LIVELY CAPITAL

EXPERIMENTAL FUTURES

TECHNOLOGICAL LIVES, SCIENTIFIC ARTS, ANTHROPOLOGICAL VOICES

A series edited by Michael M. J. Fischer and Joseph Dumit

LIVELY CAPITAL

BIOTECHNOLOGIES, ETHICS, AND

GOVERNANCE IN GLOBAL MARKETS

Edited by Kaushik Sunder Rajan

DUKE UNIVERSITY PRESS   DURHAM AND LONDON   2012

© 2012 Duke University Press

Printed in the United States of America on acid-free
paper ∞

Designed by Nicole Hayward

Typeset in Scala and Scala Sans by Tseng Information
Systems, Inc.

Library of Congress Cataloging-in-Publication Data
appear on the last printed page of this book.

CONTENTS

ACKNOWLEDGMENTS

This volume comes out of a workshop, also titled "Lively Capital: Biotechnology, Ethics and Governance in Global Markets," held in November 2004 at the University of California, Irvine. I wish to thank the National Science Foundation, the University of California's Humanities Research Institute, the Newkirk Center for Science and Society at UC Irvine, and UC Irvine's Division of Research and Graduate Studies for providing the funding that was required to conduct the workshop.

I next wish to thank the participants in the workshop for their commitment, friendship, and collegiality during the workshop; and for their patience through the long process required to convert the workshop proceedings into this edited volume. For me, thinking with, and learning from, other people is the most enjoyable aspect of being an academic, and I could not wish for a better group of interlocutors than those who are a part of this volume. In addition to the authors of the various chapters herein, I wish to thank Geoffrey Bowker, Lawrence Cohen, and Cori Hayden, who also presented papers at the workshop, but were unable to contribute their pieces to the volume.

My colleagues in the department of anthropology at UC Irvine deserve special thanks. As I planned the workshop, I received tremendous encouragement from Jim Ferguson, who was at the time chair of the department, and that support and encouragement was replicated by every one of my other colleagues, many of whom served as discussants to papers at the workshop, and all of whom contributed in significant measure to making it a success. Bill Maurer also provided a careful and critical reading of my introduction to this volume, and I have benefited greatly from his comments. In addition, people from other departments at UC Irvine and elsewhere served as discussants and interlocutors at the workshop. I wish to thank them—Bogi Andersen, Simon Cole, Jean Comaroff, Marianne de Laet, and Sharon Traweek—for adding immeasurably to the conversations. Since the workshop, Elta Smith has read (too) many drafts of my introduction as they were written, and I have benefited immensely from her patient, honest, and invaluable insight.

A few people deserve special thanks for the execution of the workshop. Caroline Melly, Guillermo Narvaez, and Neha Vora were heroic in attending to every organizational and logistical detail. Elta Smith transcribed the proceedings, while Esra Ozkan videotaped them, hence providing an invaluable archive that I was able to draw on while compiling the edited volume and writing my introduction.

Since 2002 Joe Dumit and Michael Fischer had been discussing with me the importance of providing a venue where current work on the life sciences and capital could be discussed. They both had a big role in inspiring me to organize the workshop; in providing feedback on the various grant applications that had to be written in order to get it funded; in helping me outline and structure the workshop; and in being sounding boards at various stages during the preparation of this volume. More generally, I recognize now just how much my enjoyment in collaborative work and thinking is due to having had them as my dissertation advisors. Very early in my graduate career, Joe eagerly approached me with an article that had just been written on the political economy of genomics, which is what I was planning to study in my own research. He then stopped and asked, "Are you one of those people who like it when other people are working on the same thing that you are? Because I do. I think it's so much more fun to think with other people about my work." I know that such an ethic—not just of collaboration, but of friendship and of genuine engagement with and enjoyment in the work of others—is not necessarily the norm in the academy, and is certainly not accounted for in the audit cultures that predominate in evaluating research activities and outcomes. I am grateful that I was immersed in such an ethic, because Joe and Mike taught me not just about anthropology, science studies, or social theory, but showed me by their examples the sort of academic that I wanted to be. I feel it is only appropriate that this volume is appearing in the "Experimental Futures" series, which they coedit.

I wish to thank two anonymous reviewers for careful and detailed feedback that has helped greatly in the production of this volume, and especially in the writing of my introduction. (One of them subsequently revealed himself to be Lawrence Cohen, who also, a few years previously, had the unenviable task of reviewing the manuscript of my book *Biocapital*. So I have much to thank him for, not just in terms of this volume, but in terms of my intellectual development more generally. Much of my published work has been marked, and improved, by his generous and rigorous readings.) Asya Anderson has provided invaluable editorial assistance in compiling the volume. Ken Wissoker, Courtney Berger, and Leigh Barnwell at Duke University Press have been wonderfully supportive throughout the process, and Patricia Mickelberry has been incredibly thorough and helpful in ironing out wrinkles during the volume's production. I am grateful to be working with a press that has provided such a congenial and encouraging venue for this work, for my work, and more generally, for work in the anthropology of science.

KAUSHIK SUNDER RAJAN

INTRODUCTION

The Capitalization of Life and

the Liveliness of Capital

The objective of the workshop that resulted in this volume was to investigate how new legal, social, cultural, and institutional mechanisms are emerging to regulate the global emergence of biotechnologies, and, building on that, to consider the relationship of biotechnology to ethics, governance, and markets.

The assumption underlying the workshop from the outset was that the relationship between the science in question—the life sciences and biotechnologies—and "society"—as expressed in its laws, norms, cultures, institutions, discourses, and practices—is not unidirectional. Our contemporary social realities are not simply adaptations to inexorable technological advance, as suggested by popular media renderings of science and technology. Nor can science be regulated simply by submitting it to an assumed social order, as if the latter were without history or preexisted the science. Rather the emergence of technoscience and the emergence of "the social" are simultaneous, historically constituted events.

This introduction engages some of the problem-spaces this volume inhabits, outlines intellectual trajectories within which it can be situated, and describes the particularities of individual contributions.

The Problem: Co-Production, Determination, and the Conceptualization of Lively Capital

The title "Lively Capital" is meant to point in two distinct directions. As with all workshops, this one purported to cover a topical area—broadly speaking, that of the convergence of the life sciences with systems and regimes of capital. In that register, the title refers concretely to the ways in which the life sciences are increasingly incorporated into market regimes. This is an institutional movement, away from the university and toward the market, which has been particularly marked in the American context since the late 1970s and early 1980s, and has, in the process, seen the university itself become a more entrepreneurial institutional space, one that explicitly encourages the commercialization of "basic" research conducted within its confines.

This corporatization of the life sciences can be traced back to the beginnings of the biotechnology industry, which was marked by the concomitant emergence of new types of science and technology with changes in the legal, regulatory, and market structures that shaped the conduct of that technoscience. The "new" technoscience was recombinant DNA technology (RDT), a set of techniques, developed in 1973 by Herbert Boyer and Stanley Cohen, that allows the dissection and splicing of DNA molecules in laboratories. RDT allows the life sciences to become "technological," where the product is cellular or molecular matter such as DNA or protein. Some of these proteins could, in principle, have therapeutic effects (especially for diseases that are caused by, or have as a central symptom, an abnormal amount of that protein) and be produced industrially.

While RDT was a necessary condition for the development of the biotech industry, the emergence of a new technology could hardly in itself be considered sufficient cause for the emergence of an entirely new industrial segment, which can only be understood as a consequence of the conjuncture of several events and factors. First, venture capitalists were willing to invest in a technology that had little credibility at the time as a successful business model. Second, the U.S. federal government spent an enormous amount of money on basic biomedical research through funding of the National Institutes of Health (NIH) consequent to the declaration of a war on cancer in the early 1970s.[1] Third, the Bayh-Dole Act, a piece of legislation enacted in 1980, facilitated the transfer of technology between academe and industry in the United States, and thereby enabled rapid commercialization of basic research problems. Fourth, a supportive legal climate allowed the protection of intellectual property in biotechnologies, marked, for instance, by the landmark U.S. Supreme Court ruling in *Diamond v. Chakrabarty*, in 1980, which allowed patent rights on a genetically engineered microorganism that could break down crude-oil spills. Since the 1990s, the increasingly naturalized corporatization of the life sciences has become a more global phenomenon, often explicitly or implicitly imitating events, models, or transformations that had occurred or been established in the United States, though hardly ever in seamless or homogenous fashion.

The emergence of the entrepreneurial university; the corporatization of the life sciences; the naturalization of this corporatization. Crudely speaking, this is the historical trajectory of the last four decades or so, within which this volume is located. This trajectory is neither predetermined nor uncomplicated.

I wish to illustrate this through a brief account of Stanford University's technology-transfer office, which is implicated in transformations that, over the past four decades, have helped set the stage for this volume. Stanford is often regarded as the exemplary entrepreneurial university, especially in the context of the life sciences. Its office of technology transfer was established as early as 1970, and the university's aggressive commercialization of its technologies has been seen as a huge spur to the overall commercialization of biotechnology, and specifically to the development of Silicon Valley. While Stanford, like many other universities, had before that an office of sponsored projects, which negotiated research contracts with sponsors, the development of a full-fledged technology-licensing office that focused exclusively on marketing university inventions was instigated by the entrepreneurial spirit of a particular associate director of the sponsored-projects office, Niels Reimers. Patenting university inventions was not a new phenomenon—what was at stake was institutionalizing and streamlining this process in a way that turned commercialization into an administrative priority for the university; generating expertise on patenting, often by outsourcing patent activities to patent law firms; and creating incentives for university scientists to prioritize the commercialization of their research.[2]

Stanford was not the only university to adopt a more entrepreneurial approach in the 1970s (though most technology-transfer offices in the United States were set up after the passage of the Bayh-Dole Act, in 1980). But the history of its office of technology licensing (OTL) is intrinsically linked to the history of biotechnology, and especially to the history of commercial biotechnology. Stanford's OTL was involved in getting a patent on Boyer's and Cohen's recombinant DNA technology, the most lucrative technology that Stanford licensed during that period.[3] Crucially, there were two major consequences of models such as those adopted by Stanford. First, a greater degree of collaborative research between university and industry was established (since the university retained upstream patent rights, as opposed to a "more open" or "less commercial" model, wherein patents were usually on more downstream applications and tended to reside exclusively with industries that utilized technologies in the public domain for specific product development). Second, a "spin-off" culture ensued, whereby university professors commercialized their research by starting their own companies.

By providing this potted history of Stanford's technology-transfer office, I mean to suggest that the emergence of the entrepreneurial (American) university, the corporatization of the life sciences, and the naturalization of that corporatization are complex processes. It is not the case that life sciences research before the 1970s was noncommercial, or that universities were uninterested in marketable technologies. Rather, what one sees is the changing nature and locus of "commercialization" from that time, with the university becoming a more explicit—and institutionally regulated—stakeholder in the entire process.

The consequences of commercialization are also not self-evident, because commercialization does not imply a predetermined set of actions or outcomes. Rather, the forms and modes of commercialization constitute a terrain of deeply strategic maneuver and contestation. A patent in itself does not determine the way university technologies are disseminated. All that it allows is for the patent holder to negotiate the market terrain with a certain set of exclusive rights that others in the marketplace do not possess.

The dissemination of RDT is itself instructive in this regard. Stanford decided to issue nonexclusive licenses, an approach that allowed the university simultaneously to reap enormous commercial benefit and to promote wide distribution of the technology, the latter ensuring that Stanford fulfilled the public-service function deemed appropriate for a public institution.[4] At many points along the way, Stanford pursued strategies that were in fact antithetical to a profit motive.[5] For instance, in addition to the original recombinant DNA patent that was granted in 1980, Stanford also acquired two subsequent patents, in 1984 and 1988, which covered products generated by downstream application of RDT, in prokaryotic and eukaryotic cells, respectively. In the normal course, these patents would have expired in 2001 and 2005, thereby effectively extending the length of time for which RDT patents would generate revenue for the patent holders. However, Stanford allowed the subsequent patents to expire along with the original RDT patent, in 1997. Furthermore, Stanford decided not to charge royalties to nonprofit organizations to which it licensed RDT.

The story of the RDT license suggests the potentially immense malleability of property regimes; hence, adopting a moralistic position on intellectual property is often unhelpful in responding to the actual political complexities that are enshrined within the idea and practice of commercialization.[6] But the story also points to a moment of deep ferment, wherein granting a patent on a platform technology took six years. The legitimacy of the patent was deeply debated, and the role of the university (including the private university) as existing for the public good was largely deemed non-negotiable. Thus, in referring to a certain historical trajectory that the value system of the life sciences has traversed over the past four decades, I do not mean to present a simplified narrative in which the private encroaches on the public, nor do I intend to suggest that the outcomes of such an encroachment are inevitable or known in advance. But I do want to suggest that questions about commercialization—which are, among other things, about the legitimate extent of commodification, the role of the university, and the value system of science—were deeply unsettled less than three decades ago, yet seem largely beyond argument today, certainly in the United States. What is crucial here is the changing investment in the market by the value system of the life sciences—which, in spite of numerous contingencies and open-ended, strategic negotiations, has nonetheless seen a trajectory that is resolutely more embracing of market logics over time. Even if historical trajectories are emergent and not predetermined, the fact that they have consistently unfolded in certain institutional and ideological directions over others—and not just in the United States, but in other parts of the world as well—is important to note and assess.

Stanford's RDT licensing is a useful point of departure for describing the terrain of this volume for a number of reasons. The first, quite simply, is its historical importance. RDT in many ways provided the technological conditions of possibility for biotechnology, and the mechanisms of its dissemination (which had everything to do with Stanford's strategy of nonexclusive licensing) were consequential for the development of the biotech industry. In addition, Stanford's approach to licensing exemplifies both the frictioned commercialization of the life sciences and, more broadly, the emergence of new normative systems that are more open to such commercialization. On the one hand, one sees the (at the time) contested question of the appropriate relationship of university science to commercialization, with Boyer and Cohen themselves being initially wary of the RDT patent application and thus deploying the patent, once issued, in a manner that ensured the public dissemination of the technology. On the other hand, Stanford's move toward patenting the technology in the first place, and the movement of its technology-licensing office in a distinctly entrepreneurial direction, indicates a moment of changing value systems in the sciences, away from a Mertonian disinterestedness or communism.[7]

This direction was evident in at least three ways. First, the public dissemination of a technology, while it undeniably provided access to that technology in the public domain, was also a spur to private industry. Stanford's investment as a university in the life sciences, combined with its entrepreneurial strategies of technology licensing, was undoubtedly a factor in the biotech boom in Silicon Valley. Second, self-questioning about the advisability of patenting, evident in the RDT case, eventually dissipated, as commercial enterprise came to seem like a natural course of action for university scientists. And third, Stanford to some extent emerged as a model of the entrepreneurial university, even though its mechanisms of technology transfer, which favored public dissemination of the technology, were not necessarily adopted. Over the past three decades, it has in fact become the norm for American universities to have technology-transfer offices that aggressively pursue intellectual property in the life sciences.

Stanford itself, in its actions in the RDT case, was adhering not just to the public-service function of the university, but also to the spirit of the patent regime, which is to promote the public disclosure of inventions by providing the inventor with a set of property rights. In other words, the idea (and ideal) of the public is already enshrined in the very rationale of a patent, as a particular form of intellectual property that acts in the interest of public disclosure and dissemination. That this ideal so often manifests in ways that instead protect *institutional* interests in the face of market competition, making Stanford's actions seem almost surprising or altruistic, is itself an indication of the way in which the commercialization of the life sciences has become naturalized. Exemplifying patent strategy as protectionist rather disseminating is Hoffmann-La Roche's patent on polymerase chain reaction (PCR), a technology considered in many ways as seminal as RDT for the development of biotechnology. Roche aggressively enforced the PCR patent, thereby arguably creating value for themselves without the same consideration for accessible downstream applicability that Stanford showed. Roche's approach to technology patents is perhaps by now seen as more normative than Stanford's. While both institutions engage in the commercialization of technologies, they play out market logics in decidedly different ways.

The territory of this volume is hardly limited to the playing out of intellectual-property regimes. It is more generally concerned with the nature of value systems in the life sciences, how they are perceived and deployed. What is crucial here is that, over time and more globally, the investment in market logics has been seen as variously inevitable, nonnegotiable, desirable, or even virtuous, even as market logics may themselves be contradictory. By situating the case of Stanford against that of Roche, one detects adherence both to public dissemination and to private monopolization, based on the same instruments of intellectual property. Nevertheless, whether they be more liberal or more monopolistic in inclination, both logics operate under the sign of the market, of capitalist exchange and value. Historical contingency, therefore, arises at the level of particular strategies and tactics adopted as the life sciences have been incorporated within systems and regimes of capital. This historical contingency has occurred in the shadow of an ideological embrace of the market that, while not seamless, has become increasingly uncontested over time.[8]

In other words, the fundamental nature of the market, its value systems and epistemologies, is itself shifting, in what might broadly be considered a "neoliberal" direction. *Neoliberalism* is often used as a catch-all phrase, simultaneously encompassing too much and describing too little. Yet there are uncanny overlaps between the development of life-science epistemologies and the epistemologies of neoliberal economics since the early 1970s, as Melinda Cooper (2008) has elegantly demonstrated. This is not a relationship of cause and effect—the life sciences do not change *because* of neoliberalism, nor is neoliberalism a consequence of the biotechnology revolution. Rather, what one can trace is an emergent epistemic milieu in which both economics/capitalism and the life sciences/biotechnology are undergoing radical transformation and dealing with apparently similar types of problem-spaces (such as, for instance, the understanding and management of complex systems of risk) at similar moments in time, and often drawing on one another for metaphoric or epistemic sustenance. Even the ideology of innovation, which gained enormous traction under Ronald Reagan's presidency, in the 1980s, under the influence of neoliberal thinkers such as George Gilder (1981), suggests a value system of capitalism that is in some ways quite distinct from that which Marx traced during the industrial revolution, whereby the magic of capital lies not in the creation of the surplus through the apparent exchange of equivalents, but rather in the creation of what Michael Lewis (1999) refers to as "the new new thing." Innovation as a driving force of capital draws to a large degree on technoscientific potential for value generation; technoscientific potential for value generation is fundamentally enabled by a political-economic regime that provides incentives to innovate. Rather than cause and effect, what one is confronted with are mutually implicated and emergent epistemologies and systems concerning life and value, what Sheila Jasanoff (2004, 2005a) has termed *co-production*.

While much is changing and emergent, what has remained constant, or intensified, is a set of mutual investments — the investments of the market in the life sciences, and the investments of the life sciences in the market. The co-productionist sensibility of this volume is complicated by this deep mutual investment, which *does* lead to the life sciences being increasingly "driven" by market logics, so that even when actions are apparently taken in the public good (as by Stanford with regard to RDT), the parameters of those actions make sense in terms of the market and of capital. I wish here to elaborate on this complication.

I do so, specifically, by reading Jasanoff against Marx, as an entry point into a more general problem of theory and method that this volume wrestles with. This reading, at one level, is not necessarily representative of all the chapters in the volume, which explicitly draw on or engage with either or both of these theorists to varying degrees, or not at all. But such a reading situates a problem-space that I feel all the chapters do occupy, regardless of their specific modality of engagement with it. These next reflections, therefore, are more authorial than representative, speaking to my own thinking through of problems of theory and method. I will revert to a more editorial role in a subsequent section.

On the one hand, by bringing in the notion of co-production, I want to reform a certain kind of Marxian analytics that is economically determinist. To reduce changes in the biosciences to economic causes would be inadequate. At the same time, I do not wish to suggest that co-production simply implies complexity or contingency. I am wary of a certain current anthropological move that seems to posit the establishment of contingency as the methodological solution to doing anthropology of the contemporary. This is not to deny the importance of staying attentive to contingency — Marx himself, after all, was a supremely careful theorist of the contingent. But contingency alone, I suggest, cannot adequately serve as explanation for the sorts of consistent historical unfoldings with which this volume is concerned.[9]

Both Jasanoff, with her idea of co-production, and Marx, with his historical dialectic method for studying the unfolding of systems and logics of capital, are theorists who take the contingent seriously, but refuse it as the point at which (to use a Wittgensteinian phrase) explanations run out. This leads to methodological and conceptual dilemmas though: *how* might we think causally about the trajectory of capitalization of the life sciences? How might co-production, or Marxian analytics, or indeed other kinds of conceptual frameworks, help work through the problem with which I have introduced this volume—concerning a particular political economic and epistemological trajectory that, over space and time, indicates a process of capitalization, but where the norms and forms of capitalization are complicated, and where falling prey to economic determinism would be an impoverished analytic move?

As an example of the dilemma that I am trying to elucidate, let me consider Emily Martin's work on the dual movement of capital and life toward flexibility, *Flexible Bodies* (1995). Martin's analysis is not simply diagnostic; it forces us to ask a series of analytic questions. Why flexibility? Why at this time (the 1980s and 1990s)? Martin shows that flexibility is geared toward modalities of maximizing productivity, within certain, particular political-economic conjunctures. Martin's account of "flexible bodies" most certainly shows a particular co-production of forms of life with forms of capital. But this co-production occurs under the sign of certain value systems that structure the terrain within which it happens. Melinda Cooper paints a similar canvas, showing a similar kind of dual structure, of co-production, but under the sign of particular political-economic and epistemological structures, in *Life as Surplus* (2008). This is entirely an analysis of the co-production of life sciences with neoliberal economics, but at a moment in history (the 1970s onward) wherein the appropriation of both by capital is a consistent, and increasingly dominant, feature.

My interest in this volume, inspired by works such as those just cited, is to understand the co-production of life and capital along with the consistent unfolding whereby life is increasingly appropriated (or at least appropriable) by capital. How one provides such an understanding is a fundamental methodological question, and each chapter implicitly or explicitly deals with this question in its own way and on its own terms. I myself have three suggestions in this regard, all based on a fundamental provocation: that it is analytically important to *not* abandon the question of determination, but at the same time, it is important to think of determination as *not always already implying pre-determination*.

Rather, first, one can think about *multiple determinations*. Max Weber did so throughout his work. While he never abandoned the question of causality—in many ways, it was one of the central questions that structured his work and method—he recognized that causes for social phenomena were always multiple, and establishing that multiplicity had to be an empirical task. There were a number of empirical strategies that Weber adopted: historicist; comparativist; and, in his rendering, sociological, through the device of the ideal type.[10]

Second, one can think about overdetermination. Overdetermination is originally a term of Freudian psychoanalysis, further developed and analyzed by Louis Althusser to suggest a subjectively mediated contextual relationship, which, while not causal in any simple sense, might nonetheless appear to be disproportionately important (Althusser 1969 [1965]). And so, even if a particular set of political economic formations do not in any direct and simplistic way lead to particular epistemic emergences, they could still disproportionately set the stage within which the latter take shape in particular ways and, further, *appear* to do so to various actors in ways that naturalize complicated relationships into simple causal ones. Thus, even if capitalism represents particular types of political-economic formations, in this current moment in world history, as Slavoj Žižek argues, it "overdetermines all alternative formations, as well as non-economic strata of social life" (2004, 294). Therefore, even while emphasizing the historicity and the far-from-natural emergence of capitalism as a set of political economic forms and structures, it is important to acknowledge the importance of capital as being what Žižek calls the "'concrete universal' of our historical epoch" (ibid.).

And third, one can think through Marx's phrase of the economy as being determining "in the last instance" (Marx 2009 [1858]). I wish to reflect on this phrase at some length, as an opening into thinking about Marx's method and its relevance for studying the life sciences and capital. Marx's phrase is often understood to signify his economic determinism, but it can, in fact, be read in a far more open-ended fashion. Indeed, I would argue that a proper, conjunctural reading of Marx, which is attentive to the moment and context of his writing, *must* read the phrase in a more open-ended fashion, especially considering the fact that Marx, in attempting to develop a materialist basis for understanding the economic, social, and political orders of his time, was writing against idealists of various sorts (including Young Hegelians, utopian socialists, and certain bourgeois thinkers). Frederich Engels makes this point explicitly. Even without necessarily providing a clear alternative to thinking about the problem of determination, Engels clarifies that neither he nor Marx meant themselves to be read in terms that simply implied the predetermination of the social by the economic. One can, indeed, see in Engels a quite elaborate attentiveness to contingency, but it is contingency that is operational under the sign of certain kinds of formations.

> According to the materialist conception of history, the *ultimately* deter-mining element in history is the production and reproduction of real life. Other than this neither Marx nor I have ever asserted. Hence if some-body twists this into saying that the economic element is the *only* deter-mining one, he transforms that proposition into a meaningless, abstract, senseless phrase. The economic situation is the basis, but the various elements of the superstructure—political forms of the class struggle and its results, to wit: constitutions established by the victorious class after a successful battle, etc., juridical forms, and even the reflexes of all these actual struggles in the brains of the participants, political, juristic, philo-sophical theories, religious views and their further development into sys-tems of dogmas—also exercise their influence upon the course of the historical struggles and in many cases preponderate in determining their *form*. There is an interaction of all these elements in which, amid all the endless host of accidents (that is, of things and events whose inner inter-connection is so remote or so impossible of proof that we can regard it as non-existent, as negligible), the economic movement finally asserts itself as necessary. . . . Marx and I are ourselves partly to blame for the fact that the younger people sometimes lay more stress on the economic side than is due to it. We had to emphasise the main principle *vis-à-vis* our adversaries, who denied it, and we had not always the time, the place or the opportunity to give their due to the other elements involved in the interaction.[11]

What Engels is suggesting is that the economic is, to borrow a phrase from Bruno Latour, an obligatory point of passage—it is impossible to understand the social and the political in the conjuncture in which Marx was writing, of industrial capitalism, without accounting for it. But to say that something is necessary to understand is hardly the same as saying that it predetermines the sociopolitical.

Each of these three strategies for thinking about determination as not always already predetermination—owing themselves respectively to Weber, Freud, and Marx, three of the great social theorists of the nineteenth cen-tury and the early twentieth—deploys a different explanatory modality. The first is historical and anthropological (and, relatedly, what Weber calls "sociological"); the second is ideological and subjective; and the third is, for want of a better word, structural. But that "structure," in Marx's writings, is in fact extremely complicated, and Marx is supple in his analysis of it. It is not enough to say that the economic, in Marx, is necessary but not determining. It is also important to recognize that when Marx analyzes "the economic," he is analyzing multiple things.

One can see this quite clearly in volume 1 of *Capital*, which is perhaps Marx's most schematic rendering of the structure of capital (Marx 1976 [1867]). The structural elements of capital—concerning the production of value in the relationship between the use and exchange of an object within a particular political-economic formation—are outlined at great length through most of the volume, which starts with an analysis of commodities, moves on to an analysis of money, looks at the way in which the continuous exchange of one for the other somehow leads to the generation of surplus in spite of it being an apparent exchange of equivalents, analyzes the crucial mediator (the labor of the worker) in this process, and concludes that in fact it is because this labor is always already labor power, containing within it an excess potential for work over and above that remunerated by wage, that one has the ability to generate surplus value. This much is schematic, abstract, and structural; and it is made "real," within the parameters of the schema, in the transition from analyzing absolute surplus value (which is the abstract rendering of surplus value in the hypothetical interaction between a capitalist and a worker) to relative surplus value (which is the real rendering of surplus value generated in the collective interactions between capitalists and workers, which are mediated, crucially, by machinery).

But even within this schematic rendering of the structure of capital, one sees the inescapable presence of both the ideological/subjective and the historical/anthropological. The ideological/subjective dimension is to be found in the crucial section on commodity fetishism, which speaks to the abstraction that is at the heart of an apparently thoroughly materialist analysis. It is, indeed, quite literally in the heart of the analysis, being placed carefully between the sections on commodities and on money, and at the cusp of the development of the analysis of surplus value. And it is about the commodity, on the face of it a simple material thing, being "full of metaphysical subtleties and theological niceties," and thereby appearing to individuals as a mediator of social bonds. Marx is careful to call this abstraction "fetishism" rather than "ideology," an indication of the vexed place of the concept of ideology in his writings.[12] But even if commodity fetishism is not ideology, in the specific manner in which Marx develops that concept in his early writings such as *The German Ideology* (Marx and Engels 1963 [1845]), it is still ideological; that is, it refers to the masking of real social relations through appearance, illusion, and naturalization of those illusions. Žižek, indeed, suggests that the notion of commodity fetishism is central to Marx's analysis of capital, getting as it does not "to the secret of the content of the form, but to the secret of the form itself" (Žižek 1994, 296).

The historical/anthropological dimension is to be found at the end of volume 1, in the section on "The So-Called Primitive Accumulation." In this, Marx shows the historical conditions of possibility that even make possible the establishment of the structure that is elucidated until that point. This history is a violent one, involving the enforcement of property regimes (in England, through the dispossession of peasants from the land during the enclosure movement), followed by the forcible outlawing of vagrancy that pushed these dispossessed peasants into the factories as workers for industrial capitalism. On the one hand, this section is striking in its absolute empirical specificity—the reading of particular laws that were enacted over time in England, and the care with which Marx insists on that particularity (Marx 1976 [1867], part 8). At the same time, in spite of this rigorous specificity, there is a more general structural provocation, pertaining to the constitutive relationship of violence to the history of capital, so that "capital comes dripping from head to toe, from every pore, with blood and dirt" (ibid. 926); a provocation that disputes the myth of "free" exchange of labor for wage that is propagated by bourgeois political economists, and that undergirds the entire preceding structural analysis.

In addition to these three dimensions—the structural, the ideological/subjective, and the historical/anthropological—"the economic" in Marx has a fourth dimension, in its guise as political economy, which is relevant for the concerns of this volume: its *epistemological* dimension. In order to explore this, I must back up to reiterate what Marx is doing, especially in his later works. In *Capital* (and also in the rough notes that represent his work toward it, *Grundrisse*), Marx is engaged in a project of critique that is *not* criticism or denunciation, but rather, in a Hegelian sense, a mode of analysis that attempts to explore the limits of a concept or practice. What is the concept or practice that Marx is critiquing? In his development of the concept of surplus value, Marx is analyzing capitalist exchange. But what *is* capitalist exchange? Is it concept, is it practice, or is it system? This is crucial, because often in order to grasp the "stuff of the world," as Marx is attempting to do, one has to constantly shift registers between looking at concepts, at practice, at systems, at objects. And how to do that remains a live methodological challenge for contemporary analysis, as much as it was for Marx.

I would suggest that, in fact, Marx was *not* doing systemic analysis in any simple sense; he was not trying to understand "capitalism," because capitalism as a singular is an absurdity—it did not exist in Marx's time, nor has it since. Systems of capitalist exchange certainly have existed, but they have been always fluid and emergent, taking specific organizational, institutional, and political forms in particular times and places. What Marx was trying to do, rather, was understand capital.

But what is capital, in material terms? Money and commodities. Yet money and commodities by themselves do not constitute capital; they only do so when operational in a system that is structurally geared toward the generation of surplus. Here, one is potentially confronted with a tautology, of a system that only makes sense in terms of the modalities of functioning of the objects that constitute that system, objects that themselves only function as constituents of that system when already situated within those systemic logics. After all, money and the things that circulate in exchange relations with money both existed in precapitalist societies, and systems of monetary exchange existed for centuries before capitalism. What is that magical element that breaks this tautology and mediates the exchange of money and commodity to turn them into capital? It is labor.

Who came up with this answer? Not Marx—he was playing off of this fact, which was already well-known at the time of his writing. It was well-known because political economy—an emergent but already powerful body of knowledge that explained how capital works—said so. Indeed, Marx's crucial modification to this understanding was in insisting that it was *not* labor—work that was materialized and remunerated adequately in wage—but labor power—the constitutive potential for labor over and above that remunerated in wage—that actually was the driving force of capital, by providing the conditions of possibility for surplus-value generation.

The analysis of labor power, crucially, comes not from an empirical observation of the accrual of value, but from the critique of political economy, from the fact that political economy's own analysis of capital failed to account for the fact that if labor was simply a mediator in a process of exchange and that if that labor was "free" and remunerated, then it was capital that kept growing, while labor was exploited and alienated. *This* is what political economy failed to explain; this is what *political economy* failed to explain. In other words, Marx's critique of capital *was not* a critique of capitalism, which would have been an absurdity; and it *could not have been* a critique of exchange per se, because there was no way of empirically accessing exchange per se except through the means available to know and understand that exchange.

In other words, Marx was—especially in his later writings—*critiquing the way in which we came to understand capitalist exchange*—and that way was through political economy, which was the foundational epistemology of the time that explained these mechanisms. *Capital* is, after all, subtitled "a critique of political economy"—not "a critique of capitalism," or even "a critique of capital." Political economy failed to take into account the history of capital, whose dynamics it claimed to explain; and it failed to be attentive to the abstractions with which capital was able to hide its "true" nature. Yet it functioned as authoritative knowledge.

With this elaborate exegesis on Marx's method, I wish to return to the problem of conceptualization and method that I started with. My entry point into thinking about Marx's method was to suggest that his statement that the economic is determining "in the last instance" is in fact not a determinist statement. But I have ended up by suggesting that not only is Marx not determinist, but he is also, methodologically, an epistemologist—just as he is, variously, a structuralist, a historian, an anthropologist, and a philosopher of abstraction and naturalization. Indeed, Marx cannot understand capital except through epistemology; but this not does mean, again, that political economy *determines* capitalism. What one sees, rather, is the development of an analytic that, in fact, looks rather uncannily like co-production! Marx is looking at systems and regimes of exchange and value that are made intelligible, and naturalized, through epistemology, which in turn responds to those forms of exchange and value in its own development as an authoritative form of knowledge. Value and epistemology, in this analytic, are co-produced. Yet, at the same time, Marx claims it is the economic that is determining in the last instance.

Methodologically, Marx suggests the importance of shifting registers, between the historical, anthropological, analytic, and epistemological. But when it comes specifically to the question of determination, Marx presents a challenge of method and conceptualization that is a dialectic between the co-production of value and epistemology, and the determination "in the last instance" by the economic, where "in the last instance" refers not to predetermination, but at the very least to multiple determinations and over-determinations, and where "the economic" at least has structural, ideological/subjective, historical/anthropological, and epistemological dimensions.

The question of the mutual investment of the life sciences and capital, therefore, can open itself up to a number of different explanatory schema that are attentive to contingency without reifying the contingent as itself the ultimate form of explanation; and that is what this volume does in different ways. The specification of particular relations of co-production to particular types of consistent unfoldings is, in part, the project of this book. Each chapter develops this relationship, explicitly or implicitly, in its own way; but I would suggest that all the chapters, even the ones that do not claim to be Marxist, are caught within a problematic that is shaped by the Marxian inheritance of thinking of the relationships between the co-productions and mutual determinations of the life sciences and capital.

Investment is one of those wonderful dual-edged words that points simultaneously in different directions. On the one hand, it suggests monetary investment in and by the market, speaking to regimes of valuation that fall within regimes of calculability (if not determinate, then at least probabilistic). On the other hand, it speaks to an emotional commitment—questions of obligation, passion, ethics, morality, hope, love, and pain are all at play in various peoples' investments in the life sciences and in capital.

In one register, then, *Lively Capital* is about the commercialization of the life sciences—its institutional histories, epistemic formations, and systems of valuation. In a second register, however, it refers to the lively affects—the emotions and desires—at play when technologies and research impinge on experiences of embodiment, kinship, identity, disability, citizenship, accumulation, or dispossession. In this volume, the topic of capital is specifically allowed interpretive range. The intention is not to discuss the "effects" of a particular new technological development, or the adequate ethical response to such new developments; it is, instead, to see if a potentially fresh set of approaches to looking at the life sciences and biotechnologies and their relationships to the social, to "life" as lived experience, can be facilitated.

Life Sciences and Capital as Objects of Inquiry

Three points about the objects of inquiry that this volume addresses need to be emphasized. The first is that *Lively Capital*, in its broadest conception, is about a set of interactions between science and society wherein "science" and "society" cannot be analytically purified from one another or assumed as given. The chapters in this volume examine neither the "social impacts" of science nor the impacts of society on scientific progress. Rather, as has been the method of recent work in science and technology studies (STS), both the scientific and the social are up for grabs here—none of the authors assume or believe positivist narratives that portray either the technoscience they study, or the culture, politics, or ethics within which that technoscience is completely situated, as a stable or linear process. Second, technoscience goes beyond the boundaries of the laboratory, even though the laboratory is an essential site for technoscientific production. Technoscience is equally produced in the deliberations and actions of policymakers (state and nonstate), market actors, activists (whether opposed to technoscience or fighting for its acceleration or intensification), consumers of various sorts, editors of scientific journals, and patients. One of the methodological challenges of studying lively capital is that it is not self-evident where one should look, for technoscience is always already many things—a collection of institutions, normative structures, practices, ideologies, and vocations. And third, we are not just concerned with the actions and effects of humans here. Indeed, both the actors of technoscience and the constituent elements of capital transcend human agency. Information, commodities, money, filings with the Securities and Exchange Commission, dogs, air, rice, genes, websites, and search engines, all make their appearance as central actors in the same way that scientists, policymakers, venture capitalists, bioethicists, and anthropologists do.

What, then, is lively capital as an object of study? Rather than strictly define the life sciences or capital at the outset, this volume is focused around topics that speak to their confluence, such as the promissory grammars that surround the search for therapies, the circuits of circulation of pharmaceuticals or other biologicals, questions concerning the increased recourse to information sciences in fields such as genetics or environmentalism, and the various legal techniques and political strategies deployed in order to moderate debate and construct authoritative governance regimes. Nonetheless, it is important to justify why the life sciences and capital are so central to this volume, and what is so lively about either or about their confluence.

In an attempt to theorize the contemporary, we encounter the persistence and intensification of the life sciences as one of the foundational epistemologies of our times, rather in the manner in which Michel Foucault diagnosed biology, political economy, and philology to be the foundational epistemologies upon which modernity was built in Enlightenment Europe. There have indeed been profound shifts in the understanding of "life" over the past half century. The life sciences are a shifting referent, and the biology of today often bears little apparent relation to the biology of the nineteenth century.

Simultaneously, we can see shifts in the locations of knowledge production in the life sciences, such that research is increasingly performed in corporate locales, with corporate agendas and practices. This is a corporatization, as mentioned earlier, that has been particularly marked since the 1980s and that intensified through the 1990s and the early part of the new millennium. An immediate challenge faced by work such as that contained in this volume is to trace the contours of this institutional shift, all the while situating it in the context of the epistemic and technological changes happening within the science, especially the trend in the last half century toward understanding biology at the cellular and molecular level, rather than at the level of the organism.[13] This move toward the molecular is related to the increased ease with which life gets conceived and represented in informational terms. Both commodities, such as therapeutic molecules, and "basic" scientific knowledge, which can itself be more and more easily represented as information, and packaged and commodified as, for instance, databases or diagnostic kits, are being created in this increasingly corporate environment. If STS has always been concerned with the construction of scientific facts, then what does a scientific fact mean, and how does it operate as a fact in the world, when that world is increasingly that of the market? How are these facts constituted differently in the market environment than the way in which they were constituted in a "traditional" scientific ethos, and how are objects and subjects in the world constituted in turn *by* these facts?[14] What is the relationship of the scientific representations performed by these facts to political representation? And how can we situate these changes in the technoscience and in the institutional contexts of their conduct comparatively, in different parts of the world?

At stake in this co-production of new forms of knowledge about life with new sites of such knowledge production are ways to intervene in and adjudicate on matters of health, life, and death, such as for instance through resultant emergent practices of medicine or nutritional supplementation of food crops. Therefore, the changes that this volume seeks to trace concern the changing nature and relevance of the *biopolitical*, referring to Foucault's diagnosis that one of the markers of modernity is the way in which its institutions and discourses seek to put life at the explicit center of political calculation (Foucault 1990 [1978]). Biopolitics plays out or gets transformed through questions of rights and ethics, such as privacy, informed consent, ownership, and the fluid and contested boundaries that separate the public domain from legitimate private property, and is, at least in part, *economic*. The question of the economies, mechanisms, and multiple determinations and overdeterminations of the co-production mentioned above, where "economy" refers not just to the quantitative measures of productivity or profit, but also to the *regimes of value*, both material and symbolic, that form the theater of political articulation, is in many ways the central one for *Lively Capital*.

Value is a central concept for this collection, and links questions of political economy to those of ethics and governance. It is a double-jointed word that implies not just material valuation by the market, but also a concern with meanings and practices of ethics. This is particularly salient for industries such as biotech and pharmaceuticals, which generate significant symbolic capital from being, as they are never averse to pointing out, in the business of saving lives. Systems of valuation are animated by abstractions that go beyond quantitative material indicators, and the ethical is often a shorthand that black-boxes these abstractions. Among the tasks of this collection is to open the black box—to posit ethics not as independent of political economy, but rather to see how the two are articulated, ethics being an integral part of the political economies of the life sciences and biotechnologies. Equally, ethics contains its own political economies and capacities for being institutionalized and made corporate. Therefore, parallel to emergent regimes of technoscience and capital are emergent regimes of ethics, with increasingly powerful voices in constructing discursive, normative, and ideological terrains. The analyses of these cannot be compartmentalized or kept easily separate from one another, which is why a number of the concepts concerning value in this collection are those like promise, hope, hype, obligation, and love—none of these evidently political-economic terms, yet, as they operate through this volume, absolutely central to the analysis of the liveliness of capital, of the life sciences and capital, of Lively Capital. Value also operates at multiple scales, from the local to the global, and the historical and geographic contexts from which the papers speak are essential to resist a positivist rendering of value, whether market value or ethical or symbolic value, as somehow unitary and universal. This volume, then, asks how and in what ways it becomes possible to think about ethics and politics in technocapital, and in our emergent worlds.

Intellectual Trajectories

I wish here to suggest four intellectual trajectories that operate at different scales in the work represented here, without making any claims to providing a review of the range of work done on the life sciences and society.[15] The first two are the trajectories of science and technology studies (STS) and social and cultural anthropology as separate fields; the third is their convergence; and the fourth is the impact on social theory, more broadly construed, as informed by the first three trajectories.

On the face of it, *Lively Capital* is conceived as an interdisciplinary venture that goes beyond the two disciplinary trajectories of STS and anthropology, and the volume brings together contributors housed in disciplines and areas of inquiry as broad and varied as African studies, anthropology, comparative literature, history of consciousness, public policy, rhetoric, STS, and sociology. A number of these fields of inquiry are themselves recently constituted and continuously emergent interdisciplines. And yet, the primary concern of the chapters in this volume lies in tracing particularly rapid contemporary emergence. This brings with it a set of methodological and conceptual challenges, which speak most directly to the current disciplinary concerns of anthropology perhaps more than any other. Without wishing to claim a particular disciplinary imprint for this collection, I think it is fair to say that the conjuncture in the humanities and social sciences that sets the stage for a volume such as this is an emergent and still fluid conversation between the amorphous methods of STS and the more entrenched (in institutional terms) discipline of cultural and social anthropology. This has resulted, on the one hand, in an increased turn toward ethnographic method by sociologists of science and technology, who in many ways dominated STS in the 1980s, and, on the other hand, in a serious pedagogical and theoretical engagement with STS literature that has arisen from outside the canon of disciplinary anthropology by anthropologists.

It should be noted that STS, while a small field, itself has multiple genealogies. The foundational ones, in many ways, grew out of the Edinburgh school of the sociology of scientific knowledge (SSK) and its subsequent development in France through actor-network theory (ANT).[16] But there were also other genealogies of STS represented within the United States—such as at MIT, where I received my doctorate—that grew out of other intellectual trajectories. In the case of MIT's STS program, such trajectories included, crucially, both the history of technology and the American political context of the late 1960s and early 1970s, which had seen science and technology reconstituted as intensely political sites of civic engagement.[17] Feminism provided another crucial genealogy and introduced hitherto ignored dimensions to the study of technoscience, such as by looking at the gendered and racial nature of its discourse and practice, or by generating more embodied, experiential accounts of technoscientific development, or simply by writing about science and technology to an audience committed to feminist praxis.[18] By the late 1980s, anthropology was also getting explicitly interested in science and technology as an object of ethnographic study, a seminal marker of which was Sharon Traweek's ethnography of high-energy physicists, *Beamtimes and Lifetimes* (1988).[19]

A crucial marker of a certain anthropological turn that subsequently opened the discipline up to something like STS was the "writing culture" moment of the mid-1980s. The publication of *Writing Culture* (Clifford and Marcus 1986) and *Anthropology as Cultural Critique* (Marcus and Fischer 1986) made tangible an "experimental moment" in the human sciences, and in anthropology in particular, which had been brewing for the previous two decades of the discipline's encounter with political economy, psychoanalysis, and poststructuralism. The books represented, simultaneously, a certain (particular) coming to fruition of a Geertzian interpretive anthropology and a promissory note for a modality of work yet-to-come. Calls for multisited ethnography that developed over the subsequent decade (for instance, Marcus 1995a) were elaborated out of this mid-1980s moment and represented one programmatic idea for the development of ethnographic research projects that were adequate to the contemporary intellectual and political moment of globalizing (post)modernity.

The ambition here concerned an at the time unstated recognition of the fact that ethnography had, in Douglas Holmes's and George Marcus's formulation (2005), to be "refunctioned" in a fundamental way given the changing objects of ethnographic inquiry. *Writing Culture* itself did not have the words to articulate this refunctioning, which it was already calling for, though scholarship in the subsequent two decades has argued for it in different ways.[20] It is in the context of the decentering and refunctioning of ethnography that STS comes to matter, because, regardless of whether the dialogue with anthropology is explicit or not, recent ethnographies of science and technology have contributed to methodological debates within anthropology.

What one sees, therefore, is the development of an amorphous and interesting conversation between anthropology and STS over the past two decades, though the trajectories and investments of that conversation are not uniform. This has happened at a moment when STS has turned to ethnography as a crucial modality through which to describe technoscience, supplementing and in some cases supplanting earlier foci that drew on sociology or history of science and technology. And at a parallel moment when anthropology has started engaging with a great variety of objects and processes *as* ethnographic objects and sites of study, especially engaging with globally dispersed, rapidly emergent processes that are centrally involved in the constitution of the modern world, technoscience being exemplary in this regard. It is this intellectual context—of anthropology turning to technoscience as an object of study, over the same period of time that STS turns to ethnography as a method with which to study its objects—that this volume is situated within.

I wish to suggest that the confluence of STS and anthropology is not simply a traversal of interests across distinct disciplinary spaces. STS indeed never really was a disciplinary space with clearly constituted boundaries, except in certain departments. Anthropology has been such a disciplinary space, with all the credibility, gravitas, and constraint that comes with disciplinary projects; but it has been marked by practices of inquiry that have constantly exceeded the boundaries of its own disciplinary regulation. In this sense, both STS and anthropology have been marked by an excess or surplus, which is the "stuff of the world" that has bled into their practices of inquiry to torque them. The coming together of anthropology and STS represents the coming together of two fields of inquiry that have always found it difficult to contain themselves. However, in this process, neither remains unchanged. Anthropology plus STS is not merely technoscience meets culture. Rather, it is the beginning of a profoundly interdisciplinary conversation, the likes of which can be seen in a volume such as this.

I illustrate what I mean by "interdisciplinary" by turning to the object of study here, the life sciences, which have been particularly open to being shaped by multiple disciplinary influences. When the life sciences turned "genomic," the importance of elucidating life processes in informational terms became evident. This has, indeed, been evident for at least half a century, as cybernetic understandings of life-as-code became normative in molecular biology in the 1950s and 1960s.[21] What genomics has allowed in the last decade or so is a radical change in information-generating capacity, so that the epistemic dilemma of having too little information to decode life processes at the cellular and molecular level became, almost overnight, a dilemma of having too much information to make sense of. Challenges of data generation very quickly became challenges of data mining and data management.[22] What this meant was that computational capability and informatics became central to the everyday operation of genomics research, and suddenly biologists who had not been trained in computer science had to start talking to computer scientists who often had little domain knowledge of biology. Interdisciplinary work in such a situation was not a choice—it was necessary in order to successfully keep pace with the very possibilities opened up by the experimental systems that were being studied. What changed was not just the approach to doing things, but rather the very nature of the problem; no longer was the question "How do we generate enough information to create meaning?," but rather "How do we sift through the information we have to create meaning?" Even the meaning of data came to be at stake through these interdisciplinary encounters.

For me, this has deep parallels to the challenges confronting ethnographic projects that are theoretical in scope and ambition. The problem of theory today is not simply to "make sense" of things we see as empirical observers of the world through frameworks that are already given. It is, instead, to make sense of a problem when the very nature of the problem is at stake.[23]

I wish to suggest therefore that the theoretical problem of this volume is not to come up with the "theory of" life or capital or governance or globalization or markets or neoliberalism, but is rather to come up with *forms of inquiry* that are adequate to studying a contemporary conjuncture of the life sciences and capital that contributors to the volume believe is of world-historical significance. My own investment in Marx, for instance, has nothing to do with whether he was "right" or "wrong" in a predictive sense, but has everything to do with the fact that he developed a mode of inquiry that allowed an investigation into capital in ways that political economy, because of the ways in which it had framed the problem in advance, was unable to do. Similarly, the theoretical work of this volume is to think through the modes of inquiry that might be adequate to studying the life sciences in all its historically, socially, and politically situated manifestations, at a moment when, in Michael Fischer's formulation, "life is outrunning the pedagogies in which we were trained" (Fischer 2003, 37).

Broadly speaking, there are four distinct nodes of emphasis in this volume. First, the most Marxian chapters focus directly on theorizing capital, occasionally (though not consistently) through a close reading of Marx. Second are chapters that consider questions of property, primarily through a study of intellectual-property issues in the life sciences and the ways in which these controversies have redrawn boundaries of the public and private. Third are chapters concerned most directly with questioning the spatiality of biocapital and engaging the question of what "the global" means. And fourth are chapters that are concerned most directly with the affective dimensions of lively capital—the "surplus" constituted by obligation, desire, and love, which cannot be captured by quantifiable political economic measures. Elements of all of these emphases—concerning capital, the global, property, and affect—are to be found in varying degrees in a number of the contributions.

Overview of Contributions

This volume is divided into four parts. Part I, "Encountering Value," engages biocapital through different readings of value, showing how these forms of value are always already relational and implicated in the creation of certain types of subjects. Part II, "Property and Dispossession," focuses on the constitutive place of property in the establishment of various forms of biocapital, as well as the expropriation that forms the conditions of possibility for these property regimes to be instantiated in the first place. Part III, "Global Knowledge Formations," focuses on the epistemic and institutional rationalities within which the life sciences can be understood at transnational and global scales. And Part IV, "Promissory Experiments and Emergent Forms of Life," focuses on the affective and ethical dimensions of hype and hope that are constitutive to emergent and experimental forms of biocapital.

Part 1 begins with Joseph Dumit's experimental reading of volume 1 of Marx's *Capital*, "Prescription Maximization and the Accumulation of Surplus Health in the Pharmaceutical Industry: The_BioMarx_Experiment." Dumit attempts to understand logics of contemporary pharmaceutical capital through Marx's *Capital*, in the process of which he discovers that the latter's analysis holds up surprisingly well in describing the risk logics that the former operates through (Marx 1976 [1867]). The key issue for Dumit is understanding *surplus* in pharmaceutical logic, given the fact that for Marx it is surplus value that constitutes the moment of exploitation of the worker. Dumit comes up with the notion of "surplus health," which is his analogous concept to surplus labor. He defines surplus health as "the *capacity* to add medications to our life through lowering the level of risk required to be 'at risk,'" a lowering that occurs through the setting of biomedical risk thresholds.

Dumit is tracing a particular logic, a logic that acquires its power through the fact that it speaks in a series of potentialities—the potential for future illness, the potential for future market growth, the potential for greater therapeutic consumption among ever larger segments of the world population. Like Jeremy Bentham's panopticon (Foucault 1977), this logic does not have to be realized in an actual visible structure for it to create its effects and affects. The risk threshold is the key to the operation of this potentiality; it is that which is set by biomedicine in order that its logic might unfold, in a manner analogous to the way in which the setting of wage by capital is the moment at which the conditions of possibility for the creation of surplus labor are set up and calibrated. In this logic, market size does not matter as much as market potential does, because the former is simply a retro-active indicator of events already gone by and therefore not a true indica-tor of value, which is always future oriented; just as in industrial capital, the realization of value occurs not through profit (the actual, quantifiable amount of money made over that expended), but through surplus labor (the potential for labor productivity in excess of that remunerated by wage). The logic of capital here is therefore thoroughly speculative.

The other chapters in part I locate Dumit's analysis of pharmaceutical logics of value generation next to other forms of value, thereby showing how even the surplus value that Dumit traces in rigorously Marxian fashion is animated in various ways by what Donna Haraway calls "encounter value." Haraway's "Value-Added Dogs and Lively Capital" argues that Marxian cate-gories of use and exchange value are relationally constituted and there-fore mediated by *encounters*. In other words, the fundamental categories of Marx's labor theory of value are subjective.[24] She develops this argument through an account of dog-human interactions wherein dog worlds are in-creasingly technocorporate.

Haraway focuses on dogs in part because of the constitutive history of canine labor power in the story of capital, a history that is not a part of the dominant narrative of capital as it gets told. But the insertion of dogs into systems of capital is not just as labor, but is rather threefold: in contempo-rary U.S. culture, dogs are also, themselves, commodities, as well as con-sumers of commodities. In an uncanny link to Dumit's chapter, Haraway points out that the size of the pet-food market in the United States in 2003 was equal to the size of the statin market that year.

The larger conceptual argument about exchange that Haraway is making here is one that has been made by a number of feminist science-studies scholars, which is that questions of exchange are always underwritten by those of kinship.[25] Kinship, in this context, operates at two levels. The first is at the level of the human-animal familial bond, while the second is in the importance of tracing canine pedigrees. Haraway points to the "human-animal companionate family" as a diagnostic marker of contemporary U.S. capitalism, which sees a "productive embrace of kin and brand." And so, sta-tistical issues such as the size of the pet-food market have to be considered in relation to affective issues in particular biopolitical situations involving dogs' health and illness, their life and death, where these situations often manifest *as a consequence* of the dogs' subject-position as a kin member in human households. For instance, Haraway's question "How does a companion animal's human make judgments about the right time to let her dog die?" speaks, on the one hand, to the increased salience of such questions when biomedical technologies of intervention allow canine lives to be prolonged through the diagnosis and treatment of illness in ways not possible before. But it also feels like a resoundingly familiar question to those concerned with the human impacts of new biomedical technologies, a question that (at the time of its writing, in November 2004) provided an uncanny anticipation of such debates that took political center-stage in the United States, through 2005, in media spectacles such as that surrounding the death of Terri Schiavo.

Timothy Choy's chapter, "Air's Substantiations," shows how the air in Hong Kong is simultaneously felt experience, pollution, literary imagery, and the substance of transnational capital negotiations. Choy therefore locates how the environment simultaneously becomes a health problem and an economic one. Air *matters*, and this mattering refers both to the materiality of air and to the fact that it is an object of concern, care, and investment among those who are affected by it. Choy's attempt in his piece is to layer the various ways in which it matters. He is concerned with "four forms of air . . . (1) air as medical fact, (2) air as bodily engagement, (3) air as a constellation of difference, and (4) air as an index for international comparison."

This superimposition of different forms of air points to the "warp and woof of the network being woven." These are global networks of fact, capital, affect, and lived experience. The manner in which Choy investigates air—as simultaneously material and abstract, as object and signifier, as circulating matter and as an object that we enter into constitutive lively relationships with, and as something that, while often taken for granted, functions at multiple registers simultaneously—shows an uncanny resemblance to the way in which Marx considers money, in *Grundrisse*, as, simultaneously, means of exchange, measure of exchange, universal equivalent, and as world money (Marx 1993 [1857]).

These interweaving registers produce Hong Kong as at once a global and a local site. They are saturated with the particularities of Hong Kong's historical and cultural specificity, but are also entirely dependent on locating Hong Kong's air as part of the global environmental problem that transcends boundaries and place. Choy therefore talks about air in Hong Kong in terms of what he calls a "poetics of place," where "place" does not signify *locality* as much as it signifies the particularities that are located within, and in turn both shape and defy, larger structures.

Property is the central analytic that structures part II of this volume. The chapters by Sheila Jasanoff and Elta Smith focus on intellectual-property-rights issues in biotechnology. Jasanoff's chapter, "Taking Life: Private Rights in Public Nature," looks at the way in which the law constructs the boundaries between public and private by considering landed property (through a reading of the U.S. law of takings) against intellectual-property (IP) debates surrounding new biotechnologies. The takings clause considers the contexts in which the state can appropriate land for "public" purposes. Jasanoff argues that what constitutes a public purpose in fact involves a calculus of a range of material against nonmaterial values, where the material values often pertain to property, and the nonmaterial ones concern abstract concerns of "public good" that by definition are hard to quantify. In this calculus, property rights often function as the limit to taking land for public use.

Intellectual property is a set of rights given to inventors, thereby signifying "invention," which is man-made, and has to be distinguished from a "discovery," which involves finding something that is deemed to occur naturally. The process of granting IP rights involves, as Jasanoff puts it, "a removal of the thing being patented from nature to culture." A major way in which nature is converted to culture by law is to convert nature into property, which can subsequently be rendered liquid and made to circulate. The diagnosis of these bounded categories—"invention" and "discovery," which map onto the boundary between "nature" and "culture," where what is at stake is "public" and "private"—as typical of modernity is central to Bruno Latour's analysis in *We Have Never Been Modern* (1993). Rather than leave it at diagnosis, however, Jasanoff empirically investigates the steps that need to be taken in order to shift an object through a particular legal materialization from a "natural" to a "cultural" domain.

A crucial incongruence, however, is set up between the form of law and its content. There is a fundamental formal aspect to the functioning of IP law in advanced liberal societies, one which sees it functioning consistently in certain ways to demarcate nature from culture, public from private. And yet, legal outcomes that follow the same logic of the law could vary radically, as seen in the different outcomes on the Harvard OncoMouse patent (which was held to be valid in the United States and subsequently in most advanced liberal countries, but denied by the Canadian Supreme Court).[26]

Understanding these different interpretations of patent law in the Onco-Mouse case, according to Jasanoff, involves understanding both the different conceptions of life and liveliness that are stake in the two situations, and the different imaginations of the public good that are at play—different abstractions that animate particular legal materializations.

Elta Smith's chapter, "Rice Genomes: Making Hybrid Properties," relates property to health via nutrition through a study of rice-research endeavors. The discourse of rice genomics is one of food security for the Third World as well as one of commercial value. Indeed, Smith shows that much of the primary value of rice as an object of genomic research comes from the fact that it serves as such a good model organism for more profitable food crops like maize, wheat, and soy. Hence, nutrition or food security is increasingly articulated through and constrained by commercial viability or profitability. In addition to this institutional movement is a movement within rice research toward an emphasis on genomics, where information becomes a key material object whose access and ownership becomes essential to regulate. The corporatization of biology occurs simultaneously to biology becoming more of an information science. The multiple material forms of rice—"as scientific (genetic and genomic) information, as a model cereal, as a major food staple, as a cultural icon"—are as crucial to trace as the multiple institutional forms that emerge to conduct research on rice.

Smith writes about four different rice-genome projects, of which two were publicly funded and two privately funded. She shows that in fact it is difficult to demarcate what constitutes private property from what constitutes the public domain in each of the disclosure arrangements that surround these efforts. Rather, a spectrum of property forms emerge, none of which are purely "public" or "private," but are a hybrid of the two.

As the environment of rice research becomes increasingly corporate, research priorities shift toward where the markets are. Rice, being primarily a Third World staple crop, is not particularly attractive in this regard. But this profit-making imperative is in tension with a philanthropic imperative that has become increasingly articulated out of corporate environments. Thus, while rice is an object of corporate research interest because of its value as a model organism, the corporate discourse surrounding rice research is one that constantly emphasizes the social responsibility that is being fulfilled by research that is geared toward meeting Third World nutritional needs. However, Smith shows that the public beneficiary of "public" research is never clearly articulated. None of the intellectual-property debates around rice research even touch on questions of the commodity status of new or better rice strains, questions of seed availability, or distribution mechanisms, all of which are essential to food security.

Jasanoff's and Smith's chapters look at the workings of intellectual property, in law and in practice. In the process, fundamental ontological questions—concerning nature, culture, value, the boundaries between commodities and objects in nature, or between the public domain and private property—all come to be at stake. The next two chapters, by Travis Tanner and Kristin Peterson, locate these ontological questions in terms of historical conditions of possibility; specifically, in relation to the forms of dispossession that the authors argue are necessary in order for regimes of contemporary biocapital to function. This section, therefore, sets up questions of property, which are themselves unsettled and in formation, in relation to forms of dispossession that provide the grounds upon which, as Marx also argued, capital emerges and functions.

Tanner's chapter, "Marx in New Zealand," is concerned with accumulation in the context of the genetic dispossession of indigenous peoples through projects such as the Human Genome Diversity Project (HGDP). He argues that such dispossession is not merely material (as in the extraction of human genetic material and its subsequent commodification through intellectual-property protection), but also cultural. The HGDP itself was a failed project, but similar practices of genetic archiving continue under new guises, and indigenous populations are often subject to them. These archiving practices, Tanner suggests, are not just about the production of value (whether scientific or commercial), but also of meaning. Such projects lead to the creation and reinscription of certain grand narratives—of capital and of modernity—that indigenous populations are subsumed into through their dispossession. What gets dispossessed, in part, is the possibility of alternative narratives to the grand genomic narrative, which is always already inscribed within a grand narrative of capital.

The question of narrative is crucial for Tanner. Hence, he approaches his argument not through an ethnographic study of technoscience, but rather through the reading of native literature. He focuses on Patricia Grace's novel *Baby No-Eyes* (1998), which is about Maori genetic dispossession based on a case of genetic theft that occurred in Wellington Public Hospital in 1991. This novel, and hence Tanner's argument, situates the question of biomedical ethics in the context of a larger and ongoing history, the struggle for land rights by indigenous people in New Zealand. Questions of genomic ethics, in Tanner's argument, cannot be divorced from this larger historical contextualization concerning landed property; but they also cannot be divorced from questions of narrative representation about indigenous peoples' stories. Tanner attempts to address these questions of narrative representation alongside those of legal and political representation.

The narratives of past and future get weaved together in the context of native populations as targets of genomic sampling experiments in that these experiments—seen by native populations as dispossession of something that is theirs and that is sacred—mirror the earlier dispossessions of land that were essential, in the first place, in making native populations fragile, on the verge of "extinction," and worthy of this kind of sampling effort. Therefore, Tanner's chapter has deep concerns with notions of property and ownership—are genes something that "belong" to their bearer, or can they become the intellectual property of scientists experimenting on them?

Kristin Peterson's chapter, "AIDS Policies for Markets and Warriors: Dispossession, Capital, and Pharmaceuticals in Nigeria," argues that Africa is a site that gets "emptied out," extracted of its resources, and understanding this active dispossession is essential to understanding the nature of contemporary capital flows *into* Africa. Peterson's chapter concerns the nature of capital flows, but equally of Africa's relationship to globalization. Her insistence is that Africa is not simply a marginal entity that is outside the global, or "left behind." Rather, it is constantly articulated into the global through trade, aid, development, and economic policy. In Peterson's rendering, an analysis of capital that focuses on wealth accumulation (whether through speculation or manufacturing) is not enough, because that accumulation is always undergirded by other logics—of wealth extraction and dispossession—that are simultaneously at play. Further, Peterson shows that accumulation by dispossession is not something that comes before "real" capitalism, but is an ongoing process that is a necessary condition of possibility for global capital to function.[27] This chapter provides an interesting contrast to Dumit's analysis, for what is at stake here is *access* to drugs, rather than an excess of them. And yet, access becomes a problem for Africa at the same time that excess becomes a problem in the United States through very similar neoliberal logics.

Access itself becomes a problem not just because of a constitutive lack of capacity, as is often portrayed, but because of specific, historical moments of dispossession that emptied out capacity. The implementation of World Bank and International Monetary Fund (IMF) mandated Structural Adjustment Programs (SAPS) in 1986 is a crucial moment in this regard. Peterson argues that SAPS eviscerated Nigeria's pharmacy industry, thereby destroying its drug-distribution infrastructure. This means that even when drugs now flow into Nigeria through global aid schemes, there are no adequate mechanisms for their distribution. The military must therefore be deployed to distribute drugs, hence articulating pharmaceutical economies with military ones.

Part III begins with Andrew Lakoff's piece, "Diagnostic Liquidity: Mental Illness and the Global Trade in DNA," which elaborates the theme of global trade. Lakoff traces the French biotechnology company Genset's attempts to look for genes for psychiatric illness in Argentine mental patients, showing the ways in which both institutional arrangements and epistemologies are at stake in such efforts. The key here is establishing the "potential universality of genomic knowledge about mental health," speaking to the key themes of *potentiality* and *universality* that emerge in other chapters in the volume. Establishing such universality would "render liquid" the mental illness experienced by Argentine patients as "globally" valid scientific information that can travel as a potential commodity. In the process, Lakoff shows that the very question of what constitutes a psychiatric illness is at stake.

The key issue of universality or particularity however is also an epistemic question, which, as Lakoff puts it, is one of "how to know . . . whether a case of bipolar disorder in the United States [is] the same 'thing' as a case of bipolar disorder in Argentina." These questions become constitutively ingrained in the problem of making the Argentine DNA liquid. This problem of epistemology is, further, not just a scientific experimental problem of how to identify or classify bipolar disorder: it is also linked to comparative questions of the nature of medical practice in Argentina as compared to Europe or North America, with the clinical diagnosis of bipolar disorder as such being much less common in the Argentine context. Therefore, while protocols for DNA sample collection or ethical regimes for informed consent could be quite easily standardized in the Argentine context, ventures such as Genset's also require a prior standardization of medical information, a much more difficult goal to achieve in practice.

A further complication arises from the fact that, as Lakoff shows, "bipolar disorder" is itself a biomedical category that has evolved along with the rise of a pharmaceutical-centered approach to psychiatric treatment. The purported universality of patients with bipolar disorder, therefore, hits up against the particularity of "bipolar disorder" as a disease category in the first place. And this particularity is highlighted by the Lacanian psychoanalysts who treat mentally ill patients in a ward adjacent to that in which, in the same hospital, the medical geneticists are collecting DNA samples from Genset. Complicating the epistemic difficulties faced by the medical geneticists in the story is the inherent subjectivity of psychoanalysis, which also makes scientific claims, but in opposition to the objectivity and classificatory impulses that lie at the heart of the epistemology of molecular genetics.

Wen-Hua Kuo investigates imperatives of standardization in his chapter, "Transforming States in the Era of Global Pharmaceuticals: Visioning Clinical Research in Japan, Taiwan, and Singapore." Through a comparative study of Japan's, Taiwan's, and Singapore's responses to imported drug products, Kuo looks at the moves toward regulatory standardization that occur when clinical trials become global and travel to Asia. This analysis sees the articulation of concerns with trade, with health, and with race.

Kuo's empirical material is drawn from the International Conference on Harmonization of Technical Requirements for Registration of Pharmaceuticals for Human Use (ICH). The primary attempt here is to "create a universal standard for drugs by neglecting as much as possible bodily differences," wherein "one will finally fit all." This is a logic that is deeply at odds with the personalizing logics of pharmaceutical marketing that Dumit describes, pointing to capital's contradictory impulses to universalize target populations (a move that consolidates potential markets instead of segmenting them) while individuating the subjects of therapeutic intervention (an individuation that is at the heart of neoliberal logics of self-governance).

Japan's particularity was its insistence that clinical trials be reproduced on Japanese populations in order for the results to be deemed applicable for drug marketing in Japan. Underlying this insistence was an assumption of fundamental racial or ethnic difference. In contrast to Japan, Kuo shows, Taiwan embraced ICH guidelines. This was a strategic assertion of *national* identity that showed Taiwan's ability to set regulatory guidelines for itself independent of the People's Republic of China. Meanwhile, Singapore's incentive for embracing harmonization is economic, with the burgeoning local biotech industry likely to be a beneficiary.

Kuo shows that each of these national responses attempts to authorize itself through science. Therefore, Japanese scientists are shown to embark on research that points to stratification and population uniqueness, such as population genomics. At the same time, Taiwanese researchers adopt, emphasize, and deploy other types of statistical data in order to justify their inclusion in ICH regimes. Singapore's case is tricky because of its highly multiethnic population, and so the state and researchers "just ignore racial difference" altogether. A comparison of these scientific strategies highlights the arbitrariness with which race or population or biology get accounted for or discounted in biomedical regimes. Yet it is precisely such arbitrary moves that set the conditions of possibility for subsequent "scientific" experimentation and the authority that stems from such science.

Environmentalism and scale-making are the subjects of analysis in Kim Fortun's chapter, "Biopolitics and the Informating of Environmentalism." Fortun looks at how informatics mediates political transitions at a number of scales. She does so by looking at how environmental informatics takes shape at a number of sites, such as environmental websites like scorecard. org; the deployment of environmental "worst-case scenarios" in politics and policymaking; or the ways in which informatics capacity leads to the elucidation of environmental factors involved in increased risk of asthma. In all of these sites, environmental informatics becomes a site of biopolitics. In Fortun's words, information technologies become "drivers of change at multiple scales." In this sense, "informationalism" becomes the analog to the industrialism that was at the heart of Marx's analysis of nineteenth-century capital.

The scalar dimension is key here, and is what distinguishes Fortun's analysis from linear-progress narratives of the glories of innovation. Indeed, Fortun is precisely not interested in informatics as simply a site of innovation, but rather in the ways in which it becomes one of "cultural production and ethical action." But the production of informatics deserves as much ethnographic attention as its productivity. This is because there are multiple investments in informatics, by radically different institutional actors (chemical companies as well as environmental activists, for instance), where these investments are simultaneously material, ethical, and affective. The production of informatics is "experimental," in that it is a scientific activity that is constantly being worked through and figured out, where the outcomes are not known in advance or calibrated against a readily testable hypothesis. But the production of knowledge through informatics is not just determined by the scientific experimentation that goes into the creation of websites or risk scenarios; it is also crucially dependent on how such information can be *made public*.

Fortun, therefore, maps a relationship between environmental-risk information and the public domain that is intensely political and that is, as much as anywhere else, shaped through policy initiatives and their interactions with corporate interests and grassroots mobilization. Unlike Smith, who shows the way in which informatics and public-domain issues get politically contested around the question of *property*, Fortun is concerned with the way in which *security* becomes the locus of this contestation.[28] Informatics becomes the medium through which uncertainty (in this case, environmental risk) becomes engaged. Unlike in Dumit's case of pharmaceutical marketing, however, this engagement is not yet overdetermined or controlled by corporate interests. Instead, Fortun argues that informatics leads to the creation of "discursive gaps" within environmentalism that provide the potential for various sorts of political action, including possibly revolutionary ones. While the setting of clinical risk thresholds is an act of scientific expertise in which the experts are exclusively those sanctioned by the state or by corporations, the articulation of environmental risk is an expert discourse where "expertise" is constituted in a more polymorphic and potentially democratic fashion.

Part IV begins with Michael Fortun's chapter, "Genomics Scandals and Other Volatilities of Promising." Fortun concerns himself with the temporality that is ingrained within promising, which is the temporality of "a future still to come" (in opposition to a teleological future, a future that will be).[29] The empirical material that he draws on is the story of the controversial Iceland-based genome company DeCode Genetics, which is archetypical of speculative capital.

Fortun traces the grammar of hype. The hype of genomic corporate discourse is deeply entangled with the epistemology of genomics, which also, constitutively, concerns itself with statements about the future. The promise in terms of scientific potential that genomics carries is inseparable from the promises that genome companies make about their potential as value generators in capital markets. The promises of genomics are therefore twofold, concerning themselves both with technoscientific possibilities and with expressing promises that are inscribed within the genome itself. In Fortun's words, "Being promising . . . is a fundamental aspect of the being of that favorite model organism of genomicists, a human."

On the one hand, then, the promise is "ubiquitous and unavoidable," whether in language (as Fortun shows through reading Derrida), in the market (as he shows by reading corporate filings with the U.S. Securities and Exchange Commission [SEC]), or in the body (as he shows by reading genomics). Indeed, that promising is not just a language game is evident in the ways in which promising is made a material necessity, in both genomics and in capital, *by the law*. Fortun traces how this process unfolded in high-tech in the 1990s, especially the crucial role played by changes made by the SEC in 1995 to the safe-harbor provisions of forward-looking statements. These changes made it easier to protect forward-looking statements—which are statements released by companies that speak directly or indirectly to their revenue or growth potential in ways that could significantly influence investors—from anti-fraud litigation. The materialization of hype through changes in SEC provisions made hype both essential and legally mandated. Far from being deemed an act of irrational excess or dishonesty, hype became constitutive to the discourse of high-tech capital, in ways that were nontrivial and had consequences—not least for ordinary Icelandic investors in the stock market.

Situations of commitment and indebtedness within families are the explicit themes of Chloe Silverman's chapter, "Desperate and Rational: Of Love, Biomedicine, and Experimental Community." Silverman writes about the ways in which family, kinship, and commitment are all evident in autism research. Autism constitutes a poorly understood, untreatable spectrum of disorders; therefore, when parent-advocates of children with these disorders call for treatment, their requests seem like desperation. In the process of making such calls, parents of autism-afflicted children frame the disorder as biomedical, thereby rendering it potentially treatable. This forcible rendering of a condition as biomedical has interestingly different resonances in the case of autism than it does in the context of bipolar disorder as addressed in Lakoff's chapter. While in the latter case, biomedicine constitutes the hegemonic discourse of global capital and reductive, corporate-driven genome science, in the setting of autism it is biomedicine that forms the "alternative" discourse.

Silverman's chapter traces the way in which this biomedical knowledge is produced in parent-practitioner communities. Empirically, it is an account of a parent advocacy project, Defeat Autism Now! (DAN!). If Michael Fortun's chapter described a political economy of hype, then Silverman's describes what Carlos Novas and Nikolas Rose (2004) refer to as a "political economy of hope." What Silverman shows is the parents' faith and investment in biomedicine, a particularly desperate and irrational kind of faith in reason and rationality. They hold a belief that biomedical experimentation will provide therapeutic outcomes for their suffering children, but this can only come about if parents—many of whom are outside the circuits of expertise that constitute the institutions of biomedicine—are involved in bringing about such experimentation themselves (either through funding, or advocacy, or collaborations with researchers, or in some cases, through getting involved themselves in science). Their faith in science is entwined with the love that animates this faith. Faith and love here are not opposed to objectivity, but rather are involved in the creation of new forms of objectivity.

Such familial investment is hardly innocent or painless. Indeed, it involves making the child an object of therapeutic experimentation, with all the risks that attend such subjection. Questions concerning such experimentation are simultaneously epistemic and political. Choices are being made and research agendas are being set, but by agents who do not fall into the categories of actors—scientific or corporate—who normally are vested with the authority to do so. One of the promissory terrains here therefore involves the reconstitution of expertise, thereby potentially leading not just to new therapies for autism, but also possibly reconstituting the very grounds of biomedical knowledge production.[30]

Michael Fischer's chapter, "Lively Biotech and Translational Research," concerns itself with experimental epistemology, institutional development, and the "peopling" of technoscience. It is written as a memorial to the late Harvard researcher Judah Folkman, one of the pioneers of angiogenesis research, which came to have significant applications in the development of anticancer therapies. Fischer begins by narrating Folkman's "Decalogue," an account of the multiple challenges involved in translating experimental research into the clinic.

But there are other sorts of translations, and translational challenges, which are at play in the different tracks of Fischer's narrative, for example, interdisciplinary translations between experimental and informational science; institutional translations between laboratory and market; the translation of skills between mentors and students, or between laboratories; global circulations of knowledge through the mobility of graduate students and postdoctoral researchers; and transnational translations that are occurring through these circulations, as well as through institutional endeavors (especially in Asia) to create new types of collaborative alliances that are not necessarily centered on the traditionally powerful research centers of the West. Hence, while earlier chapters in the volume (such as Tanner's and Peterson's) focus on the dispossession that is constitutive to the operation of technocapital, the Fortuns's, Silverman's, and Fischer's chapters open up the question of technocapital's experimental nature and promissory potential. This promise is not simply articulated as a glorious narrative of inevitable innovation, but rather is worked through in terms of the constitutive opportunities and challenges that are inherent in working within these promissory arenas for its practitioners.

Fischer shows that the investments in emergent epistemic, institutional, and "peopled" forms of the life sciences are not just financial or epistemic, but also speak to individual biographical histories of migration, reflect different national sentiments and scientific priorities, and call not just on structural constraints, but also on affect, aesthetics, and a whole range of tacit skills—what might be considered, following Choy's chapter, a "poetics" of lively biotech and translational research. But these stories are not merely contingent; they speak to larger structural and historical issues. For instance, why are certain research initiatives privileged over others? When? By whom? What are the implications, for instance, of Fischer's account (through a conversation with a Chinese postdoc based in the United States) of agricultural biotechnology falling "out of the loop" in the United States, such that cutting-edge work, where the fame and profit are to be found, ends up being almost entirely in the domain of biomedicine? If the life sciences, among other things, claim to provide us with knowledge about what it means to be human, then who gets to decide what kinds of work end up enabling or representing those claims? And how do these decisions operate in the spaces and double-binds that exist between the structural and the improvisational, the expropriative and the promissory, between "what we have been or what we will be" (Derrida 2002), the lively spaces of technocapital, which are also the spaces for the emergence and contestation of the epistemological, the ethical, and the political?

Notes

1. According to Cynthia Robbins-Roth (2000), 11 percent of all federal research-and-development money was allocated to basic biomedical research, and the National Cancer Institute alone was spending nearly a billion dollars annually on basic research by 1981.

2. For an account of the commercialization of the research university in the 1970s, with a specific focus on the Cohen-Boyer recombinant DNA patents and Stanford's licensing strategies, see Hughes 2001.

3. The RDT patent was jointly held by Stanford (where Cohen was employed) and the University of California (where Boyer was employed), but Stanford handled the licensing of the technology.

4. Maryann Feldman, Alessandra Colaianni, and Kang Liu document the actual commercial value of the Cohen-Boyer patent. The value is measured not just in terms of the market value of the technology itself, but also in terms of its function as a platform for downstream product development. The patent generated $35 billion in sales for 2,442 downstream products. The technology was licensed to 468 companies. By the end of 2001, Stanford and the University of California had generated $255 million in revenue from the patenting and licensing of the technology. The patent application was filed in 1974, and the patent was granted in 1980, the year in which *Diamond v. Chakrabarty* authorized the patenting of life-forms and in which the entrepreneurial university received legislative sanction through the Bayh-Dole Act. The patent expired in 1997. See Feldman, Colaianni, and Liu 2007 for an elaboration of this case,

which is regarded in many ways as a gold standard in strategic university technology licensing.

5. I draw extensively on Feldman, Colaianni, and Liu 2007 in this account of Stanford's RDT licensing strategy.

6. In *Biocapital*, I have argued for the importance of staying attentive to what Rosemary Coombe refers to as an "ethics of contingency with respect to commodified social texts" (Coombe 1998, 5; Sunder Rajan 2006). This was in response to the unfolding politics around DNA patents, which were broadly condemned as the ultimate signifier of the commercialization of the life sciences (reflecting, as it were, the commercialization of "life itself"). What did such a position signify, I wondered, when major proponents for gene patenting included patient-advocacy groups for rare genetic diseases, who argued that being about able to take out patents on gene sequences expressed in such diseases was one of the few mechanisms to ensure that they could have some control and direction over research into these diseases; and when major opponents of gene patenting included big multinational pharmaceutical companies, who did not want their downstream product development fettered by upstream property rights and licensing arrangements?

7. For Merton's normative structure of science, consisting of the four norms of universality, disinterestedness, communism, and organized skepticism, see Merton 1942.

8. Indeed, one can see this trajectory, wherein deep contestation regarding the very grounds of capitalization gives way to a more naturalized acceptance of such capitalization (even if the mechanisms and strategies of capitalization are varied), in other parts of the world more recently. In the late 1990s and early 2000s, as the Indian state pushed its scientific establishment to become more entrepreneurial in its approach, I noted debates in India similar to those that had taken place in the United States in the 1970s.

9. For Paul Rabinow's development of an idea of method adequate to the anthropology of the contemporary, see, for instance, Rabinow 2003. Rabinow's use of the concept of assemblage, derived from Gilles Deleuze, is relevant here (Rabinow 1999). Rabinow defines assemblages as contingent articulations of heterogenous elements; and it is the mapping of these contingencies that Rabinow sees as important for what he calls concept work. An elaboration of thinking with assemblages as a methodological solution to what they call "anthropological problems" is provided by Aihwa Ong and Stephen Collier in their introduction to *Global Assemblages* (Ong and Collier 2005). Jasanoff's idea of co-production is something that forces us to go beyond the contingent as explanation or solution; and Jasanoff herself, while deeply attendant to the contingent and the particular, should not be misread in a manner that simply allows one to substitute complex contingency for co-production.

10. The best-known example of Weber's method in action as relevant to this argument is, of course, his analysis of the relationship between religion and capitalism in *The Protestant Ethic and the Spirit of Capitalism* (2008). For Weber's elucidation of the ideal type and its role in sociological method, see Weber 1978.

11. Engels letter to Bloch, 1890 (Engels 1972 [1895]). I am grateful to Dominic

Boyer for conversations on the relationship of ideology to materialism, which have helped me think through the question of determination "in the last instance."

12. Indeed, the word *ideology* virtually drops out of Marx's later writings. See Balibar 1994 for a reflection on this fact.

13. The idea that life is information has been very much a part of the central dogma of molecular biology since the 1950s, and sets the stage for more recent emergences in the life sciences, such as recombinant DNA technology in the 1970s, or genomics in the 1990s. For historical and philosophical reflections on the life sciences becoming more molecular or informational, see Doyle 1997; Doyle 2003; Jacob 1993 [1973]; Kay 2000; Keller 1995; Keller 2002b.

14. For the seminal essay that speaks to this "traditional" ethos, see again Merton 1942, which outlines the normative structure of science at that time—norms that differ radically from the corporate scientific ethos that prevails today in the life sciences.

15. Such a review would undoubtedly be a useful task, but it would be a different task from introducing *this* volume, which is a particular reading, representation, inflection of such work writ large. Apologies, therefore, that crucial bodies of work in the life sciences and society remain uncited. A useful resource to consult for such a review would be Stefan Helmreich's overview of what he calls "species of biocapital" (2008).

16. On SSK see Bloor 1991 [1976]. On ANT see Latour and Woolgar 1986 [1979]; Latour 1987; Latour 1988; Callon 1986.

17. For work in this early MIT STS vein, see, for instance, Noble 1979; Weiner 1987; Winner 1978; Winner 1988.

18. The list of relevant works in this genre is too extensive to cite here. But some key works include Keller 1984, concerning women in science; Keller 1992, Martin 1991, and Cohn 1987, about the gendered nature of technoscientific discourse; Rapp 2000a, for an exemplary account of women's experiences of new reproductive technologies; and Haraway 1991, for essays explicitly situating technoscience within feminist praxis.

19. See Martin 1998 for an overview of the role of anthropology in the cultural study of science and technology.

20. Particularly important in this regard is programmatic writing on anthropology and ethnography by key contributors to *Writing Culture*, such as Michael Fischer, George Marcus, and Paul Rabinow. See for instance Fischer 2003; Fischer 2007a; Fischer 2007b; Marcus 1995a; Marcus 2005; Marcus 2007; Rabinow 2003. Over the past two decades Fischer and Rabinow have become central figures in STS, not just through their own work, but through their mentoring of a generation of graduate students doing ethnographic work on science and technology. Marcus has been a curious, interested, and partisan onlooker of and interlocutor with the anthropology of science.

21. See for instance Kay 1996; Kay 2000; Keller 2002a; Monod 1971.

22. I can illustrate this through personal experience of gene-expression studies. Just a little more than a decade ago, I was engaged in a master's research project which involved studying gene expression in the slime mould *Dictyostelium dis-*

coideum. Around the time I was finishing my degree, in 1996–97, the lab I worked in adopted a new method called RNA differential display. What differential display allowed was the ability to study gene expression in two different samples (say, between a wild-type and mutant slime mould) in a single experiment. This study of differential gene expression could allow a comparison of the functioning of certain genes in different states (for example, between different tissues, or between different stages of development within a single cell type, or to look at whether a gene is turned on or off in particular situations of stress or disease). However, one already had to know which gene one was interested in to start with; and such a project could constitute an entire doctoral dissertation. Within the next three years, however, high-throughput technologies such as the Affymetrix DNA chip had been developed, which allowed one to compare two entire genomes of interest for differential DNA expression in a single experiment. For example, a major early publication that showed the utility of the Affymetrix chip used it to differentiate acute myelogenous leukemia from acute lymphocytic leukemia (Golub et al. 1999). This was a landmark paper because it enabled the classification of these two cancers based on the differential gene expression patterns of fifty genes, without any prior biological knowledge. Normally, tumor classification would require clinical, pathological, and cytological analysis. Classifying these cancers is of crucial importance in choosing the right treatment regimen, and the regimens for these two types of leukemia vary considerably. Often cells that follow different clinical courses look similar in biopsies, and traditional diagnosis of one or the other form of leukemia requires a complicated battery of tests. The DNA chip enabled this form of cancer diagnosis to move away from systems based on visual analysis to molecular-based systems, and in this experiment, allowed the comparative measurement of activities of nearly seven thousand genes expressed in bone marrow samples from thirty-eight patients. In other words, one moved very quickly from a stage of studying the expression of a single gene that was limited by how much prior information one had about that gene, to a stage where one could study the expression of seven thousand genes without any prior information about each of them. This scaling up was entirely consequent to the development of informatics capabilities that could generate and process large amounts of data very quickly.

23. See Rabinow 2003 and Rabinow 2004 on how problematization is the problem to be tackled. This is developed in conversations in Rabinow's Anthropology of the Contemporary Research Collaboratory (ARC), for which see www.anthropos-lab .net. See also Aihwa Ong's and Stephen Collier's introduction to *Global Assemblages*, for the development of the notion of "anthropological problems" (Ong and Collier 2005, 3–21).

24. Jason Read (2003) develops this argument for the constitutive place of subjectivity in Marx's political economy in a more philosophical vein.

25. For examples of such work, see for instance Franklin 1997; Franklin 2007; Franklin and Ragone 1997; Franklin and MacKinnon 2002; Franklin and Lock 2003; Ginsburg and Rapp 1995; Haraway 1997; Rapp 2000a; Strathern 1992a; Strathern 1992b; Strathern 2005.

26. This speaks to Jasanoff's crucial departure from Latour's actor-network

methodology. While Latour's diagnosis of the moves of purification that underlie the modernist enterprise remain valid in both the U.S. and Canadian cases that Jasanoff traces, the fact remains that the status of biotech patenting has emerged as significantly different in the two locales. This reflects the importance of Jasanoff's insistence on comparative methodology even within studies of advanced liberal societies, which has threaded through all of her work but is clearly explicated in the context of biotechnology policy in *Designs on Nature* (Jasanoff 2005a). See also Sperling 2006 for similar methodological moves in his comparative analysis of bioethics in the American and German contexts.

27. "Accumulation by dispossession" is David Harvey's term (2003).

28. Michel Foucault argues that security is one of the pillars of modern biopolitical governmental rationality, what he calls "governmentality." See Burchell, Gordon, and Miller 1991.

29. The distinction is by Derrida (1994).

30. See Epstein 1996 for a similar argument in the context of patient activism around HIV-AIDS.

PART ONE

ENCOUNTERING VALUE

1

JOSEPH DUMIT

PRESCRIPTION MAXIMIZATION AND THE
ACCUMULATION OF SURPLUS HEALTH IN
THE PHARMACEUTICAL INDUSTRY

The_BioMarx_Experiment

The late Roberto Goizueta transformed Coca-Cola in the early 1980s. He had an insight—a simple but stunningly powerful one that he shared with his senior executives. What, he asked almost casually, was the average per-capita daily consumption of fluids by the world's 4.4 billion people? The answer was: 64 ounces. And what, he asked, is the daily per-capita consumption of Coca-Cola? Answer: less than 2 ounces (Charan and Tichy 1998). "We remain resolutely focused on going after the other 62," Mr Goizueta said (Coca-Cola Company 1996, 6). However absurd Goizueta's redefinition of the Coca-Cola Company's market might seem, it has been taken as a key transformative insight throughout the business world, demonstrating that "every business is a growth business" (Charan and Tichy 1998, 435). And that "virtually infinite growth" is a matter of finding the right formulation for the virtual and then actualizing it. One pharmaceutical parallel to global human liquid consumption is illness risk and the capacity to take drugs to ward it off. The more I read about markets and talk with pharmaceutical marketers, the more it seems as if this unbelievable growth in therapeutic consumption to ward off the risk of illness is happening. In this chapter, I elaborate the logics within the pharmaceutical industry that naturalize such growth.

Over the last twenty years, many scholars in medical anthropology and allied fields, including myself, have studied pharmaceutical resistance—pill diversion, strategic noncompliance, complementary and alternative medicine, and organized antimedicalization movements—as positive, creative, agentive challenges to the U.S. health system. While such resistance is widespread, the scale of our analyses has been enveloped (though not eclipsed) by the more general, macro scale of continued growth of pharmaceutical prescriptions.

According to the data,[1] which I have reviewed extensively with statisticians and economists, the average insured American purchases ten to thirteen different prescriptions per year. Over 10 percent of all Americans, and 50 percent of those over forty-five years of age, are buying cholesterol-lowering drugs, a number comparable to that for antihypertensives. The growth rate in users of chronic treatments has been and is conservatively projected to be over 10 percent per year, and the total number of pills taken grows 3 percent per year (Express Scripts 2007). Antidepressant use, despite negative publicity on suicidal side effects, is projected to increase 5 percent a year, and the rates of use remained steady in children even where such drugs were more or less banned. Stimulants, especially attention-deficit drugs, are growing 15 percent per year overall and 30 percent per year among those under nineteen years old. The scale of the market in prescription drugs is $500 billion per year. This does not include over-the-counter drugs, vitamins, nutraceuticals, alternative medicines, and so on.[2] The challenge is to account for the sheer amount of drugs being consumed and the mechanisms of their continued growth.

The numbers arrest me, even if I don't know how to believe them. What has abducted my interest is the problem of accumulation—the state of biochemical accumulation in the bodies of Americans continues at a rate that seems unbelievable, absurd, and unsustainable. At the center of the prescription growth in the United States are clinical trials as the key modality for determining facts about health and treatment, and guidelines that use those trials to redefine illness as a threshold. In this manner, health is reframed in the way that Goizueta reframed the potential consumability of soft drinks as something that can be grown in a virtually unlimited manner.

> "We want to recommend more aggressive treatment to people who are at very high risk," said Dr. James I. Cleeman, the coordinator of the group that issued the guidelines, the "National Cholesterol Education Program" of the National Heart, Lung and Blood Institute.
>
> "And," he added, paraphrasing Shakespeare, "there are more of them out there than are dreamt of in your philosophy." (Kolata 2001, 1)

Cleeman's suggestion is that clinical-trial implications exceed not only what one thinks, but also one's very imagination. To imply a failure of imagination is to propose a need to reframe the problem. The people-at-risk (patients-in-waiting), he intimates, are not visible, even to themselves.[3] If Americans want to be healthy in the future, they must necessarily trust clinical trials and treat the numbers that they propose. Cleeman is talking about guidelines, introduced in 2001, which lowered the recommended level of bad cholesterol, thus tripling the number of people defined as high risk. Within his declaration is an order of reality in which epistemology (where clinical trials redefine high risk) determines that people who fall into the newly identified category *are* now ontologically at high risk, and being at high risk, they should (as in need to, ethically, imperatively) be put on treatment. Even as Cleeman pronounced these words, though, new clinical trials were under way that three years later, in 2004, would take the undreamed of numbers of people he referred to in 2001 and triple *them*, to 200 million. And in 2006 the thresholds were lowered even further.[4]

In a paragraph taken from an article in the *Wall Street Journal* published in 2004, the author emphasizes a set of population statistics that intensify an argument about the dangers of not listening to doctors and clinical-trial data: "Only a fraction of people with high cholesterol are on statins, despite a barrage of drug-company advertising backed up by guidance from public-health officials. About 11 million Americans currently take one of the statins, while some public-health experts say that at least 36 million should probably be on one. Globally, the discrepancy is even more dramatic: About 25 million are taking the pills while an estimated 200 million meet guidelines for treatment" (Winslow 2004, 1). The new target number, of 200 million people worldwide, represented one out of every thirty persons on the planet.

Universal screening programs and mass pharmaceutical regimes continue to regularly appear in the news, with the line between good use and abuse being increasingly hard to draw. The twenty-first century has already seen recommendations for mandatory cholesterol screening starting at age twenty for all Americans and for prescribing standard pharmaceutical treatments for the approximately 30 percent of the population expected to be at high risk when tested. Children are subject to screening for obesity and other risk factors for heart disease in similar ways. Each of these screens works by setting a number, a threshold, which, when crossed, triggers a diagnosis of risk or disease and a recommendation for treatment. Underlying the controversies surrounding mammograms, PSA prostate-cancer tests, and other screens is a concern as to whether, in light of evidence suggesting that lower thresholds might help more people, there could be *any reason* not to make the test more sensitive.

Looking at the growth rates and the projected rates of growth, I am confused. How does "our" pill-taking continue to grow? My conversations with colleagues, doctors, and economists invariably end with the interim conclusion that an ever-increasing prescription rate makes no sense. However, despite actively looking for projections that prescription rates will taper off in the future, I cannot find them. Instead, rates are predicted to grow 8–15 percent per year. While the number of prescriptions must surely, logically, stop growing at some point, it seems there are other logics at work.

Rereading Marx, I heard echoes of this pharmaceutical logic, where increases in productivity paradoxically create more work: "Hence, too, the economic paradox, that [machinery,] the most powerful instrument for shortening labour-time, becomes the most unfailing means for placing every moment of the labourer's time and that of his family, at the disposal of the capitalist for the purpose of expanding the value of his capital" (Marx 1976 [1867]). Could it be that health has become expandable? Doctors, government health officials, pharma marketers, all pronounce the most mystical, teleological sentences, as if they were channeling Marx about an infinite imperative for accumulation, substituting pharmaceuticals in the body for hours of labor: "We want to maximize the number of new prescriptions"; "We want to identify people at risk at the earliest possible point."

Marketers want to maximize the number of prescriptions in order to maximize profits. They see clinical trials as investments whose purpose is to increase sales of medicines: "Important clinical studies to conduct from a scientific or medical perspective are sometimes not important studies to conduct from a drug development perspective" (Spilker 1989, 372). Pharmaceutical researchers openly express their unhealthy predicament: "One of the significant problems for the Pharma industry is that of the 400 disease entities identified, only 40 are commercially attractive by today's requirements of return on investment" (Bartfai and Lees 2006, 14). They see patients as points of resistance: "Pharma's New Enemy: Clean Living" (title on cover of *Forbes*, 29 November 2004).

Just substitute a few words and these marketers and researchers could be quoting *Capital*. This is not surprising, for Marx was, after all, quoting the industrialists of his day. What has shifted are the terms: it is Illness as Value that is now being maximized, and the Health of Patients rather than their Labor that is being exploited. There is a parallel in form: perhaps marketers *see* unproductive health the ways capitalists saw unproductive labor. That is, marketers may see clinical trials as investments that increase the extent and intensity of prescriptions the way capitalists saw machinery as an investment that increased the extent and intensity of labor hours. The grammar and logic of capitalists that Marx studied in *Capital*, in other words, seem to be mirrored by strategies of pharmaceutical executives and marketers.

Step-by-step, the logic is impeccable. Everyone agrees with the basic points and the underlying framework: first, since medicine is so expensive, pharmaceutical companies are required to fund much of the research; second, as companies, they must of course be able to earn a return on these "investments." This framework is not scandalous. If the analogy holds, then it makes clear a strange dynamic: health as a growth field through treatments; surplus health growth via clinical trials. Mickey Smith, the author of a dozen classic works on pharmaceutical marketing, describes this indefinite resource of health as growth. "For as long as everyone is destined to die from some cause, a decline in one can only come at the expense of an increase in another. This is an inescapable truth, yet there seems to be some failure to recognize it. What society, and the pharmaceutical industry to some degree, is doing is making conscious or unconscious decisions about 'tolerable' causes of death" (Smith, Kolassa, Perkins, and Siecker 2002, 32).

Smith is pointing out that if health is defined as reducing risk, then health is an infinite phenomenon, since for every risk you reduce or eliminate, you still have a 100 percent risk of dying from something else. The limit to health research is not, then, a realizable healthy body, but a risk-free body, which instigates a virtually infinite process. There is always room for another study and another treatment, until patients can't take any more treatment due to side-effects, costs, or effort. To put this back in Coca-Cola terms, where Goizueta took the prospective soft-drink market from a measure of people's desires for Coke to the limit of their *capacity* to consume liquid, pharmaceutical companies have redefined health from a measure of symptom reduction to the limit of our body's *capacity* to consume treatments. How many drugs could we be mandated to take?

It looks therefore as if pharmaceutical companies have found a way to grow health through clinical trials, redefining health as treatment, in part by expropriating the means of diagnosing illness, through screening tests that tell us and the doctor that we need treatment. Consequently, the interests of the pharmaceutical industry lie not in reducing treatments, but in increasing them. No matter how obvious this might seem now, I didn't immediately see the connections, even when pharmaceutical researchers said it directly: "No one is thinking about the patients, just market share" (Bartfai and Lees 2006, 73).

The dilemma might be summarized this way: clinical trials are by and large conducted in order to test new treatments for healing a disease state or reducing the risk of future disease. Clinical trials designed to *reduce* the amount of medication people take and still save lives sounds like a win-win solution—the company will have a better, more targeted drug to sell, and people will get better faster—but in practice this kind of trial is remarkably rare, even counterintuitive. If successful, such a trial would remove a large number of people from a risk category, essentially assuring them that they had less risk than they had thought, and the drugs they had been taking for health would no longer be understood to provide such. As I have talked with doctors as part of my fieldwork, they, too, have registered a sense of how odd this dilemma is. Most trials are set up so that either they are successful and a new, more intensive treatment regimen is indicated, or they fail and the status quo prevails. Only the trials that backfire and find excessive side effects result in reduced treatment. Doctors are particularly struck by how easy it is to put people on medication because they meet guideline criteria and how difficult it is to get them off. There are often no studies conducted to determine when it would be better or safer to stop giving a medication to a patient, while at the same time there are very few studies of the long-term effectiveness or safety of those medications (Klein et al. 2002). Such studies do not interest drug companies, because, again, they could conceivably shrink the market for treatments. The general trend, therefore, is for the industry to conduct only trials that would grow the market by increasing the amount of medication in our collective lives, and the empirical data for U.S. pharmaceutical consumption bears this out.

run The_BioMarx_Experiment.prog in all [marx.works]: replace [Capital] with [Biomedicine]

Marx's categorical analysis seeks to explain some of the apparent anomalies of modern social life as intrinsic aspects of its structuring social forms: the continued production of poverty in the midst of plenty, the apparently paradoxical effects of labor-saving and time-saving technology on the organization of labor and social time, and the degree to which social life is controlled by abstract and impersonal forces despite the growing potential ability of people to control their social and natural environment.
MOISHE POSTONE, *TIME, LABOR, AND SOCIAL DOMINATION*

Rereading *Capital*, I found what I felt to be remarkable parallels between the pharmaceutical growth process and especially the chapters on machinery. Machinery both multiplied labor power and therefore seemed to be a reason to labor less, and yet in the industrial system, it had the paradoxical effect of increasing the amount of labor needed, in order to continue to produce surplus labor. Theoretically this was worth pursuing. Thus, in order to better understand the pharmaceutical industry and the logic and political economy of treatment maximization, I have conducted an experiment: I have attempted to channel Marx, twenty-first century Marx, using twenty-first century technology. Karl BioMarx is the author of a future automatic translation of *Capital* into *Biomedicine*.[5] I use *biomedicine* as the general term because I think this analysis has relevance to the broad set of health industries that are science, statistical, information-based, and for-profit.

Methodologically, Marx operates with a fascinating strategy: he reads newspapers, government reports, factory guidelines, and economists. One of the analytic tactics he deploys regularly is to point to how much of what he would like to say as *critique* has already been said *openly*, in public and in reports, by capitalists. That is, exposé alone is not critique; one must show how the system reinforces its worst tendencies despite being conscious of them. Furthermore, most of what Marx found scandalous had been publicly scandalous at the time it was first being perpetrated. Yet critique and scandal had been assimilated, naturalized, and their very scandalousness and systematicity had been forgotten. What makes it hard to keep the intolerable nature of a particular phenomenon in focus is twofold: first, economists cover it up with logical explanations; second, capitalists and workers may perceive many aspects of it to be necessary and even desired, but still consider it to be in need of tweaking. Indeed, Marx himself notes that he is not claiming that the capitalist system, for all of its horrors, is worse than what went on before—only that it needs considerable improvement.

For this chapter, I checked out a number of pharmaceutical-industry textbooks from the University of California libraries. Written by pharmaceutical researchers and marketers, as well as by management consultants who work for or have worked for pharmaceutical companies, these books are intended to teach readers about the workings of the pharmaceutical industry. They are practical, orienting books. They include *Drug Discovery: From Bedside to Wallstreet* (2006), by Tamas Bartfai, a long-time pharmaceutical researcher at Hoffman la Roche and now chair and professor of neuropharmacology at Scripps Research Institute, and Graham V. Lees, a scientific editor and publisher; *The New Medicines: How Drugs Are Created, Approved, Marketed, and*

Sold (2006), by the researcher Bernice Schacter; and *A Healthy Business: A Guide to the Global Pharmaceutical Industry* (2001), by Mark Greener. They are concerned, above all, with helping scientists and laypersons understand *how* it matters that pharmaceutical research and sales is always a business.

With this in mind, I used an electronic copy of *Capital* (from the Marxists Internet Archive, www.marxists.org) and began to systematically substitute key words. The aim was to create a theoretical fission between contemporary healthcare shifts and nineteenth-century industrial shifts. What became immediately apparent was how careful a writer Marx was. Systematic substitution actually works. After many attempts to find an appropriate program of substitutions, the result is a text in which many of the sentences seem uncannily prescient, and many more present surprising and challenging formulations.

I treat this as an experiment. It is ongoing. The words I've come to select are quite specific to the logic of surplus health and derived through interaction with the pharmaceutical marketing literature. In this substitution, *value* becomes *illness, machinery* becomes *clinical trial, employment* becomes *treatment*, and so on. What this produces is an experimental logic of medical growth through increased diagnosis and the magic of symptom-fetishism.[6] An examination of the passages that use the substitute terms helps one think through possible articulations of how thresholds work in patients and for biomedical growth, and how the system of biomedicine has logics that easily outstrip local analysis of pharmaceutical action.[7] This chapter is a preliminary read through some of the consequences of *Biomedicine* alongside some passages from my forthcoming book.[8]

My pocket origin story of this biomedical form of capital is as follows: it could be said to start when medicine as an arm of capital (charged with maintaining workers for work) became an industry itself, beginning in the 1930s and picking up steam after the Second World War. The healthcare industry has its own imperatives for growth that on the face of it are contradictory to capital; that is, healthcare grows by treating more illnesses, yet it should not remove workers from the workplace. The solution is to appropriate that part of health that is not needed for work. This surplus health includes those persons who are too young or too old to work, and it includes illnesses that can be treated "on-the-job," so to speak, without keeping the worker from working. The latter encompasses both illnesses of the everyday (like mild depression) and illnesses of the future (like risk factors and symptomless illnesses like cholesterol). Each of these areas of illness can easily be shown to be major targets of diagnostic and therapeutic development, and each has had phenomenal growth in the last fifty years, intensifying especially in the last decade.[9]

Comparing pharmaceutical textbooks with *Biomedicine* has allowed me to explore the ways in which mass medicine functions as a regime of capital, and hopefully to better understand why medicine continues to be developed in such promising ways (more knowledge of health and illness, and more treatments) that are nonetheless tragic (we are taking more medicines, and not necessarily dying less). Working through BioMarx alongside contemporary writings also helped me understand more about how Marx struggled to make sense of the capitalists of his day. Moreover, the uncanny prescience of BioMarx provides insight into the strange ways in which pharmaceutical analysts talk about health. In the conclusion I will discuss more about my relation to Marx and our relation to the logic of biomedicine. In the meantime, I will walk through this logic, learning how pharmaceutical companies have come to see the population. Perhaps it will also become evident how the public has come to accept a notion of quantitative health, as investment, and how this might not lead to the best future.

replace all [commodit(y/ies)] with [Symptom(/s)]

Where *Capital* begins with a discussion of the commodity, *Biomedicine* begins with the symptom, and this allows us to see how strange health and illness have become.

> The Healthy Life of those societies in which the Biomedical mode of Medicalization prevails, presents itself as "an immense accumulation of Symptoms," its unit being a single Symptom. Our investigation must therefore begin with the analysis of a Symptom.

> Felt-Illness become a reality only by use or Healing: they also constitute the substance of all Healthy Life, whatever may be the social form of that Healthy Life. In the form of society we are about to consider, they are, in addition, the material depositories of Measured-Illness.

> As Felt-Illness, Symptoms are, above all, of different qualities, but as Measured-Illnesss they are merely different quantities, and consequently do not contain an atom of Felt-Illness. (C1: ch. 1)

The distinction described here is between, on the one hand, experiencing Felt-Illness (feeling sick) and going to the doctor to change how you feel, and, on the other hand, watching an advertisement or getting screened (measuring your illness) and being told you may be at risk and should talk to your doctor about possible treatments. In this substitution, symptom is shown to be two-sided in the way that Marx showed the commodity to be two-sided. It had both a use value and an exchange value. The exchange value renders the commodity quantitatively comparable to every other commodity. With biomedicine, we see a similar shiftiness. A symptom seems to be both felt and measured. As measured, the symptom begins to take on a life of its own. A screening test for high cholesterol tells us we have been ill without knowing it, and we begin to experience ourselves as *having high cholesterol*. Or a test suggests we have a high risk for prostate cancer and we feel the need to do something. Even a question may suggest that though we think we feel fine, we *actually feel ill* without knowing it. For example, here is a transcript of an early unbranded commercial by Lilly to help promote Prozac.

> VOICE-OVER: Have you stopped doing things you used to enjoy? Are you sleeping too much, are you sleeping too little? Have you noticed a change in your appetite? Is it hard to concentrate? Do you feel sad almost every day? Do you sometimes feel that life may not be worth living?
> VOICE-OVER: These can be signs of clinical depression, a real illness, with real causes.
> SCREEN-TITLE: Depression strikes one in eight
> VOICE-OVER: But there is hope, you can
> SCREEN-TITLE: Get your life back
> VOICE-OVER: Treatment that has worked for millions is available from your doctor. This is the number to call for a free confidential information kit, including a personal symptoms checklist, that can make it easier to talk with a doctor about how you're feeling. Make the call now, for yourself or someone you care about.

This direct-to-consumer (DTC) television commercial begins as a checklist in the form of an interrogation, with simple questions that are very general: are you sleeping too much or too little? But the seriousness of the questions is contained in the follow-up: "These can be signs of clinical depression." This conclusion converts the questions into a medical algorithm, a logical process following a series of steps. The checklist of symptom questions *measures* your potential illness. But the grammar arrests: "These can be signs" is a peculiar phrase. It is retroactively transformative, inscribing aspects of one's life as symptoms. What you had previously thought of—if at all—as personal variations in mood and habit are brought into heightened awareness; if you had not considered them to be symptoms, now you might. The first implication is that you are, maybe, suffering from a serious disease and do not know it. But this is not a presymptomatic form of awareness. Unlike the situation in Nelkin's and Tancredi's *Dangerous Diagnostics* (1989), where a brainscan or genetic test reveals a disease before it manifests symptoms, here you find out that you have been suffering from symptoms without knowing it.

The grammar of the phrase "These can be signs of x" or "You could be suffering from x" are not simple performatives (see Austin 1962). They do not assert that you *have* depression, nor do they diagnose. For legal, marketing, and health reasons, the grammar is explicitly modalized as possibility: "These *can* be," "You *could* be," "You *might* be." But such suggestions do give you a new potential. You cannot, morally, ignore the possibilities they raise, because your status has changed via this information. You really might *be* suffering (see Sacks and Jefferson 1992). You are *now* at risk (for being at risk), you now *know* that you have been at risk, you *have* to try to do something about it. And the commercial draws you out with "There is hope." Why is there hope? Because treatments are available.

The dynamic here shifts symptoms away from being *felt* toward being *measurements* controlled by others. Even though you self-diagnose, you do so via an algorithm, converting your embodiment into quantitative signs. Where Marx described the odd situation of the worker who must be free to choose to work, but nevertheless must work, BioMarx details the parallel requirement that a person must both be free to accept a diagnosis and yet is obliged to accept it.

> The Patient instead of being in the position to Submit Symptoms in which his Health is incorporated, must be obliged to offer for Submission as a Symptom that very Healthiness, which exists only in his living self.

> For the conversion of his Test-Scores into Biomedicine, therefore, the Diagnoser of Test-Scores must meet in the market with the Fearful Patient, Fearful in the double sense, that as a Fearful man he can dispose of his Healthiness as his own Symptom, and that on the other hand he has no other Symptom for Submission, is short of everything necessary for the realisation of his Healthiness. (C1: ch. 6)

The dynamic goes as follows. A person watches a commercial and goes to a doctor, or gets screened and finds out she has a risk factor. She is supposed to take action to reduce it: change her lifestyle or take a pill. This is for her health, here defined not in terms of illnesses whose suffering she feels, but in terms of a measure whose threshold she has crossed. By taking the treatment she reduces her risk, and this generates health in herself. This health, one should note, is based entirely on clinical trials whose facts have set the guidelines that determine her risk. Felt illness still exists—those times when her body drives her to see a doctor—but more and more of her "illnesses" are detected only through measurements, checklists, and biomarkers. These measures are her symptoms, which she fearfully and proudly tries to reduce. As one patient shouts to the world in a commercial: "I've lowered my cholesterol!"

From a marketer's point of view, the question is how to get you to add depression, breast cancer, cholesterol to *your* lived anxieties, to your personal agenda, enough so that you attend to it, find more information, and talk to your doctor about it.[10] The problem marketers must solve is how to get their particular facts into your head as facts that you come to depend on. For instance, another commercial begins with a scene of middle-aged people on exercise bikes in a gym, working out but looking tired. The only sound is of a ball rolling around, and superimposed above the exercisers is a spinning set of numbers. Finally the ball is heard dropping into place; the number is 265. The cholesterol roulette is over. The text on the screen: "Like your odds? Get checked for cholesterol. Pfizer."

The form of such commercials draws on a public-health logic of awareness: the unaware consumer-at-risk must be made into a patient-in-waiting. In order to achieve this, the consumer's felt-sense of health must be attacked as not simply mistaken, but dangerous. A campaign for colorectal cancer exemplifies this, first asking, "Are you the picture of health?," then warning, "You may look and feel fine, but you need to get the inside story."[11]

> Once adopted into the Medicalization process of Biomedicine, the means of Health passes through different metamorphoses, whose culmination is the Threshold, or rather, an automatic system of the Clinical Trial . . . so that the Patients themselves are cast merely as its conscious linkages. In the Threshold, and even more in the Clinical Trial as an automatic system, the Felt Illness, i.e., the Body quality of the means of Health, is transformed into an existence adequate to fixed Biomedicine and to Biomedicine as such; . . . it is the Threshold which possesses skill and strength in place of the Patient, is itself the virtuoso, with a soul of its own in the mechanical laws acting through it. . . . The Patient's activity, reduced to a mere abstraction of activity, is determined and regulated on all sides by

the movement of the Clinical Trial, and not the opposite. The science . . . does not exist in the Patient's consciousness, but rather acts upon him through the Threshold as an alien power, as the power of the Threshold itself. The appropriation of living Health by objectified Health.[12]

The process described here is the turning over the state of your health-in-general or healthiness to biomedicine so that it is something you *must* look up; it is fully expropriated from your own experiences, since you can't trust your senses, even though "you may look and feel fine." In turn, you come to experience *risk itself.* "Lipitor caught everyone by surprise since it did not have competitive outcomes data, but it shifted what counted as success beyond long-term clinical trials (hope) to short-term biomarker reductions (signal, as in lowered cholesterol numbers), which in turn became experiencible" (Moss 2007, 31).

Converting hope into a signal, a biomarker (cholesterol) that becomes a type of felt-illness (experiencible) completes the transformation of measured illness into the two-sided symptom that takes on a life of its own. Bio-Marx calls this symptom-fetishism, where the number becomes embodied. Risk is more real than lived healthiness. The question is thus: how did measured illness and risk come to be the definition of health?

replace all [Cooperation] with [Preventive Health]

The shift to measured illness took place during the twentieth century, at the intersection of public health and clinical studies, as *preventive health.* Public health recognized that some illnesses needed to be treated collectively in order to prevent repeated outbreaks. Vaccinations are the prototype. Large-scale clinical studies, beginning in the 1950s, approached illnesses collectively even if they were not spread like infectious diseases. The Framingham Heart Study was foundational as a prospective (future-oriented) study that examined the health of over five thousand people in Framingham, Massachusetts, over their lifetimes and eventually across three generations. It aimed to discover connections between ongoing behaviors (like smoking) or biomarkers (like cholesterol) and future events (like heart attacks).

Clinical studies like the large-scale Framingham Heart Study enabled pioneers in epidemiological medicine like Geoffrey Rose to articulate the need for a comprehensive notion of "preventive health." In his now classic treatise, *Strategies for Prevention* (1993), Rose described how large-scale studies slowly transformed our definition of health from the traditional, simple models of disease—in which the patient first suffers, then calls on the doctor—to an epidemiological, measured model of diseases like hypertension and high cholesterol. These represented "a type of disease not hitherto recognized in medicine in which the defect is quantitative not qualitative" (Pickering 1968, in Rose, Khaw, and Marmot 2008, 7). Rose describes the traditional model of diseases as one of felt-illness, whose treatment aims at removing the felt-illness. The quantitative model, however, is one of measured deviance, whose treatment aims at reducing the risk of future adverse events.

The most significant difference between these two models is in the form of diagnosis as we move away from "Has he got it?" to "How much of it does he have?" (Brayne and Calloway 1988, quoted in Rose, Khaw, and Marmot 2008, 9). Given a continuum of measurements (blood pressure tends to be shaped like a bell curve in populations), Rose asks where the diagnostic line should be drawn. Our current administrative medical system demands clear decisions; "this decision taking underlies the process we choose to call 'diagnosis,' but what it really means is that we are diagnosing 'a case for treatment,' and not a disease entity" (Rose, Khaw, and Marmot 2008, 10). In other words, given the continuum of scores where everyone has some blood pressure, the only meaningful reason to draw a line is because that line makes a difference in what we do about it. Rose's public-health perspective allows him to see and state clearly consequence of quantitative disease models, that *diagnosis equals treatment.*

Rose's argument thus tracks that of chapter 13 of *Capital*, on "Cooperation," in which Marx examines how man's collective labor is much more productive than the sum of individual efforts—it becomes social labor. Rose, in turn, offers the concept of population illness. Rose shows how important it is to consider many illnesses, like hypertension, as population illnesses, requiring population-based treatments. To study population illnesses though, for which the actual events (like heart attacks or deaths) are rare, one needs to run very large clinical studies. In one kind of clinical study, the *clinical trial*, the population being studied is randomized at the beginning into often two groups, with one being given a specific treatment, like a drug, and the other group being given a placebo or a competitor treatment. Clinical trials study the effect of a treatment on a large group of people over a period of time (two weeks to ten years), and it is the large scale of the trial that allows a small treatment effect to be multiplied enough to be visible. For example, it might require a trial of ten thousand people over the age of thirty taking a beta-blocker for more than five years to determine whether the risk of a heart attack has been reduced by that particular treatment. If ten fewer fatal heart attacks are found (forty among the five thousand taking the drug versus fifty among those not), then the study might reach statistical significance and be used to get FDA approval for the drug.

If the trial is well-designed, its result then becomes a type of public-health fact. If you are over thirty and take the pill every day for five years, you reduce your chances of a heart attack by 20 percent (this is a big percentage, as the original chance of having a heart attack was 1 percent, now reduced to 0.8 percent). Of course, another way to look at it is that 500 people must take pills every day for five years in order to prevent one heart attack; the number needed to treat (NNT) is therefore 500. Of those people, 495 would not have had heart attacks in any case, and four of them would have had heart attacks despite taking the pill. Thus there is a lot of unnecessary treatment (another way to think of that NNT). This kind of result is known as the "prevention paradox" in which "many people must take precautions in order to prevent illness in only a few" (G. A. Rose 1992, 12).

Rose's analyses were hailed as critical insights that reformulated how to treat coronary heart health and brought public health and clinical medicine back together. Rose uses the word *precautions* because treatments for him mean lifestyle, diet, and behavioral changes first, and he displays extreme caution regarding prescriptions, since he points out that large NNTs greatly amplify the number of long-term side-effects, many of which would be all but impossible to detect without other massive, long-term, and prohibitively expensive clinical trials.

replace all [value(s)] with [Illness(es)]
replace all [labor] with [Health]

Rose's book is fascinating because it clearly demonstrates how preventive logic based on clinical trials does not have to result in more medicine. Rose is acutely aware, however, that it could. I have dwelled on Rose's arguments for preventive medicine not only because they've been persuasive in policy and research, but also because they show the true power, insight, and innovation of population (or mass) health. Since he literally marks off population health from traditional felt-illness, he provides an image of measured, quantitative disease that is not economically beneficial, but is plausible and rational on humanitarian grounds. Rose in fact begins his book with a meditation on the *uneconomic* consequences of preventive health, noting that it does not save the state money, since it tends only to postpone, rather than truly prevent, health problems. He also points out that the longer one lives, the more treatments one tends to require. In the end, Rose argues for preventive medicine solely on humanitarian grounds, that more life with less illness is better.

Though he hints at the fact that preventive health has no inherent limit and that thresholds of treatment must be carefully and socially debated, Rose does not confront or even consider the use of clinical trials and preventive-medicine logic by for-profit pharmaceutical companies. The shift from clinical trials primarily for health to clinical trials for profit has been taking place in pharmaceutical companies since the 1950s (Greene 2006). The historian Steve Sturdy summarizes the process:

> The scene was set for the rapid institutionalization of clinical trial procedures in the postwar years. Drug companies, government agencies and charitable organizations now realized that, by exerting strict controls over the supply of new drugs, they could force clinicians to participate in standardized clinical trials. But at the same time, clinicians came to recognize that they too could benefit from participating in such trials. The development of dramatically effective new drugs, including the antibiotics and subsequently such molecules as cortisone (Cantor 1992; Marks 1992), did much to raise public expectations of the power of modern medicine. By acting as gatekeepers to potentially beneficial new therapies, clinicians could do much to enhance their own professional prestige and authority over patients. As a result, large scale clinical trials became one of the defining features of the postwar medical landscape. Drug companies and administrative bodies were now able to conduct large scale clinical experiments to measure the therapeutic effects of a wide range of novel substances. . . . Doctors had now ceded much of their clinical autonomy to the administrative demand for standardized forms of medical practice. (Sturdy 1998, 283)

The importance and centrality of prevention and clinical trials for advancing healthiness is disputed by no one. Current spending on clinical trials exceeds $14 billion per year. According to governmental and nongovernmental studies, in 2004 around 50,000 clinical trials took place in the United States, involving 850,000 people in industry-funded preapproval testing, and another 725,000 in postmarketing (Phase IV) trials. In addition, 750,000 more people participated in government-funded trials. While these numbers may seem large, within the health industry they represent a crisis of under-enrollment. Four out of every five clinical trials are delayed due to recruitment problems. "The number of trials has doubled in the past 10 years, forcing companies to seek trial participants in emerging markets outside of the saturated areas in the United States and Western Europe. Emerging markets such as India, China, and Russia offer drug companies a volume of potential subjects, and trials can often be executed at reduced costs" (Ernst and Young 2006). The pressure to rapidly complete clinical trials has led to doctors being paid to enroll their own patients, and in a number of countries, people participating as experimental subjects in exchange for healthcare.[13]

At the same time, the ever-increasing scale of clinical trials, the sheer number of them, and the size of each one has put them more or less out of even government's financial reach. Across the board, the pharmaceutical industry, government officials, and even critics agree that only corporate institutions have the resources to conduct most clinical trials. For example, in examining the Celebrex and Vioxx discussions at the Food and Drug Administration (FDA), the pharmaceutical researcher Bernice Schacter noted, "Lengthy discussion about what kind of trial or trials are needed to clarify the issue of the relative cardiovascular safety of the NSAIDS [nonsteroidal anti-inflammatory drugs], triggered both by the FDA's question and a suggestion by Dr. Robert Temple, Director of CDER's [Center for Drug Evaluation and Research's] Office of Medical Policy, that what he called an ALL-HAT trial [Antihypertensive and Lipid-Lowering Treatment to Prevent Heart Attack Trial] be done to compare the cardiovascular effects of NSAIDS using naproxen and diclofenac as controls. Whether such a megatrial could be done and who would fund it remained unclear, though the enthusiasm among the members of the committee was high" (2006, 219). In other words, the proper questions that needed to be asked with regard to the drugs probably "could not" be investigated, since the clinical trial would be too expensive for government funding and since the direct-comparison questions would be too risky for a pharmaceutical company to ask, given that corporate research funding is tied to investments.[14]

The problem is that biomedical companies are first and foremost companies; they exist to make profits, and therefore they must run clinical trials as *investments* whose purpose is to grow returns. This fact is the key form through which marketing takes over research design in pharmaceutical companies. The insight we get from BioMarx is that the return on investment is calculated not solely on labor of workers or clinical-trial subjects, but that value is seen to accrue from the patients via *treatment numbers* (even speculative ones). Hence clinical trials become machinery for generating evidence for generating prescriptions. In other words, the flipside of an evidence-based marketing strategy is that markets are made through evidence, and potential marketable evidence (from a clinical trial) is the determining factor in running the clinical trial in the first place.

The mystery of value for Marx lies in the fact that value is not in fact tied to material wealth; instead capitalists are fixated on labor as value. For the capitalist, the machinery, factory, raw materials, and distribution are all sunk costs, or "fixed capital." The only "variable capital" is the worker, whose full day of labor power the capitalist forces him to sell. If he could, the worker would sell his labor only dearly, but since there is no better work elsewhere and the worker is easily replaced, he must sell a whole day's labor in order to work at all. The wages the capitalist pays the worker is enough to allow him to survive and reproduce. After the worker has put in enough hours to cover his cost (and the cost of the fixed capital), everything else is *surplus labor* resulting in *surplus value*. Capital brings this into being by striving (through competition) to bring necessary labor time to a minimum (so as to maximize surplus labor time), and the result is science, but in a misshapen form.

> "Hence it posits the superfluous in growing measure as a condition—question of life or death—for the necessary."
>
> 1. capital "calls to life all the powers of science and of nature as of social combination and of social intercourse, in order to make the creation of wealth."
> 2. "on the other hand it wants to use labor time as the measuring rod for the giant social forces thereby created, and to confine them within the limits required to maintain the already created value as value." (1993 [1857]: 706)

Science and technology become tragedy because capital has a peculiar measuring system. Marx is quite clear that capitalism brings this science into being, but because it insists on labor as the measure of value, even when labor isn't necessary, it uses science and technology to produce more surplus labor, rather than to produce material wealth.[15]

In working with marketers and working through BioMarx's version of surplus, I have come to understand the pharmaceutical development process better. The pharmaceutical company sees the clinical trial, the pills, and marketing as sunk costs; the only variable capital is the total number of prescriptions (TRX) that are filled, which is the number of patients times the number of prescriptions they purchase. Health research therefore is measured by the number of projected total treatments.

Biomedicine thus calls to life the powers of science (though cooperation and social force) in order to create wellness, but

> on the other hand it wants to use Treatments as the measuring rod for those giant health forces, and confine them within the limits required to maintain the already created Health as Health. (BioMarx)

In other words, a pharmaceutical company thinks of health directly in terms of prescriptions, so that, as Rose concluded, treatments are the "meaning"— that is, use of—Health. Therefore a patient is valuable to pharma to the extent she takes treatments and continues to take them. A "healthy" person who is not on or does not like to be on medicine is, from the perspective of this economy, not valuable. In other words, from the perspective of value, healthiness is antithetical to biomedicine—only Health, abstracted and valorized, is valuable.[16]

Each clinical trial is evaluated first by whether and by how much profit it will generate for the company. Thus Bartfai and Lees take pains to spell out to their readers: "The company's order of priorities is extremely clear. The major factors in selection of a clinical candidate in the company's own priority order are: (1) marketing . . . (2) internal economics . . . (3) scientific, technical and legal issues . . . The regulatory and marketing groups, and then the clinicians, can always override scientific considerations; they 'call the shots.' . . . Under current circumstances this is unavoidable . . . Decisions of this caliber are so expensive and so delicate for the companies' future that they cannot be left to scientists and clinicians alone" (2006, 71–72). *Unavoidable* priorities. Companies are only doing what they have to do in order to survive. This is stated in the same manner as Marx: the executives, like the capitalists, are possessed by the circumstances. The result is that each clinical trial must be designed so that it increases the number of prescriptions purchased. It might seem that a steady state—keeping the population healthy and improving drug efficacy—would be enough to keep an industry alive, but the pressures on biomedicine to grow are enormous, leading to the need to accumulate prescriptions.

Marx described how "Capital as such has to grow" (G III:317). This accumulative sentiment pervades pharmaceutical industry discussion. "In order for Pharma and biotech companies to maintain double-digit growth rates through 2005, they need to multiply their productivity by a factor of five" (Perkins 2002, 148). Similarly, Mark Greener, a former research pharmacologist and editor of *Pharmaceutical Times*, notes, "The stock market expects the pharmaceutical sector to grow at a healthy rate. A survey of 15 analysts in 2000 found that they expected the large pharmaceutical companies to grow between 12% and 15% per year between 2000 and 2005. They also expected sales to increase by between 8% and 10% each year, with the market increasing between 6% and 8% annually. However the U.S. market—the last unfettered, free pharmaceutical market—accounts for some 75% of growth worldwide, reflecting in part the impact of pricing controls" (Greener 2001, 36). Thus, not only is profit an unavoidable priority, but massive growth is too. The problem is precisely that pharmaceutical companies are *expected* to run clinical trials, and even critics like Jerome Kassirer, the former editor of the *Journal of the American Medical Association* (*JAMA*), concede that they *legitimately want profits* (Kassirer 2005, 188)

For Kassirer and many critics, the answer is better regulation in order to define ethical bounds for rule-binding the system to enable profitable pharmaceutical health. The fight between regulators and profits often leads to bizarre encounters, as described by Leonard Weber, healthcare consultant and former director of the Ethics Institute at the University of Detroit Mercy. "Drummond Rennie, a *JAMA* editor interviewed by Peter Jennings on Bitter Medicine 2002 agreed that drug companies 'are intent on keeping consumers on drugs, which are not as good as older drugs, for the simple requirement of profit.' 'Rennie responded yes, absolutely, and it would be strange if they didn't. "They've got to be prevented." Rennie's point was that the pharma industry needs to be understood as part of the for-profit business world" and "will do whatever they can in the pursuit of profits' limited only by legal restraints" (Weber 2006, 13). Although Weber goes on to suggest better business ethics, he leaves untouched the fundamental transformation of health value as measured by treatments. Indeed, better regulations would help curb the abuses, like withholding information on side-effects, but it does not address a deeper, structural concern, which is the dynamic shift that takes place when clinical trials are run by industry in order to grow itself.

replace all [machinery] with [Clinical Trials]

Increasingly, large companies need the mature sales . . . generated by several blockbusters—drugs that achieve sales of more than $1b annually—to fund R&D programmes and meet shareholders' expectations of growth.

MARK GREENER, *A HEALTHY BUSINESS*

A significant problem for the FDA is that there are too many me-too drugs submitted. . . . The companies see it as a way of generating profits, through establishing a new market share, and it is also seen as a safe way to introduce a new drug . . . [since the competition] has already validated the target. . . . But no one is thinking about the patients, just market share.

TAMAS BARTFAI AND GRAHAM V. LEES, *DRUG DISCOVERY*

The *solution* to growth is clinical trials. They alone can increase the productivity of prescriptions, creating more drugs for more people for longer periods of time. Their dynamic in healthcare parallels that of machinery in capitalism. In *Capital*, machinery occupies a pivotal role, unleashing collective productivity and thus allowing one man working one hour to produce more than what two or ten or a hundred were able to produce without the machine. Machinery's paradox is the most tragic for Marx. The tremendous increase in productivity enabled by machinery would seem to liberate man from endless toil; with more wealth produced from less effort, it would seem to follow that less work would need to be done. But, historically, the opposite happened: machinery led to longer working days and required more workers, since capitalists came to see *not* using the machines as wasteful and therefore wanted them to be used at all possible times. Furthermore, machinery's relative ease of use meant that women and children could also be employed, expanding the labor pool tremendously and cheaply.

The analog in biomedicine has a similar paradox: clinical trials are an amazing way to increase the healthiness of the population. Although they have promised to reduce the amount of time we spend attending to our health, they have instead increased the time, energy, money, and treatments we apply to our health, and have extended treatment to children and the elderly in increasing numbers. This may seem overstated, but bear with me. Machinery under capitalism did increase productivity (and successful clinical trials do offer ways to increase healthiness), but capitalists employed machines not for that purpose, but instead to increase their profits. If a machine could save labor but would not increase profits, for example, then the capitalist would not use it. Marx discussed a case in England where women were sometimes used instead of horses for hauling barges because the women as "surplus population [were] beneath all calculation. Hence we nowhere find a more shameless squandering of human labor-power for despicable purposes than in England, the land of machinery" (C1:517). Cases like this, in which the decision to use a machine turns on whether it is "worth" it to replace humans, speak to the core of the labor theory of value. Increase productivity—but only if it does not cost more than employing labor and increases the surplus value of the remaining labor. Thus, only those machines which increased profits (e.g., by making a product that could be sold at a higher profit) would be installed.

The problem is eerily the same in biomedicine. "Today, 62% of the sales of Pharma is in the United States. But how many know that there are over 800 compounds that are sold in Europe, and which are highly efficacious, therapeutically wonderful, but which have not yet been registered in America? And they won't ever be. Because by now the patent life is so short, and the FDA so slow, that they cannot be. . . . [T]he marketers in the U.S. won't be interested. And it is nonnegotiable; this is how it is" (Bartfai and Lees 2006, 138). America as a pharmaceutical contradiction is all too familiar. Many critics echo Ken Silverstein (1999) in decrying a world where "millions [are spent] for Viagra, pennies for the poor." It may seem straightforward to hold drug companies responsible for the choices they make regarding which diseases to research, but as Bartfai and Lees indicate, this critique is internal to corporate logic. Almost every pharmaceutical industry textbook I found narrates an ongoing debate over precisely this issue, of whether a pharmaceutical company can afford to care about medicine and people, rather than about profits. Passing along this debate to future industry scientists is precisely the purpose of pharmaceutical industry textbooks in bringing it up. "Should they develop for this specific use and that one, but not the one unlikely to succeed or unlikely to generate a sufficiently large market?" (Schacter 2006, 116).

Especially today, under the pressure to maintain growth, apparently vile decisions are driven by a clear perception of "waste." One is literally "throwing money away" if one is not making "as much as one could" compared with other investments. "One of the significant problems for the Pharma industry is that of the 400 disease entities identified, only 40 are commercially attractive by today's requirements of return on investment. . . . Society needs to find a way to make more diseases commercially attractive if it wants Pharma investment in treating any of the other 350 diseases affecting hundreds of millions of people" (Bartfai and Lees, 2006, 14). Here Bartfai and Lees, a former pharma researcher and a publisher, seem to be calling for regulation to save them from their own structural violence. This call echoes an industrial practice, which Marx registers in a chilling footnote, wherein even kidnapping raids of children, who were forced to work in factories for more than twelve hours per day, were unstoppable as long as competition with other capitalists existed. One group of factories actually submitted a petition to the British government in 1863, pleading, "Much as we deplore the evils before mentioned, it would not be possible to prevent them by any scheme of agreement between the manufacturers. . . . Taking all these points into consideration, we have come to the conviction that some legislative enactment is wanted" (C1:297). Bartfai and Lees state that the question is not one of choices, but of structural pressure. Their mode is one of enlightened attack: it is society that needs to find ways to make thing better, to make unprofitable diseases profitable.[17] Similarly, Marx describes how under the regime of machinery, "Capital takes no account of the health and length of life of the worker, unless society forces it to do so" (C1:381).

Across the board, pharmaceutical-industry analysts are unabashed about these constraints and presuppositions. "Pharmaceutical companies tend not to invest in tropical medicines because they are unlikely to recoup their investments. . . . Given the pressure on pharmaceutical companies to maximize their return on investment, this attitude is unlikely to change without a major change in shareholders' attitudes" (Greener 2001, 122). This is what Marx was striving to explain about the interactions of Capital, how once you see it in process, then the entailments of that process have a type of force to them—not a deterministic force, but influential nonetheless. From inside the pharmaceutical industry, one feels that one's life is at stake, and certainly one's company is. The pressure from "other possible investments" (such as other disease research at the same company) is key, since this means that there does not even need to be competition from other companies for the process of value generation to exert its force. The very fact that disease research is an investment establishes an equivalence between investments, such that they become comparable along one dimension: quantitative health, or treatments.

> To the out-cry as to the physical and mental degradation, the premature death, the torture of over-Treatment, it answers: Ought these to trouble us since they increase our profits? But looking at things as a whole, all this does not, indeed, depend on the good or ill will of the individual Pharma. Free competition brings out the inherent laws of Pharma Experience, in the shape of external coercive laws having power over every individual Pharma. [82] (Ch. 10)

The problem of comparing possible treatment research within pharmaceutical companies is that saving one set of lives through research and development, marketing, and sales must be compared on return-on-investment profit grounds with saving other lives who may return more net profit.

"Products that are not able to limp along must be eliminated. They are a drain on a business unit's financial and managerial resources, which can be used more profitably elsewhere" (Perkins 2002, 122). Most critics do not begrudge pharmaceutical companies this attitude, because they understand and have naturalized corporate funding of research.

Here is the twist so peculiar in capitalism and biomedicine: the company that one loves because it makes healing medicines becomes secondary (logically) to the money it returns. The disease one wants to cure becomes secondary to its market size. It comes to appear that it has to be this way. "Pharmacoeconomics plays a pivotal role. Drug development is very capital intensive and even big indications such as malaria and tuberculosis are affected. The cost means that small indications suffer, regardless of how good the science is. If drug discovery were a science-driven activity, one would expect scientists to be running drug companies. However, since Roy Vagelos of Merck retired [in 1995], no Big Pharma has been run by a scientist; they are all run by people who were trained in economics" (Bartfai and Lees 2006, 71). Bartfai and Lees suggest that because drug development is capital intensive, economic value comes naturally to supplant scientific or health value. The reason why this can be justified has its roots in Geoffrey Rose's insight that once disease comes to be defined as on a continuum with health, the only meaningful diagnosis is that which indicates treatment. Treatment therefore equates with diagnosis, and the market indicated by a diagnostic threshold is both a measure of profit and the very definition of "health." As health is an a priori "good," comparisons of two possible clinical trials turn on their relative profitability.

Inside a pharmaceutical company, this comparison is a source of continual negotiation, where clinical research directed at healthiness can clash with market research, leading to struggles over who should really be deciding clinical directions. Bert Spilker, the head of project coordination at Burroughs Wellcome and author of many pharmaceutical textbooks, writes of this struggle in his six-hundred-page *Multinational Drug Companies: Issues in Drug Discovery and Development*. Note how "medical value" retains only a ghost of its apparent persuasiveness: "The cooperation of research and development and marketing groups may be severely tested when an investigational drug has a high medical and low commercial value and the project draws resources (or would draw resources) away from projects that the marketing group believes have greater commercial value and are of high or medium medical value" (Spilker 1989, 427–28). There is a defensiveness in the qualifying phrase "of high or medium medical value," as if a me-too drug with low medical value would not be chosen no matter how commercially valuable it was.

Hidden (and assumed) within these debates over medical and commercial value is the fact that, like machinery, clinical trials seen from the point of view of investments become a different sort of beast than those seen from a medical point of view. The very innovative power of science and technology, productivity, and intensity comes to be transformed, mutated into profit and growth monsters. As Marx puts it, "This process of separation . . . is completed in large-scale industry, which makes science a potentiality for production which is distinct from labor and presses it into the service of capital" (C1:482). Similarly, Steve Morgan, Morris Barer, and Robert Evans, writing in 2000, eerily repeat the same insight: "Science and objectivity are of interest to a private, for-profit corporation only insofar as they further the quest for profits" (660). Looking closer at how clinical trials are implemented via BioMarx will let us see how surplus health continues to be expanded.

replace all [productivity] with [Knowledge]

Using financial, contractual and legal means, drug manufacturers retain a degree of control over clinical research that is far greater than most members of the public (and, we suspect, many members of the research community) realize.

STEVE MORGAN, MORRIS BARER, AND ROBERT EVANS,

"HEALTH ECONOMISTS MEET THE FOURTH TEMPTER"

If growth is achieved through choosing to study those diseases that have the biggest markets, those markets can be stretched wider through choosing how to design clinical trials so they indicate more of the population for treatments. This is possible because clinical trials were designed to compare treatments for existing diagnoses that had known outcomes, such as cures. When clinical trials are used to define a diagnosis along a continuum, they turn out to be remarkably flexible. In *Rose's Strategy of Preventive Medicine* (2008), Geoffrey Rose uses as an example the potential benefits of serum-cholesterol reduction on coronary-heart-disease deaths. In a table—which I have reproduced here, since it makes clear just how open clinical trials can be—he breaks down risk by age and sex.

Rose points out how a screening program for men 55–64 would require 230 men to be screened and 100 of those screened to be treated for five years, and would prevent one death on average. And this, he says, "relative

TABLE 1 Estimates of potential reduction in coronary heart disease deaths from screening for raised serum cholesterol (> 6.5 mmol/l) in different age and sex groups

Age in years	25–34	35–44	45–54	55–64
Percentage with raised level				
Men	20	35	40	45
Women	15	20	50	70
5-year deaths per 1,000 in this group				
Men	1.2	5.8	21.3	48.1
Women	0.2	1.1	4.5	15.9
*Number screened to prevent 1 death in 5 years**				
Men	21,100	2,500	600	230
Women	137,300	23,200	2,200	450
Number treated for 5 years to prevent 1 death				
Men	4,200	860	230	100
Women	20,600	4,560	1,100	320

*Assuming 20% reduction in deaths among all eligibles.
SOURCE: Based on Rose, Khaw, and Marmot 2008.

to other preventive or therapeutic measures . . . would be reckoned a good value" (Rose, Khaw, and Marmot 2008, 37). Among women 25–34, however, such a program would require screening 137,300 women and placing 20,600 of those screened on treatment for five years to prevent one death. The number needed to treat (NNT) is so high in this instance because that demographic has so few coronary deaths to begin with (only 0.2 per 1,000, or 1 in 5,000, over five years). Rose's response illuminates a crisis in health prevention. Despite appearing objective, one must engage in a relative valuing of lives: "Unless one takes the extreme and wholly unrealistic view that the saving of a life is worth any price at all, then it is hard to justify" (ibid.). Rose's need to increase the hyperbole, saying "extreme and wholly unrealistic," reveals the dilemma, since he is nonetheless talking about a program that would save lives.

Empirically, a clinical trial could be designed to show a population goal for putting everyone over twenty-five on cholesterol-lowering drugs, despite the incredibly huge NNT. Which clinical trial to run, where to draw the line, is thus a social and political dilemma. The dilemma becomes a full-blown contradiction when one considers the structural economic constraints. First, only a few of the many possible population and cholesterol groups can be studied; and second, due to the large size of the groups and the expense of clinical trials, it is pharmaceutical companies who are allowed to choose the groups. Make no mistake, the clinical trial will generate legitimate and true facts about health; it will indicate treatment for whatever population subset it successfully studied. And since very few similar clinical trials are likely to be conducted, the facts generated by the first trial may very well be the *only* facts available about this kind of health.[18]

Now ask yourself: if you were running a pharmaceutical company and had to choose between a study that could show a high treatment benefit for men over forty-five, or one that would show a low benefit for (but still save lives among) everyone over twenty-five, which would you fund?

> As Pharma, he is only Biomedicine personified. His soul is the soul of Biomedicine. But Biomedicine has one single life impulse, the tendency to create Illness and surplus-Illness, to make its constant factor, the means of Medicalization, absorb the greatest possible amount of surplus-Health. (Ch. 10)

Since the problem for pharmaceutical companies is how to keep growing despite the constant pressure of stockholders, competitors, and the time-bombs of their own patents running out, clinical trials *must* be designed to maximize their markets in order to maximize their investment return. The point is that there is corporate awareness of the variety of possible clinical trials, and conscious selection of those trials that meet the desired profile (long-term, large population, etc.). Bernice Schacter's book, *The New Medicines*, written by a researcher to train future pharmaceutical researchers, explains the dilemma this way: "If the team elects to seek approval for a narrow subset of patients with a certain condition, then the market for the drug may be too small to make financial sense for the company. If they seek the widest use, for example, for everybody with arthritis, they are at a greater risk of failing to demonstrate safety and efficacy and therefore failing to get approval. This is based on biology" (115).

Here again one discerns the parallel to machinery in capitalism: machinery can only be implemented when it increases surplus value. In this admonition, Schacter makes clear why clinical trials cannot be designed with healthiness as a priority, why companies must see health as a *means* to profit through increasing treatments. And this is how clinical trials that result in larger NNTs (i.e., less efficient drugs) come to be valued more than smaller ones. The dynamic of surplus health is at work in the fact that the larger NNT means that more people will be taking the drug without benefitting from it, but since there aren't more facts available about who will actually benefit, it appears as if everyone taking the pill is benefitting a little by reducing their risk. "Ideally the study would look at the drug's effects on the so-called 'hard' endpoints—death or a heart attack, for example. However such studies tend to be large, expensive, and lengthy. So many studies rely on 'surrogate' endpoints. These predict the risk of suffering a hard endpoint either for each patient or from a population perspective. . . . Taken across the whole population these are associated with, for example, a risk of stroke asthma or heart attack. However that does not show that any particular patient will develop the disease" (Greener 2001, 61).

> In one word, surplus-Illness is convertible into Biomedicine solely because the surplus-Risk Diagnosis, whose Illness it is, already comprises the material elements of new Biomedicine. (BioMarx)

A recent article by D. G. Manuel et al. in the *British Medical Journal* comparing eight cholesterol treatment guidelines in four countries made this dynamic visible (see 1). The researchers designed a graph that mapped the population indicated by each of the guidelines and the lives that would be saved, assuming that the clinical-trial evidence was correct and the guidelines were scrupulously followed. The first thing to note about the graph is that different guideline committees in different countries and at different times made very different choices about how to implement the facts at their disposal. They came up with vastly different percentages of the population indicated for treatment and numbers of lives potentially saved. The curve on the graph is the researchers' extrapolation of the ideal treatment-to-saving rate. The fact that it is a curve and not a point shows something very important about the graph as a whole. Any point under the curve is a potential guideline. And any point would save lives. The graph thus illustrates that there are thousands of potential clinical trials that could be run and thousands of consequent treatment indications.

In order to compare the different guidelines, the authors describe the horizontal distance from the curve for each guideline as the efficiency gap, which implies that the same number of lives could be saved while treating fewer patients. The vertical distance to the line they call the efficacy gap: how many more lives could be saved with a different guideline targeting the same percentage of the population. They thus suggest that prudent committees design clinical trials and guidelines to move the points up and to the left. Yet such trials would be predicated on decreasing the numbers of people on medication. Like Rose, these researchers fail to consider the countervailing pressure of companies, for whom the value of a clinical trial

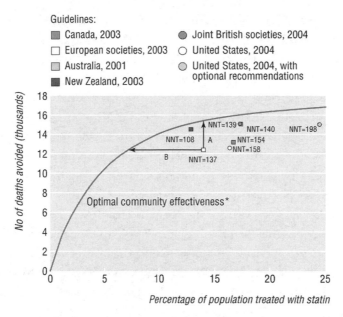

Guidelines:

■ Canada, 2003   ● Joint British societies, 2004
□ European societies, 2003   ○ United States, 2004
▨ Australia, 2001   ◐ United States, 2004, with
■ New Zealand, 2003   optional recommendations

*The optimal community effectiveness curve shows the number of CHD deaths avoided
if the highest risk people were treated first
A=The effectiveness gap or the difference between the potential deaths avoided by the
guideline recommendations compared with the optimal number of deaths
avoided if the highest risk people were recommended treatment.
B=The efficiency gap or the difference between the percentage of the population
recommended statin treatment compared with the minimum population that could be
treated to avoid the same number of deaths.
NNT=Number needed to treat to prevent 1 CHD death over 5 years

1. Number of deaths from coronary heart disease (CHD) prevented
over five years by percentage of Canadian population aged 20–74
years treated with statins for different national guidelines for the
management of dyslipidaemia. From Manuel et al. 2006, used
by permission.

and of consequent guidelines would be based instead on how far to the right
they would appear on the graph, which translates into vastly more treatments. Thus Schacter states clearly, "The challenge for the project team is to
design the most efficient development plan, within the regulatory and ethical constraints, that will provide the largest market, and the best return on
investment. The trials should be no larger, nor run longer, than required to
provide evidence for efficacy and safety" (2006, 117).

It comes as no surprise, perhaps, that the U.S. guidelines are the furthest
to the right and getting more so with each guideline revision. The medical historian Jeremy Greene notes that once Merck had started conducting
clinical trials for its drug Mevacor so that they reinforced the then controversial cholesterol guidelines, subsequent trials by many companies "came
to exert a formative influence on the guidelines themselves" (2006, 197). At

the same time, there is nothing inherently wrong with any of these guide-lines. The revised U.S. guidelines are projected to save approximately 300 more lives per five years than New Zealand's (15,000 versus 14,700), at the mere cost of putting 12 percent more of the population on statins (24 per-cent versus 12 percent).

Rose's struggles with this entailment of his preventive-health proposal reflects the inability of the logic of prevention to stop the biomedical ap-propriation of clinical trials. Almost ironically, Rose notes the "folly" of those health-service managers and policymakers who get so caught up in mistaking "people treated" for improved health that they say things like, "This has been a good year for the National Health Service. . . . [W]e have treated more patients than ever before." Rose's assessment is that they are managing health services "according to the principles of the market" (Rose, Khaw, and Marmot 2008, 38). Precisely! Rose nowhere seems to recognize that what he calls this "blinkered attitude" is the very goal of pharmaceutical marketing.

> Drug companies commonly control the research question (with what products and what doses, and for what patients and conditions, is the new drug compared?), they control the selection of patients for the trials, they control how drop-outs and side-effects are reported and treated in the analysis, and they control what information makes its way into scientific presentations and peer-reviewed publications. Drug companies often use surrogate endpoints to establish a product's efficacy (and to establish a market for the product), despite absence of evidence that the surrogate outcome and health status are in fact correlated, and sometimes, in the face of evidence that they are not. (Morgan, Barer, and Evans 2000, 661)

In this manner, biomarkers undergo a transformation from being additional signs of an illness or risk of one into being the means of defining a new ill-ness for treatment.[19] Symptoms become commodities not because they are paid for, or even because they involve biomarkers, but because that is the only way to decide on illness. The person is dependent on the clinical-trial evidence for knowing whether or not he is ill and needs treatment. And the patient has no way of knowing whether or not a test finds a real thing, or whether the treatment works. The switch to "preventive, population medi-cine" makes it possible for the biomarker to become the "fact" that defines an illness for treatment, with its attendant consequences.

Rose's argument is cited by pharmaceutical companies in their explana-tions of what they have been up to! "Defining illness is somewhat arbitrary.

Blood pressure and lipid levels, for example, have a normal 'bell-shaped' or 'n-shaped' distribution. In such cases, the abnormality is quantitative rather than qualitative. As a result, the point at which clinicians decide that something is abnormal and therefore warrants treatment, is an arbitrary decision usually based on population risk (Rose 2008, 6–8). This means that different clinicians can—and sometimes do—draw different conclusions about the point at which they will intervene. Clearly such factors can influence the success of a particular medication" (Greener 2001, 47). It was in fact this citation of Rose in Greener's guide to the pharmaceutical industry that led me to his *Strategies for Prevention* in the first place. Rose's lack of understanding of the political-economic context of clinical-trial operations causes him to overlook the true function of clinical trials in biomedical capitalism in the same way that economists missed the function of machinery in capitalism. It is Rose, then, who wears blinders, for marketing must insist that the clinical trial be designed such that success will generate a bigger and more profitable market in prescriptions, whereas Rose assumes that the point of designing a clinical trial is to maximize healthiness in society and that this requires careful discussion of the tradeoffs between the size of the population indicated by the clinical trial and the costs of treating that population.

All of Rose's assessment modes are moot in the context of pharmaceutical companies, whose criteria of value is number of prescriptions. Those companies are quite explicit in calling for the maximum market size that can be reliably diagnosed. The pharmaceutical management consultant Arthur Cook provides an example in his *Forecasting for the Pharmaceutical Industry*.

> For the patient-based forecaster success-stories revolve around diseases such as benign prostatic hypertrophy and HIV. Benign prostatic hypertrophy (BPH) is a disease that affects men, usually in older age. Cadaveric epidemiological studies suggested that the prevalence of BPH was as high as 95 per cent in men over the age of 65. However, the number of men treated for BPH was significantly lower. With the advent of new diagnostic technologies, physicians were able to monitor for an enzyme associated with BPH and were able to diagnose patients earlier in their disease. This led to market growth through an increase in diagnostic rates. (Cook 2006, 41–42)

Cook here shows how since pharmaceutical companies have begun to operate the clinical trials that define the facts about illnesses and risks, patients are quite literally forced to rely on those facts and to submit their healthiness to screening and diagnosis. The healing and *actual* risk reduction the patient receives is the equivalent of wages in *Capital*. But what of the excess or surplus? Say the NNT is 50, then for every 50 patients who are diagnosed and treated, 49 are treated without needing that treatment. For BioMarx, their treatments are surplus and the healthiness used (that is spent in screening, purchasing prescriptions, and side-effects) is "surplus health."

But isn't this the public-health argument and the prevention paradox? Yes, and no. What makes the difference in machinery is not in the machinery-worker relation, but in the capital dynamic that makes it the *means* for profit on investment.

> Now in order to allow of these elements actually functioning as Biomedicine, the Pharmaceutical Pharma requires additional Health. If the exploitation of the Patients already Treated do not increase, either extensively or intensively, then additional Healthiness must be found. For this the mechanism of Pharma Medicalization provides beforehand, by converting the Patient Class into a class dependent on Risks, a class whose ordinary Risks suffice, not only for its maintenance, but for its increase. It is only necessary for Biomedicine to incorporate this additional Healthiness, annually supplied by the Patient Class in the shape of Patients of all ages, with the surplus means of Medicalization comprised in the annual Medicalize, and the conversion of surplus-Illness into Biomedicine is complete. From a concrete point of view, accumulation resolves itself into the Return to Healthiness of Biomedicine on a progressively increasing scale. The circle in which simple Return to Healthiness moves, alters its form, and, to use Sismondi's expression, changes into a spiral. (B-C1: ch. 24)

In biomedicine, the current market for a drug is the limited viewpoint, and to reframe it we need to take into account everyone who might *possibly* be able to take the drug. In this worldview, the first step is not to look at those who already visit their doctors, but to take a potential threshold diagnosis and calculate how many people *would* be part of that threshold and therefore *should* be consumers of the relevant drug. Interviewed in the industry journal *Pharmaceutical Executive*, an Aventis executive, Thierry Soursac, describes this process: "'Up until now,' he says, 'when we were looking at the size of the market, we tended to open this market data bible called IMS and say, Okay, the market of proton pump inhibitors is that much, and the market of hypertension is that much, and this is the size of the market we have to tap into.' The problem, Soursac explains, is that IMS data represent a 'rearmirror view' of markets, a view of the past, not the potential" (Shalo 2004). Soursac's reference to the problem of the rear-mirror view of markets is almost an exact quote from *Every Business Is a Growth Business*, this time referring to Robert Nordelli, the chief executive officer of General Electric Power Systems. Pharmaceutical executives, in other words, have already been translating capital into biomedicine, substituting health for value and illness for labor.

The question thus arises: how big is the market for statins or other drugs? One approach, used by most drug companies, has been to measure the number of diagnoses for the indication and use this as a benchmark for market size. One company, IMS Health, is the acknowledged leader in this type of information gathering. IMS tracks almost every prescription written by every doctor, then sells this information to marketers and drug companies so they can track exactly how well their campaigns are going. "[Soursac's] entire marketing strategy hinges on his belief that pharma companies need to 'look at how many human beings on the planet have specific diseases that can be addressed by our drugs; this is the market. Whether this market has translated into any sales of drugs in the past is irrelevant.' Soursac cites the possibility that a market with low sales may be suffering from underdiagnosis of a condition or poor documentation of disease epidemiology in certain geographic areas" (Shalo 2004). When Soursac suggests looking at who on the planet "has the disease," he means to estimate the number of people who could be determined to be below a specified threshold. Arguing then from this "potential," he outlines a strategy for achieving it, beginning with changing how the disease is diagnosed and how it is documented.

One way to grow a megamarket is to emphasize underdiagnosis by identifying a hidden epidemic. In one instance of market production in Japan, for example, targeted epidemiological studies were designed specifically to show undetected, even unimaginable levels of deep venous thrombosis (DVT) and thus to literally create a market for its diagnosis and treatment. "'People in the company said there are too few patients in Japan,' [Soursac] says. 'But I looked at the U.S. and Europe. . . . And thought this is sure to be a big market'" (ibid.). This approach is a textbook business-growth tactic, emulating Goizueta's admonition that Southern California's average monthly Coke consumption could be tripled because Hungary's consumption was three times greater per month. Growth for Goizueta was premised on the notion that ideals, and not averages, were appropriate target norms. So Soursac commissioned a third-party epidemiology study that found rates of disease in Japan to be identical to those in the United States. "'Suddenly, by having that data in your hand and being able to share it with the health authorities and medical institutions, you certainly create the market for diagnosis and treatment of DVT which didn't exist before'" (ibid.).

National populations and potential markets are made equivalent through epidemiology and international-standards bodies. Factual humanitarian claims of disease prevalence are mobilized to invoke nation-state ethical responses, opening up markets. Health facts, in Soursac's view, are highly contestable. If epidemiological data suggests one conclusion, another study might counter it. The result would seem to be a contradiction or even a controversy. But when the data are properly shared, emphasized, and amplified, a new patient-population-in-waiting can be created whole cloth, where "one didn't exist before."

> Pharma does not know that the normal Prevalence of Health also includes a definite quantity of Unneeded Health, and that this very Unneeded Health is the normal source of his gain. The category of surplus Health-time does not exist at all for him, since it is included in the normal Treatment-Time, which he thinks he has Needed for in the day's Healthiness. (Ch. 20)

Here we can return to the unimaginable and virtually infinite numbers of people at high risk prophesied by Cleeman, and inflect these numbers with the materialization of theory by Soursac. One reason why we cannot imagine how many people are at risk and need treatment is because the population does not exist until the questions are posed.

In this remarkable perspective shift, we see how pharmaceutical marketers and executives envision health: as a growth opportunity and a virtually unlimited one. Could they be right? Could we be headed for a world in which greater numbers of drugs are taken for life, such that it comes to be our bodily intolerance to multiple drugs taken simultaneously, rather than our lived health, which provides some sort of limit? Can the number of drugs that Americans consume continue its double-digit growth for another century?

replace all [hour(s)] with [Treatment(s)]

Intensification: From Mass Patients to Chronic

Health care has changed dramatically in the past 35 years, as treatment has increasingly migrated from the doctor who directed care in the hospital to patients who now prevent illness through medication use in an unsupervised community setting. . . .

[M]edications [now] treat illnesses early in their natural history, long before painful or disabling symptoms are apparent. . . . With [these] asymptomatic conditions, patients are often unable to determine if they need treatment at all and/or whether the product is working. . . . This difficulty is only likely to grow. As researchers unravel the molecular basis of an illness, *manufacturers increasingly turn incurable diseases into merely chronic ones.*

WINDHOVER INFORMATION INC., "MOVING BEYOND
MARKET SHARE" (EMPHASIS ADDED)

Thus we see, that Clinical Trials, while augmenting the human material that forms the principal object of Biomedicine's exploiting power, at the same time raises the degree of exploitation.

If Clinical Trials be the most powerful means for increasing the Knowledge of Health—i.e., for shortening the Treatment-Time required in the Medicalization of a Symptom, it becomes in the hands of Biomedicine the most powerful means, in those industries first invaded by it, for lengthening the Treatment-Time beyond all bounds set by human nature. It creates, on the one hand, new conditions by which Biomedicine is enabled to give Fearful scope to this its constant tendency, and on the other hand, new motives with which to whet Biomedicine's appetite for the Health of others.

BIOMARX, *BIOMEDICINE*

Soursac's policy of finding the epidemiology to back up projections is an example of efforts to increase the number of people on a treatment as much as possible, but it still runs into limits. Maximizing markets by choosing the most profitable diseases and maximizing the number of people indicated by clinical trials are only the first steps; the next is to maximize the length of time people stay on a treatment by increasing the number of prescriptions per diagnosis. One way to achieve this is to study younger pools of risk patients. Bartfai and Lees explain how biomarkers enable this extension: "Drug companies do not have the time to wait for the actual therapeutic effect to manifest itself. . . . That is why the designers of clinical trials are always looking for 'surrogate endpoints'; that is you look for something which indicates the therapeutic effect indirectly" (2006, 121).

They use the example of chronic slowly progressing diseases like osteoporosis, rheumatoid arthritis, and Alzheimer's.

If you want to look at the disease progression of say a neurological disease such as alzheimers (AD), the length of the study becomes a major financial and marketing issue. Since the disease can be detected much earlier, one can elect to perform the trial on patients with mild to moderate symptoms, as defined neuropsychologically, and try to show efficacy against a slow decline, but that can take 24 to 36 months. The market size will of course be much bigger; they are younger and there are more of these patients who are less likely to die from other causes during the trial. (Bartfai and Lees 2006, 156–57)

In this passage, Bartfai and Lees demonstrate how one can choose to study the mild, earlier forms of a disease, which will vastly increase both the number of people in the market (as well as the NNT) and the length of time each of those people are on the drug. Pharmaceutical companies face enormous pressure to redefine diseases this way, by studying them with the purpose of identifying a market that is as large as possible.

> Hence that remarkable phenomenon in the history of Modern Pharma, that the Clinical Trial sweeps away every moral and natural restriction on the length of the Treatment-Time. Hence, too, the economic paradox, that the most powerful Test for shortening Health-time, becomes the most unfailing means for placing every moment of the Patient's time and that of his family, at the disposal of Pharma for the purpose of expanding the Illness of his Biomedicine. (B-C1:532)

The source of this expansion possibility lies in the fact that treatment value is counted via the total number of prescriptions. This was brought home to me when, in talking with a group of marketers about chronic illness and poring over a large flowchart of patient decision points, I was directed to a loop in one corner of the chart where repeated prescriptions were encapsulated. "We would love to increase the number of prescriptions a patient takes," said the marketer, "because the profit is the same if one patient takes a drug for four months, as it is for four patients taking the drug for one month." This interchangeability of patient numbers and prescription consumption is reflected in the Express Scripts report under "utilization," which is prevalence (the number of people indicated for the drug) times intensity (the average length of prescription per patient).

The effect of quantitatively extending risk in this way not only places more people in the category of "at risk," but essentially changes the *quality* of the disease, rendering it chronic. This illustrates the fact that, from the marketer's perspective, "the economic driver in health care has shifted from the physician to the patient. While physicians continue to control episodes of short-term, acute illness, such as hospitalizations, patients increasingly drive the financial and clinical outcomes for chronic diseases through the simple daily act of taking a pill, often over a long period of time. In financial terms, the shift from acute to chronic care medicine means that between 75–80% of a prescription's value is now concentrated in the patient's return to the pharmacy for refills" (Windhover Information Inc. 2002, 64). The point being made here, in a pharmaceutical report published in 2002, is that if one were to compare two clinical trials, one for a cure and one for a chronic treatment approach to a disease, one would find the latter to have a four to five times better chance of becoming a blockbuster.

Similarly, Michael Kremer and Christopher Snyder investigate why drugs are more profitable than vaccines.

> In a simple representative consumer model, vaccines and drug treatments yield the same revenue for a pharmaceutical manufacturer, implying that the firm would have the same incentive to develop either ceteris paribus. We provide more realistic models in which the revenue equivalence breaks down. . . . The second reason for the breakdown of revenue equivalence is that *vaccines are more likely to interfere with the spread of the disease than are drug treatments, thus reducing demand for the product.* By embedding an economic model within a standard dynamic epidemiological model, we show that the steady-state flow of revenue is greater for drug treatments than for vaccines. (Kremer and Snyder 2003, emphasis added)

Kremer and Snyder make explicit that in too much drug research, cures get in the way of repeat revenue. The corollary of seeing clinical trials as instruments or means for maximizing prescriptions especially when used to lengthen treatment time is that everything that gets in the way of those treatments becomes a *loss*. Since the expected return-on-investment for a clinical trial is the total possible prescriptions, everything which impedes their realization is perceived as a barrier to overcome. BioMarx says something very similar after stating that "the development of this objective Healthy Life is in opposition to, and at the cost of, the human individual."

> The productivity of Health in general = the maximum of Risk Diagnosis with the minimum of Health, hence the greatest possible cheapening of Symptoms. This becomes a law in the Biomedical mode of Medicalization, independently of the will of the individual Pharma.

A version of the flowchart the marketers showed me, with patients' return for prescriptions, appears in almost every pharmaceutical textbook.

What I didn't understand at the time is that what the marketers call "utilization" (Express Scripts) is, from the user's point of view, "bioavailability," the overall availability of an individual's metabolism for the maintenance of pharmaceutical flows.[20] The consequence of this formulation is that marketers envision patients literally as points of resistance (rather than of consumption). Their physiological rejection of many drugs, their desire to stop taking different treatments, even their sense of their own wellness are all obstacles to be overcome. "Applying such metrics to a variety of chronic disease states reveals that a marketer's real enemy is less the share lost to competitors than the cumulative effects of patient attrition over time" (Windhover Information Inc. 2002, 69).

Note that this means that marketers are directly opposed to your decision not to continue taking a prescription because you feel better or want to try an alternate form of medicine. The business magazine *Forbes* reinforced this battle image with a cover story on *Pharma's New Enemy: Clean Living*. Recalling Goizueta's redefinition of the Coke market toward human liquid consumption, which he declared as a war on coffee, tea, and tapwater, the point here is that for good business reasons, Pharma has found a way to grow through declaring war on living without drugs. Or as BioMarx puts it: "*If the Patient Heals his disposable time for himself, he robs Pharma*" (B-C1: ch. 10).

What seems absurd here, that medical research could define healthiness as its enemy rather than a goal is precisely the absurd paradox that confronted Marx regarding the factory. Extending the analogy in which machinery is equivalent to clinical trials, we see that the logical counterpart to hours is treatments or prescriptions. As investments, factories and machinery come to be seen by capitalists as in need of maximization; any time those machines are not in use comes to be experienced as a loss, no matter how absurd that perception may be. Marx describes this transmogrification of factories into entities that require laborers, which in turn requires laborers to work as long as possible.

> Furnaces and workshops that stand idle by night, and absorb no living labour, are "a mere loss" to the capitalist. Hence, furnaces and workshops constitute lawful claims upon the night labour of the work-people. The simple transformation of money into the material factors of the process of production, into means of production, transforms the latter into a title and a right to the labour and surplus labour of others. An example will show, in conclusion, how this sophistication, peculiar to and characteristic of capitalist production, this complete inversion of the relation be-

tween dead and living labour, between value and the force that creates value, mirrors itself in the consciousness of capitalists. (C1: Ch11)

Marx goes on to quote a "grotesque" passage in which a mill owner considers his "property damaged" when workers don't work long enough to keep the machines running at all times and thus fail to maximize the capitalist's return on investment.[21] The counterpart in biomedicine is that, from the perspective of pharma, a drug's possible market becomes its *expected* market, and every person for whom that drug could be indicated is seen as a loss of revenue if he or she is not in fact taking the drug.

Mickey Smith's *Pharmaceutical Marketing* includes a chart on the "Decomposition of the Market," which lists the math through which patients with chronic condition X (1,000,000) translates eventually into only 7,350,000 prescriptions whereas there was an "original potential" of 12,000,000 prescriptions (1 million patients × 12 prescriptions). The remainder is termed "Prescriptions 'Lost'" with "Lost" in quotation marks, as if Smith knows he is treading on ethically suspect grounds; but in the summary chart, defined in terms of the potential market, the remainder is simply listed as "Total Prescription Loss." Smith uses this chart to make the point that increasing compliance may be cheaper than increasing market share.

This type of paradox arises, according to Marx, because capitalists do not employ machines *in order to* produce things and generate material wealth for society, but perceive material wealth *merely as a means* to make more money.

> This contradiction comes to light, as soon as by the general employment of machinery in a given industry, the value of the machine-produced commodity regulates the value of all commodities of the same sort; and it is this contradiction, that in its turn, drives the capitalist, without his being conscious of the fact, to excessive lengthening of the working-day, in order that he may compensate the decrease in the relative number of labourers exploited, by an increase not only of the relative, but of the absolute surplus-labour. (C1:531, ch. 15)

For BioMarx, it is no less tragic.

> This contradiction comes to light, as soon as by the general employment of the Clinical Trial in a given Market, the Illness of the Threshold-Medicalized Symptom regulates the Illness of all Symptoms of the same sort; and it is this contradiction, that in its turn, drives Pharma, without his being conscious of the fact, to excessive lengthening of the Treatment-Time, in order that he may compensate the decrease in the relative num-

ber of Patients exploited, by an increase not only of the relative, but of the absolute Surplus-Health. (B-C1:531)

The phrase "without his being conscious of the fact" is stunning, and intriguing. Marx does constantly attend to the psychology of the capitalist, whom he sees as structurally produced ("this contradiction in turn drives Pharma"). He points out, for instance, that what may appear at first glance to be miserly greed, the impulse to accumulate, must in fact be understood as a kind of possession: Capital possesses the capitalist, such that seeing through Capital's eyes—from Capital's point of view—comes to be so deeply embedded that it blinds the capitalist to all else. Thus, Marx says, Capital inhabits the soul of the capitalist.

It is only the notion of possession by capital that prepares me to understand the following claim, which is otherwise, to me, incomprehensibly callous.

> Looking at the business of mental disease objectively, but without cynicism, a common denominator of these indications is that they share the distinction of not being cured by these pharmacological treatments. This makes the market even more attractive. The patients have to take the drugs chronically. Not only are the diseases not cured, but there are few treatments that give 100% relief to those who have a syndrome. All usual response rates are 60 to 70% for a really good drug. . . . This gives a double opportunity: (1) one can enter a partially saturated market with a drug that works on patients unresponsive to existing treatments; and (2) one can improve on the side effect profile or the efficacy in terms of the time required for the onset. (Bartfai and Lees 2006, 221)

"Objectively, but without cynicism": it is as if the authors dimly recognize how outrageous the rest of the paragraph will seem, yet they cannot stop, because the point they are making *is* objective. The world they live in, that we live in, is a world in which medical research *is* a financial investment demanding returns. In our world, it is objectively true that drugs that cure people or stop the spread of a disease (like vaccines are supposed to do) "reduce revenue." Chronic treatments are more valuable to research, and drugs that work only on a subset of people (i.e., those with a large NNT) generate more prescriptions (by indicating a larger market) than those that treat everyone they are indicated for (i.e., those with an NNT of 1). This is the reality of biomedical capitalism within which we and the pharmaceutical companies must operate and survive.

Bartfai and Lees claim their view is "without cynicism." That is, while it might appear as if they are speaking in a scornful and bitterly mocking attitude, as if they were motivated only by selfish interests, they are not. What they are saying, in essence, then, is that unlike Marx's capitalists they are *consciously* possessed and understand clearly that they have no choice but to "excessively lengthen treatment time . . . beyond all bounds set by human nature," to let BioMarx rephrase them. "Cynicism" would imply that they had based their analysis on selfish motives and that they therefore believe it to be the product of human nature. But they rest in the knowledge, instead, that their conclusions simply reflect the objective contradictions of healthcare.

What is a Treatment-Time? What is the length of time during which Biomedicine may Heal the Illness whose daily Diagnosis it buys? . . . [I]n its blind unrestrainable passion, its were-wolf hunger for surplus-Health, Biomedicine oversteps not only the moral, but even the merely physical maximum bounds of the Treatment-Time. . . . It reduces the sound sleep needed for the restoration, reparation, refreshment of the bodily powers to just so many Treatments of torpor as the revival of an organism, absolutely exhausted, renders essential. It is not the normal maintenance of the Illness which is to determine the limits of the Treatment-Time; it is the greatest possible daily expenditure of Illness, no matter how diseased, compulsory, and painful it may be, which is to determine the limits of the Patients' period of repose. Biomedicine cares nothing for the length of life of Illness. All that concerns it is simply and solely the maximum of Illness, that can be rendered fluent in a Treatment-Time. (B-C1: ch. 10)

A different sort of logic was proposed by two researchers in the *British Medical Journal* in 2004. Based on a meta-analysis of existing risk, biomarker, and threshold trial data, the authors proposed a single multipill that would save lives to such an extent, they argued, that everyone over fifty-five should be mandated to take it. Their logic is an extension of Rose's prevention analysis. In a nod to cost, but not to consent, they suggest that a low-cost version of this polypill, using generic components off patent, would work, even if "10% of the users were intolerant" (Wald and Law 2003, 4). "Intolerance" here is a formulation of the literal limit of the body's resistance to too many drugs, that is, when it throws them up. Their proposal thus involves calibrating the drug to the maximum number of effects, side effects, and cost that society will tolerate before rebelling (the NNT of the polypill was estimated to be between 600 and 800).

With "bodily intolerance" we confront the same ultimate physiological barrier that Marx's capitalists found with labor, and that Goizueta found with Coke: the surprisingly expandable but not unlimited elasticity of the human body. Where Goizueta suggested that the target for Coke's growth was human liquid-consumption capacity, Marx locates it in the labor humans can be pushed to do, and BioMarx in the body's tolerance for pills.

> And this law is only realized because it implies another one, namely that the scale of Medicalization is not determined according to given needs, but rather the reverse: the number of Risk Diagnoses is determined by the constantly increasing scale of Medicalization, which is as much Undiagnosed health as possible, and this is only attained by engaging in Medicalization for Medicalization's sake. (B-PEM 1038)[22]

Their article concludes with a call for the end of thresholds altogether by taking them to their natural limit: "It is time to discard the view that risk factors need to be measured . . . everyone is at risk" (Law and Wald 2003, 4). This is a naturalized form of the suggestion to "put statins in the water supply," no longer even a half-joke, but a policy proposal.

Conclusion
replace all [laborer(s)] with [Patient(s)]
replace all [wage] with [Risk]
replace all [wage-laborer(s)] with [Patient(s)-at-Risk]

We see then, that, apart from extremely elastic bounds, the nature of the Measure of Symptoms itself imposes no limit to the Treatment-Time, no limit to surplus-Health. Pharma maintains his rights as a purchaser when he tries to make the Treatment-Time as long as possible, and to make, whenever possible, two Treatment-Times out of one. On the other hand, the peculiar nature of the Symptom Submitted implies a limit to its Healing by the purchaser, and the Patient maintains his right as Submitter when he wishes to reduce the Treatment-Time to one of definite normal duration. There is here, therefore, an antinomy, right against right, both equally bearing the seal of the law of Measures. Between equal rights force decides. Hence is it that in the history of Pharma Experience, the determination of what is a Treatment-Time, presents itself as the result of a struggle, a struggle between collective Biomedicine, i.e., the class of Pharma Companies, and collective Health, i.e., the Patient Class.

BIOMARX, *BIOMEDICINE*

In using Marx to construct BioMarx, I did not assume that he was correct about the economy. Rather I attended carefully to the careful way in which he read the capitalists and the economists of his day. He attempted to make explicit the logic of their way of valuing the world, which was via labor. He found this directly in their writings: in the way they kept their account books, the way they complained about wasted labor, and the way they chose to invest in machinery (or not). In this manner he executed a close reading of their practices as they were enabled by their perspective—the material-ization of their theories of value. By following the logic of the competing demands of this perspective—the way in which growth, for instance, be-comes critical for a company's survival—he tried to show that apparently contradictory results, like constant crises, were results of this logic. And in turn he was able to show how the actions of capitalists made sense in light of this logic.

I, too, have had to adopt this convoluted approach, as I have been con-fronted with apparently absurd and sometimes vile practices by pharma-ceutical companies, and with corporate statements that have, in a calm, ob-jective, and at times defensively complaining voice, logically justified those practices. I did not attend to the scandalous practices of cheating in a clinical trial, suppressing research, or ghostwriting results, but rather to the need to develop drugs for diseases that affect the insured middle class and to avoid focusing on diseases that largely affect the poor, and to the need to define illness as risk and increase as much as possible the number of treatments, even if this means that most of the people being treated would not benefit from the drug.

When I've assessed these latter practices without the context and logic unearthed by BioMarx, it has been impossible to understand pharmaceu-tical companies as anything but predatory. Using Marx in this way helps clarify the logic that drives pharmaceutical companies as a logic that is not exclusive to that particular corporate culture, but is shared by all of us—and that it is a historically specific logic, a way of seeing health qua mass health as risk reduction, that results in health being defined via treatments.

When I presented a talk on this material (without mentioning Marx) as grand rounds to the psychiatry department at Alta Bates Medical Center, the first question I got was "Is there any hope?" The second was "What do you recommend we do?" BioMarx suggests we are only at the beginning of a transformation of health. In analogy to the fight over the length of the work-ing day, BioMarx suggests that there will be increasing struggles over how much medicine we can be mandated to take. Perhaps the various debates regarding the new "vaccines"—like a human papilloma virus vaccine for all girls ages nine to thirteen, and a meningitis vaccine for all children—are signs of this. And will the same happen with the polypill? Will there be a fight to determine the level of drug intake we are forced to tolerate?

> One of the most important problems, therefore, which the Diagnoser of a Health Market has to solve is to find out the maximum speed at which he can run, with a due regard to the above conditions. It frequently happens that he finds he has gone too fast, that breakages and bad Treatment more than counterbalance the increased speed, and that he is obliged to slacken his pace. (B-C1: ch. 10)

Reading BioMarx this way, I have started having little out-of-body experiences. The future is calling me. This nightmare future where, when we wake up, we first check the latest clinical trials, then order our pack of pills for the day. Some people—the "intolerants" we call them—have negative reactions to the pills. This is not their fault, as they protest, but their insurance goes up all the same.

In this future, there are still two parties: the More-Lifers rule on a platform of immortality. Life extension is a reality, they say, as long as we all do our part and participate in enough clinical trials. The opposition party—More-Choicers—complain that health is not a formula, that we should not be forced to take so many drugs with so many unknown interactions. Current guidelines require monitoring of seventy-five biomarker measures like cholesterol levels (six types), the body-mass index, a composite behavioral-emotional score, and a steady-attention test. In order to receive health coverage, all of these must remain above the acceptable level or one must be taking the appropriate preventative pharmaceutical therapy. There is growing unrest over the legal intolerance rate of 10 percent, meaning that 10 percent of the population has detrimental reactions to the required medications. If the rate isn't lowered below 3 percent, many predict that the More-Choice party may take back the house.

That is my objective paranoid self speaking. My analytic self and the pharmaceutical analysts begin with the tragedy. The more "health" we gain, the more medicine we consume. Key to this is the fact that most clinical trials are designed by the pharmaceutical industry, and in the most mundane capitalist way, this means that they are designed such that if they are successful, they will increase the market in drugs. The corollary is startling: almost all of the facts that we have gathered about our mass health over the last twenty years tell us to take more drugs. They are not bad facts, but they are limited.

We are not asking—and tragically, in our current system, we cannot afford to ask—what sort of "health" we might have if we took fewer drugs. Those facts have not been produced. Until we rethink the infrastructure for the design of large-scale clinical trials and screening tests, this trend will continue—even if we clean up the abuses of clinical trials.

The disturbing analytic problem is that the pharmaceutical executives and marketers seem to be shouting the same thing. Don't blame us, they say. We know the problems better than you do, but we are trapped within them, too, even as we perpetuate them. Thus Mickey Smith explains, "Society has medicalized human problems, it appears. Medicine has perhaps been an accessory, and the pharmaceutical industry, certainly, has provided both with the means. To expect either of the latter parties to do, or have done, otherwise bespeaks a considerable naïvete" (2002, 35).

Similarly, Bartfai and Lees blame two groups "as the real root causes of all the problems the industry and society face" (2006, xvii), including why there are not enough drugs for the people who need them. The groups are "the lawyers who litigate and the venture capitalists who may want too much return from too short an investment and can switch their investments and allegiances on a whim" (ibid. xvii). To point to lawyers distracts from the real insight: the structure of drug companies is that of a capital company whose chief allegiance is to shareholder returns.[23] Whether by venture capitalists or mutual funds, drug companies must always increase returns, and that means more treatments.

Thus it seems we are in a bind, even as we have an ever-increasing number of treatments available, which continually reduces our risk. We call for more regulation, better surveillance of drug companies, and even structure incentives better, like orphan drug laws and granting companies six additional months of patent protection if they study the effects of their drugs on children. We get more information on relative safety, companies get more money, and, predictably, children end up taking more and more drugs (reducing their risks more).

But this use of *we* marks me as a medically insured U.S. citizen most deeply. My suspicious ears hear another trend being traced by the ethnographies of clinical trials conducted globally. There is a much bigger, global story to tell about clinical trials that is the object of ongoing research by Kaushik Sunder Rajan, Adriana Petryna, Kris Peterson, myself, and others.

The insidiously banal logic of modern-day pharmaceutical industry growth drives clinical trials in ways that have become so naturalized that it is hard to imagine health research in any other way. How is it that we live with, love, and ingest medicines produced through this same experimental zone? How is it that our privileged consciousness sees this as all so natural, our bodies willingly given over, and other bodies so easily made invisible?

replace all [materials] with [Bodies]
replace all [property] with [Life]
replace all [price] with [Prevalence]
replace all [profit] with [Market-Increase]

Notes

1. IMS Institute for Healthcare Informatics, *The Use of Medicines in the United States: Review of 2010* (Parsippany, N.J.: IMS Health Incorporated, 2011); Office of the Actuary, Centers for Medicare and Medicaid Services, *National Health Expenditures, Forecast Summary and Selected Tables.*

2. This scale allows for tremendous variation. Huge disparities exist in prescription rates by state, and by counties in states, and within counties. Often a single prescriber can increase the prescription consumption in an area five- to tenfold.

3. Formulation from Sunder Rajan 2006.

4. The old guidelines held that blood levels of low-density lipid (LDL) cholesterol, the bad kind that clogs arteries, should stay below 100 milligrams per deciliter (mg/dL) and, ideally, below 70 mg/dL for very high-risk patients. According to the new advisory, issued in 2006, those guidelines are now recommended for all people with established heart disease. See Edelson 2006.

5. See https://sites.google.com/site/biomarxexperiment/home. The original is at http://www.marxists.org/archive/marx/works/1867-c1/index-l.htm.

6. See appendix A for a list of the substitutions. I have not thought through even in passing the relations of biocapital in the sense that Kaushik Sunder Rajan (2006) uses it; the capitalization of biology in the sense used by Hannah Landecker (2007); the commodification of bodies in either Adriana Petryna's take on health or clinical-trial trafficking (2005) or in Lawrence Cohen's take on organs (1999); or the proposed rewriting of *Capital*, vol. 3, by Sarah Franklin and Margaret Lock (2003).

7. Other words may be more appropriate for other areas of analysis. If anyone has terms they would like to experiment with, I can easily code them and produce a substituted *Capital* in less than a minute. Automation does save labor!

8. All quotes from *Biomedicine* I have left in the raw automatic-substitution format. Please see the websites for the full text of *Biomedicine.*

9. The inspiration for this origin story lies in the work of the Socialist Patients Kollectiv (SPK) (Sozialistisches Patientenkollektiv 1993).

10. See Greene 2006 on the history of this since the 1960s.

11. "Are You the Picture of Health?" campaign for colorectal-cancer screening. Available at the Centers for Disease Control and Prevention website, http://www.cdc.gov.

12. [Fixed Biomedicine and the Development of the Knowable Forces of Society] NOTEBOOK VII, End of February, March. End of May–Beginning of June 1858, The Chapter on Biomedicine (continuation).

13. Compare Petryna 2005, Sunder Rajan 2007, Fisher 2009, Petryna, Lakoff, and Kleinman 2006, and Peterson, chapter 7 in this volume.

14. See for instance J. Urquhart, "Some Key Points Emerging from the COX-2 Controversy." *Pharmacoepidemiology and Drug Safety* 14, no. 3 (2005).

15. The work of Moishe Postone, in *Time, Labor and Social Domination* (1993), was extremely useful in helping me to understand this dynamic.

16. Pastoral care or biopolitics in terms of care of the population, from this perspective, is only meaningful when such "care" involves an expanding and constant domain of prophylaxis. Normalization in a Foucaultian sense is meaningless in this regard—the only state of "normality" that generates value is one that expresses an aggregate potentiality of future illness.

17. They continue: "A more profound 'revelation' [than the fact that drug companies deceive us] is that members especially of the U.S. society are prepared to take too many drugs with little provocation" (Bartfai and Lees 2006, xv). This statement speaks fundamentally to why citizens should in no way expect drug companies or any other company to look out for the public interest or even to tell the truth. As Althusser (2006) suggests, the bourgeoisie lies easily because the lies are naturalized, continuous, "common-sense" discourses all by themselves.

18. Ironically, cholesterol-lowering drug trials are actually quite numerous because the market is so large that almost every company competes for it (Greene 2006).

19. See Linda Hogle (2001) and Jennifer Fosket (2002) on clinical-trial entry criteria determining "high risk" as a result.

20. See Cohen on bioavailability (Cohen 1999, 2004), as well as the federal definition: "the rate and extent to which the active ingredient or therapeutic ingredient becomes available at the site of drug action" (section 505[j] of the Food, Drug, and Cosmetic Act as amended by Title XI of the Medicare Prescription Drug, Improvement, and Modernization Act of 2003 [505(j)(8)(A)(ii)]).

21. During the revolt of the English factory lords between 1848 and 1850, "the head of one of the oldest and most respectable houses in the West of Scotland, Messrs. Carlile Sons & Co., of the linen and cotton thread factory at Paisley, a company which has now existed for about a century, which was in operation in 1752, and four generations of the same family have conducted it," this "very intelligent gentleman" wrote a letter in the *Glasgow Daily Mail*, on 25 April 1849, with the title "The Relay System," in which among other things the following grotesquely naïve passage occurs: "Let us now . . . see what evils will attend the limiting to 10 hours the working of the factory. . . . They amount to the most serious damage to the millowner's prospects and property. If he (i.e., his "hands") worked 12 hours before, and is limited to 10, then every 12 machines or spindles in his establishment shrink to 10, and should the works be disposed of, they will be valued only as 10, so that a sixth part would thus be deducted from the value of every factory in the country."

22. From author's "BioMarx" substitution of the *Economic Works of Karl Marx 1861–1864* at chapter 6: http://www.marxists.org/archive/marx/works/1864/economic/ch02a.htm (at [450]).

23. Lawyers are problematic in that their litigation puts drug companies on the defensive (Smith, Kolassa, Perkins, and Siecker 2002).

2

DONNA J. HARAWAY

VALUE-ADDED DOGS AND LIVELY CAPITAL

Marx dissected the commodity form into the doublet of exchange value and use value. But what happens when the undead but always generative commodity becomes the living, breathing, rights-endowed, doggish bit of property sleeping on my bed—or giving cheek swabs for your genome project, or getting a computer-readable ID chip injected under the neck skin before the local dog shelter lets my neighbor adopt her new family member? Canis lupus familiaris, indeed; the familiar is always where the uncanny lurks. Further, the uncanny is where value becomes flesh again, in spite of all the dematerializations and objectifications inherent in market valuation. Marx always understood that use and exchange value were names for relationships; that was precisely the insight that led beneath the layer of appearances of market equivalences into the messy domain of extraction, accumulation, and human exploitation. Turning all the world into commodities for exchange is central to the process. Indeed, remaking the world so that new opportunities for commodity production and circulation are ever generated is the name of this game. This is the game that absorbs living human labor power without mercy. In Marx's own colorful, precise language, which still gives capitalism's apologists apoplexy, capital comes into the world "dripping from head to toe, from every pore, with blood and dirt" (1976 [1867], 926).

What, however, if human labor power turns out to be only part of the story of lively capital? Of all philosophers, Marx understood relational sensuousness, and he thought deeply about the metabolism between human beings and the rest of the world enacted in living labor. As I read him, however, he was finally unable to escape from the humanist teleology of that labor—the making of man himself. There are, finally, no companion species, reciprocal inductions, or multispecies epigenetics in his story.[1] But what if the commodities of interest to those who live within the regime of Lively Capital cannot be understood within the categories of the natural and the social that Marx came so close to reworking, but could not finally do under the goad of human exceptionalism? These are hardly new questions; but I propose to approach them through relationships inherent in contemporary U.S. dog-human doings that raise issues not usually associated with the term *biocapital*, if, nonetheless, crucial to it.

First, however, a caveat about the "specificity," in many senses, of my arguments: co-constitutive relatings between taxonomically recognizable dogs and people are at least tens of thousands of years old and reach across and through the earth (and the near reaches of outer space in U.S. and Soviet space programs) to almost every habitat and lifeway, past and present, and, with luck, to come. "Dog is my co-pilot" is more than a joke or a slogan to put on one's car along with the Darwin fish. Multispecies, trans-species becoming with each other in naturecultures is the name of the game of life on earth; the partners do not precede their dynamic knottings. Critters, human and not, make each other up in flesh and sign, literally and figuratively. There is no such thing as the universal human or the universal dog. Therefore, it makes no sense to talk of human-dog relating as a singularity of any kind, including capitalist sorts. That does not mean that living with, thinking with, "one" complex tangle in its interlaced times and spaces does not shape attachment sites for many other connections—not generalizations, but possible connections to other muddy worlds that nonholistically extrude and intrude in the fibrous slime of situated becomings with. Ahumanist temporalities infuse all the tissues of all the critters; history is not a human monopoly. In that sense, "we have never been human" is my motto. We are, rather, "companion species"—that is, mortal mess mates at a terran repast.[2]

So, in this chapter, I explore value tangles of mainly late-twentieth- and early-twenty-first-century, Euro-American-U.S. dog-human doings, some of which shape consequential and quite specific kinds of "globals" in commodity networks, queer and normative kin-formations, and other sorts of worlding familiar to analysts of biopower and biocapital. It would be a mistake to read my analyses as paradigmatic for other situated dog-human doings, but also a mistake not to grasp attachment sites for analysis and action.[3] I may be telling a tale of a decadent late imperial power, where lively capital runs on doggy investments, but it is a tale with many other storied and lived times—past, present, and to come.

There is no shortage of proof that classic rabid commodification is alive and well in consumer-crazy, technoscientifically exuberant dog worlds in the United States. I will give my readers plenty of reassuring fact-packages on this point, sufficient to create all the moral outrage that we lefties seem to need for breakfast and all the judgment-resistant desires that we cultural analysts seem to enjoy even more. However, if a Marx-equivalent were writing *Biocapital*, volume 1, today, insofar as dogs in the United States are commodities, as well as consumers of commodities, the analyst would have to examine a tripartite structure: use value, exchange value, and encounter value—without the problematic solace of human exceptionalism.[4]

Trans-species encounter value is about relationships among a motley array of lively beings, where commerce and consciousness, evolution and bioengineering, and ethics and utilities are all in play. I am especially interested here in "encounters" that involve, in a nontrivial but hard-to-characterize way, subjects of different biological species. My goal is to make a little headway in characterizing these relationships in the historically specific context of lively capital. I would like to tie my Marx-equivalent into the knots of value for companion species, especially for dogs and people in capitalist technoculture in the early twenty-first century, where the insight that to be a situated human being is to be shaped by and with animal familiars might deepen our abilities to understand value-added encounters.

Valuing Dogs: Markets and Commodities

Like a 1950s TV show, companion-animal worlds are all about family. If European and American bourgeois families were among the products of nineteenth-century capital accumulation, the human-animal companionate family is a key indicator for today's lively capital practices. That nineteenth-century family may have invented middle-class pet keeping; but what a pale shadow to today's doings that was! Kin and brand are tied in productive embrace as never before. In 2006 about 69 million U.S. households (63 percent of all households) had pets, giving homes to about 73.9 million dogs, 90.5 million cats, 16.6 million birds, and many other critters.[5] As an online report on the pet food and supplies market from MindBranch, Inc. for 2004 stated, "In the past, people may have said their pet 'is like a member of the family,' but during 1998–2003 this attitude has strengthened, at least in terms of money spent on food with quality ingredients, toys, supplies, services, and healthcare."[6] The consumer habits of families have long been the locus for critical theory's efforts to understand the category formations— like gender, race, and class—that shape social beings. Companion-species kin patterns of consumerism should be a rich place to get at the relations that shape emergent subjects, not all of whom are people, in lively capital's naturecultures. Properly mutated, the classics like gender, race, and class hardly disappear in this world—far from it—but the most interesting emergent categories of relationality are going to have to get some new names, and not just for the dogs and cats.

The global companion-animal industry is big, and the United States is a major player. I know this because I have dogs and cats who live in the style to which I and my whole post-Lassie generation have become indoctrinated. Like any scholar, however, I tried to get some hard figures to go with the coming examples. The Business Communications Company publishes an annual analysis of market opportunities and segments, company fortunes, rates of expansion or contraction, and other such data dear to the hearts of investors. So for the first draft of this chapter, I tried to consult *The Pet Industry: Accessories, Products, and Services* for 2004 online. Indeed, I could have downloaded any of the alluring chapters, but all of them are proprietary, and so to peek is to pay. To get access to the whole package would have cost me over $5,000, a nice piece of evidence all by itself for my assertion in the first sentence of this paragraph. An alternative data source, Global Information Inc. (the self-described online "vertical markets research portal"), offers twenty-four-hour, five-day-per-week updates for pet marketers on forecasts, shares, research and development, sales and marketing, and competitive analysis. Ignore these services at your peril.

In the end, I settled for training-sized statistical tidbits from Business Communications and from the free summaries for the 2006 survey available on the website of the American Pet Products Manufacturing Association. In the United States alone in 2006, pet owners spent about $38.4 billion overall on companion animals, compared to $21 billion in 1996 (constant dollars). The global market for pet food and pet-care products for 2002 was U.S. $46 billion, which is an inflation-adjusted increase of 8 percent over the period 1998–2002. The inflation-adjusted growth rate for 2003 alone was 3.4 percent, driven, we are told, by pet owners' demand for premium foods and supplies.

Consider just pet food. ICON Group International published a world market report in February 2004. The report was written for "strategic planners, international executives and import/export managers who are concerned with the market for dog and cat food for retail sale."[7] The point was that "with the globalization of the market, managers can no longer be contented with a local view." Thus, the report paid special attention to which countries supply dog and cat food for retail sale, what the dollar value of the imports is, how market shares are apportioned country to country, which countries are the biggest buyers, how regional markets are evolving, and how managers might prioritize their marketing strategies. Over 150 countries are analyzed, and the report makes clear that its figures are estimates of potential that can be drastically altered by such things as "'mad cow' disease, foot-and-mouth disease, trade embargoes, labor disputes, military conflicts, acts of terrorism, and other events that will certainly affect the actual trade flows." Indeed. Nonetheless, the report neglected to state the underlying obvious fact: industrial pet food is a strong link in the multispecies chain of global factory farming.[8]

The *News York Times* for Sunday, 30 November 2003, is my source for the $12.5 billion for the size of the U.S. pet food market in 2003 ($15 billion in 2006) (Koerner 2003). I did not know how to think about how big that was until I read another story in the *New York Times* (2 December 2003) telling me that in 2003 the human cholesterol-lowering statin market was worth $12.5 billion to the pharmaceutical industry (Kolata 2003). How much human blood-lipid control is worth how many dog dinners? I'd throw away my Lipitor before I shorted my dogs and cats. Marx told us how the purely objective nature of exchange value evacuates the trouble springing from such use-value comparisons. He also told us how such things as statins and premium dog food become historically situated bodily needs. He didn't pay nearly enough attention for my taste to which bodies, in the multispecies web linking slaughter labor, chicken cages, pet dinners, human medicine, and much more.

I cannot now forget these things as I decide how to evaluate both the latest niche-marketed dog food purported to maximize the sports performance of my agility dog and also how her nutritional needs are different from those of my older but still active pooch. A large and growing portion of pet-food products addresses specific conditions, such as joint or urinary tract health, tartar control, obesity, physiological demands, age-related needs, and so on. I cannot go to an agility meet to run with my dog without tripping over brochures and booths for natural foods, scientifically formulated foods, immune-function-enhancing foods, homemade-ingredients foods, foods for doggy vegans, raw organic foods that would not please vegans at all, freeze-dried-carrot fortified foods, food-delivery devices to help out dogs who are alone too much, and on and on. Indeed, diets are like drugs in this nutritional ecology, and creating demand for "treatment" is crucial to market success. Besides diets, I feel obligated to investigate and buy all the appropriate supplements that ride the wavering line between foods and drugs (chondroitin sulfate and glucosamine sulfate or omega-3 fatty acids from flax seed or fish oil, for example). Dogs in capitalist technoculture have acquired the mixed blessing of the "right (obligation) to health," and the economic (as well as legal) implications are legion. Dog nutraceuticals also land them as well as their humans in the midst of controversies over depleting shark populations for chondroitin from their cartilage or hyping consumers on the benefits of omega-3s from cheap plants when the fatty acids from disappearing fish stocks are probably the ones needed for joint and heart health.

Food is not the whole story. The Business Communications Company stressed that there was growth in all segments of the companion-animal industry, with rich opportunities for existing players and new entrants. Health is a giant component of this diversifying doggy version of lively capital. Small-animal veterinarians are well aware of this fact, as they struggle to incorporate the latest (very expensive) diagnostic and treatment equipment into small practices in order to remain competitive. A special study done in 1998 revealed that vets' income was not growing at the rate of comparable professionals because they did not know how to adjust their fees to the rapidly expanding services they were routinely offering.[9] My family's credit-card records tell me that at least one of the vet practices we frequent got the point in spades. In 2006 U.S. citizens spent about $9.4 billion for vet care for pets. As a reality check, I turned to the *World Animal Health Markets* of 2010, a report that profiled animal health markets in fifteen countries, accounting for 80 percent of the world share.[10] The conclusion: in the affluent parts of the globe, the pet-health market is robust and growing.

Consider a few figures and stories. Mary Battiata wrote a feature article for the *Washington Post* in August 2004 that followed her search for a diagnosis for her aging family member, her beloved mutt, Bear, who showed troubling neurological symptoms. After the first sick visit to the vet cost $900, she began to understand her situation. She was referred to the Iams Pet Imaging Center in Washington for an MRI. Or rather, Bear was referred, and his guardian/owner Mary wrestled with the ethical, political, affectional, and economic dilemmas. How does a companion animal's human make judgments about the right time to let her dog die—or, indeed, to kill her dog? How much care is too much? Is the issue quality of life? Money? Pain? Whose? Does paying $1,400 for Bear's MRI add to the world's injustice, or is the comparison between what it costs to run decent public schools or to repair wetlands and what it costs to get Bear diagnosed and treated the wrong one? What about the comparison between people who love their pet kin and can afford an MRI, and people who love their pet kin and can't afford annual vet exams, good training education, and the latest tick and flea products, much less hospice care (now available in a few places for dogs and cats)? What comparisons are the right ones in the regime of lively capital?

Other high-end treatments now available for pets include kidney transplants, cancer chemotherapy, and titanium joint-replacement surgeries. The University of California, Davis, recently opened an up-to-the-minute treatment and research hospital for companion animals with the kind of cancer care expected in the best human medical centers. New veterinary drugs— and human drugs redirected to companion animals—emphasize pain relief and behavior modification, matters that hardly appeared on Lassie's people's radar screens, but involve serious money and serious ethical dilemmas today.[11] In addition, vets-in-training today take courses in the human-animal bond, and this diversifying region of the affectional family economy is as richly commodified and socially stratified as is any other family-making practice, say, for example, assisted reproduction for making human babies and parents.[12]

Pet health insurance has become common, as is malpractice insurance for vets, partly fueled by the success in court of arguments that companion animals cannot be valued like ordinary property. "Replacement value" for a companion dog is not the market price of the animal. Neither is the dog a child nor an aged parent. In case we missed the point in all the other aspects of daily life, efforts both to establish money damages and to pay the bills for our companions tell us that parent-child, guardian-ward, and owner-property are all lousy terms for the sorts of multispecies relationships emerging among us. The categories need a makeover.

Besides vets, other sorts of health professionals have also emerged to meet companion-animal needs. I get regular professional adjustments for my Australian Shepherd sports partner, Cayenne, from Ziji Scott, an animal-chiropractic-certified practitioner with magic hands. No one could convince me that this practice reflects bourgeois decadence at the expense of my other obligations. Some relationships are zero-sum games, and some are not. But a central fact shapes the whole question: Rights to health and family-making practices are heavily capitalized and stratified, for dogs as well as for their humans.

Beyond the domains of dog medical services, nutrition, or pedagogical offerings, canine consumer culture of another sort seems truly boundless. Consider vacation packages, adventure trips, camp experiences, cruises, holiday clothing, toys of all kinds, daycare services, designer beds and other animal-adapted furniture, doggy sleeping bags and special tents and back-packs, and published guides to all of the above. On 24 September 2004, the *New York Times* ran ads for dog shopping that featured a $225 raincoat and $114 designer collar. Toy dogs as fashion accessories to the wealthy and fa-mous are a common newspaper topic and a serious worry for those who think those dogs have doggish needs.[13] The American Kennel and Boarding Asso-ciation in 2006 reported that the significant industry growth is in the high-end pet lodgings, such as the new San Francisco hotel, Wag, which charges $85 per night and offers massage, facials, and swimming pools. Webcam TV for traveling humans to watch their pets in real time in communal play areas is standard at San Francisco's middle-of-the-market $40-per-night Fog City Dogs Lodge.[14] For those whose commodity preferences are more book-ish, look at the companion-animal print culture. Besides a huge companion-species book market in categories from anthropology to zoology and the whole alphabet in between, two new general-audience magazines make my point. *Bark* is a Berkeley, California, dog literary, arts, and culture rag that I read avidly, and not just because they favorably reviewed my *Companion Species Manifesto*. The East Coast finally faced its responsibilities in this mar-ket segment; and so, with articles on such matters as how to win a dog-custody battle and where to find the best ten places to walk with your dog in Manhattan, the *New York Dog* appeared in November–December 2004, aiming to rival *Vogue* and *Cosmopolitan* for glossy values.[15] And all of this hardly touches the media markets crucial to hunting with dogs, playing dog-human sports, working with dogs in volunteer search and rescue, and much more. It seems to me that it is all too easy in dogland to forget that resistance to human exceptionalism requires resistance to humanization of our part-ners. Furry, market-weary, rights bearers deserve a break.

Enough, or rather, almost enough—after all, in lively capital markets "value-added" dogs aren't just familial co-consumers (or co-workers, for which you must go to the next section of this chapter). In the flesh and in the sign, dogs are commodities, and commodities of a type central to the history of capitalism, especially of technoscientifically saturated agribusiness. Here

I will consider only kennel-club registered "purebred" dogs, even though those surely aren't the canines that come first to mind in connection with the term *agribusiness*, no matter how much pedigree-packing dogs return us to crucial nineteenth-century economic and cultural innovations rooted in the biosocial body. In *Bred for Perfection* (2003), Margaret Derry explains that the public data keeping of lineage—the written, standardized, and guaranteed pedigree—is the innovation that fostered international trade in both livestock like sheep and cattle and fancy stock like show dogs and chickens. And, I might add, race- and family-making stock. Institutionally recorded purity of descent, emphasizing both inbreeding and male lines that made female reproductive labor all but invisible, was the issue. The state, private corporations, research institutions, and clubs all played their roles in moving practices for controlling animal reproduction from pockets of memory and in local endeavors of both elites and working people to rationalized national and international markets tied to registries. The breeding system that evolved with the data-keeping system was called scientific breeding; and in myriad ways this paper-plus-flesh system is behind the histories of eugenics and genetics, as well as other sciences (and politics) of animal and human reproduction.

Dog breeds, not variously differentiated and stabilized kinds, but breeds with written pedigrees, were one result. Across continents, dogs with those credentials could command very nice prices, as well as fuel amazing practices of heritage invention, standards writing and maintenance, sales-contract development, germ-plasm trading, health surveillance and activism, reproductive-technology innovation, and the passionate commitment of individuals, groups, and even whole nations.[16]

The proliferation of dog breeds, and their movement into every social class and geographical region of the world, is part of the story. There are many breeds specifically produced for the pet market, some quite new, like the cross of Borzois and Longhaired Whippets to make the little sighthound called Silken Windhounds. Witness today's explosion in toy breeds and teacup breeds as fashion accessories (and too often, medical disasters). Or the popularity of the puppy-mill-produced dogs because they carry an American Kennel Club (AKC) purebred dog pedigree. Or, as I move away from outrage to love affair, I am reminded both of the knowledgeable, talented, self-critical dog people whom I have met in performance dog worlds, as well as in conformation show dog scenes, and also of their accomplished, beautiful dogs. And of my dogs, including Roland, the one with the fraudulent (that Chow-Chow dad) AKC Australian Shepherd registration, gotten so that he can play with agility in the shepherds' sandbox, as long as he is reproduc-tively sterilized.

But is he necessarily reproductively silenced? What happens when pedi-gree, or lack of it, meets petri dish? Consider the Dolly technique so in-sightfully written about by Sarah Franklin in *Dolly Mixtures* (2007). Dolly the pedigreed sheep might have been the first mammal who was the fruit of somatic cell nuclear transfer cloning, but she was at the head of a growing parade of critters. By tracing the many biosocial threads in Dolly's geneal-ogy across continents, markets, species, sciences, and narratives, Franklin argues that emergent ways of fleshly becoming are at the heart of biocapital, as both commodities and as mode of production.[17] Franklin maintains that breedwealth was the crucial new kind of reproductive wealth in the late eighteenth and nineteenth centuries; and control over the reproduction (or generation by other means) of plants and animals (and, to varying degrees, people) is fundamental to contemporary biocapital's promises and threats. The traffic between industrialized agriculture and scientific medicine for people and animals is especially thick in Dolly mixtures and spillovers. Cur-rent innovations and controversies in stem-cell research and therapeutic as well as reproductive cloning are at the heart of the transnational, trans-specific action.

Stem cells and dogs take us inevitably to Hwang Woo-Suk and the Seoul National University. The international scandal surrounding Hwang's an-nouncement in *Science* magazine, in 2004 and 2005, of achieving the globalized biomedical grail of human-embryonic-stem-cell clones and the subsequent revelation, in December 2005, of fabricated data, bioethics vio-lations in egg donation, and possible embezzlement have a more authen-tic canine backstory that only makes sense in light of *Dolly Mixtures*. In the United States, the well-hyped dog-cloning Missyplicity Project was di-rected to the affectional commodity pet market.[18] Not so the biomedical dog-cloning efforts of Hwang and his nine South Korean associates, plus Gerald Schatten, a stem-cell researcher at the University of Pittsburgh, who announced Snuppy, an Afghan Hound puppy cloned with the Dolly tech-nique, in August 2005.[19] A biotechnical splice to his core, Snuppy is fab-ricated of S(eoul) N(ational) U(niversity) and (pu)ppy. Hwang's research career must be understood in the context of agribusiness animal research moved to human biomedicine. His professorship is in the department of theriogenology and biotechnology in the College of Veterinary Medicine at Seoul National University. Before Snuppy, Hwang reported having cloned a dairy cow in 1999, and he was widely regarded as a world leader in the field.

There is a great deal about Hwang's dramatic rise and fall that is not clear, but what is clear is the thick cross-species travel between agribusiness research and human biomedicine often obscured in the U.S. "ethical" debates over human-stem-cell technologies and imagined therapies or reproductive marvels.

Pricey U.S. dog cryopreservation services, university-private company collaborations for canine cloning research geared to the pet market, and Korean national efforts to become first in a major area of biomedical research are not the only arias in this lively capital opera. However, even if freezing my AKC mutt Roland's cells in anticipation of making a nuclear clone of him could only happen over the dead bodies of my whole polyspecific and polysexual family, these Dolly spillovers, especially Snuppy, do suggest just the right segue to the second part of this chapter.

Valuing Dogs: Technologies, Workers, Knowledges

Referring to advertisements for the sale of working sheepdogs, Donald McCaig, the Virginia sheep farmer and astute writer on the history and current state of herding Border Collies in Britain and the United States, noted that categorically the dogs fall somewhere between livestock and co-workers for the human shepherds.[20] These dogs are not pets or family members, although they are still commodities. Working dogs are tools that are part of the farm's capital stock, and they are laborers who produce surplus value by giving more than they get in a market-driven economic system. I think that is more than an analogy, but it is not an identity. Working dogs produce and they reproduce; and, in relation to lively capital, they are not their own "self-directed" creatures in either process, even though enlisting their active cooperation (self-direction) is essential to their productive and reproductive jobs. But they are not human slaves or wage laborers, and it would be a serious mistake to theorize their labor within those frameworks. They are paws, not hands. Let's see if we can sort through the implications of the difference even in the face of the evolutionary homology.

To do so, I turn to Edmund Russell's arguments about the evolutionary history of technology in his introduction to the collection *Industrializing Organisms* (2004). Far from keeping organic beings and artifactual technologies separate, putting one in nature and the other in society, Russell adopts recent science and technology studies insistence on the co-production of natures and cultures and the interpenetration of bodies and technologies. He defines organisms shaped for functional performance in human worlds as biotechnologies, that is, "biological artifacts shaped by humans to serve human ends" (2004, 1). He goes on to distinguish macrobiotechnologies, such as whole organisms, from microbiotechnologies, such as the cells and molecules that get all the press as biotechnology itself in the current science and business press.

In that sense, dogs deliberately selected and enhanced for their working capacities, for example as herders, are biotechnologies in a system of market farming that became contemporary capital-intensive agribusiness through a welter of nonlinear processes and assemblages. Russell is interested in how the ways in which human beings have shaped evolution have changed both human beings and other species. The tight boxes of nature and society do not allow much serious investigation of this question. Russell's major efforts are directed at analyzing organisms as technologies; and he looks at biotechnologies as factories, as workers, and as products. In spite of the fact that Russell gives almost all the agency to humans—who, I admit readily, make the deliberate plans to change things—I find his framework rich for thinking about valuing dogs as biotechnologies, workers, and agents of technoscientific knowledge production in the regime of lively capital.

Leaving aside such critters of the past as spit-turning dogs or cart-hauling dogs, whole dogs are simultaneously biotechnologies and workers in several kinds of contemporary material-semiotic reality. Herding dogs are still at work on profit-making (or more likely, money-losing) farms and ranches, although job loss has been acute. Their work in sheep trials is robust, but located in the zone between work and sport, as is the labor of most sled dogs. Livestock guardian dogs have expanding job opportunities, in part because of the reintroduction of ecotourism-linked, heritage predators (wolves, bears, and lynxes) in sheep-raising areas of the French Alps and Pyrenees, as well as predator control on U.S. ranches no longer allowed to use poisons. Dogs have state jobs and jobs franchised out to private providers as airport-security laborers, drug and bomb sniffers, and pigeon-clearing officers on runways.

The popular television show *Dogs with Jobs*, using newspaper-style classified ads for jobs as the visual icon for the show, is a good place to get a grip on dogs as workers.[21] Most of the dogs seem to be unpaid voluntary labor, but not all of them are. Jobs include epilepsy warning, cancer detection, hearing-aid work, guiding the blind, serving as psychotherapeutic aides for traumatized children and adults, visiting the aged, serving as aides for wheelchair-bound people, providing rescue services in extreme environments, and more. Dogs can be and are studied and specifically bred to enhance their readiness to learn and perform these kinds of jobs. For all of them, dogs and people have to learn to train together in subject-changing ways. But more of that later.

Part dogs (or delegated dog wholes or parts in other material bases than carbon, nitrogen, and water) might have more work in lively capital than whole dogs. Consider, in addition to Snuppy's stem-cell scene, dog-genome projects. Canine archived genomes are repositories useful for research for product development by veterinary pharmaceutical enterprises and by human biomedical interests, as well as by—a gleam in researchers' eyes— behavioral genetic research. This is "normal" biotechnology. Sequencing and data-basing the complete dog genome was made a priority of the U.S. National Human Genome Research Institute (NHGRI) in June 2003. Based on a Poodle, the first rough dog-genome sequence, about 75 percent complete, was published that year. The first full draft of the dog genome was published and deposited in a free public database for biomedical and vet researchers in July 2004. In May 2005 a 99-percent-complete sequence of the genome of a Boxer named Tasha, with comparisons to ten other kinds of dogs, was released. Dogs belonging to researchers, members of breed clubs, and colonies at vet schools provided DNA samples. The team that got this draft, in the process developing procedures that might speed the deposition of many more mammalian genomes, was headed by Kerstin Lindblad-Toh of the Broad Institute of MIT and Harvard as well as the Agencourt Bioscience Corporation. Part of the NHGRI's Large-Scale Sequencing Research Network, the Broad Institute got a $30 million grant for the work. These are the kinds of public-private arrangements typical of microbiotechnology in the United States and, with variations, internationally.[22]

Further, once the genome was published, the Center for Veterinary Genetics at the University of California School of Veterinary Medicine called for individual dog people and clubs to contribute to a full repository of many of the different breeds of dogs in order to address the needs of different domains of dogdom. The goal was to enlarge the DNA data bank from its then current sampling of the genetic legacy of a hundred breeds to more than four hundred international canine populations. Many research projects involving dog genes, organs, diseases, and molecules could be used to illustrate these kinds of things. The part-dogs are reagents (workers), tools, and products, just as whole dogs are in macrobiotechnological kinds of knowledge and production projects.

There is also another sense in which dogs are valuable workers in technoculture. In laboratories, they labor as research models both for their own and for human conditions, especially for diseases that could be "enclosed" for medical commodity production, including previously unknown sorts of services, to address newly articulated needs. That, of course, is what their archived genomes are doing; but I want to look more closely at another mode of this scientific medical canine labor in the context of lively capital. Stephen Pemberton explores how dogs suffering from hemophilia became model patients, as well as surrogates and technologies for studying a human disease, over the course of years, beginning in the late 1940s, in the laboratory of Kenneth Brinkhous at the University of North Carolina, Chapel Hill. This research is what made human hemophilia a manageable disease by the early 1970s, with the availability of standardized clotting factors.[23]

Bleeder dogs did not just appear at the lab doorstep as ready-made models and machine tools for making things for humans. The canine hemophiliac was made through representational strategies, dog-care practices, breeding and selection, biochemical characterization, development of novel measurement devices, and semiotically and materially joining hemophilia to other metabolic-deficiency disorders (especially diabetes and pernicious anemia, both treatable by administering something functionally absent in the patient, and both being diseases in which dogs played a large role in the research, with crucial payoff in techniques and devices for working with dog organs and tissues). The principal problem Brinkhous faced when he brought into the lab male Irish Setter puppies who showed the stigmata of bleeding into joints and body cavities was keeping them alive. The puppies had to become patients if they were to become technologies and models. The entire labor organization of the laboratory addressed the priority of treating the dogs before anything else. A bleeding dog got transfusions and supportive care. Lab staff could not function as researchers if they did not function as caregivers. Dogs could not work as models if they did not work as patients. Thus, the lab became a clinical microcosm for its research subjects as an essential part of the last century's revolution in experimental biomedicine. As Pemberton put it, "We cannot understand how scientists discipline their experimental organisms without understanding how these organisms also discipline scientists, forcing them to care" (2004, 205).

In the late twentieth century, drugs developed for people (and surely tested on rodents) came to be agents of relief for dogs, too, in a kind of patient-to-patient cross-species transfusion. This kind of dogs-as-patients scene is part of my own adult origin tale in dogland. My middle-class childhood tale had more to do with confining the multispecies civic commons by leash laws in the 1950s than with biomedicine. Toward the end of her sixteenth, and last, year of life, in 1995, my half-Lab mutt, Sojourner—that grace-giving whelp of an irresponsible backyard breeder whom we named for a great human liberator—and I began to frequent her vet's office in Santa Cruz. I had read Michel Foucault, and I knew all about biopower and the proliferative powers of biological discourses. I knew modern power was productive above all else. I knew how important it was to have a body pumped up, petted, and managed by the apparatuses of medicine, psychology, and pedagogy. I knew modern subjects had such bodies, and that the rich got them before the laboring classes. I was prepared for a modest extension of my clinical privileges to any sentient being and some insentient ones. I had read *Birth of the Clinic* and *The History of Sexuality*, and I had written about the technobiopolitics of cyborgs. I felt I could not be surprised by anything. But I was wrong. I had been fooled by Foucault's own species chauvinism into forgetting that dogs, too, might live in the domains of technobiopower. *Birth of the Kennel* might be the book I needed to write, I imagined. *When Species Meet* is the mutated spawn of that moment.

While Sojourner and I waited to be seen by her vet, a lovely pooch pranced around at the checkout desk while his human discussed recommended treatments. The dog had a difficult problem—obsessive self-wounding when his human was off making a living, or engaging in less justifiable non-dog activities, for several hours a day. The afflicted dog had a nasty open sore on his hind leg. The vet recommended that the dog take Prozac. I had read *Listening to Prozac* (2003), so I knew this was the drug that promised, or threatened, to give its recipient a new self in place of the drab, depressive, obsessive one that had proved so lucrative for the nonpharmaceutical branches of the psychological professions. For years, I had insisted that dogs and people were much alike, and that other animals had complex minds and social lives, as well as physiologies and genomes largely shared with humans. Why did hearing that a pooch should take Prozac warp my sense of reality in the way that makes one see what was hidden before? Surely, Saul on the way to Damascus had more to his turnaround than a Prozac prescription for his neighbor's ass!

The canine patient's human was as nonplussed as I was. She chose instead to put a large cone, called an Elizabethan collar, around her dog's head so that he couldn't reach his favorite licking spot to suck out his unhappiness. I was even more shocked by that choice—what, I fumed internally, can't you get more time to exercise and play with your dog and solve this problem without chemicals or restraints? I remained deaf to the human's defensive explanation to the vet that her health policy covered her own Prozac, but that the pills were too expensive for her dog. In truth, I was hooked into the mechanisms of proliferating discourse that Foucault should have prepared me for. Drugs, restraints, exercise, retraining, altered schedules, searching for improper puppy socialization, scrutinizing the genetic background of the dog for evidence of canine familial obsessions, wondering about psychological or physical abuse, finding an unethical breeder who turns out inbred dogs without regard to temperament, getting a good toy that would occupy the dog's attention when the human was gone, accusations about the workaholic and stress-filled human lives that are out of tune with the more natural dog rhythms of ceaseless demands for human attention: all these moves and more filled my neo-enlightened mind.

I was on the road to the fully embodied, modern, value-added dog-human relationship. There could be no end to the search for ways to relieve the psychophysiological suffering of dogs and, more, to help them achieve their full canine potential. Furthermore, I am convinced that is actually the ethical obligation of the human who lives with a companion animal in affluent, so-called First World circumstances. I can no longer make myself feel surprise that a dog might need Prozac and should get it—or its improved, still-on-patent offshoots.

Caring for experimental dogs as patients has taken on intensified meaning and ambiguities in twenty-first-century biopolitics. A leading cause of death for older dogs and people is cancer. Enabled by comparative post-genomics tying humans and dogs together as never before, in 2006 the National Cancer Institute set up a consortium of over a dozen veterinary teaching hospitals to conduct drug tests of possible benefit against human cancers on pet dogs living at home who are subject to the same malignancies. The proposal was for a parallel nonprofit group to collect tissue samples and DNA from pet dogs to pinpoint genes associated with cancer in dogs and people. Companion dogs would be clinic patients and not kenneled lab pooches, who might then be relieved of some of their burden; and grants and companies would pay for the experimental drugs. Dogs might benefit from the drugs, but they would get them with lower standards of safety than would be required in human testing. That's the point, after all, for enlisting dogs in National Cancer Institute state-of-the-art testing in the first place. Pet owners may have to pay for things like biopsies and imaging, which could be very expensive. Researchers would have neither the animal-rights scrutiny nor the financial burden of caring for lab dogs, including paying for those MRIS.[24] Pet owners and guardians would have the power to call a halt to further experimental treatment, based on their sense of the experiences of their dogs. This system of drug testing seems to me superior to the current one, because it places the burden of suffering (and opportunity of participating in scientific research) on those specific individuals, humans and dogs, who might reap the benefit of relief. In addition, experimentation would take place much more in the open than can ever be possible or desirable with lab animals, perhaps encouraging deeper thinking and feeling by a diverse human population of pet owners, as well as clinicians and scientists.[25]

What I find troubling here is a growing ethos that subjects pet dogs to the same search for "cures" that human cancer patients endure, rather than continuing to work within and improve current standards of care in vet practice to reduce cancer burdens and provide supportive care guided by quality-of-life criteria and not by maximum prolongation of life. Chemotherapy that dogs currently get rarely aims to eliminate the cancer, and dogs consequently generally do not experience the terrible sickness from drug toxicity that most people, in the United States at least, seem to feel obligated to accept. How long can that moderate veterinary approach to dog illness, and acceptance of death as profoundly sad and hard but also normal, endure in the face of the power of comparative postgenomic medicine and its associated affectional and commercial biopolitics?

So dogs became patients, workers, technologies, and family members by their action, if not choice, in very large industries and exchange systems in lively capital: pet foods, products, and services; agribusiness; and scientific biomedicine. Dogs' roles have been multifaceted, and they have not been passive raw material to the action of others. Further, dogs have not been unchangeable animals confined to the supposedly ahistorical order of nature. Nor have people emerged unaltered from the interactions. Relations are constitutive; dogs and people are emergent as historical beings, as subjects and objects to each other, precisely through the verbs of their relating. People and dogs emerge as mutually adapted partners in the naturecultures of lively capital. It is time to think harder about encounter value.

Valuing Dogs: Encounters

Why not start with prisons, since we have been touring other large industries in lively capital and this one is immense? There are many places we might go—dogs terrorizing detainees in Iraq, for example, where the encounters shaping enemies, torturers, and attack dogs made use of the social meanings of all the "partners" to produce definite value in lively capital.

International human-rights apparatuses (and where were the animal-rights outcries on this one?); scrutiny of franchised-out interrogation functions; and the moral, psychological, and financial economies of contemporary imperialist wars: who could deny that all these are at the heart of enterprise and investment? Or we could travel to the high-security, high-technology, soul-destroying prison in California's Pelican Bay to track the attack-dog production, dog-fighting culture, and Aryan gang operations run from the prison, resulting in the dog-mauling death of a young woman in her apartment hallway in San Francisco and an outcry for exclusion of dogs from public space in general (but not apartment hallways).[26]

All of these prison dog-human encounters depend on the face-to-face meeting of living, meaning-generating beings across species; that is the encounters' power to terrorize and to reach into the core of all the partners to produce dogs condemned to euthanasia when their usefulness has ended and people fit to carry on the profitable enterprise of the prison-industrial complex, as inmates, lawyers, and guards. However, I want to think about co-shaping dog-human encounters in another prison context, one that makes me pay a different kind of attention to coming face-to-face across species, and so to encounter value. Therefore, let's go to Animal Planet television again, this time to watch *Cell Dogs*.[27] If dogs became technologies and patients in the world of hemophilia, they become therapists, companions, students, and inmates in the world of cell dogs. It's all in the job description.[28]

Every Monday evening in 2004, Animal Planet turned its attention to a different prison work project that had reforming prisoners teaching reforming pooches their manners in order to place them in various occupations outside the prison. The narrative and visual semiotics are fascinating. First, the entering dogs have to be made into inmates in need of pedagogy if they are to have productive lives outside. Fast frame cuts show cell doors clanging behind the dogs, who then get assigned to one prisoner-apprentice teacher, to live in the same cell with this individual human inmate for the duration of their joint subject-transforming relating. Dog trainers teach the prisoners to teach the dogs basic obedience and sometimes higher-order skills for placement as assistance dogs or therapy dogs, or even just as household family members as pets. The screen shows the incarcerated dogs preparing for life outside by becoming willing, active, achieving obedience subjects. The pooches are obviously surrogates and models for the prisoners in the very act of becoming the prisoners' students and cellmates.

The technologies of animal training are crucial to the cell-dog programs.

These technologies include the postbehaviorist discourses and the equipment of so-called positive training methods (not unlike many of the pedagogies in practice in contemporary schools and child-counseling centers); some older technologies from the military-style, Koehler training methods based on frank coercion and punishment; and the apparatuses and bodily and mental habits crucial to making family members and happy roommates in close quarters. There is another sense of technology operating here, too: in their personal bodies themselves, the dogs and people are freedom-making technologies for each other. They are each other's machine tools for making other selves. Face-to-face encounter is how those machines grind souls to new tolerance limits.

The canines must be modern subjects in many senses for the cell-dog program to work. The dogs require and model nonviolent, non-optional, and finally self-rewarding discipline and obedience to legitimate authority. That is the route to freedom and work outside—and to survival. That death awaits the failed dog is a leitmotif in many of the programs, and the lesson for their teachers is not subtle. The traffic between performing and modeling is thick for both the humans and the dogs, teachers and students, docile bodies and open souls to each other. Life and death are the stakes in the prison-industrial complex. Prison-reform discourse has never been more transparent. Arbeit macht frei.

Leaving the prison through mutual self-transformation of dogs and people is the nonstop theme. The humans must stay behind to finish their sentences (some are lifers); nonetheless, when their dogs are successful canine citizen-workers outside, the human inmates leave jail in two senses. First, through their dog students, the convicts give themselves to another human person, to someone free, someone outside; and so they taste freedom and self-respect both by proxy and in the flesh. Second, they demonstrate their own reformed status as obedient, working subjects who can be trusted with freedom in a society divided into the outside and the inside. Part of the proof of worthiness is the human prisoners' act of surrendering, for the benefit of another, the companion and cellmate with whom they have lived for weeks or months in the only physically intimate, touching, face-to-face relationship that they are allowed. The graduation scenes, where the human inmates sacrifice themselves to give their intimate companions to another to achieve a better life for both, are always intensely emotional. I dare you to be cynical, even if all the knives of critical discourse are in your hands. Maybe it's not all "arbeit macht frei" here, but something more like "touch makes possible." Since I can't be outside ideology, I'll take that one, face-to-face and eyes open. The rhetoric that connects categories of the oppressed in these programs is not subtle (prisoners, animals, the disabled, women in jail, black men, strays, etc.); all belong to categories that discursively need much more than remedial training. However, there is potential in these projects for much more promising entanglements that question the terms of these tropes and the conditions of those who must live them.

Perhaps it would be possible to rethink and retool cell dogs to work their magic to build subjects for a world not so fiercely divided into outside and inside. Marx understood the analysis of the commodity form into exchange value and use value to be a practice crucial to freedom projects. Maybe if we take seriously encounter value as the under-analyzed axis of lively capital and its "biotechnologies in circulation"—in the form of commodities, consumers, models, technologies, workers, kin, and knowledges—we can see how something more than the reproduction of the same and its deadly logics-in-the-flesh of exploitation might be going on in what I call "making companions."

In *Making Parents: The Ontological Choreography of Reproductive Technologies*, Charis Thompson compares and contrasts capitalist production to what she calls a "biomedical mode of reproduction," which I think of as core to the regime of lively capital. Thompson is studying the making of parents and children through the subject- and object-making technologies of biomedically assisted reproduction, a very lively area of contemporary investments of bodily, narratival, desiring, moral, epistemological, institutional, and financial kinds. She is acutely alert to the classical processes of production, investment, commodification, and so on in contemporary human-assisted reproduction practices in the United States. But she is adamant that the end of the practices makes a difference; that is, the whole point is to make parents by making living babies. *Capital*, volumes 1–3, did not cover that topic. *Biocapital*, volume 1, must do so.

In two columns, Thompson sets out the following lists, which I borrow, abbreviate, and abuse.[29] In practice, parents-in-the-making selectively seek out, endure, elaborate, and narrativize various objectifications and commodifications of their body parts. Women do this much more than men do because of the fleshly realities of assisted conception and gestation. Many sorts of social stratification and injustice are in play, but they are often not of the kinds that those seeking their fix of outrage find whenever they smell the commodification of humans or part humans. Living babies properly assigned make living parents content with their objectifications. Other actors—technicians, doctors, janitors, friends, many more—in this mode of

TABLE 1 Capitalist production and biomedical reproduction

Production	Reproduction
Alienated from one's labor	Alienated from one's body parts
Capital accumulated	Capital promissory
Efficiency/productivity	Success/reproductivity
Life course finite and descent linear	Loss of finitude/linearity in life course and descent
Essentialism of natural kinds, social construction of social kinds	Strategic naturalization, socialization of all kinds

SOURCE: Adapted from table 8.1 in Thompson 2005, 249.

reproduction may be made invisible to render them non-kin and reproductively impotent. The lure of kin-making is the name of this promissory game of reproduction.

I am interested in these matters when the kin-making beings are not all human and children or parents are not the issue, literally. Companion species are the issue. They are the promise, the process, and the product. These matters are mundane, and this chapter has been replete with examples. Add to those many more proliferations of naturalsocial relationalities in companion-species worlds linking humans and animals in myriad ways in the regime of lively capital. None of this is innocent, bloodless, or unfit for serious critical investigation. But none of it can be approached if the fleshly historical reality of face-to-face, body-to-body subject-making across species is denied or forgotten in the humanist doctrine that holds only humans to be true subjects with real histories. But what does "subject" or "history" mean when the rules are changed like this? We do not get very far with the categories generally used by animal-rights discourses, where animals end up permanent dependents ("lesser humans"), utterly natural ("nonhuman"), or exactly the same ("humans in fur suits").

The categories for subjects are part of the problem. I have stressed kin-making and family membership, but rejected all the names of human kin for these dogs, especially the name of children. I have stressed dogs as workers and commodities, but rejected the analogies of wage labor, slavery, dependent ward, or nonliving property. I have insisted that dogs are made to be models and technologies, patients and reformers, consumers and breed-wealth; but I am in need of ways to specify these matters in nonhumanist terms, where specific differences are at least as crucial as continuities and similarities across kinds. Parts do not add up into wholes, and inners do not sort well from outers. Critters and companion species, not humans and animals, seem to be the ontological denizens of my terra.

Biocapital, volume 1, cannot be written just with dogs and people. I face up to my disappointment in this sad fact by rejoicing in the work of my fellow animal (and other critters) studies and lively capital analysts across lifeworlds and disciplines.[30] Most of all, I am convinced that actual encounters are what make beings; this is the ontological choreography that tells me about value-added dogs in the lifeworlds of biocapital. Dogs are not figures to analyze nor resources to exploit; they are co-making partners in the multispecies, naturalcultural tangles of living value.

Notes

Revised from chapter 2 in *When Species Meet* (University of Minnesota Press, 2008).

1. Marx came closest in his sometimes lyrical early work, "Theses on Feuerbach" and "The Economic and Philosophic Manuscripts of 1844" (Tucker 1978). He is both at his most "humanist" and at the edge of something else in these works, in which mindful bodies in inter- and intra-action are everywhere. I follow Alexis Shotwell's subtle analysis of Marx's near escape from human exceptionalism, implicit in his discussions on how labor power becomes a commodity, sensuousness, aesthetics, and human species being (Shotwell 2006, 111–21).

2. For the argument that "companion species" might be a better way to characterize the material-semiotic, historical, philosophic, and scientific knottings at stake than posthumanism, see Haraway 2008. Companion species are all of us—not just dogs and people, not just humans and animals, not just macrospecies and microcritters.

3. For example, consider the evidence for the dog-human affectional and ritual ties in the archaeologist Darcy Morey's analysis (2006) of earth-wide ancient and recent dog burial sites. For complex semiotic and power tangles between people and dogs crucial to understanding past and present colonial and state relations for the upper Amazonian Runa of Peru and the nonhuman animals (dogs, panthers, insects, etc.) they interact with, see Kohn 2007. For a multispecies, canine-human "becoming with" (and dying with) approach in quite another idiom in contemporary Aboriginal Australia's Northern Territory, see Deborah Bird Rose 2006. Contemporary dog-people affectional and economic pet relations outside Euro-America are also multiple and exuberant, not to mention politically contentious. For example, the zeal with which officials in the People's Republic of China periodically repress pet dogs in cities is only matched by the inventiveness of middle-class Chinese citizens to conceal and keep their cherished dogs. Some estimates puts pet dogs in China today

at 150 million—one for every nine citizens ("Pets Contribute to China's Economy" 2005). That seems to push the bounds of the middle classes! None of that keeps dogs off the menu (except around Olympic sites) any more than pet love keeps large numbers of dogs out of puppy mills in the United States. My point is that past and present dog-human doings are ubiquitous, important, and not amenable to typological, ahistorical generalization.

4. An early interdisciplinary effort to write that missing Marxist volume is Franklin and Lock 2003. Then came the following abbreviated but crucial list, which I take from the graduate seminar I taught in the winter of 2007, "Bio[X]: Wealth, Power, Materiality, and Sociality in the World of Biotechnology": Sunder Rajan 2006; Mander and Tauli-Corpuz 2006; Strathern 2005; Waldby and Mitchell 2006; Mbembe 2001; Franklin 2007; Petryna, Lakoff, and Kleinman 2006. The course grew partly from thinking about a "figure" in the sense introduced in chapter 1, "When Species Meet: Introductions."

Consider a fictional multiple integral equation that is a flawed trope and a serious joke in an effort to picture what an "intersectional" theory might look like in Biopolis. Think of this formalism as the mathematics of science fiction.

$$\int_{\alpha}^{\Omega} Bio\,[X]n = \iiint \ldots \iint Bio(X_1, X_2, X_3, X_4, \ldots, X_n, t)\, dX_1\, dX_2\, dX_3\, dX_4 \ldots dX_n\, dt = Biopolis$$

X_1 = wealth, X_2 = power, X_3 = sociality, X_4 = materiality, $X_n - ??$

α (alpha) = Aristotle's & Agamben's bios

Ω (omega) = Zoë (bare life)

t = time

Biopolis is an n-dimensional volume, a "niche space," a private foundation committed to "global is local" biocracy (http://www.biopolis.org/), and an international research and development center for biomedical sciences located in Singapore (see http://en.wikipedia.org). How would one solve such an equation?

5. The figures, provided by the American Pet Products Manufacturers Association (APPMA), are from a free online teaser for their "2005–2006 APPMA National Pet Owners Survey," available for purchase by non-APPMA members for $595. See http://www.americanpetproducts.org. The APPMA annual Global Pet Expo, the industry's largest trade show, is a real eye opener for any remaining sleeping romantics about pet commodity culture. It is not open to the general public, but only to retailers, distributors, mass-market buyers, and "other qualified professionals." By not shelling out $595 for the pet owners survey, I lost my chance to get the lowdown on such things as details on where U.S. pet dogs are kept during the day and at night, groomer visits and methods of grooming used, methods used to secure dogs in the car, type of food and size of kibble purchased, number of treats given, types of leash or harness used, type of food bowls used, information sources consulted and books or videos owned, dog-care items purchased in the last twelve months, pet-themed gifts purchased, holiday parties for dogs given, expressed feeling about benefits and

drawbacks of dog ownership, and much more—all duplicated for every common species of pet. Not much in the practice of capital accumulation through the lives of companion animals is left to chance.

6. MindBranch, Inc., "U.S. Pet Food Market," Market Research Report for June 2004, http://www.mindbranch.com/listing/product/R567-0062.html, accessed 22 June 2011.

7. A brief, free, PDF-format summary is available online from MindBranch, Inc. To learn more, you have to pay. To get my limited commercial facts for this chapter cost only my phone number inscribed on an online form, followed by an advertising call or two—much more easily resisted than the new liver cookies at Trader Joe's.

8. The links of pet food within the transnational capitalist food industry are so elaborate and involve so many kinds of products that the supply chain gone awry can threaten national economies and international public health across species. See Nestle 2008.

9. Mary Battiata (2004) tells us that a four-year veterinary education in the United States costs about $200,000. Setting up a small vet practice starts at about $500,000. Battiata cites a study of vet fee structures and lagging salaries published in 1998 by the consulting firm KPMG.

10. www.pjbpubs.com/cms.asp?pageid=1490, accessed May 2007.

11. In the mid-2000s the slogan "One Medicine" emerged to signal the convergence of human and vet medicine across the globe in the context of bird flu, terrorism, global agribusiness, comparative biomedical genomics, wildlife-human relations, and much else. The military has not been slow to see the relevance of a one-medicine approach to bioterrorism, biomedicine, and food supply. In 2008 the Sixth Annual "One Medicine" Symposium featured a one-medicine approach to climate change (http://www.onemedicinenc.org, accessed 17 August 2008).

12. Thompson 2005. See also Haraway 2003b.

13. See, for example, La Feria 2006.

14. Berton 2006. The marketing in all of the examples discussed was entirely directed to affluent human beings' ideas and fantasies, and paid scant heed to anything like biobehavioral assessments of how dogs and other boarded species would do best in unfamiliar surroundings. Paying for a "training vacation" might go a long way toward increasing civil peace, say, compared to paying for suites appointed with color-coordinated humanesque furniture and Animal Planet TV shows.

15. Lavery 2004.

16. For their place in complex nationalisms and ethnic identity discourses, consider the Karelian Bear Dog, the Suomen-pystyykorva (Finnish Spitz Dog), the Norsk Elghund Grå (Norwegian Elkhound), the Kelef K'naani (Israeli Canaan Dog), the Australian Dingo (an Eora Aboriginal word), the Islandsk Farehond (Iceland Sheepdog), the Korean Jindo Dog, and the Japanese Shiba inu, Hokkaido inu, Shikoku inu, Kai inu, and Kishu inu—and I have hardly started. Comparing the fascinating histories, discourses, and cultural politics in which Canaan Dogs and Dingoes figure would require another book. Both kinds of dogs scavenge and hunt in the "pariah" or "primitive" dog categories, made over for globalized breed club standardization. Re-

constituted or reinvented dogs of the hunting elites of European feudalism also have a fascinating contemporary story. Check out the Irish Wolfhound in this regard, complete with the breed's Celtic origin story, from the first century B.C., along with the details of the dogs' nineteenth-century "recovery," enabled by a Scot: Captain George Augustus Graham's breeding of dogs called Irish Wolfhounds who still remained in Ireland with Borzoi, Scottish Deerhounds, and Great Danes. The popularly recited details of the Great Rescuer's craft seem never to pollute the pure origin story of ancient nobility, or to disturb the keepers of the closed stud books in the breed clubs. "Value-added" seems the right term for these breeding operations!

Probably the world's most important collection of Southwest Indian art, including weaving, pottery, Kachina figures, and much else, is housed at the School of American Research in Santa Fe, New Mexico, in exquisite adobe buildings commissioned by two transplanted, wealthy, eccentric, New York women, Elizabeth and Martha Root White. The sisters also raised many of the most famous Irish Wolfhounds of that breed's early period in the United States, between the 1920s and the Second World War, on this rugged and beautiful property. The land and buildings now serve as a major anthropological research and conference center. Rathmullan Kennel's Irish Wolfhounds are buried in a little graveyard on the grounds, marking the value-added encounter of wealth, gender, aestheticized and reinvented tradition in dogs and human beings, white people's collection of indigenous artifacts on a grand scale, philanthropy, activism in support of Pueblo Indian land rights and health, patronage of the arts of Europe, the United States, and Indian nations, as well as scholarship of a kind that reaches across generations, nurturing some of the best twentieth- and twenty-first-century anthropology in all subfields. When I visited the dogs' graves at the School of American Research in 2000, after writing the first versions of "Cloning Mutts, Saving Tigers" for Sarah Franklin's and Margaret Lock's workshop on "New Ways of Living and Dying," the bones of the Whites' Irish Wolfhounds seemed like fleshy, fantasy-laden, Euro-American ancestors in this complex colonial and national tangle. See Stark and Rayne 1998. For photographs of people, grounds, and dogs — including a re-creation, organized by the White sisters for a Santa Fe festival, of a sixteenth-century hunting party with Irish Wolfhounds — and for a detailed description of the myriad practices that sustained these upper-class show dogs, see Jones 1934, online at www.irishwolfhounds.org. The interlaced constructions of Native American arts and crafts and Irish Wolfhounds in Santa Fe is perceptively analyzed in Mullin 2001.

17. Franklin 2007. See also Clarke 2007.

18. Haraway 2003a; Haraway 2008, chapter 5, "Cloning Mutts, Saving Tigers," 133–57, 355–60. Genetic Savings and Clone, the private corporate labs where the never-successful Missyplicity Project came to rest after the researchers at Texas A&M lost heart, went out of business in October 2006, leaving its frozen companion-animal tissue bank to the livestock-cloning firm, ViaGen. Genetic Savings and Clone did announce the live birth of two cloned cats in 2004, and mounted its Nine Lives Extravaganza, the world's first commercial cloning service for cats, with an advertised price, in February 2006, of $23,000 plus sales tax. CopyCat, one of the kittens born

in 2004, cost $50,000. No sequel called "Cheaper by the Dozen" followed. The president of the Humane Society of the United States could only have been called ecstatic at hearing of Genetic Savings and Clone's departure; he was quoted by Reuters news service, on 13 October 2006, for calling the business failure a welcome "spectacular flop," given the need for spending resources on addressing pet overpopulation. That was my reaction, too; I had just read the local newspaper's list of shelter dogs and cats needing homes in my small town that month. Nonetheless, predictably, in 2008, a commercial spin-off of Hwang's (successful) dog-cloning work, Seoul-based RNL Bio, sold five cloned puppies to an American woman named Bernann McKinney for $50,000 (Kim 2008). McKinney's Pit bull Bsooger had died of cancer in 2006. That the first commercial cloned pups were Pit bulls is at best ironic, given the widespread hysteria (ideology) about dangerous breeds.

19. Hwang et al. 2005. Somatic-cell nuclear transfer—the Dolly method—was the technology employed. In view of the faked data on human-embryonic-stem-cell (hESC) clones, Snuppy's authenticity was questioned; but in January 2006 an independent investigator pronounced him to be a definite clone of Tel, the DNA donor, and a major advance for stem-cell research. See the Wikipedia page for Snuppy for an introduction to this story. Over 1,000 dog embryos were transferred into 123 bitches to get three pregnancies and one living dog. The special difficulties involved in cloning dogs compared to other animals are detailed in Kolata 2005. On the hESC controversy, Hwang still has supporters in South Korea, and many scientists elsewhere acknowledge the extraordinary international competitive pressures at play in the whole field.

20. From McCaig's posting on CANGEN-L, a Canine Genetics Internet discussion list, around 2000. To understand the work of Border Collies and the way they are regarded by their people, see McCaig 1992 [1984]; McCaig 1998a; McCaig 1998b.

21. Track the show through http://www.dogswithjobs.com, accessed May 2007. For the history of dogs as behavioral genetics research subjects, see Scott and Fuller 1965; Paul 1998; Haraway 2003c. The early hopes for the first U.S. Canine Genome Project, which was led by Jasper Rine and Elaine Ostrander, included connecting dog genes and behaviors, using crosses of purebred dogs identified for different behavioral specializations, like Newfoundlands and Border Collies. Some of the talented fruits of those odd crosses play agility at the same trials that Cayenne and I frequent. The ideas about behavioral genetics in some of the early pronouncements of the Canine Genome Project were the butt of jokes among dog people and also other biologists for their simplistic formulations of what different kinds of dogs do and how "genes" might "code for" "behaviors," formulations that are rare in postgenomic discourse. Check out Polly Matzinger "Finding the Genes that Determine Canine Behavior," http://www.bordercollie.org/health/k9genome.html, accessed 22 June 2011, for an explanation to dog people of what the Canine Genetic Project was about. Research into behavioral genetics is not necessarily simplistic or unimportant for people as well as other species. However, old-fashioned ideology dressed up as research plays a big role in the history—and probably future—of this field. Ostrander

mainly concentrated on comparative cancer genomics in dogs and humans at the Fred Hutchinson Cancer Research Center in Seattle. In 2004, the National Human Genome Research Institute (NHGRI) named her as the new chief of its Cancer Genetics Branch, one of the seven research branches in the Division of Intramural Research. Related to psychopharmacogenetics, comparative behavioral genetics remains a long-term research commitment in the NHGRI.

22. Lindblad-Toh et al. 2005. Elaine Ostrander was one of many prominent (and not so prominent) coauthors on this paper. Several international labs also had canine genetic mapping projects of various kinds dating from the 1990s. Comparative mammalian genomics, including ever better drafts of whole-genome canine sequence data, is a rich interdisciplinary field of current research. With the guidance of Mark Diekhans, a software engineer and fellow member of the University of California, Santa Cruz, Science and Justice group, I spent a wonderful afternoon browsing dog sequence data. See http://genome.ucsc.edu, accessed 17 August 2008.

23. Pemberton 2004.

24. See Pollack 2006.

25. Comparative oncology is the name of the very active research and clinical work that tracks similar cancers across biological species, including dogs and people. To follow progress in the alliance of pet dogs and their people in undertanding and treating cancers that affect both, see Melissa Paoloni and Chand Khana, "Translation of new cancer treatments from pet dogs to humans," http://www.nature.com/nrc/journal/v8/n2/abs/nrc2273.html, accessed 22 June 2011; for current open trials studying the feasibility of personalized medicine approaches to similar pet dog and human cancers, see https://ccrod.cancer.gov/confluence/display/CCRCOPWeb/Clinical+Trials, accessed 22 June 2011. Recent dog-focused publications from the Comparative Oncology Program of the National Cancer Institute include: "A Compendium of Canine Normal Tissue Gene Expression," PLoS One, May 2011 (http://www.plosone.org/article/info%3Adoi%2F10.1371%2Fjournal.pone.0017107); "Guiding the Optimal Translation of New Cancer Treatments from Canine to Human Cancer Patients," Clinical Cancer Research 15:5671–77, 15 September 2009. Osteosarcoma has been an early focus in this research; see M. C. Paoloni, C. Mazcko, E. Fox, T. Fan, S. Lana, et al., "Rapamycin Pharmacokinetic and Pharmacodynamic Relationships in Osteosarcoma: A Comparative Oncology Study in Dogs," PLoS ONE 5, no. 6 (2010): e11013, doi: 10.1371/journal.pone.0011013. My dog Cayenne's best friend Willem, a Great Pyrenees dog, died of metastatic osteosarcoma at seven years of age; maybe that is why I notice all of this with love and rage, the only affects that seem up to the task of staying with the trouble of both cancer and its research apparatuses.

26. This awful story can be tracked in Southern Poverty Law Center 2001. In the year in which two large mastiff-type dogs mauled Diane Whipple to death in a San Francisco apartment building, the incidence and severity of dog bites in San Francisco in all public places was significantly lower as a result of effective public-education programs. That did not stop the public demand to remove dogs from public areas or greatly restrict their freedom in the wake of the mauling. About twenty

dog-bite-related human deaths occur in the United States per year in a dog population of over seventy million. That does not justify any of the deaths, but it does give a sense of the size of the problem. See Bradley 2006.

27. For details on the series, which aired in 2004, see the Internet Movie Database, http://www.imdb.com.

28. See also Neal 2005. Go to the Pathways to Hope website, http://www.path waystohope.org, for more on the Prison Dog Project. Canine Support Teams is the project at the California Institute for Women. The Pocahontas Correctional Unit in Chesterfield, Virginia, is a women's facility that trains inmates in dog grooming. Gender assumptions seem well groomed here. The Second Chance Prison Canine Program, in Tucson, is "a group of advocates for people with disabilities, prison inmates, and animal welfare in Arizona [who] coordinate a prison pet partnership program to address issues common to these three groups" (http://www .secondchanceprisoncanine.org). Go to the Jayne Cravens and Coyote Communications website, http://www.coyotecommunications.com, for a partial list of active prison dog-training programs, which include institutions with projects for training stray dogs and cats as well as dogs for people with disabilities. Websites accessed 14 May 2007. Also see Harbolt and Ward 2001. Canada and Australia also have programs.

29. Thompson 2005, 8.1.

30. For example, besides the texts already cited in note 4, see Hayden 2003; Helmreich 2003; TallBear 2005; Hirsch and Strathern 2005.

3

TIMOTHY CHOY

AIR'S SUBSTANTIATIONS

The writer Xi Xi, from Hong Kong, opens her experimental short story "Marvels of a Floating City," a mixed-media piece that intersperses brief narratives with reproductions of paintings by René Magritte, with a fantastic image of a metropolis—a thinly veiled Hong Kong—emerging from the sky.

> Many, many years ago, on a fine, clear day, the floating city appeared in the air in full public gaze, hanging like a hydrogen balloon. Above it were the fluctuating layers of clouds, below it the turbulent sea. The floating city hung there, neither sinking nor rising. When a breeze came by, it moved ever so slightly, and then it became absolutely still again.
>
> How did it happen? The only witnesses were the grandparents of our grandparents. It was an incredible and terrifying experience, and they recalled the event with dread; layers of clouds collided overhead, and the sky was filled with lightning and the roar of thunder. On the sea, myriad pirate ships hoisted their skull and crossbones; the sound of cannon fire went on unremittingly. Suddenly, the floating city dropped down from the clouds above and hung in mid air. (Xi 1997, 106)

I love this image. It transforms a city that can at times feel dense and overwhelming into a thing of quiet and delicacy. Xi Xi shows Hong Kong as a place moved by the slightest touch of a breeze, as a place that can become absolutely still. It reminds me of the Hong Kong I sometimes encountered on late-night walks past Hong Kong's government buildings, while taking the slow ferry between Hong Kong and Lantau Island, and at times while sitting on MTR subway trains when, following the example of many others around me, I would listen to music through headphones and take a nap.

Xi Xi's conceit also turns Hong Kong into something like a natural object, something nearly elemental. Hong Kong's mercantile and military origins become almost atmospheric—a storm depicted by layers of clouds and a sky filled with flashes and roars. The pirates themselves—the British Lord Palmerston and the others—are absent in this picture (their presence is marked only by the crossed flag which is raised into the sky), but the meteorological impact they had in birthing the floating city is made clear.

Xi Xi's pairing of city and sky is fanciful and metaphoric—the images of dangling and floating recall the questions about an uncertain future that preoccupied Hong Kongers in the late 1990s—but for me, Xi Xi's image of the floating city is particularly compelling because it also invokes something profoundly literal. Air is central to the understanding and experiencing of Hong Kong.

To explain what I mean by this I need to tell you another story of city and sky, this one just slightly less fantastic. In April 1999 Tung Chee-hwa visited the headquarters of the Walt Disney Corporation in Los Angeles, California. The visit was perhaps intended as a triumphant exercise of social capital, meant to perform and to buttress a relationship forged through a controversial agreement Tung had signed earlier that year, between the Walt Disney Corporation and the Hong Kong government.

The agreement amounted to a joint business venture. Disney would build a theme park in the Special Administrative Region, a park which not only would serve as a draw for international tourists, but also (Tung hoped) provide service-sector jobs to the increasing—and increasingly vocal—ranks of unemployed people in Hong Kong. In return, the Hong Kong government would be the primary investor. The agreement was criticized roundly, for its environmental oversights as well as for the economically vulnerable position it forced on Hong Kong. At least in the Disney Corporation, though, Tung had a supportive ally. They were in agreement: a world-class park for a world-class city was exactly what Hong Kong needed.

Unfortunately, Tung's visit to Los Angeles was marred by another voicing of doubt and criticism, this time from within the Disney Corporation itself. During the visit, Michael Eisner, Disney's chief executive officer, took the opportunity to express concern about the poor air quality in Hong Kong, adding that the smog did not mesh particularly well with the family image that Disney so prided itself on cultivating. Eisner never said explicitly that Disney's continued participation in the theme park hinged on an improvement of Hong Kong's air. But people with whom I later spoke—shopkeepers, environmental activists, and taxi drivers alike—nevertheless interpreted the event as something of a threat, as though Eisner had taken Tung aside and whispered in his ear that Disney would pull out if Hong Kong's air quality did not improve.

One could have remarked on the irony inherent in this moment, when a corporation based in—and associated so strongly with—the smoggy city of Los Angeles faulted another city for its poor air, but Tung made no attempt to do so. Instead, he returned to Hong Kong and sheepishly reported the exchange to his advisors and to the Hong Kong public through the news media.

The newspapers had a field day. Hong Kong had just coughed its way through the worst winter of air pollution in its recorded history. Many Hong Kong residents had checked themselves into hospitals citing respiratory problems. The winter of poor air had also forced my partner, Zamira, and me to relocate from our apartment in Sai Ying Pun, an aging urban district in Western Hong Kong where we had been living since arriving in the city, to a flat in a village house in Mui Wo, a rural town on the coast of Lantau Island. Zamira had suffered three sinus infections in six months while we lived in the city—it was time to move.

I remember feeling a guilty sense of relief when I read the news. The extremity of the air pollution—the worst in history, remember!—made my partner's illness, and our move from city to village, count now as a moment of participation in a genuinely Hong Kong experience. Until then, I had sought to cultivate indifference toward air and air pollution. Although we, like our friends, routinely avoided waiting or walking on busy streets because the air stung our eyes and throats, and though we often charted a course for rural areas on weekends to get away from the city pollution, I consistently refused to comment on or even to notice the air. My justification was simple, if not simple-minded: the people I met in my first months in Hong Kong who were most vocally critical of the air quality were almost without exception expatriate businesspeople from the United States. I did not want to be associated with them. The air pressed upon me, for instance, at a cocktail party celebrating the publication of a book by the renowned Hong Kong landscape photographer Edward Stokes. I was chatting with a representative from the American Chamber of Commerce and his wife when the air pressed upon me. Hong Kong has to see, she told me, that the environment is an economic problem. Hong Kong wanted to build this Cyberport, for instance, but who would want to come to Hong Kong to work if the air was bad? If you could not even see? This was the first time, but certainly not the last, that I heard Hong Kong's air coupled with the future of its economy.

At the same time, many of my Cantonese-speaking, Hong Kong–born friends often vocalized their suspicions that politicians who built campaign platforms on the topic of air pollution were motivated by selfish and middle-class interests. Such politicians were only trying to preserve real-estate values for the properties of elites, they said. So, in what I then considered an ethnographer's effort to become immersed in an ethics more grounded in Hong Kong's particularity, I tried hard to act as if the air stinging my throat were a commonplace, not worthy of notice.

But the air persisted. Zamira's illness, the record-breaking winter pollution, and the Disney debacle together forced me to take notice of the air that had been swirling everywhere around, above, and through me and everybody else the entire time I had been in Hong Kong. I remembered then that during my first field visit to Hong Kong, in 1996, when I had asked various environmental officials about the pressing environmental issues, one of the first issues mentioned had always been air quality. Not only that, but air had mediated ruminations about Hong Kong's impending political handover to Chinese sovereignty. *The real concern is transborder pollution, the official at the* EPD *tells me during an interview months before the 1997 handover. How will we deal with the air and water pollution that comes down from the Mainland? The air is framed as a threat from the north in these pre-post-colonial months.* What remained to be seen, they said, was how the Chinese government would respond to Hong Kong's attempts to reduce air and water pollution in Mainland China. We will soon see, they seemed to be telling me, what the implications of the political handover will be. One activist told me explicitly that they were trying to lie low, and that rather than making any political demands, they would concentrate on building relationships with Mainland bureaucrats before the transfer of power.

This account of my gradual awakening to the significance of air mimes a standard trope in ethnography, that of the epiphany-in-and-of-the-field. But it is also something else, or it can be if you shift your attention away from my eventual ethnographic realization and look more closely at my initial attempts to disavow my difficulties with the air. That disavowal was plainly an endeavor to distance myself from expatriates; it was a localizing and nativizing enterprise, one whose motivations were analytically untenable but nonetheless impossible for me to resist. If I avow that at stake in my initial disavowals was a naïve dream of being a Chinese American anthropologist more able to stomach an everyday, everyman's Hong Kong life than my imagined doppelgängers, the well-paid expatriates—including those of Chinese descent—it is only to point out that whatever lines of distinction I imagined—and whatever means I saw available to identify with some people and to distance myself from others—themselves point to the key issue. Air mattered powerfully in Hong Kong. It mattered in deeply felt, variegated, and variegating ways.

All That Is Air

Air matters too little in social theory. Marx famously described the constant change that he saw characterizing a "bourgeois epoch" as a state in which "all that is solid melts into air," and that provocative phrasing served in turn as a motif for Marshall Berman's diagnosis of "modernity" as a shared condition in which all grand narratives were subject to skeptical scrutiny. Yet, aside from signifying a loss of grounding, air is as taken for granted in theory as it is in most of our daily breaths. This is unfortunate, because thinking harder about air, that is, not taking it simply as solidity's opposite, might offer some means of thinking about relations and movements—between places, people, things, scales—means that obviate the usual traps of particularity and universality. These traps themselves, it turns out, are generated through an unremarked attachment to solidity.

To understand this attachment, it is helpful to revisit the context and afterlife of Marx's commonly cited line "All that is solid melts into air." The passage where it appears is about a sweeping change.

> The bourgeoisie cannot exist without constantly revolutionising the instruments of production, and thereby the relations of production, and with them the whole relations of society. Conservation of the old modes of production in unaltered form, was, on the contrary, the first condition of existence for all earlier industrial classes. Constant revolutionising of production, uninterrupted disturbance of all social conditions, everlasting uncertainty and agitation distinguish the bourgeois epoch from all earlier ones. All fixed, fast-frozen relations, with their train of ancient and venerable prejudices and opinions, are swept away, all new-formed ones become antiquated before they can ossify. All that is solid melts into air,

all that is holy is profaned, and man is at last compelled to face with sober senses his real conditions of life, and his relations with his kind. (Marx 1978a [1867], 475–76)

Marx argues here that with capital as such comes a constant revolutionizing of society. This is one of the livelinesses of capital. When surplus value is a motivating abstraction, what once were means to generate differential value—the instruments of production—can become a fetter to that project once those instruments are fixed and ubiquitous. A technology might at one time lower the costs of production or enable new forms of goods and markets, but if that technology becomes ubiquitous in a given market through others securing similar means, the advantage it offered disappears. One might try to revive dead capital through new markets, as many of the other chapters in this volume illustrate, but if it cannot be resuscitated, something livelier must take its place.

Marx's rendering of this process of endless dynamism hinges on a remarkable figuration of solidity. On the one hand, solidity stands for fixity and reliability. The phrase "all that is solid" renders firm industrial society and the long-standing nature of relations among people and between people and land. On the other hand, this very fixity is itself *historical*. Solidity, in other words, is not fixed at all. Marx materializes this paradox of fixity-nonfixity in his language, through his images of relations being "fast-frozen" or "ossified," for these images beg the question of what existed before the freezing and ossification. His images of solidification as a process imply a prehistory, one of presolidity.

There are typically two responses to an image of the world wherein solidities dissolve. A philosopher might strive for some contingent conceptual fixities to make sense of this swirling about. In fact, Marx does precisely this at this moment in his analysis, and it requires an unavoidable universality. We hear in the passage a mantric, assonant, repetition of "all." "All social conditions," "all fixed, fast-frozen relations," "all new-formed [relations]"— together they aggregate, yielding an image of a whole that in turn gives way to the epochal atmospheric world of capital. Similarly, social theorists since Marx have sought to develop general terms like *flexible capital*, *postmodern condition*, and *neoliberalism* to grasp and contain a world of dynamism and change.

Meanwhile, another response, one common among cultural anthropologists today, is to refuse the universalizing gesture, and perhaps even to refuse the very project of the concept. This might take the form of repudiating either the claim that "everything" is melting or the idea that there can be "whole relations" in the first place. Such abstractions kill, this response goes, doing violence to particular human lives and practices that lie outside the terms of the analysis, and such lives are accessible only through empirical work.

The first response is the one usually charged with being up in the air, with not being concerned with concrete details, particular conditions, specific lives on the ground; but in fact, both responses are of a piece. Both responses, whether universalizing or particularizing, seek solid analytic ground, and both find their ground through resort to a "one." This is so whether the one is the unifying one of the "all" or the irreducible particular one refusing subsumption into the general. The conceptual one and the empirical one are a conjoined pair, and both suffer vertigo without firm footing.

Air is left to drift, meanwhile, neither theorized nor examined, taken simply as solidity's lack. And there seems at first to be no reason not to let it. When solidity is unconsciously conflated with substance, when only grounding counts for analysis, air can only be insubstantial. And we are stuck with the twinned ones—universal and particular—grounded, fixed, and afraid.

Environmentalists in Hong Kong, however, would press us on this attachment to the ground, as would Marx himself. The environmentalists would ask, "Is not this stuff floating above and around us itself deeply substantial?" As for Marx, we should remember that his claim is ultimately about a dialectics of solidity. Solidities all have a presolid past, and air lies in solidity's future. As he declares in a speech during the anniversary of the *People's Paper*, "The atmosphere in which we live weighs upon everyone with a 20,000 pound force. But do you feel it?" (Marx 1978b [1867], 577–78).[1] It would be a mistake, in other words, to search only for ground when above and around us is substance aplenty. Our living with this substance, furthermore, is neither universal nor particular. Air is not a one; it does not offer fixity or community; but it is no less substantial. The question is whether we can feel it.

Hong Kong might help us feel it. From a certain point of view, there is no "air" in itself. Still, in the following pages I follow the lead of people in Hong Kong in attributing (imposing?) a loose category, air, to encompass a number of things, among them dust, oxygen, dioxin, smell, particulate matter, visibility, humidity, heat, and various gases. There is no air in itself; air functions instead as a heuristic with which to encompass many atmospheric experiences. The abstraction of air does not derive from asserting a unit for comparison or a common field within which to arrange specificities, but through an aggregation of materialities irreducible to one another (including breath, humidity, SARS, particulates, and so on). Thinking about the materiality of air and the densities of our many human entanglements in airy matters also means attending to the solidifying and melting edges between people, regions, and events.

This might help us to imagine a collective condition that is neither particular, nor universal—one governed neither by the "all," nor through the "one nation, one government, one code of laws, one national class-interest, one frontier, and one customs-tariff" that Marx envisioned, nor even the "one planet" of mainstream environmental discourse. Instead, it orients us to the many means, practices, experiences, weather events, and economic relations that co-implicate us at different points as "breathers." I like this term, *breathers*, which I borrow from environmental economics; it refers to those who accrue the unaccounted-for costs that attend the production and consumption of goods and services, such as the injuries, medical expenses, and changes in climate and ecosystems. I like the term because its very vacuousness constantly begs two crucial questions that are both conceptual and empirical: what are the means of counting costs? And who is not a breather?

THE STORY OF AIR's substantiation in Hong Kong hinges on acts of condensation, and this chapter engages in parallel acts to condense that story. Consider how the air pollution monitoring stations dotting Hong Kong yield a measurement for RSP, or respirable suspended particulate. Enclosed machines on rooftops and streets ingest millions of mouthfuls of wind a day, calming it so that what it holds can fall out and so that enough of those particles can be collected for counting—to accumulate enough of the particular for it to register as weight, as substance worth talking about. Miming this method, this chapter collects the details in a diffuse set of contexts: the production of air pollution as a local and global medical concern, the material poetics of *honghei* (ambient air) in daily discourse and practice, the acts of large- and small-scale comparison signaled by air, and the transformations that condense Hong Kong's air into measurable particles and then further into a particular, yet internationally recognized, metric for risk.

In short, four forms of air concern me: (1) air as medical fact, (2) air as bodily engagement, (3) air as a constellation of difference, and (4) air as an index for international comparison. Ultimately, my aim is to gain a deep understanding of all of them and to move seamlessly between their methods and registers. Rather than focus on just one, I make a start in each of them because conveying the dispersal of air's effects and its substantiations is one of my chief aims. This has produced a text that can seem diffuse; its argument requires some work to condense. But that is exactly what people concerned with air must do: turn the diffuse into something substantive.

Air and Dying

Climatologically, there are two Hong Kongs. Beginning in May and June, the air in Hong Kong swells as winds blow in from the tropical south, bringing heat and humidity. The air temperatures will range from the mid-80s to the high 90s Fahrenheit, while the humidity hovers around 95 percent. The air sticks to you as you walk, forms a sheen on your skin as you move from an air-conditioned bus, taxi, or building to the outside. In the late summer, there are the typhoons, great oceanic whirlwinds that occasionally batter the small island with wind and rain as they spin through the Pacific. In colloquial Cantonese, typhoons are called *da fung*, the beating wind.

Then, around late September, the winds begin to shift. Cooler and drier air gradually blows in from the north, across Mainland China and Asia. The temperatures can drop into the mid-40s Fahrenheit—as they did in the winter of 2000, when the streets filled with puffy North Face jackets—while the humidity drops to 70 percent. In these drier months, Hong Kong can feel temperate. In the summer, the air in Hong Kong is heavy with heat and water, but in winter months its weight comes from a different kind of load, as the cool, dry winds sweep the smoke and soot from the skies above China's industrial factory zones into Hong Kong.

It is these sooty, winter, months that most likely motivated Michael Eisner to pull Tung Chee-hwa aside during Tung's visit to Los Angeles. If Eisner's criticism of Hong Kong's air was indirect and vague, the critiques voiced a few years later by Hong Kong doctors were specific and direct. In 2001 and 2002, faculty from the departments of community medicine at the University of Hong Kong (HKU) and the Chinese University of Hong Kong (CUHK) published separate articles in internationally known scientific journals linking Hong Kong's air pollution and declining health. The first of the two, published by Chit-Ming Wong, Stefan Ma, A. J. Hedley, and T. H. Lam in the journal *Environmental Health Perspectives*, was entitled "Effect of Air Pollution on Daily Mortality in Hong Kong" (2001). The second, published in *Occupational and Environmental Medicine* by T. W. Wong, W. S. Tam, T. S. Yu, and A. H. S. Wong of CUHK's department of community and family medicine, was entitled "Associations between Daily Mortalities from Respiratory and Cardiovascular Diseases and Air Pollution in Hong Kong, China" (2002). The articles' findings were chilling. Both studies concluded that there were significant short-term health effects of acute air pollution. More people died of cardiovascular or respiratory illness on days with bad air quality than they did on days with good air quality. The HKU study also compared warm and cool weather data and found that the chance of pollution-correlated mortality was statistically higher in the cool season.

Both articles take pains to locate themselves in a citational network. I mention this not to argue that citational networks are invoked to confer authority on the articles; that point about scientific articles has been well-argued by others already, and though certainly applicable in this case, it is not what interests me most.[2] Instead, I am interested in the warp and woof of the network being woven, for it lends a specific character to the objects and political substances emergent in it. One way to see how is through the titles of some of the citations which form the network:

"Particulate Air Pollution and Daily Mortality in Detroit" (Schwartz 1991)
"Air Pollution and Mortality in Barcelona" (Sunyer, Castellsague, and Saez 1996)
"Particulate Air Pollution and Daily Mortality in Steubenville, Ohio" (Schwartz and Dockery 1992)
"Air Pollution and Daily Mortality in London: 1987–92" (Anderson et al. 1996)
"Air Pollution and Daily Mortality in Philadelphia" (Moolgavkar et al. 1995)
"PM_{10} Exposure, Gaseous Pollutants, and Daily Mortality in Inchon, South Korea" (Hong, Leem, and Ha 1999)
"Daily Mortality and 'Winter Type' Air Pollution in Athens, Greece: A Time Series Analysis within the APHEA Project" (Touloumi, Samoli, and Katsouyanni 1996)
"Air Pollution and Daily Mortality in Residential Areas of Beijing" (Xu, Dockery, and Gao 1994)

There is a remarkable, almost numbing uniformity to the titles. They share a syntactic structure, differing from one another through a paradigmatic substitution of terms within that structure. In each, a compound subject is first offered through a conjunction of air pollution with mortality, later to be positioned through a locating "in." Though there are minor variations in the first half of the titles—"air pollution" might be modified as "particulate air pollution" or "winter type air pollution"—the most significant transformations are take place in the second half, the prepositional phrase naming a particularity of place.

In this structure, we discern something about the workings of exemplarity as political method. Through the mustering of a network of almost identical examples, and by naming Hong Kong's air through a similarly almost identical name, the doctors make Hong Kong an example in a much larger problem. At the same time as that example draws power from the network, though, it also lends stability to that network. The co-examples as a whole, as a network, substantiate a conjunction of objects—air pollution and death—differentiated only by place.

One thing to notice here is the play of particularity in the formation of political substance. Rather than jeopardizing its stability, the proliferation and accumulation of particulars is key to the citational network's existence. The production of Hong Kong air is both a localizing and a globalizing project. It is localizing because it carves out the uniqueness of Hong Kong. It lends it specificity—the hallmark of that last phrase is place-based specificity. At the same time, it is globalizing because it performs membership in an international community of atmospheric and medical science, and in an international, global problem.

Equally important, the common form of the titles signals common method. Both articles were "retrospective ecological studies" employing "time-series analysis," a method which amounts to statistically correlating the "number of people dying on a particular day" (or a day or two later) with meteorological data and air-pollutant concentrations over a long-term period.[3] The statistical method used was a Poisson regression model "constructed in accordance with the air pollution and health: the European approach (APHEA) protocol" (Wong, Tam, Yu, and Wong 2002, 2). The near-identity of the titles in this particular citational network, in other words, is premised on a near-identity of technique. Hong Kong's air cannot be simply asserted to be one example among many of deadly airs in the world; co-exemplarity is actualized through the standardization of technique.[4]

This generation simultaneously of general problem and specificity is resonant with dynamics analyzed in other chapters in this volume. It bears comparing, for instance, with the collecting, formatting, and iterating of data in environmental informatics, as Kim Fortun explores in this volume (chapter 10), in that it generates a general problem precisely by arraying and juxtaposing particularities, though as Fortun observes, environmental informatics enable this process to be iterated across sites and types of information not possible in the space of one study. In fact, the Hong Kong daily-mortality studies discussed here would be but two among a vast library of data sets for information engineers like those Fortun describes, raising the question of the extent to which such studies might be produced in anticipation of themselves being informatted and networked. The carving out of specificity through geographic location also underscores Sheila Jasanoff's observation, in this volume (chapter 4), that specificity plays a vital role today in legitimating claims of intellectual innovation and ownership. What becomes clear looking across these topoi is that while specificity is at play in all these moments, one cannot take for granted what specificity means. There is no specificity in general, and the real work of specificity must be gleaned from the pragmatics of the specific knowledge practices in which specificity as a concept is figured.

To understand this, let us examine the specific conditions in which the citations appear in the Hong Kong articles. Consider this excerpt from the HKU study's conclusion.

> In setting air pollution control policy from a public-health viewpoint, it is important to identify the health effects of air pollutants from local data. Because of the lack of data, there are few studies based on daily hospital admissions and mortality in the Asian Pacific region. For hospital admissions there has been only one study in Australia (36) and two in Hong Kong (30, 37). For mortality studies, there have been one in Beijing, China (38) based on 1-year daily data, two in Australia (36, 39), and two in Korea (40, 41). Our report should contribute to the understanding of the effects of air pollutants in this region and may clarify the differences in effects and mechanisms between Western and Eastern populations.
>
> Local data on health effects of air pollution are required for setting standards and objectives for air pollution controls. When local data are not available, foreign data may be helpful, but they may not be relevant or applicable because of a difference in climate or other conditions. Our findings in this study provide information to support a review of air quality objectives with consideration of their effects on health. (Wong, Ma, Hedley, and Lam 2001, 339)

Here, the network (plotted by the integers corresponding with the citations at the end of the article) is invoked through a naming of its holes, "the lack of data." The naming of the general problem is indistinguishable from the claim for the primacy of the specific.

The explicit value of the Hong Kong study is marked as clarifying differences in effects "between Western and Eastern populations." By identifying Hong Kong's warm, humid summer and cool, dry winter, the Hong Kong University study reminds us that we are in the subtropics; and the specific ways in which it cites its network of relevant citations give that reminder a certain freight. It identifies and locates the work of Hong Kong doctors within the terms of a center and periphery of scientific practice. As scientists in the periphery, the researchers must negotiate a double bind not unlike the one Lawrence Cohen describes facing gerontological organizations and authors in India in the 1970s, who, in appending "India" to their names and publication titles, "claim[ed] local autonomy from internationalist [gerontological] discourse, but [did] so through a reassertion of epistemological subordination" (Cohen 1998, 90).

The Hong Kong doctors navigate this bind through an appeal to local appropriateness: "When local data are not available, foreign data may be helpful, but they may not be relevant or applicable because of a difference in climate or other conditions." Note, they do not say that the category does not apply "here" or that air pollution is a Western problem; they simply maintain that better, more local data is needed. This is a supplementary strategy, one that has the potential to disturb, even while leaning on, the centrality of temperate studies: "This study provides additional information for our previous study on hospital admissions (21), and the many time series studies on air pollution and mortality in temperate countries (1–11, 13, 15, 17–19, 28, 29, 33, 35, 38, 39)" (Wong, Tam, Yu, and Wong 2002).

The Hong Kong studies "contribute to" and provide "additional information for" the networked assemblage of other conjunctions of air and mortality "in temperate countries," and in doing so, they help it to grow. Yet at the same time, their very act of "adding to" articulates through implication an inadequacy in the apparently whole original to which they contribute.[5] The Hong Kong doctors' exemplification of Hong Kong names geographic unevenness in—even while extending the reach of—an emerging coalescence of scientific and political substance.

This emergent substance is fragile stuff. Daily-mortality studies face criticisms that they establish no causal link or proof of impact in the long term. Some epidemiologists, for instance, argue that even if one can show that the number of people dying on a day with high air pollution is significantly greater than on a comparable day with lower pollution, the early deaths might be of people who had little time left to live anyway.[6] Those most vulnerable on high-pollution days are those in fragile health or in advanced stages of terminal illness, the argument goes. This is termed a "harvesting effect." Those who died were going to die soon; they were simply harvested early. Long-term cohort studies are needed to determine precisely how many, if any, person-years have been lost. Only with such data, this argument concludes, can the extent to which air pollution decreases life be understood.

Such a refusal to recognize air's daily effects by scaling time out seems absurd at first, but we should recognize it as a logical side-effect of rendering illness and health into prognosis. As Sarah Lochlann Jain illuminates in her analysis of "living in prognosis," a prognosis, which assigns people a certain percent chance of being alive in the next number of years based on how long others considered to be in comparable medical and demographic categories have lived, puts one in the mind-wrenching position of living counterfactually, always juxtaposing one's living against aggregated odds of dying.[7] The analytic of harvesting simply takes this head-wrench to the extreme—by deeming your death today unworthy of note simply because most others in your position, whether good air day or bad, did not survive much longer than you.

Substantiating Hong Kong air as a dangerous substance will require crunching not only numbers. It will require grappling with how to think about a cause of death when causes are multiple and overlapping, and how, when lives and causes are complex, to say when it matters that a person dies today—and not tomorrow or next year. These efforts are crucial if air pollution's effects on health are to be grasped.

At the same time, they run a risk of narrowing our sense of what matters in human-atmospheric relations. When we ask how many more people die on particularly polluted days than would have died if the air were clear, death becomes a proxy for air's effects, and death itself is rendered a problem of lost time—which in turn prompts the demand for more accuracy in counting the time in person-years lost. (How many person-years will be spent counting person-years?) But it bears remembering that air's human traces are found not only in those who die, their times of death, or total person-years lost, but in the fabric of living.

Air and Living

One day I collapsed when I got home, my stomach somersaulting like it had at the Tung Chung Citiplaza, where I had needed to stop at the public washroom instead of catching my connecting bus. A fever hit me that night, leaving me weak and useless. The next morning, I called Wong Wai King, my collaborator and informant in Tai O village, with chagrin to cancel our appointment.

"You're sick, eh? Yeah, the honghei these days has been really bad."

I found this strange. The air hadn't seemed that bad. But I spoke to others, who nodded knowingly and recalled that the air had been particularly wet on that hot, muggy day.

THE LINK FORGED by the doctors between air and health was not novel or isolated. Nor, as Wong Wai King and others helped me to see, was air's impact on health in Hong Kong limited to its particulate load. Already circulating was an existing discourse of *honghei* (Cantonese: ambient air) and health. Reviewing my notes back in San Francisco, I noticed this entry from August 2000: "Ah Chiu has been sick. She got a cold or something. It's a common thing to get colds out here in the summer. Nobody thinks it's strange, because they all know that when going in and out of air conditioning, you can get really cold and then sick." My notes and memories are dotted with such commentaries. Sometimes, I was told, it was too hot. Other times, it was cold, dry, or wet. Intrigued, I consulted Traditional Chinese Medicine texts and doctors for mention of honghei, but I was unable to find any. There was plenty, however, about *feng* (Mandarin for "wind"; *fung* in Cantonese).

In Traditional Chinese Medicine texts, honghei (Mandarin: *kōngqì*) denotes one of two sources of acquired *hei*. The other source is food. Hei, widely recognized in its Mandarin pronunciation, *qi*, is the fundamental life force in TCM, often translated as breath. Honghei is thus a breath in two senses; it is a source of vital breath, and it is breathed. In everyday use, honghei refers to the air in one's surroundings.

Though breath is vital, wind is dangerous. "Wind is the first evil," my acupuncturist back in California, Marliese, explained to me. "It opens the body to secondary ills."

The historian of science Shigehisa Kuriyama offers a beautiful account of the central role played by wind in the history of Chinese medical conceptions of the body. He highlights the tension that existed between, on the one hand, feelings of an ultimate resonance between the body's breath and the surrounding winds, and, on the other, anxieties about human subjection to chaos, where humans were opened to irregular and volatile winds by their skin and pores. Through close study of medical and philosophical texts, Kuriyama shows clearly that "meditations on human life were once inseparable from meditations on wind" in both Chinese and Greek medicine (1999, 236).

What most strikes me in Kuriyama's account is his attention to language, both in the ancient texts he studies, and his own. Winds and air whistle through his writing as much as they do through the texts he analyzes. Listen, for instance, to his discussion of the connection that the philosopher Zhuangzi drew between earthly winds and human breath.

> The winds of moral suasion, the airs that rectify the heart, and now the heavenly music of gaiety and sadness. All these bespeak a fluid, ethereal existence in a fluid, ethereal world. A living being is but a temporary concentration of breath (*qi*), death merely the scattering of this breath. There is an I, Zhuangzi assures us, a self. But this self is neither a shining Orphic soul imprisoned in the darkness of matter, nor an immaterial mind set against a material body. Anchored in neither reason nor will, it is self without essence, the site of moods and impulses whose origins are beyond reckoning, a self in which thoughts and feelings arise spontaneously, of themselves, like the winds whistling through the earth's hollows. (Kuriyama 1999, 245)

By allowing the airs to permeate his own account's figurations and similes, Kuriyama conveys to his readers Zhaungzi's theorization of human permeability and impermanence more vividly and viscerally than a less writerly account could. Later, Kuriyama will show how much more dangerously the winds are figured in subsequent texts, and it is this later sense of wind's danger that my acupuncturist in California inherits through her study of TCM.

Air's meanings in Hong Kong seem to exceed this classical medical genealogy. Among people I have known in rural and urban Hong Kong, good fung characterizes good places. Wind's ubiquity, however, and the way it wends its way into everyday talk recall the inseparability of wind and life that Kuriyama describes and the lyrical trace of an imminently atmospheric sense of the self and health. Meditations on life through wind are as present as ever.[8]

IN TAI O, the air is on the tips of people's tongues. "Hello, good day. Nice feng today, isn't it?" The old men sit on the benches by the Lung Tin Housing Estate, Dragon Field, facing the road that connects Tai O to the rest of Lantau Island, watching the hourly bus come in with visitors. Their shirts are loose. The breeze curls through Lung Tin, finds Wong Wai King sitting on the concrete steps outside her bottom-floor apartment. She sips some sweet water, closes her eyes, and plays her guzheng. "Wah, hou shufuhk," she tells me. "Ah, it's so very shufuhk."

The word *shufuhk* means "comfortable," but also much more. When people say they're not shufuhk, they mean they're not well. Conversely, when Wong Wai King and others tell me that they're shufuhk, they mean to tell me that they are experiencing a crisp, pure pleasure that saturates their being. Like a cool breeze on a hot sticky day. Clean sheets on a bed. Or the way a cup of tea might warm you from the inside when you're cold. The word is ubiquitous.

Places are made into living things through a blend of landmark and language, as anthropologists of place have taught us to see, and the air in Hong Kong is undeniably part of the rhetoric of its place.[9] But air, polluted and otherwise, is a daily materiality as well as symbolic field. To explore a material poetics of place, and air's function within it, we need to ask after the material and meaningful ways in which air enters into human and geographic life as such. For the notion of a poetics of place to have any teeth, for it to do more than simply legitimate linguistic study as a study of something linked to the material world, we must also go after the nonverbal ways that air operates poetically. How does air serve as a meaningful and material unit in the building of Hong Kong? Let us take an atmospheropoetic tour.

Some of the neighborhoods I choose for this tour are among Hong Kong's most famous. Central is the financial center of Hong Kong and the heart of its government. Central's illuminated towers, set against a foreground of the green waters of Victoria Harbor, adorn most of the stereotypical tourist images of Hong Kong. Less celebrated internationally but well-known both in Hong Kong and in Hong Kong tourist literature is Mong Kok, a district on the Kowloon Peninsula. To many, Mong Kok is the antithesis of Central. Mong Kok is commonly held to be more Chinese than Central. While the English language appears on shop signs and restaurant menus in Central and sometimes comes out of shopkeepers' mouths, it is rare in Mong Kok. Whereas Central offers at least some Western comforts, Mong Kok caters to Hong Kong Chinese and to tourists seeking a flavor of Chinese alterity within Hong Kong.[10]

Tai O, described just a few pages ago, should be considered a part of this tour, along with Lung Kwu Tan and Ha Pak Nai. Tai O is a popular destination for domestic and foreign tourists, though not long ago it was considered a dirty backwater. Lung Kwu Tan and Ha Pak Nai, meanwhile, are relatively less well-known villages in Hong Kong's New Territories, hemmed in by a power station and a landfill and facing the impending construction of a municipal-waste incinerator.[11] With their inclusion, another axis of difference becomes clear. Central and Mong Kok might in isolation evoke an imagined opposition between Western and Chinese in Hong Kong, but when Tai O, Lung Kwu Tan, and Ha Pak Nai become stops on our tour, Central and Mong Kok find themselves partners in urbanity set against the rural New Territories.

CENTRAL. In the winter the air in Central sweeps in dark swirls through Connaught Road, blowing under squealing double-decker trolley cars before whirling up Pedder Street toward Lan Kwai Fong, Central's famed restaurant and bar area. It chases the heels of trundling buses and racing taxis, and flings gusts of soot at the ankles of the pedestrians waiting at the crosswalk, who, almost in unison, lower their heads and cover their mouths and noses with a hand or handkerchief—a loosely synchronized nod and an almost instinctive gulp of held breath—as the wake of air washes over them.

Lung Kwu Tan. In Lung Kwu Tan and Ha Pak Nai, two villages in Hong Kong's northwestern New Territories, the air smells cleaner at first—that is, it doesn't smell of diesel. There are fewer buses out here. Fewer taxis. But it does smell of garbage, of the garbage water that leaks from refuse trucks. People talk about the dust that settles on their vegetables from the cement factory's smokestack. And then there are the flies that fill the air, making it so that you might want to keep your mouth more tightly closed while you're breathing. Now people are worried about what else might come from the air, as the government plans to build their incinerator here. Dioxins, says Rupert Yu, a Greenpeace campaigner working to halt the incinerator's construction, the most poisonous substance humans have ever created.

Still, the air is on the water, and this yields cool breezes. On weekends, it fills the sails of windsurfers, and it carries the scent of visitors' barbecue, even if the occasional atmospheric shift wafts reminders of the cement factory, power station, and landfill nearby.

Mong Kok. In Mong Kok, a neighorhood on the Kowloon Peninsula that has been called the most densely populated area in the world, the winter winds are as sooty as those in Central. Dust expelled from the backs of abundant buses, trucks, and taxis barely settles before it is stirred up once again. Pedestrians cross the street with the same nodding gestures. Off the street, though, the winter wind might find itself broken by a crowd, trapped and thawed by the press of people gathered to shop and play.

The same is true a bit farther north, in Yau Ma Tei, where there is also the night opera. There are two women performing, one middle-aged with glasses, wearing a skirt, leaning deliberately toward her microphone under bright incandescent lights. The musicians sit to the left, one smoking a cigarette while he plays his erhu. The music, the voice, they quaver. They sound like old radio. The air is full, too, with shells, with black beans, the slightly sticky smell of cow parts being stewed, durian, skewers of pork, oyster omelets. The smell of diesel fades into memory, and the cold air, defeated, rises to the overlooking skyscrapers in warm ripples.

WE HAVE TAKEN a slight detour from the issues of health that first brought us to consider the air. But we have retained the issues of the body, the question of immediacy—the coughs, instinctive intakes of breath. Part of air's substantiability in Hong Kong comes from the fact that it is always breathed.

The poetic mattering of Hong Kong's atmosphere encompasses not only Wong Wai King's rhapsodic "Wah hou shufuhk," but also her sip of sweet water, the placement of her chair, and the coughs and nods of the pedestrians aiming to cross the street in Central. Air's poesis, the co-productive engagements between people and air, range from commentary, to breath, to avoidance, to the flip of an air-conditioner switch. Put another way, air is not only an object of cultural commentary, and not only a nonhuman materiality always already enmeshed in webs of social and cultural practice. It is something embodied that engages with humans through bodily practices. The smell, breath, wind, weather, typhoon, air conditioning, air pollution, height, verticality, science, sound, oxygen, smoking. The tactility of the atmosphere.

The anthropologist and musician Steven Feld (1996) has argued that sound and voice provide a useful entry for the anthropological study of relations between person and place. He identifies the sonic resonance of the human chest cavity as a central feature of the links and feedback loops between people and their environments. How similarly fruitful might an anthropology of air be, an anthropology of this stuff sensed in and through the moment of bringing breath into the interiority of the body, or at the moment when wind opens the body to ailments? Air muddies the distinction between subjects and environments, and between subjects. This thickness and porosity rendered by air is part of what makes the air and the airborne such deeply felt elements. Bodies may be, as the geographer David Harvey (1996) argues, intersections of large- and small-scale spatial practices; but if bodies are an intimate location of effects and agencies, air is the substance that bathes and ties the scales of body, region, and globe together, and that subsequently enables personal and political claims to be scaled up—to global environmental politics—and down—to the politics of health.

Air's Comparisons

In August 2000 a feature entitled "A Breath of Fresh Poison" was published in the *South China Morning Post*. In the article, readers are introduced to a sympathetic character, Fred Chan Man-hin, who had recently returned to Hong Kong from Canada to start a company. Although he initially "planned on being here forever," he tells the Post, "the pollution has affected my decision. I can't work and be sick all the time." Today, Chan "avoids his office in the Central business district because the pollution gives him dizzy spells and migraine headaches. He has spent tens of thousands of dollars on doctors and tests to find a cure for the allergies, viruses, and exhaustion that he cannot seem to shake" (Ehrlich 2000, 13).

The article throws into relief a signature feature of air's substantiation as a problem in Hong Kong. It does not merely recount Chan's unshakeable health woes; it makes a pointed comparison. Chan had initially left Hong Kong for Canada, we are told, and had later returned to make his fortune, but now the pollution might affect his decision to "be here forever." If air constitutes a danger in Hong Kong, part of its threat derives from its capacity to serve as an index for comparing Hong Kong with Canada and other places.

This capacity of air as a point of comparison first became evident to me through my family, particularly through jokes about how predictably those who do not live in Hong Kong get sick when they visit. My mother's cousin, Ling, playfully chided her when she fell ill, for instance, when my parents visited Hong Kong near the end of my fieldwork. "You, your cousin Maggie, and your brother To—you all get sick whenever you come back to Hong Kong." My mother falls ill almost every time she visits Hong Kong, as do I. Ling knows this well, as we usually go to her or her husband for antibiotics. "You're not *jaahppgwaan*, not accustomed, to the air," Ling says. "Will you still visit?"

Will we still visit? This simple question draws us back to the landscape photographer's cocktail party, to my conversation with the American Chamber of Commerce representative and his wife, who wondered aloud how investors could be expected to come to Hong Kong if the air quality continued to deteriorate. It echoes Disney's admonishment to the chief executive. Air is not only an index of health. It is an index for comparing livability, well-being, global attractiveness.[12]

I cannot leave the matter of air's comparability at this level of global comparison, for it misses some of the subtle comparisons and distinctions that operate within the city-state. We are now acquainted with the air of some of Hong Kong's neighborhoods, its qualities, and its dangers; now questions of justice and equity beg to be asked. How are Hong Kong's air spaces distributed? Who gets to occupy those with the cleanest air? Who breathes the street? Who breathes mountains? Who breathes the sea? Who breathes flies?

A FEW WEEKS after moving to Mui Wo, I returned to Sai Ying Pun to visit with the fruit vendor, Mrs. Chau. Ah, you've come back, Mrs. Chau said, loudly enough for passers-by to hear. I smiled, a bit embarrassed, and replied that the oranges looked good. I asked her to pick some for me, and for a glass of juice, and we chatted for a while there on Mui Fong Street.

I missed Sai Ying Pun, I told her. Mui Wo was nice, but it wasn't as convenient. There were also all the mosquitoes, I continued. Expecting some sympathy, I offered my arms to show her my mosquito bites, but Mrs. Chau dismissed them with a wave and a laugh.

Sure, there are mosquitoes, she said. But I'm sure the honghei is much better there.

Of course. Of course honghei mattered to Mrs. Chau, who worked every day on the busy corner of Mui Fong Street and Des Voeux Road, just down the street from one busy bus stop, where diesel buses pulled in nearly every minute, and across the street from another. Hillary, the stationer down the street, at least had a door between the street and his shop, and his shop was air-conditioned.

FAR FROM BEING UNIFORM, Hong Kong consists of pockets. Studies in the loosely Marxist or critical geographic tradition take as a premise that there are social inequities, mapped and realized through spatial distinction.

Through their lenses, we discern a geographically uneven distribution of environmental harm, where the rich have access to good air, while the poor are relegated to the dregs, to the smog and dust under flyovers or on the streets.[13] One can, in other words, discern a political-economic geography of air. The poorest air quality was initially in the urban areas, in the industrial zones. Now, the bad air is being exported, as Hong Kong companies relocate their factories in the Guangdong Province, where labor costs are lower and environmental standards are more lax. But then the pollution comes back in those notorious winter winds.

These arguments help to ground the air in a solid sociological critique of social and geographic stratification; for this reason they are politically vital. At the same time, such fixings need less rigid company. When mapping the spatial distribution of social inequity, an account of air must at some point leave land-based maps, for they can divert us from the movements of air and breathers alike—not to mention mobile pollution sources, such as the taxis, buses, airplanes, and cargo ships crucial to the circulations of Hong Kong's industries. To the geography of air and the dialectics of air and capital, I simply add three corollaries: (1) air is made not only in emissions, but also in the respiration and movements of breathers; (2) neither those who emit particulate, the winds that carry it, nor those who breathe it sit still in places; and (3) as Kuriyama reminds us, there has always been more to air than particles.

THE STRATIFICATION OF AIR SPACES in Hong Kong has been loosely tied to income, and incomes and occupations have also been racially marked. White-collar expatriates, with their generous compensation packages, have to a greater extent than most people in Hong Kong been able to choose to live somewhere clean and central. Air spaces have been constituted in part by the racialized and classed bodies that live, work, and play in them.

The Peak and the Mid-Levels have long served Hong Kong's elite as airy refuges. Almost from the moment British colonists occupied the small island off China's southern coast, they turned toward the peaks that formed the dramatic backdrop for the small harbor they so desired, looking upward for some respite from Hong Kong's summer heat and humidity. If for mountaineers the staggering heights of snowcapped peaks presented a dream of sublimity and transformation, the Peak in Hong Kong offered to colonists a more mundane, more everyday, yet perhaps equally treasured transcendence of place, time, and air.[14] Even relatively recently, civil servants have had privileged access to apartment buildings high up.

In colonial times, people cared mostly about the heat and humidity. The winter winds, whose passage through the landmass of greater Asia lent them coolness and dryness, were greeted with great pleasure. Today, that dryness and that passage through China have made winter less popular than it used to be. Real estate up high continues to be prized, but now it is valued not only for respite from summer heat and humidity, but also because it promises at least some relief from roadside pollution and congestion, as well as convenient access to work and play.

The Mid-Levels, known in Cantonese as *zhong saan kui*, or the "mid-mountain area," are found a bit downhill from the Peak, and they, too, serve as something of a refuge from the soot below. The apartment towers are spaced farther apart than in the neighborhoods at lower altitudes, and there are fewer cars. Commercial skyscrapers are less prevalent up here, and the common mode of commuting here is the escalator, the longest covered outdoor escalator in the world—the same one that stars in Wong Kar Wai's film, *Chungking Express*. The escalator descends into Central from the top of the Mid-Levels in the morning, carrying not only local and expatriate professionals on their way to the office, but also domestic helpers heading down to the markets to buy the day's groceries. Later, at 10:00 a.m., the escalator will reverse itself so that the domestic helpers won't have to climb the many flights of stairs back to their employers' homes. Scores of restaurants and bars have sprung up around the escalator. The escalator and the easy commute it offers into Central have made the terraced streets of the Mid-Levels a pocket of real estate that is even more highly valued today than it was in colonial times.

Much of Hong Kong seems designed to get off the ground—into the air, and out of it. In colonial times, the English built their mansions in the Mid-Levels and the Peak. Today, when I walk with Hemen, a representative of the Tsing Tao Beer Company, he wends his way expertly through Wanchai, a government and nightlife district on Hong Kong Island, without ever touching the ground. We spend the day on the walkways that link this hotel to that shopping center. Some walkways are covered, others enclosed. Up here, we avoid the cars and the exhaust. My grandmother and I got lost once in these walkways. I remember how she pointed down to the street. There, she said, that's where I want to go. How do we get there? We never made it—we were lost in the flyovers.

AIR IS LIKE FOOD, essential to human life. Any anthropology worth its salt, however, asks after the meanings of the essential and its manifestation in material and semiotic constellations of power. Writing of food and eating, Judith Farquhar observes that "[a] political economy of eating emphasizes the uneven distribution of nutritional resources, while a political phenomenology of eating attends to the social practices that make an experience of eating" (2002, 46). For an adequate account, both ends of the analytic pole are necessary, as is everything in between. Air similarly calls simultaneously for an understanding of its distribution and an emic analysis of its presence and distinction in acts of living. Like foods and tastes, air is enrolled into projects of social, racial, ethnic, cultural distinction. When diasporic Chinese find the air in Hong Kong or China unbearable, their coughs, comments, and airplane tickets distinguish person and region. Consider also how atmospheric qualities figured in colonial poetics of difference. The Chinese "do not suffer from the oppressive heat of the lower levels during the summer months as Europeans do," theorized the signatories to a petition in 1904 to create a "Hill District" for Europeans.[15] Air marked the moments when colonists grasped for something concrete to say to concretize their deep unease—a sense that all around them, permeating everything, was difference.

Air's Index

We have seen that people in Hong Kong have a number of techniques for reading the air—dirtiness, wetness, heat, breeze, height. And we have seen how threats and health are substantiated through air's breezing and breathing. In this section, I want to look at one of the state's measures. Air's substantiations, as we have seen them thus far, present a mess for a planner or politician. To facilitate communication and policy, they need something easier to evaluate—a measure that can be translated back into coughs and particles, if need be, but that is simpler and more encapsulating. Little wonder that air, an index of so much, should have an index of its own.

The Air Pollution Index (API) in Hong Kong is calculated in a manner similar to that of other countries, such as the United States, Australia, and Mexico. Air pollution monitoring stations throughout Hong Kong collect data on several target pollutants: sulfur dioxide (SO_2), carbon monoxide (CO), nitrogen dioxide (NO_2), and respirable suspended particulate (RSP). The raw data for each pollutant, usually measured in micrograms per cubic meter within a given period of time (1 hour, 8 hours, 24 hours), is turned into a sub-index calibrated so that an index of 100 will correspond with a density of pollutant that is dangerous to health. That reading of 100 corresponds to different densities for different pollutants. For instance, for SO_2, an index of 100 is calibrated to 800 micrograms per cubic meter of air (800 $\mu g/m^3$) in a 1-hour period, while for NO^2, the 100 is calibrated to 300 $\mu g/m^3$. For the general Hong Kong API, the highest of the 5 sub-indices (measured in different locations) for a given hour or day is taken as the API for that hour or day.

The clarity of the number 100 — so metric! — in the index is what grabbed my attention; it brought to mind the history of the kilogram.[16] In 1799, in an effort to standardize measurements in France, the French National Assembly decreed that a "kilogram" would be defined as the mass of a decimeter of water at 4 degrees Celsius. Brass and platinum weights were made with equivalent mass, and the platinum one, the "kilogramme des archives," would eventually become the standard mass for twenty other countries in Europe through a treaty known as the Convention du Mètre. A more durable copy of the kilogramme des archives, made of platinum and iridium, was later fashioned as the international standard and called "K." Twenty copies of K were then apportioned to each of the signatories of the Convention du Mètre. Was this 100 of the API a universal measure — like the kilogram — calibrated across national and cultural difference through an ultimate standard?

It seems so at first. Common methods and machines internationally unite those who seek to measure air's load. These methods and machines serve as paths of translation — along them, air can be turned into vials of dust, which can in turn be transformed into indices. These are "circulating references," organizations and transformations of matter that allow material to assume more mobile forms.[17] The reversibility of these translations ensures the indices' stability and rigor, assuring their users and proponents of a pathway back to the dust. It takes an apparatus of techniques and methods — not simply the calibration of danger to the integer 100, but also the replicability and reversibility of the translations between air and number — to qualify Hong Kong's API as an index among others. There is a standardization, then, to the techniques for measurement, as well as to the form of the API.

When I reviewed the Air Pollution indices of several other countries, however, I was surprised to find that an API of 100 is calibrated to different amounts of dust in different places. For instance, for carbon monoxide, the one-hour objective in Hong Kong is 30,000 $\mu g/m^3$, while in California the equivalent objective is 23,000.[18] If the air in California had 24,000 $\mu g/m^3$

of carbon monoxide in it during a one-hour period, the API would read over 100 and be considered unhealthy, while in Hong Kong the API might hover only around 80 and be considered acceptable.[19] Between the final API form and the standard methods for measurements lies a space for governing what will register as risk or danger.

Most striking is the difference in objectives for RSPS (PM_{10}). The twenty-four-hour target in Hong Kong is 180 $\mu g/m^3$, while the federal standard in the United States is 150. The California standard is lower still, at 50 $\mu g/m^3$, which is the same as levels deemed acceptable by the World Health Organization.[20] That is, the Hong Kong threshold at which the air is considered to contain an unhealthy level of RSPS is almost four times as great as the threshold in California, exemplifying the fact that standards for danger are different in different places.

Calibrating the API is a technique for managing the public perception of risk—for a public that includes vendors like Mrs. Chau, sick entrepreneurs like Fred Chan, corporations like Disney, and residents weighing arguments that a more democratic government could care better for its people.[21] The API can be read alongside the adjustment of risk thresholds that Joe Dumit analyzes, in this volume (chapter 1), in the context of pharmaceutical marketing, where marketers aim to lower the published thresholds so that more people will feel and be deemed unwell and, therefore, fit for medication. It also has resonances with the novel iterations of data in environmental informatics explicated by Kim Fortun, also in this volume (chapter 10). Together these examples illuminate a common situation, where the ongoing tuning, tweaking, and reiterating of numbers, graphs, and maps becomes central to affective and aesthetic work—the making visible and experienceable (or invisible and unexperienceable) of risks that are difficult to articulate.[22] A symptomless biomarker becomes felt as disease; an intuited tie between social difference and health verges on presence. Through the API's calibration, the smell of diesel drifts in, then out; a breath feels alternately thick and thin, clean and dirty, invigorating and debilitating. It is not simply that the API is deployed for persuasive ends, but that the very technical practice of its generation—as much as commentaries on the breeze, held breaths, and treatises on the effect of southerly versus northerly winds—brings air into sense and sensibility. This is an aesthetic technology with serious stakes.

Air's Poetics

First of all the enveloping hot air, ungiving, with not a flicker of movement, a still thermal from which there is no relief. You are surrounded by hot air, buoyed up by hot air, weighed down by hot air. You inhale hot air, you swallow hot air, you feel hot air behind the ears, between the legs, between the toes, under the feet.

Many hours later, a very slight stir, followed by the suggestion of a breeze. The thermal remains.

Yet more hours later, a sudden tearing gust of wind, and the storm has arrived.

LOUISE HO, "STORM"

What kind of substance is Hong Kong's air at the end of the day? One shared, particular, and comparable, one realized in bodily, sensory, practical engagements of breath and movement, as well as through the material and mathematical transformations of medical method. One fixed in the whorls between buildings, mobile as it blows across town, across borders, across disciplines—one that signals a global political economy, postcolonial anxiety, as well as concerns about health and well-being.

Air's qualities are coupled with Hong Kong's industries. Think of the smokestacks of industrial factories making goods and the cargo ships moving freight; the carbon footprints of the jets and taxis moving finance workers; the mark on the air from the coal- and gas-burning power plants that send electricity to Hong Kong's skyline and to the electronics shops, bursting with gleaming toys to be bought and powered with leisure money or credit. Think of the combustion at the end of consumption's lifecycle, where discarded things are incinerated. Air pollution is both condition and effect of capital. We burn in making, we burn in consuming, we burn in discarding, and the smoke has nowhere to go but up. Once up, this smoke constitutes its own threat to Hong Kong's place in financial circuits.

Meanwhile, Hong Kong doctors work to locate their concerns about the atmospheric load in Hong Kong within broader concerns about health, as well as within international science. Pedestrians and environmentalists worry about the winter shift in the wind that brings China's air into Hong Kong. Air's capacity to hold many forms of substance helped solidify a village-NGO collaboration mobilized to halt construction of an incinerator in Hong Kong's New Territories.

Air disrespects borders, yet at the same time is constituted through difference. Neighborhoods have different atmospheres; nations generate and apply different pollution standards; leaders worry about the state of their air compared to that of other nations. The winds themselves derive from differences in air pressure between regions, and similar relativities allow our lungs to inhale and exhale. Gradients, whose foundations are the contact and bleeding of difference, move air through the spaces we live in and through our bodies.

HOW DO WE THEORIZE this shifting substance bound up in processes of production and consumption that also holds and touches much more? What manners of thinking about scales, distinctions, and connections does it open to us? My answers to these questions remain preliminary, but let me outline for now an argument for air's potential to reorient discussions of political universalism.

Recent efforts in post-Marxist political philosophy to retheorize universalism can be brought fruitfully to bear in air's analysis, but they also meet a limit.[23] As exemplars of such efforts, consider the interventions made in *Contingency, Hegemony, Universality*, a series of provocations and exchanges between Judith Butler, Ernesto Laclau, and Slavoj Žižek. The authors in this exchange agree that there are no self-obvious political or ethical universals unstained by particularity, and that the concepts of the universal and the particular are best understood in relation with each other and with their deployment in historically specific political acts. On the question of how precisely to understand the relation of the universal and the particular, however, the authors differ strongly.

For Laclau, the universal is an "impossible and necessary object" in the constitution of any political articulation, in both theoretical and political terms. "From a theoretical point of view," he argues, "the very notion of particularity presupposes that of totality. . . . And, politically speaking, the right of particular groups of agents—ethnic, national or sexual minorities, for instance—can be formulated only as *universal* rights" (Laclau 2000, 58). The particular is thus for Laclau never outside of, or prior to, a field of relative and necessary universality within which particulars come to be known as such. The universal, in its very impossibility and necessity, grounds politics (and analytics) of particularity.

Butler, meanwhile, argues almost the reverse point. "If the 'particular' is actually studied in its particularity," she writes, "it may be that a certain competing version of universality is intrinsic to the particular movement itself"

(Judith Butler 2000a, 166). That is, a close study of particular political move-ments might reveal that they actually are the universals that they seemed to rely on. Universality, for Butler, rather than simply preceding the particular, is in fact generated and iterated through particular visions of the universal.

Žižek, following Hegel and Marx, invokes the concepts of oppositional determination and the concrete universal to solve the paradox of the uni-versal's and the particular's simultaneity. Of all species within a genus, he argues, there is always one that is both member of the genus and determiner of the terms defining that genus. Furthermore, the historically specific con-dition of global capital structures the situation of political particularisms; and class politics, he maintains, while one among multiple forms of politics, serves as the model for politics in general.

Any of these positions could ground air's analysis to good effect. We might lean on Butler's concept of "competing universalities" to argue that the daily mortalities substantiated by Hong Kong's doctors not only buttress a universalizing claim of air pollution's link with dying, but also instantiate a particular, competing, version of this universality that questions the pe-ripheralization of Hong Kong scientists and Hong Kong health in interna-tional science. We could borrow a page from Žižek to argue that in air's en-tanglement with capital we encounter the air relation determining all other air relations. Or, twisting somewhat Laclau's characterization of the relation between universalism and contingently articulated political blocs, we could see air emerging as an empty, yet always necessary, universal—to be filled in with honghei, RSP, typhoons, buses, breezes, science, flies—making en-vironmental politics, rather than class politics, a primary field for political claims.

Before long, however, air would push back. Each approach offers a theory of politics through a solution to the universal-particular paradox; but to do so each leans on an initial opposition between the universal and the particu-lar to render their coexistence paradoxical in the first place, in need of a solu-tion. As I hope to have conveyed, however, air's encompassment of universal and particular does not present itself as a paradox. It is a banality. Rather than a solution to a paradox of scale, then, air asks for a theoretical language that does not find its movement through multiple scales and political forms remarkable in the first place.

Can we, following Kuriyama, learn to hear air whistling through the hollows of theory? Doing so means making permeable the grounding dis-tinction drawn between the unruly manifold of matter and putatively prior conceptual forms.[24] For ethnography, it also means adopting a different relationship than usual with the concrete. Listening to air, thinking through this diffuse stuff in the thick of becoming, requires less literal materialism.

The poet and critic Charles Bernstein remarks on the relation between poetry and philosophy: "Poetry is the trump; that is to say, in my philosophy, poetry has the power to absorb these other forms of writing, but these other forms do not have that power over poetry. . . . When I think of the relation of poetry to philosophy, I'm always thinking of the poeticizing of philosophy, or making the poetic thinking that is involved in philosophy more explicit" (1992, 150–51). Thinking, for Bernstein, is always a poetic act. Poetry is always thinking. This figuring of always poeticized philosophy pushes me to make explicit the poetic thinking involved in theorizing problems of universality and scale.[25] What are the "universal" and "particular" but conventionalized figures for theory's poetics? Their ossification should be clear when those most ardently debating their definition declare the inadequacy of their terms, and then return to rest on them again and again. Some tropic invigoration might help—a poetic revival through the activation of examples, where details yield not simply particularity, but the potential for mobile metaphors. Might the material poetics of Hong Kong's air—with its whirlings, its blowing through scales and borders, its condensations, its physical engagements, its freight of colonial, economic, and bodily worries about health and well-being, its capacity to link and to divide, its harnessing for simultaneously local and cosmopolitan projects—provide that reviving breath theory needs?

Notes

1. Cited in Berman 1988, 19.

2. On citational networks, see Latour and Woolgar 1986 [1979].

3. The CUHK group's study spanned the four-year period from 1995 to 1998. The HKU study used data for the period 1995–97. Meteorological data was obtained from the Hong Kong observatory, and air-pollutant concentrations were obtained from the Environmental Protection Department.

4. On the work of standards and standardization, see Bowker and Star 1999. Also see Andrew Lakoff's contribution to this volume (chapter 8). Lakoff insightfully shows the ways in which the specific freights of psychiatry and Lacanian psychoanalysis in an Argentine hospital bear on the possible substantiation of bipolar disorder as a viable diagnosis and thus on the liquidity or potential flow of genetic information about Argentine patients as information—globally liquid information—about a globally common mental disorder.

5. Homi Bhabha, following Derrida, elaborates the disturbing power of "being additional" in a postcolonial situation. "Coming 'after' the original, or in 'addition to' it, gives the supplementary question the advantage of introducing a sense of 'secondariness' or belatedness into the structure of the original demand. The supplementary strategy suggests that adding 'to' need not 'add up' but may disturb the calculation" (Bhabha 1994, 155).

6. For an example of this form of argument, see McMichael, Anderson, Brunekreef, and Cohen 1998.

7. Jain 2007.

8. My thanks to an anonymous reviewer and to Rachel Prentice for helping me draw this connection.

9. See, for instance, Feld and Basso 1996; Rodman 1992; and Raffles 2002.

10. Mong Kok's role in imaginations of Hong Kong can be illuminated by Mong Kok's selection as a challenge in an American reality-television show. Contestants were asked in other parts of Hong Kong to complete tasks such as lowering a shipping container with a crane or finding the tallest building in Central. In Mong Kok, however, their task was simply to find a certain tea shop where they would be asked to drink a bitter tea. Mong Kok's tightly packed and sporadically marked streets drove at least one contestant to tears.

11. For a discussion of an attempt by village residents and Greenpeace activists to halt the construction of the incinerator, see Choy 2005.

12. For an analysis of state efforts in East and Southeast Asia to craft exceptional spaces attractive to foreign-capital investment, see Ong 2006. Ong adopts the term *ecologies* metaphorically to refer to the desired labor and financial conditions that are predicted to be conducive to a city's insertion in global trade circuits. I would merely add that in the pursuit of such desires, the ecologies of landscapes, airscapes, and waterscapes can be equally subject to concern and government.

13. For a strong and crucial argument for the recognition of air pollution as a social class issue in Hong Kong, see Stern 2003. Stern points out that, despite the fact that Hong Kong's lower classes suffer greater exposure to air pollution — in their occupations and in their homes — Hong Kong elites have generally set the anti-air-pollution political agenda.

14. For a discussion of the historical racialization of urban space in colonial Hong Kong, see Bremner and Lung 2003.

15. Quoted in Bremner and Lung 2003, p. 244.

16. The following draws heavily on material presented in "What Is the History of Weighing (FAQ—Mass and Density)," 8 October 2007, available on the National Physical Laboratory website, http://www.npl.co.uk, accessed 23 December 2007.

17. See Latour 1999.

18. Air Pollution Index, available at the website for Hong Kong's Environmental Protection Department, http://www.epd-asg.gov.hk/english/backgd/backgd.html, accessed 27 June 2011. The objective was 30,000 µg/m3 both in 2002, when the first version of this chapter was drafted, and again in 2007, when it was revised.

19. The World Health Organizations's acceptable NO_2 annual mean is 40 μg/m³; Hong Kong's target is 80 μg/m³.

20. The World Health Organization's target for PM_{10} is 50 μg/m³ and for $PM_{2.5}$ is 25μg/m³.

21. For instance, the former Hong Kong legislative council member Christine Loh made air quality and the state of Victoria Harbor central issues of her tenure in the Hong Kong Legislative Council (LegCo). During her work in LegCo, as well as in Civic Exchange, a think tank she founded after leaving government, she has consistently figured environmental issues as examples of how important it is for Hong Kong's government to heed the needs and voices of its public.

22. Also relevant to this point is the account by Adriana Petryna (2002) of Ukranians' struggles after the Chernobyl accident to substantiate their debilitations as radiation-caused ailments in order to receive state aid, while confronted with narrow and fluctuating definitions of radiation sickness.

23. For instance, see Butler, Laclau, and Žižek 2000. Also see Balibar 1995a; Badiou 2003.

24. On the distinction between the a priori and a posteriori, Judith Butler remarks, "We might read the state of debate in which the a priori is consistently counterposed to the a posteriori as a symptom to be read, one that suggests something about the foreclosure of the conceptual field, its restriction to tired binary oppositions, one that is ready for a new opening" (2000, 274). With this, Butler argues that strict distinctions between the a priori and a posteriori signal a devolution of discussion rather than an elevated state, an argument that such oppositions indicate a foreclosed and unfruitful practice of theorizing. In the context of the exchanges with Laclau and Žižek in which Butler makes this argument, this can be read as a claim that theory would benefit from an infusion of historical materiality. One should not mistake Butler's argument, however, for a call for empiricism against theory. This would repeat the same mistake of strict distinction. I take her statement, instead, as an invitation to think about theory's poetics.

25. Poetry is an act of creation for Bernstein, a hopeful writing with ambition to forge something new. "Poetry is aversion of conformity in the pursuit of new forms, or can be. By form I mean ways of putting things together, or stripping them apart, I mean ways of accounting for what weighs on any one of us, or that poetry tosses up into an imaginary air like so many swans flying out of a magician's depthless black hat so that suddenly, like when the sky all at once turns white or purple or day-glo blue, we breathe more deeply" (Bernstein 1992, 1).

The pursuit of new forms and the quest for new arrangements of things so that the skies change color around us and so we all may breathe more deeply—such poetics are in tune with the most critical and ambitious of theoretico-political projects. I see the critical creativity of such poetics as kin with the serious play of Joseph Dumit's BioMarx substitutions and Kim Fortun's iterative informatics (chapters 1 and 10, this volume).

PART TWO

PROPERTY AND DISPOSSESSION

4

SHEILA JASANOFF

TAKING LIFE

Private Rights in Public Nature

Once out of nature, I shall never take
My bodily form from any natural thing . . .
W. B. YEATS, "BYZANTIUM"

When does something in nature become property? When may it be owned, exchanged, manipulated, given away, entirely used up, or even destroyed? Does it matter if that thing is alive? I will show in this chapter that the answers given by the law depend not on the nature of the thing in question but on the culture that conditions our understandings of what it means for something to be natural.

Takings: Public and Private

Let us begin with land, the bedrock from which the juncture of nature and property springs. The Fifth Amendment to the U.S. Constitution provides in its "takings clause" that no private property "shall be taken for public use, without just compensation." That clause circumscribes the state's power of eminent domain, which allows governments to claim land for public purposes such as building highways and railroads, protecting environmental amenities, and fostering neighborhood renewal. Private ownership of land is not in itself a barrier to takings. The law focuses instead on interpreting the key terms of the constitutional guarantee, that is, on defining what is a legitimate "public use," entitling appropriation by the state, and what constitutes "just compensation." For instance, a controversial Supreme Court decision of 2005, *Kelo v. City of New London*, held that a local authority may take private property for economic redevelopment, even though the proposed use would benefit some private parties, and even though the condemned property included private homes that were neither derelict nor abandoned. All that the municipality needed to show was that it had formulated a carefully considered plan to benefit the entire community. In a sternly utilitarian calculus, it was enough justification that many would gain from a decision that disadvantaged only a few.

That logic can undermine public takings that fail to meet a mathematical test. Protecting the environment, for example, was at one time unquestionably a valid public purpose under the takings clause; indeed, the public value of environmental protection was so taken for granted that it hardly called for explicit justification. Governments imposed restrictions on land use, and even blocked development altogether, in order to prevent environmental degradation or improve the quality of urban life. Measures such as reducing traffic congestion or creating green spaces still are seen as legitimate public purposes, but recent case law makes it clear that the state's power to mandate such measures is not unlimited.

Beginning in the 1980s, property owners gained increasing political influence and the federal courts began restricting the power of eminent domain in environmental cases. In a decision signaling that the tide was turning, the Supreme Court held that a law depriving property of all its economic value would always count as a taking, triggering a right to compensation (*Lucas v. South Carolina Coastal Council* 1992, 505 U.S. 1003). In other landmark cases, the court instructed communities that they had to demonstrate an "essential nexus" between imposed conditions and legitimate regulatory objectives (*Nollan v. California Coastal Commission* 1987, 483 U.S. 825), as well as a "rough proportionality" between those conditions and the scope of the planned development (*Dolan v. City of Tigard* 1994, 512 U.S. 374). Most controversially, the court ruled that a city wishing to protect a designated wetland had to pay a property owner for any resulting losses, even though the owner had acquired his property knowing that it included a legally protected wetland.[1]

The logic in all these cases favors uses of real property that generate greater economic value over uses that promote only nonmaterial values, such as the integrity or aesthetic quality of nature. The relative value of alternative uses is measured, in part, by their capacity to circulate.[2] Economic development overflows the territorial limits of land; it converts place, which sits still, into money, which moves. A mall generates more wealth than a park. Aesthetic or other intrinsic value by definition does not circulate, or so we imagine; it stays bound up in the land, permitting at best private enjoyment by those who own or have access to it. The inverse principle is that removing land from imagined, or even imaginable, economic productivity—making it sit still, lie fallow, or be seen by only a few—requires compensation, even if the taking serves a recognized public good. Not surprisingly, the first rule has tended to benefit mainly private property developers, although in *Kelo* it was a city (to be sure, acting in concert with a private developer) that put forward the more gainful land-use plan, and thus overrode the interests of private dwellers who could not so readily translate land into money.

Inanimate land, then, can be owned, developed, mined for private gain, or be otherwise made to circulate in commerce; and it can be "taken" out of circulation for noneconomic public purposes only on adequate compensation. Undeveloped land offends U.S. law's predisposition toward use and commerce. It is as if in the eye of the law the right of ownership *naturally* belongs to the party who can do most to render land fluid, enabling a bounded, immovable territory to overcome its inert condition. Land becomes lively through the commerce it enables or sustains. That logic explains why seeking to maintain spaces in static or undeveloped form, whether for preservation or for individual use as in *Kelo*, has fared less well in recent legal conflicts than efforts to open up those spaces to the presumptively greater mobility of commerce. In the hierarchy of utility, the user who does most to overcome land's earthbound limits gains priority, though it may be at the cost of compensating a less entrepreneurial prior owner.

A very different imagination regulates the ownership of living things and of nature's laws. Here, the assumption in Western legal systems has been that lively nature belongs to all: it is the taking of nature for private purposes that should be banned or strictly regulated. Nature is considered to be the common property of humankind. Private claims would corral a part of universal nature, secreting it away from open access or enjoyment. In particular, one may not stake an intellectual-property claim based on the mere discovery of a natural object or law. Einstein could no more copyright the equation "$e = mc^2$" than Edmund Hillary or Tensing Norgay could own Mount Everest, or Neil Armstrong, the Apollo astronaut who in 1969 took the famed "giant step for mankind," could lay claim to the moon whose surface he first set foot on. The logic again is utilitarian: no private use, the law imagines, could possibly outweigh the public benefit of a nature whose works and workings remain available, in equal measure, to everyone. In order to claim an intellectual-property right in natural objects, the claimant has to do more than merely find it, picture it, publicize it, or celebrate it. The would-be patent holder has to change the quality of the thing itself, so that it no longer partakes of the character of the "commons" that we see in nature. The object of the patent claim has to be clearly marked, indeed set apart from nature, as a creation of human ingenuity and enterprise. It has to be a result of invention, not discovery. It has to remove the thing being patented from nature to culture.

How have these understandings of inanimate land versus living nature, and of public versus private benefit, intersected with, disrupted, and to some degree refashioned intellectual-property rights in the products of modern biotechnology? What steps, specifically, are required to move something from the domain of nature to the domain of culture? "Taking life" by asserting private-property rights in natural objects or phenomena involves two kinds of moves that are tacitly, though not explicitly, granted controlling status in the law: specificity and circulation. Put differently, the property claim has to involve *both* taking a specific, characterizable, and reproducible bite (and, today, perhaps as much byte as bite) out of nature *and* a capacity to make the excised element circulate widely in commerce. In this way, patentable, human-made, "natural" objects escape the conditions of unruliness, complexity, and nongovernability that environmental historians have associated with American visions of wild nature (Nash 1982).

I elaborate this argument in five parts. I first look at varied treatments of the nature-culture boundary to illuminate how that divide is constructed and maintained while things move back and forth across it. I then offer the historical example of slave ownership to illustrate the importance of ontological stability and circulation in legal constructions of life as capital. I next provide a brief account of U.S. patent law and the framework it lays out for asserting intellectual-property claims in biological materials. Turning then to case law, I compare the strategies for culturing nature, or failing to do so, in two leading North American patent decisions: the judgments on patenting life forms by the Supreme Courts of the United States and Canada in 1980 and 2002, respectively. I conclude with reflections on the law's role in constructing the metaphysics, the value, and the moral economy of nature in the era of biotech patenting.

Nature/Culture: Intertwinings

It hardly needs stating that ideas of nature stem from culture. Nevertheless, the intensity of engagement between scientific and other imaginations—literary, artistic, and legal, for example—in crafting biological categories deserves comment. Modern biotechnology builds on long histories
of human preoccupation with the natural. Can art emulate nature? Explora-
tions abound, from Galatea, the marble image of perfect womanhood awak-
ened to life by Pygmalion's prayers, to Oscar Wilde's happy prince, who
gave away his precious apparel to the poor and hungry until his leaden heart
was revealed to be worthy of god's grace (Wilde 1990). How *un*natural is
nature? Collectors such as Rudolf Virchow in Berlin and Peter the Great in
St. Petersburg sought answers to that question by amassing biological mal-
formations and monstrosities in their cabinets of curiosities. Where does
the human end and the nonhuman begin? Hybrids that cross that categorical line are both revered and feared. The potbellied, elephant-headed god
Ganesha is one of India's favorite deities, whereas in Western legends part-
human creatures were seen as slippery, treacherous beings—mermaids and
centaurs, vampires and werewolves.

Long predating the genetic turn, the possibility of tampering with
human nature presented forbidding possibilities. In the Gothic tales of the
nineteenth century, attempts to compete with nature usually led to disaster. Mary Shelley's *Frankenstein* became the iconic cautionary tale of human
overreaching. H. G. Wells and Robert Louis Stevenson offered in *The Invisible
Man* and *Dr. Jekyll and Mr. Hyde*, respectively, their own, near mythic glosses
on the hazards of tinkering with human nature. Leon Kass, the first chair of
the President's Council on Bioethics, established by George W. Bush, asked
his fellow committee members to read Nathaniel Hawthorne's short story
"The Birthmark" (1843), in which an overzealous scientist attempts to rid
his beautiful wife of a hand-shaped mark on her cheek. He cures the defect
but kills the beneficiary.

Biomedicine today uses material technologies to reinscribe the nature-
culture boundary yet again. These developments challenge our intuitions
about when human life begins and ends, which states of being are worth
preserving, and who should decide. The case of Karen Ann Quinlan in New
Jersey spotlighted the uncertain legal and ethical status of persons in a per-
sistent vegetative condition, kept "alive" through technological devices supporting nutrition and respiration (*Matter of Quinlan* 1976, 70 N.J. 10). In
2004 the fate of Terri Schiavo, a brain-dead woman in Florida, plunged

America into a media-saturated controversy about who should have power, in disputed cases, to bring an end to liminal states of existence. While the family fought bitterly over the removal of Schiavo's feeding tube, the U.S. president and members of Congress took up the cudgels in support of continued treatment in a misguided attempt to placate America's pro-life religious Right. Disrupting kinship categories, new reproductive technologies, beginning with the first use of in vitro fertilization, in England in 1978, have allowed childless couples to conceive, postmenopausal women to bear children, same-sex couples to become parents, procreation to occur between dead and living spouses, and babies to be preselected to donate healing tissue to diseased siblings (Thompson 2005). By introducing new biological entities into the world—frozen gametes, cell lines, induced pluripotent cells—these techniques have also tested the limits of our understanding of the human community.

Curiously, it was the birth in Edinburgh's Roslin Institute of a sheep named Dolly, cloned from a mammary-gland cell of an adult ewe, that provoked some of the deepest reflections on human nature. The announcement of Dolly's birth in an issue of *Nature* in February 1997 sparked intense debate about the ethics of human cloning. At the epicenter of controversy was the legal and moral status of human embryonic stem cells. Pregnant with possibility, these biologically undifferentiated entities excited biomedical researchers as much as they disturbed Christian fundamentalists and others who believe that human life begins with the fusion of egg and sperm. In their moral universe, the extraction of stem cells so as to prevent an embryo from developing into a full human being was equivalent to murder. President George W. Bush, explaining why he would restrict federal funds for working with embryonic stem cells, ratified this reading with characteristic bluntness. His administration would not, he pronounced, promote research "which destroys life in order to save life" (Stolberg 2005).

In all of these stories, nature manifests itself through disagreements about what should be seen as natural. As Bruno Latour famously observed, "The settlement of a controversy is the *cause* of Nature's representation, not the consequence" (1987, 99). What we see as belonging to nature is, in other words, the result of all kinds of prior commitments to ways of seeing, studying, and classifying life. Natural settlements are contingent on established ways of knowing.[3] If nature appears devoid of social influences, then that, too, is the result of concentrated labor, which Latour (1993) in a bow to biotechnological processes labeled "purification."

We accept that traditional or premodern societies did not see nature as cleanly bounded off from society. The Achuar of the Amazon, for example, up until the late twentieth century, still conceived of plants and animals as persons because they possess the ability to communicate with people (Descola 1996, 224). But our own nature-culture divisions are no less the products of cultural predispositions and institutional histories. They are experience congealed into material-semiotic systems and actors.[4] The new technologies of life naturalize new self-definitions and forms of sociality (Rabinow 1992). For instance, reproductive technologies reinscribe stereotypical gender relations by constructing technologically mediated conception as "natural," and hence desirable (Hartouni 1997). *Not* using the available technologies—by choosing to remain childless, for example—then becomes the marked, or unnatural, behavior.

Nonhuman entities in some sense actively participate in technological innovation. Blurring hard distinctions between agents and things acted on, technological systems come into being as products of complex enrollments and translations among their heterogeneous components, both animate and inanimate.[5] Recognizing this interplay, the proponents of actor-network theory (ANT) have ascribed a kind of agency to the "actants" (nonhuman agents) in such systems, denying any *a prioristic* divide between nature and culture. The inanimate components are capable of resisting, and so redirecting, human agency; as Latour pointed out with his famous example of the traffic bump, or "sleeping policeman," inanimate technological artifacts can take over human functions, with varying normative and social, as well as physical, consequences (Latour 1992).

Even in modern societies, then, distinctions between person and property, actor and object, human and nonhuman should not be seen as given in advance. Our definitions of things as they are and things as we would like them to be are not independent but are frequently established together through processes of co-production (Jasanoff 2004). It follows that ontological settlements reached in one social or cultural context, with its particular normative commitments, need not be universal, though they may become so through concerted attempts to harmonize differences. As I have shown elsewhere, Britain's epistemic respect for empiricism and common sense logic, coupled with a political history of deferring to elite public servants, produced a legal regime that treated pre-fourteen-day-old embryos (sometimes called pre-embryos) as not yet human (Jasanoff 2005a). Parliament and, by extension, the British public, accepted the view that the formation of the primitive streak, precursor to the central nervous system, in embryos around the fourteenth day marks a meaningful rupture in human biological development. There was no scientific consensus that the embryo under fourteen days is any less human than after that date. The fourteen-day limit was not, to start with, universally accepted. In Germany and the United States, for example, dominant cultures of public knowledge, or civic epistemologies, kept alive questions about the embryo's moral status and could not produce a pragmatic bright line for research comparable to Britain's. Only gradually has the fourteen-day rule been smuggled into U.S. policy, largely through the voluntary, self-regulatory activities of scientific bodies interested in furthering research with stem cells.

Many controversies about the right way to draw the line between nature and culture illustrate the centrality of the law as a device for performing what I call "ontological surgery" in modern political systems. Courts, legislatures, and regulatory agencies routinely grapple with conflicts about the nature and meaning of natural objects. How we define and characterize boundary-crossing objects, and how we choose to interact with the resulting things, are worked out as much through law as through scientific research and development. Such concepts as the environment, clean air, brain death, DNA fingerprint, or even "natural mother" are located in webs of meaning crucially shaped by the law. The law constructs both life and capital and, more specifically, demarcates those aspects of life that can be owned from those that cannot.

Ontological Border-Crossing:
Naturalization, and the Law

A key function of the law is to produce inevitability. A legal decision takes a social issue that is uncertain, disordered, or contested and reorders it within a system of preordained rules and norms. Questions and ambiguities are temporarily eliminated, and the fuzzy boundary between lawful and lawless conduct is constituted once again as sharp and discernible—in a word, natural. In today's text-based legal systems, judicial opinions naturalize the restored order, so that, in the ideal case, no one challenges the rightness of the winning argument. Before a definitive judgment is handed down, there are many possible ways to think about the rights and wrongs of the case, as well as many reasons for choosing between possible outcomes; the law itself appears unsettled. Afterward, only one reading prevails. When a court succeeds, its reading looks like the only one that could have been reached under the circumstances. As if naturally all other rules and reasons fall away, and the law appears to dictate the outcome rather than the outcome rewriting the law. Like scientific writing, legal text-making produces its own authority by erasing the contingencies from which the dominant ruling emerges.

Of course, such erasures can never be complete, and some of the moves by which contingency is backgrounded can be easily recovered within the textual practices of the law. American law writing, for example, offers space for dissenting opinions, so that, in highly contested cases, any reader can follow the debates behind the ultimate legal ruling.[6] Dissenting opinions, however, are consigned to history, like discarded scientific theories. It is the holding in the case, and the arguments justifying it, that circulate as law. All else is considered *dicta*, mere sayings without the force of law. The most authoritative decisions are those that provoke no dissent, but even powerful dissenting voices in the law go underground unless, in the rarest of cases, they colonize the imagination of succeeding generations and ultimately get resurrected as what *should* have been the law all along.

A noted example of such a shift in U.S. law occurred around the infamous *Dred Scott v. Sandford* decision (1857), which is especially interesting here because it sheds historical light on the law's ability (or lack of it) to produce definitive ontological settlements. The case tested at its most basic level the stability of the legal compromise that had allowed slave states and nonslave states to coexist within the Union before the Civil War. Technically, it raised questions about whether Scott, a black man, had the right to sue in federal courts; whether his years of residence in free northern states had made him a free man; and whether Congress had the right to ban slavery in the northern territories acquired through the Louisiana Purchase. Chief Justice Roger Taney, an eighty-year-old, patriotic son of the South and a former slaveowner, ruled on all three questions in a 7–2 decision that has reverberated through time as a moment of shame in American constitutional jurisprudence.[7] The dissenting voices of Justices John McLean and Benjamin Curtis won the day morally, but legally it was the seven-member coalition of proslavery Democrats who prevailed.

Taney devoted nearly twenty-four pages of his fifty-five-page opinion to the first legal issue, which he phrased in this way: "The question is simply this: can a negro whose ancestors were imported into this country and sold as slaves become a member of the political community formed and brought into existence by the Constitution of the United States, and as such become entitled to all the rights, and privileges, and immunities, guaranteed by that instrument to the citizen, one of which rights is the privilege of suing in a court of the United States in the cases specified in the Constitution?" (*Dred Scott v. Sandford* 1857, at 403). He concluded, in what historians have rejected as a perversion of facts and logic, that blacks were not entitled to be regarded as citizens of the United States (McPherson 1990, 174).

There is little doubt that Taney's was a deeply felt, "visceral" opinion, written by an aging, bereaved man who saw a whole way of life that he valued threatening to disappear before his eyes, and was determined to fight that outcome with the most powerful instrument at his command: his nation's constitutional law. But was there no logic to Taney's position? At stake in the case was not only the political composition of the United States but the ontology of a group of persons who constituted a sizeable part of its population. Could American blacks claim rights of citizenship (and thus of personhood) by suing in federal courts, and yet be treated as property, not persons, in some parts of the country? For Taney, this specter presented an intolerable contradiction. An early and ardent defender of Jacksonian free enterprise, Taney was perhaps especially troubled by the notion of a person who could act as a free agent and fully autonomous citizen in some contexts, but elsewhere would function only as useful chattel, devoid of agency. Logic, for him, demanded a resolution of this ambivalent, boundary-disrupting identity, and he opted for a characterization of Dred Scott that would remain unalterably on the side of chattel, regardless of where in the country his masters chose to transport him. As chattel, Scott and others like him could circulate smoothly across state borders, an integral and ontologically stable component of the economic and cultural system that Taney had given his life to defending.

Dred Scott proved to be a pyrrhic victory for the South. In less than five years the slave states and the free states were at war, in a conflagration that put paid to the notion of dual ontologies—as goods and as persons—for human beings of any color living within the United States. Abraham Lincoln, who presided over the Union's victory, bought into Taney's logic of singular ontologies, but famously not into his normative settlement. For Lincoln, the confusion over where to place black humanity within the framework of the U.S. Constitution could be resolved in only one way. *All* men, as the Declaration of Independence stated, were created equal. In a famous speech, just a year after *Dred Scott*, Lincoln said, "I believe this government cannot endure, permanently half slave and half free."[8] A question for modern biotechnology, and particularly for intellectual-property law as it relates to living things, is whether the time has come for similar ontological clarification with respect to manipulated biological entities.

Invention and Its Rewards

In contemporary industrial societies, an inventor, author, or artist is entitled to retain exclusive rights for a period of time in the products of intellectual or artistic creativity, through patents or copyright. For the founders of the American republic, the importance of scientific and technological innovation was important enough to deserve constitutional support. It was not inquiry in and of itself that the founders prized, but useful inquiry, promoting economic and social well-being. Accordingly, the word *science* explicitly appears in the U.S. Constitution only in the clause governing intellectual property. Article 1, section 8 provides that Congress shall have the power "to promote the progress of science and useful arts, by securing for limited times to authors and inventors the exclusive right to their respective writings and discoveries."

The U.S. Patent Act, originally drafted in 1790 by Thomas Jefferson, among others, implements that constitutional grant of authority, and it still remains, with minor revisions, the governing text for assigning intellectual-property rights in the United States. Canada, too, regulates intellectual property under an almost identically worded statute. The operative provisions in both nations specify what sorts of things can be patented ("patentable subject matter") and under what conditions. Thus, patents can be issued for "any new and useful process, machine, manufacture, or composition of matter, or any new and useful improvement thereof."[9] As the words *new* and *useful* make clear, the most important objective of patents is to reward novelty and utility. Things already discovered and things that contribute nothing to human welfare deserve no acknowledgment. The inventive step, moreover, must be a real advance and not "obvious" to persons "skilled in the art."[10] If a discovery lies too close to what knowledgeable inventors deem to be "prior art," then the claim to exclusivity is unfounded and deserves no special legal protection.

Part of the justification for granting patents is that they encourage inventors to put into the public domain knowledge and know-how that would otherwise be held in confidence and would not circulate for the benefit of would-be inventors and the public. In order for useful inventions to circulate, they must be exactly reproducible, and this requirement is secured by the law's demand for specification. To this end, the law calls for "a written description of the invention, and of the manner and process of making and using it, in such full, clear, concise, and exact terms as to enable any person skilled in the art to which it pertains, or with which it is most nearly connected, to make and use the same, and shall set forth the best mode contemplated by the inventor of carrying out his invention."[11] Biological materials that are too complex to be specified through written description can meet the patent law's specificity requirement in a more physical way: through deposit in a storage service like the American Type Culture Collection located in Virginia.

Patents confer a temporary monopoly on the holder, contrary to the spirit of free circulation that is central to a market economy. The conventional rationale for nevertheless granting such a monopoly is that the inventor compensates society by placing the know-how and the invention itself in the public domain. Any subsequent person wishing to make the same product or use the same process must acquire a license from the patentee; but anyone with sufficient resources can in principle obtain a license, just as anyone with sufficient ingenuity can alter a patented process to make it more efficient and productive, or combine it in novel ways with other processes to produce new entities that can in their turn be patented. The key to making productive use possible is that the patent system must be capable of precisely describing what the inventor did to stake out a claim. Just as owned land has to be precisely surveyed or described, so a patented invention must be described in ways that render it reliably, consistently reproducible. Accidental inventions, those for which there is no surefire recipe for others to follow, cannot be patented.

Intellectual-property laws were not written with modern biotechnology in mind, but a concern for protecting property rights in biological materials dates back at least to the U.S. Plant Patent Act of 1930. The Plant Variety Protection Act of 1970 extended similar rights to nonsexually propagated plant life. Behind these laws was a growing consensus that breeding new plants satisfies the patent regime's desire for the new, the useful, and the reliably reproducible. Utility, as noted, is firmly tied to the notion of replication: to circulate, a patented product or process has to move about as a formula, package, or set of characteristics that can be accurately described, recognized, and exchanged. Even an organism has to be reproducible in predictable ways in order to merit a patent. Thus, Golden Rice, the first and most vigorously debated product of agricultural biotechnology's turn to high-value-added staple commodities, is a highly specific modification of one of the world's most widely consumed grains.[12] It results from incorporating into rice a new trait that no rice ever had before: genetically engineered beta-carotene that confers on the resulting "golden" rice the power to spur vitamin A production in the consumer's body. It is *novel* in that it crosses the line between food and pharmaceuticals, and builds into ordinary rice a new characteristic that makes this rice nutritionally richer than the natural grain. It is *useful* in that it caters to a new and presumably needy market—people suffering blindness through malnutrition in impoverished regions of the developing world. It is *non-obvious* in that many ingenious extensions of existing knowledge and craft were needed to imagine, create, and standardize this product for widespread cultivation. Indeed, so complex was the translation process that moved Golden Rice from idea to object that it required, according to Ingo Potrykis, of the Swiss Federal Institute of Technology, who is widely regarded as the "father of Golden Rice," seventy separate patents to acknowledge all the inventions that he and his colleagues drew on in producing their new product (Potrykis 2001).

In its landmark 1980 decision, *Diamond v. Chakrabarty*, the U.S. Supreme Court laid out a very broad interpretation of the patent law as it applies to biological products. At issue in the case was whether Ananda Chakrabarty, a research scientist working for General Electric, could patent a new form of the pseudomonas bacterium that he had created. The bacterium was capable of breaking down the components of crude oil and thus was considered potentially useful for cleaning up oil spills. Up to that time, patents had not been granted for living organisms other than plants, and some observers feared that extending patent rights to manufactured life would start society down the slippery slope to commodifying and thus reducing the integrity of life itself. By a 5–4 majority, however, the Supreme Court rejected these fears as irrelevant. Quoting a congressional committee report, the court concluded that the legislature had intended patentable subject matter to "include anything under the sun that is made by man." Chakrabarty's bacterium clearly met that test since it was a product of laboratory manipulation and had never previously existed in nature. Though the decision concerned a bacterium, *Chakrabarty*'s authority was soon used to underwrite patents on many higher animals, including mice, pigs, cats, and cattle.

As the center of biotechnological research and development moved toward drug discovery and the search for genetic links to diseases in the 1990s, scientists and drug companies saw increased value in owning property rights in genes and even gene fragments. Two questions immediately arose. Are genes, entities that (unlike Chakrabarty's bacterium) do occur in nature, encompassed within the law's definition of patentable subject matter? And do they meet the law's additional constraints of novelty, utility, and non-obviousness? Despite some controversy, the courts and the United States Patent and Trademark Office (PTO), in its 2005 guidelines, ruled positively on both issues.

The decision to allow patenting of genes illustrates again the inclination of property law to favor moves that put otherwise unproductive inanimate matter into circulation, creating economic value. Rejecting arguments that genes are natural objects, and therefore unpatentable, the patent office stated: "An isolated and purified DNA molecule that has the same sequence as a naturally occurring gene is eligible for a patent because (1) an excised gene is eligible for a patent as a composition of matter or as an article of manufacture because that DNA molecule does not occur in that isolated form in nature, or (2) synthetic DNA preparations are eligible for patents because their purified state is different from the naturally occurring compound" (U.S. Patent and Trademark Office 2005, 1093). The patent office cited by way of historical precedent a patent in 1873 to Louis Pasteur for purified yeast, free from disease germs, as a composition of matter. In both cases, purified yeast and purified genes, the inventive step consisted of removing a biologically occurring "composition" from any constraining material matrix. Purification, in effect, was a process of *de*naturing, of taking something out of its natural context. In pure and isolated form, genes are no longer nature's instruments, subject to the vagaries of natural law, but are amenable instead to human intentions and purposes. They are ripe for entering the cultural worlds of sociality and commerce. Indeed, as the PTO guidelines stipulate, a gene patent may be granted only if the claimant provides a description of how the gene will be used.

A number of U.S. biotech patent decisions begin to make sense if reexamined within the specificity-circulation framework. Most instructive, perhaps, is the much-discussed 1990 decision of the California Supreme Court in the dispute between a patient named John Moore and his doctors at the University of California, Los Angeles (UCLA), over who owned the cells excised from Moore's spleen during his treatment for hairy cell leukemia (*Moore v. Regents of the University of California* 1990, 793 P.2d 479). The court held that Moore could sue for lack of informed consent, but that he was not the proprietor of his own cells and tissues, and hence could not pursue a claim of "conversion" or, colloquially, theft. The case has been analyzed by a number of scholars who stress its internal contradictions from the standpoint of legal reasoning.[13] For example, Moore's cells were deemed to be present in all human beings, thereby ruling out his claim to uniqueness; but they were at the same time held to be novel enough as generators of lymphokines to justify the researchers' patent claim. Equally, the court concluded that granting Moore ownership rights in his own tissues might hamper research, but that granting patent rights to the UCLA researchers would not create a similar impediment.

These apparent contradictions fall away when one views the actions of the UCLA medical researchers as a project in mining nature for extractable entities that can freely circulate. Lodged in Moore's body, the leukemia cells were part of a complex organism and were rendering no value to society at large—indeed, they were producing a potentially incurable disease in the host's own body. Cut loose from context and allowed to exist in their own right, they became the raw material for an "immortal" cell line of possible therapeutic value, which could be instrumentally used to cure other people. Structurally and functionally, the spleen cells were the same, whether in or out of Moore's body. But by excising them from their unruly context, the UCLA researchers rendered this specific piece of nature more tractable to human ends. Though not guilty of *legal* conversion (theft), they in this way effected an ontological conversion, turning nature into property.

Instructive, too, is the outcome of the lawsuit brought by the prostate-cancer specialist William Catalona against his employer, Washington University, to acquire ownership of a biorepository containing some thirty thousand patient samples that the physician had collected during his research. Catalona wished to take the samples with him when he joined the medical faculty of Northwestern University. He argued that the samples were originally the property of his patients, who had donated them for research, so that he now had exclusive control of them. In a unanimous opinion, the Eighth Circuit Court of Appeals upheld the rights of the university, denying the claims of both Catalona and his patient-donors (*Washington University v. Catalona* 2007, 490 F.3d 667). Superficially, the judgment appears to line up against circulation, since the court did not allow the repository to move with Catalona. Consistency reappears, however, when we look below the surfaces, to the deep structures of capital accumulation. Faced with a choice between the individual researcher's abstract claims of public benefit—"a publication enriches the scientific community, is consistent with the wishes and consent of the patients, contributes to the progress of medicine by furthering research, and in some cases may bring grant money into the university" (Lori Andrews 2006, 399)—and the university's superior wealth-generating and circulation-enabling power, the court opted for the latter.

Another instructive case involved the attempt by a Mississippi farmer named Homan McFarling to circumvent a licensing agreement with Monsanto, the manufacturer of a genetically modified soybean crop that McFarling had cultivated. In order to boost sales of its popular herbicide Roundup, Monsanto had produced a variety of so-called Roundup Ready crop plants, with growth enzymes resistant to glyphosate, the herbicide's active ingredient. The modified crops would grow in fields treated with Roundup, thus prompting more farmers to use the two products together as a unified technological package. As part of its marketing strategy, the company required all farmers to sign an agreement with their seed distributor, promising not to store or replant Roundup Ready seeds from one growing season to another. Bridging the turn of the century, from 1999 to 2000, McFarling decided to test this new move in patent law by breaking the terms of the agreement and replanting seed that he had saved from his first planting. He alleged that Monsanto's patent did not cover the second-generation product. His legal argument consisted of two parts: first, by restricting use of the second-generation seed, Monsanto had impermissibly broadened the scope of its patent, constituting "patent misuse"; second, by making it impossible to separate the patent on the genetically modified trait from the seed that contained it, Monsanto was performing an impermissible act of "tying" the two products, in violation of antitrust law.

McFarling lost, but not until he had pursued his claim up to the Supreme Court, which refused to hear his appeal on 27 June 2005. That decision let stand the adverse judgment of the Court of Appeals for the Federal Circuit (CAFC), the specialized tribunal that hears all first-round patent appeals. The CAFC ruling, together with the amicus curiae (friends of the court) brief filed by the U.S. government in support of Monsanto, against McFarling, sheds light on the metaphysics of the transition from nature to property in the American legal imagination (Supreme Court of the United States 2005).

With respect to McFarling's claim of patent misuse, the Federal Circuit held that extending Monsanto's patent to the second-generation seed, which also contained the patented anti-glyphosate growth gene, did not constitute an illegal broadening of the patent's scope. This was because the seed in the next generation was, in effect, a nearly identical copy of the original seed, because the plant was "self-replicating" (ibid. 14). This was not, in other words, a case of a manufacturer attempting to keep a subsequent inventor from using the original invention to produce a newly useful product. It was, as the government's brief also argued, a case of a downstream user illegitimately trying to undermine an inventor's lawful right to restrict how its patented product, and all more or less identical products, should be used.

The term *self-replicating* smoothly elided any distinction between one generation and the next of a genetically altered seed. It foregrounded the patented gene and rendered pragmatically immaterial the seed that contained it. It also, incidentally and without fanfare, elided the farmer's labor in cultivating a generation of crops capable of bearing new seeds. McFarling, by this reckoning, was not and could not be party to Monsanto's inventive step; his work of propagation needed no special acknowledgment from the standpoint of patent law. To recognize him in any respect as a party to the invention would only have muddied the waters and gotten in the way of those with more power to make the product circulate. In advocating this result, the government's position bore a striking conceptual resemblance to *Johnson v. Calvert* (1993), the gestational-surrogacy case in which the California Supreme Court decided that the genetic mother of a child should be seen as its natural mother. The surrogate mother, who had provided nine months of labor to bring a genetically unrelated child to life, was held to have no rights to the being she had nurtured, but whose identity she could not legally claim to have shaped.

With respect to the tying issue, the government again argued that no law had been broken. McFarling had claimed that he was, in effect, being forced to buy unwanted new seed as a result of having invested in Monsanto's genetically altered trait. The Federal Circuit considered this claim invalid. McFarling, the court held and the government agreed, was not "entitled to purchase respondent's patented invention without also honoring limits imposed on the use of the product in which that invention finds its useful, tangible expression" (Supreme Court of the United States 2005, 16). The invention (the modified trait) and the product in which it found expression (the seed) became in this way a single, indissoluble package, part of culture not nature. Therefore, as the government counterargued, keeping McFarling and other farmers from storing and reusing seed from earlier plantings was simply within Monsanto's broad legal right to refuse to license its product (Roundup Ready seed of any generation)—a right enjoyed by all patentees. Further, illustrating the potency of economic arguments, the government claimed that the restriction on reuse was not, in any case, likely to be anticompetitive. Monsanto could, after all, have charged an additional fee for reuse, and this plus the cost of monitoring and enforcing any such relicensing system would arguably have driven up costs to a degree that would not have benefited consumers.

As these U.S. cases illustrate, the history of granting patents on biological inventions has tended toward expansion of property rights, though occasionally a case sends a reminder that limits still exist. Such a point was reached in a case involving a diagnostic test for deficiency in B vitamins. An American company named Metabolite Laboratories held patents on both the correlation between elevated levels of the amino acid homocysteine and the vitamin deficiency, and a blood test based on that fact; and it sued a licensee, LabCorp, when it stopped paying royalties because it had switched to using other tests (*Laboratory Corp. of America v. Metabolite Laboratories* 2006, 548 U.S. 2006). Metabolite claimed that LabCorp was infringing its patent simply by permitting physicians to see the homocysteine–vitamin deficiency correlation without paying licensing fees. The suit attracted considerable media attention. In an op-ed column in the *New York Times*, the science fiction writer Michael Crichton attacked Metabolite's claim: "Basic truths of nature can't be owned" (2006, WK 13). Perhaps Crichton was defending his own right to speak of such discoveries in his writing without incurring possible charges of infringement. The case reminds us, in any event, that specificity (as in the statement of a natural law) is not in itself enough to establish a patentable claim. To circulate effectively—usefully—in society, the claim still has to be materialized in some way (in a seed or a blood test), just as money historically achieved circulation through embodiment in shells, feathers, precious metals, and other products of nature.

Legal Metaphysics: Nature or Composition of Matter?

Diamond v. Chakrabarty marked in its way the first day of creation for the U.S. biotechnology industry. With that decision, the Supreme Court threw open the door to patenting any living things that were the creation of human hands and human ingenuity. In effect, nature became a mine for those who could satisfy the authorities that they had extracted or synthesized something that no longer properly belonged in the realm of the natural, and that was at the same time new, non-obvious, and useful. Although the case concerned only the microorganisms produced by Ananda Chakrabarty, most observers believed that the court's logic extended equally well to higher organisms, including mammals. In fact, the PTO waited until 1987 to apply the ruling in this fashion, stating in its guidelines of that year that patents could be granted for any "non-naturally occurring, non-human multicellular living organisms, including animals."[14] The first patent granted for a transgenic mammal in the United States was for the so-called Harvard OncoMouse, a mouse strain that had been modified with a gene to increase its susceptibility to cancer. The resulting construct was useful for cancer research, since the genetic alteration made the OncoMouse more suitable as a model for studying the development of the disease.

Environmental, religious, and animal-rights groups protested the extension of patents to higher organisms, but the PTO's broad reading of *Chakrabarty* remained intact. That human intervention had produced an entity not otherwise occurring in nature was sufficient to define the altered thing as a patentable "composition of matter." Other possible readings of the nature-culture boundary were accordingly ruled out. Over time, though with some hiccups along the way, patent offices in Europe, Japan, and other countries mostly fell in line behind the American decision. But in 2002, in *President and Fellows of Harvard College v. Canada*, the Canadian Supreme Court became the first judicial body to rule that higher animals could not be viewed as compositions of matter under the Canadian Patent Act, whose wording is almost identical to that of the corresponding U.S. law.[15]

In neither the U.S. nor the Canadian case was there a monolithic position to which we unproblematically attach the label of "legal culture." Indeed, each was a 5–4 decision, with a vigorous, almost winning dissent. The significant point is that the opinions split along very different intellectual lines. The U.S. debate centered on an imaginary of progress: in science, in law, and in national life. The Canadian decision, by contrast, occupied itself with the difference between life and matter, even in the case of such a lowly creature as a mouse. The differences of opinion in the United States mapped roughly onto the liberal-conservative divide on the court, a divide historically associated with opposing views of the free market and the utility of centralized regulation. All five of the majority justices in this case were Republican appointees, hence presumably more in sympathy with economic interests; the dissenters included the only two Democratic appointees (Thurgood Marshall, Byron White) and two moderate-to-left Republicans (William Brennan, Lewis Powell).

In Canada, the court's make-up reflected the fault line of national identity, which runs along linguistic and religious divisions between Anglophone and Francophone, and between Protestant and Catholic. The judges' names may tell a part of the story. On the side of the cautious majority were Michel Bastarache (the lead author), Claire l'Heureux-Dubé, Charles Doherty Gonthier, Frank Iacobucci, and Louis LeBel. On the side of the acerbic minority were all three judges with identifiably English names, including the lead author William Binnie, together with Louise Arbour, John C. Major, and Beverly McLachlin. Some (l'Heureux-Dubé, LeBel, Arbour) had received a part of their education in Canadian Catholic institutions or in

France (Bastarache, Gonthier). Bastarache himself was an authority on language rights, a federalist by political inclination, and, perhaps most important, a man who had experienced genetic tragedy in his own life: both of his children suffered from a congenital convulsive disorder, and both had died before he was appointed to the court, one at three and a half and one at seventeen (Corelli and Bergman 1997).

In comparing the two decisions, it is helpful to read the *Chakrabarty* opinion together with two contrasting amicus curiae briefs. Routinely submitted in important Supreme Court cases, such briefs supplement the arguments advanced by the parties and often provide conceptual and rhetorical resources that the justices use in crafting their written opinions. Key amicus briefs were submitted in *Chakrabarty* by the biotechnology company Genentech and the People's Business Commission (PBC), an antibiotechnology group founded by the author and political activist Jeremy Rifkin. Genentech devoted a considerable part of its argument to debunking the claims put forward by PBC, dismissing its opponent as having an "essentially Luddite philosophy" (Supreme Court of the United States 1979a, 11). Appealing to an American myth of self-fashioning through technology, the strategy succeeded. The majority opinion relied substantially on the Genentech brief and, except for a few dismissive words, almost entirely ignored the PBC's.

On Liveliness

The most revealing points of comparison between the two decisions, as well as between the two *Chakrabarty* briefs, have to do, first, with the characterization of life and liveliness and, second, the definition of the public good. Genentech's intention was to diminish as far as possible the distance between Chakrabarty's microorganism and the kinds of objects for which patents had been granted in the past. To this end, the brief stressed the similarity between the novel bacterium and inanimate matter, such as chemicals and even carburetors. There could be no legal dispute worth commenting on, Genentech argued, if the question before the court were the patenting of a plasmid, for plasmids are "absolutely inanimate" structures, each building block of which "is an absolutely dead bench chemical" (Supreme Court of the United States 1979a, 15). Curiously, however, these very "dead chemicals" (a term used several times in Genentech's text) are endowed with the capacity to "cough the bacterial engine [that contains them] into useful life"—a rhetorical move that at once converts the dead chemical into an agent of life, and its living container into a mere machine, like a car engine (ibid. 16,17). Such admixtures of living and dead substances, Genentech suggested, should be treated in effect as utilitarian objects whose composition should not in any way bear on their patentability: "Can it be said that Congress intended patents on living organisms inside inanimate bits of straw but prohibited them in the case of inanimate bits of chemical inside microorganisms, or are we beginning to draw distinctions that border on the silly?" (ibid. 16).

To PBC these questions seemed anything but silly. For this intervenor, the very attempt to equate a living organism with a mere inanimate composition of matter raised profound ethical issues. After all, the PBC argued, "the thing which sets living organisms apart from nonliving entities is their very 'aliveness'" (Supreme Court of the United States 1979b, 5). Granting patents on life, they suggested, was the first step down an all too predictable slippery slope toward turning human beings into objects of manipulation and design, in violation of the human spirit. In support, they quoted the American physician and ethicist Leon Kass, later head of George W. Bush's ethics commission, and the French philosopher of technology, Jacques Ellul. Over and over, the PBC brief insisted that once the line between life and nonlife was breached, there would be no way to hold on to the most meaningful distinctions, between natural and artificial reproduction, between human and machine, and between higher and lower organisms. Neither law nor science would be in a position to stop the slide, for "if patents are granted on microorganisms there is no scientific or legally viable definition of 'life' that will preclude extending patents to higher forms of life" (ibid.). Notably, the brief did little to unpack the difference between nature and artifice. It was fundamentalist in its assertion that human life should be held apart from manipulation and ownership. It raised no serious metaphysical questions.

In advocating for Chakrabarty's patent, Genentech, too, bought into PBC's humanist concerns, but not into their implications for patenting life. Its brief repeatedly sought to distance the chemical-like microorganism from any connection with forms of life that could give rise to deeper ethical concerns. Thus, the company noted that "animal cloning, test tube insemination and other extravagances have nothing to do with the minute concerns of Chakrabarty, and those in turn have nothing to do with gene-splicing, which alone has generated all the controversy" (Supreme Court of the United States 1979a, 10). It would be perverse, the brief went on, to impede potentially life-giving research by showing altruism toward "invisible bacteria that can be freeze-dried to a powder having no semblance of living-ness" (ibid. 12). The question before the court, accordingly, was not, as PBC had argued, "the rapid proliferation of genetic technologies in the areas of energy, agriculture, medicine, industrial processes and many other aspects of the nation's economic life" (Supreme Court of the United States 1979b, 3). Rather, the court's obligation was simply "one of statutory interpretation, of grammar leavened with reason" (Supreme Court of the United States 1979a, 11). The majority went along with this narrow construction of its role, announcing in the opinion's first line that its task was only "to determine whether a live, human-made micro-organism is patentable subject matter under 35 U.S.C. §101" (*Diamond v. Chakrabarty* 1980, 447 U.S. 305).

PBC's fears turned out to be partly justified. In spite of its focus on one of the most minute and micro forms of life (a bacterium), *Chakrabarty* was read within a few years to authorize the expansion of patent protection across the full domain of genetically altered life forms, including all nonhuman animals. In Genentech's conceptually groundbreaking brief, the living container for a patentable "composition of matter" (a gene, a plasmid) came to be seen as mere matter—analogous to an automobile engine needing to be coughed into life, or even to straw. Once that sleight of mind was accomplished, the same reasoning was smoothly extended by administrative decree to genetically altered higher animals, oysters, mice, and eventually larger mammals. All these met the test, in Genentech's words, that they were "called into being solely by the hands of man" (Supreme Court of the United States 1979a, 4), and in the Supreme Court's echoing language, were "not nature's handiwork," but the inventor's own (*Diamond v. Chakrabarty* 1980, 447 U.S. 310).

This was precisely the move that the five-member majority of the Canadian Supreme Court refused to make twenty-two years after *Chakrabarty*. That court's leavening of grammar with reason led to a very different metaphysical resolution than did the U.S. Supreme Court's seemingly unproblematic construction of the law.[16] The situation confronting the Canadian justices was, of course, crucially different from that in *Chakrabarty*. Canada had extended patent rights to microorganisms and fungi without debate or litigation. In the *Harvard College* case, however, the Canadian court had before it a mammal that could by no stretch of the imagination be likened to a mere composite of inert chemicals.

History mattered. By 2002, after Dolly and the cloning wars, some of the Canadian jurists clearly felt that the specter of the slippery slope was more real than it had seemed in 1980. In particular, the PBC's warnings about the march toward depersonalizing and commodifying human nature seemed to have more substance than it had two decades earlier. Allowing patents on higher organisms, the majority concluded, would create problems in a time when the boundary between animals and humans was becoming blurred through biomedical advances such as xenotransplantation. As Justice Bastarache wrote for the majority: "The pig receives human genes. The human receives pig organs. Where does the pig end and the human begin?" In such an environment, it was imperative for metaphysical lines to be redrawn and clarified through legislative action. "In my view," Bastarache observed, "it is not an appropriate function for the courts to create an exception from patentability for human life given that such an exception requires one to consider both what is human and which aspects of human life should be excluded" (*President and Fellows of Harvard College v. Canada* 2002, SCC 76, para. 181).

It was, however, in what the dissent dismissed as "murine metaphysics" that Bastarache's opinion most strikingly parted company from *Chakrabarty* (ibid. para. 45). For the Canadian dissenters, who in essence followed *Chakrabarty*'s logic, classifying the OncoMouse as a composition of matter was thoroughly unproblematic because every cell in its body had been changed through the addition of an oncogene: "The oncogene is everywhere in the genetically modified oncomouse, and it is this important modification that is said to give the oncomouse its commercial value" (ibid. paras. 68, 69, 96). By contrast, the majority was substantially less impressed by the inventor's degree of control over the whole mouse. To them, it was almost common sense that altering one small bit of a complex organism's genetic code does not produce an altogether different entity, a human invention that is no longer part of nature.

> Although some in society may hold the view that higher life forms are mere "composition[s] of matter," the phrase does not fit well with common understandings of human and animal life. Higher life forms are generally regarded as possessing qualities and characteristics that transcend the particular genetic material of which they are composed. A person whose genetic make-up is modified by radiation does not cease to be him or herself. Likewise, the same mouse would exist absent the injection of the oncogene into the fertilized egg cell; it simply would not be predisposed to cancer. *The fact that it has this predisposition to cancer that makes it valuable to humans does not mean that the mouse, along with other animal life forms, can be defined solely with reference to the genetic matter of which it is composed.* The fact that animal life forms have numerous unique qualities that transcend the particular matter of which they are

composed makes it difficult to conceptualize higher life forms as mere "composition[s] of matter." It is a phrase that seems inadequate as a description of a higher life form. (ibid. para. 163, emphasis added)

In this way, the Canadian court repudiated the logic that had, in the United States, so smoothly subordinated and made immaterial the container of the patentable genetic trait, regardless of whether that container was a mouse or a microorganism. Specificity and circulation were not enough, in the Canadian jurists' minds, to confer property rights over the mammalian matrix within which the inserted oncogene found expression.

The Public Interest

The two North American patent decisions also differed in their understandings of the public interest and the respective roles of legislatures and courts. Chief Justice Warren Burger, writing for the *Chakrabarty* majority, picked up on two themes that Genentech had highlighted and that resonated well with America's founding myth of progress through discovery. First, the court sustained the company's view that the patent law, as an instrument for furthering invention, should be given an expansive reading. Genentech had played on the theme of exploration and advancement—"The system seeks not to catalogue the past, but rather to compass the future" (Supreme Court of the United States 1979a, 4)—which it backed up with evidence from legislative history. Quoting congressional committee reports accompanying the law's 1952 reenactment, Genentech observed that "patent laws are written in large and prospective terms, so as to include 'anything under the sun that is made by man'" (ibid. 6). This formulation proved influential with the court. Burger in turn sustained the widest possible application of the law with a quotation from Thomas Jefferson—"Ingenuity should receive a liberal encouragement" (*Diamond v. Chakrabarty* 1980, 447 U.S. 308)—and he incorporated into his opinion the same bit of 1952 legislative history that Genentech had cited. From 1980 onward, the rubric "anything under the sun that is made by man" became identified with the subject-matter provision of the Patent Act, now ratified by the highest law of the land (ibid. 309).

The court's second theme, again echoing Genentech, was judicial deference to the will of Congress. The company's brief had asserted that it was up to the legislature, not the courts, to decide what to include in or exclude from "the broad compass of patentability" (Supreme Court of the United States 1979a, 7). The field was too complex for judicial evaluation, and the "surgical precision" of legislative policy discriminations was to be preferred to the "meat ax" approach of the petitioner, who was advocating the curtailment of advances across a vast field of technological development (ibid. 8). The Supreme Court agreed that restrictions on patenting, if there were to be any, had to come from Congress. Courts were institutionally incapable of making the right sorts of judgments. On the one hand, fears such as those voiced by the PBC (referred to only obliquely as "the [petitioner's] *amicus*") could be resolved only "after the kind of investigation, examination, and study that legislative bodies can provide and courts cannot" (*Diamond v. Chakrabarty* 1980, 447 U.S. 317). On the other hand, any attempt to put brakes on innovation would be fruitless anyway, because "legislative or judicial fiat will not deter the scientific mind from probing the unknown any more than Canute could command the tides" (ibid.). Even this shopworn metaphor of helplessness was not the Court's own. In its brief, Genentech had written that it was not for the Court "to attempt, like King Canute, to command the tide of technological development" (Supreme Court of the United States 1979a, 12).

For the *Chakrabarty* majority, then, all the relevant lines—legal, metaphysical, institutional—were bright lines, admitting no ambiguity. In particular, there was and could be no question whether an object for which a patent was sought existed in nature or was the work of human hands. Anything on the nonhuman side of the boundary deserved a patent in accordance with the broad purposes of the law. Limiting patentability—hence limiting the circulation of inventions—was the step that required justification, and courts moreover lacked the institutional capacity to carve out exceptions. If there were slippery slopes and special dangers inherent in patenting life, those were matters that only lawmakers could competently address.

The Canadian Supreme Court, too, decided that it was not in a position to make the sorts of policy judgments that the OncoMouse case called for, but the starting point for invoking judicial restraint was the reverse of that adopted in the United States. Unlike the U.S. court, which took the ontological line between natural and manmade as clearcut, the Canadian justices saw scientific and technological practices as blurring important boundary lines: between human invention and nature's handiwork, especially with regard to gene-altered complex organisms; between human and nonhuman organisms and entities; and between permissible innovation and protection of a valued status quo. Accordingly, the court viewed the patenting of higher life forms as "a highly contentious and complex matter that raises serious practical, ethical and environmental concerns that the Act does not contemplate" (*President and Fellows of Harvard College v. Canada* 2002, SCC 76, para. 155). The justices refused to undertake such a "dramatic expansion of the traditional patent regime" (ibid.); instead, they held that higher life forms did not constitute a "manufacture" or "composition of matter" within the meaning of the term *invention* in the Patent Act.

The principles of judicial construction and deference were essentially the same for both courts: it was their application that differed. In both common law cultures, it was concededly the legislature's job to make policy through law. Courts could determine whether or not a particular question fell within the law's intended purview, but they could not redraw the legislated lines to fit new facts in the world. The point of divergence between *Chakrabarty* and *Harvard College* lay in their treatment of the ontological challenges posed by biotechnology. The U.S. court saw life patents as unproblematic in the light of an analysis that stressed the manufactured character of bioengineered entities and the inevitability of technological advances. Applying the classic "but for" test, the Supreme Court concluded that the bacterium would not have existed but for Chakrabarty's ingenuity. The Canadian court, faced with a mouse rather than a microorganism, trained its analytic sights on an altogether different question. That human agency had endowed a commercially valueless animal with high economic value did not, in that court's judgment, turn a mouse into a "composition of matter." It merely raised difficult questions about where this entity, and others like it, stood in relation to the objectives of the Patent Act, and to human values more broadly. In the public's interest, *that* complexity had to be resolved by Parliament, not the judiciary.

Conclusion

Patent law is often described as a highly technical area of legal practice, mainly oriented to resolving questions of priority (who is the first mover), as well as whether the claimed invention meets the tests of novelty, non-obviousness, and utility. It is seen as a site of technical assessment and narrow legal construction, not for high matters of ethics, politics, or philosophy. For U.S. patent law in particular, as the rhetoric of litigants and of courts constantly reminds us, the most important policy choice was that written into the Constitution and the first Patent Act: "Ingenuity should receive a liberal encouragement." Patenting is only a means to an end, namely, that the nation should continue to renew itself through invention, and so remain true to its founding imaginary.

That reading, when applied to claims relating to novel biological constructs, has led to a systematic favoring in U.S. law of moves that isolate specific bits of nature and put them into economic circulation. In this respect the law of life patents is logically consistent with the law of "takings" as it relates to real property. Whether nature resides in inanimate land or in living things, what the law rewards is the act of economic agency that takes something that was fixed, embedded, and immovable and makes it specific, dynamic, and commercially value-laden. In short, *lively*. Like precious metals mined from a mother lode, genes and other biological constructs extracted from living nature can be patented by anyone ingenious enough to detach them from their physical surroundings and make them useful in commerce or industry. That logic of the market, grounded in Lockean notions of value creation through labor, helps to explain why John Moore, whose diseased body produced cells of potential therapeutic value, could not patent his own tissues, but why such a patent was granted to the physicians and the company that manufactured from his spleen a cell line for use in biomedical research and development. It explains why William Catalona could not act as an agent on behalf of his sample-donating patients. It also explains why Homan McFarling, the farmer who wanted the right to replant Monsanto's genetically altered soybean, failed in his bid to restrict the company's rights in the second-generation seed that contained the manipulated genetic trait. Backtracking a century, it may also help explain why Chief Justice Roger Taney, confronted by the ambiguous ontology of the slave body, opted for the legal reading that would allow it to circulate freely without losing value.

Even market logic, however, has metaphysical ramifications when it is applied to "private takings"—taking things out of public nature for private gain. As *Chakrabarty* and *Harvard College* dramatically illustrate, granting patents on living things involved decisions about where to draw the line between life and matter. Silently and with little fanfare, the U.S. Supreme Court decided in *Chakrabarty* that any circulating commodity containing within it a part altered by human invention is eligible for a patent. The court's interpretation of invention seamlessly carried over to the patenting of higher animals, blurring the line between living genetically engineered cows or soybean seeds and mechanical contraptions such as cars with newly designed engines.

The Canadian Supreme Court's refusal to classify the OncoMouse as a "composition of matter," and hence to treat it as patentable, drew a line that U.S. patent law does not recognize between patentable microorganisms and nonpatentable higher life forms. But it also put back on the agenda of public debate previously glossed-over questions about the degree of intervention, innovation, and control that must be shown in order to move something across the boundary from nature to culture, or from life to capital. That question is unlikely to recede as life and capital come to be bonded together in ever more intricate assemblages. In the lively, global landscape of intellectual-property law, it remains to be seen whether the division between the aggressively materialist commodity culture of the United States and Canada's more cautious respect for the ungovernable complexity of life can long endure.

Notes

1. *Palazzolo v. Rhode Island et al.* 2001, ruling that acquisition of land after an environmental regulation's effective date does not bar a takings claim, since otherwise future generations could not assert a right that the present generation was unable to assert through lack of will or resources (533 U.S. 606).

2. On theories of valuation, see Graeber 2001; Singer 2000. Singer argues, consistently with the broad argument of this chapter, that the ownership model of property rights is deficient because it includes no element of obligation.

3. Even philosophers of science who do not embrace a radical skepticism about the reality of nature accept the contingency of specific representations. See Kitcher 2001; Hacking 1999.

4. See particularly Haraway 1991, "A Cyborg Manifesto: Science, Technology, and Socialist-Feminism in the Late Twentieth Century," 149–81. Also see Latour 1993. Unlike Haraway, Latour has not written specifically about the socially constitutive properties of the life sciences and technologies, but much of his work on hybridization and purification of material objects bears importantly on our understanding of the contemporary metaphysics of living things.

5. For a classic exposition of actor-network theory, including a definition of the concept of translation, see Callon 1986.

6. One can see this practice as similar in some ways to the German ideal of *Nachvollziehbarkeit* (follow-through-ability) that Stefan Sperling describes in "Science and Conscience: Bioethics, Stem Cells and Citizenship in Germany" (2006). Only in the U.S. case what seems to matter is the fact of a debate and not the transparency of the reasoned argument itself.

7. For a detailed account of the contorted politics of the *Dred Scott* decision, see McPherson 1990, 170–81.

8. Abraham Lincoln, speech before the Republican State Convention, Springfield, Illinois, 16 June 1858.

9. 35 USC Sec. 101.

10. 45 USC Sec.103.

11. 35 USC 112.

12. On the concerns expressed about Golden Rice in relation to food security and public health, see Jasanoff 2005b.

13. See Boyle 1996, 97–107; Jasanoff 2005a, 213–24.

14. Donald J. Quigg, Assistant Secretary and Commissioner of Patents and Trademarks, "Animals — Patentability," 7 April 1987, available at U.S. PTO, Consolidated Notices, 3 December 2008, 1337 CNOG 487, http://www.uspto.gov/web/patents/patog/week53/OG/TOCCN/item-115.htm#cli115, accessed 23 June 2011.

15. Patent Act, R.S., c. P-4, s. 1.

16. Of course, the Canadian court too saw its task as a matter of statutory construction rather than policymaking. Both the majority and the dissent agreed that the words of the Patent Act should be read "in their grammatical and ordinary sense" (*President and Fellows of Harvard College v. Canada* 2002, SCC 76, paras. 8, 11, 154, 155).

5

ELTA SMITH

RICE GENOMES

Making Hybrid Properties

The field of genomics is one of the busiest sites of research and debate at the nexus of technoscience and policy, and also the most upstream. The work of genomics involves developing scientific knowledge and technologies to produce new information about organisms, the implications of which stretch from basic research to improved pharmaceuticals and agricultural products. In this emerging arena, intellectual-property rights in genes and genome-related information are highly contested. For example, in early 2000 two world leaders, President Bill Clinton and Prime Minister Tony Blair, announced the completion of genome maps for the human species. Compared in importance with the moon landing, and called the scientific breakthrough of the twentieth century, the sequenced genomes became the focus of intense media attention and public speculation. At the heart of the announcement by Clinton and Blair was a call to private genomics companies to make their information public, leading many to speculate whether they intended to change the policy landscape for gene patents (Abate 2000). The Clinton-Blair announcement, following similar announcements by the French and Japanese governments, also incited the sell-off of biotech stocks and initiated debates among business, legal, and political leaders around the world over the direction in which this breakthrough would take humankind.

When two draft rice genomes, one generated by a public sequencing consortium and the second by the private biotechnology firm Syngenta, were published in the journal *Science*, in April 2002, they were hailed as the most exciting genomic efforts since the completion of the Human Genome Project, announced only two years earlier.[1] The rice-genome projects were described as a solution to food-security concerns in the Third World and as key to mapping more commercially valuable but more genetically complex cereal grains. The publication of the genomes was a controversial move for *Science*. The editorial on 5 April answered charges that *Science* was violating the norms of scientific publication and potentially threatening prospects for humanitarian aid by allowing Syngenta to publish its sequence without releasing its genome information to the public domain, as the journal typically requires. The *Science* publication followed five years of mapping and sequencing efforts by four different groups, two publicly and two privately funded. While publication made the property debates surrounding these projects public for the first time, these debates were simply the culmination of long-standing and ongoing negotiations over the property rights that would emerge with the new scientific information. From the time an international consortium of scientists began the first mapping effort, in 1997 — followed in these efforts by a private biotechnology firm, in 1998 — new conceptions of property rights arose alongside the generation of genomic data.

In articles and websites discussing the four projects, genome maps were debated and described by many actors as necessarily public. These arguments contrast with the private patent claims that could be and were made on individual genes and with the trade-secret status of private biotechnology firms' genome projects. In genome research, intellectual-property protection takes the form of trade secrets and patents. Trade-secret laws protect commercially valuable knowledge so long as that knowledge remains secret. Trade-secret status is both more and less protective than a patent: it is more restrictive than a patent for information that remains completely confidential, but information that becomes available either legally or through guesswork is no longer protected. Patents allocate de facto monopoly power to a group or individual "creator" to prevent unauthorized use of their inventions for a limited period of twenty years. This includes sale, distribution, and use without license or other agreement. In exchange, inventors must release information that would allow others to replicate their creation. To obtain patent rights, the invention must meet three requirements: novelty, inventiveness, and utility in the United States or industrial applicability in the United Kingdom.

Debates over property rights permeate every discussion of the scope and import of new biotechnologies. The imagined benefits include drug development, therapeutics, increased agricultural production, and nutritional improvement. But genomics is a set of upstream technoscientific practices, producing large volumes of information with as yet little practical significance, and with marketable products in most cases at least several years down the road. As can be inferred from the human- and rice-genome examples, this is a highly contested terrain for property rights at all of the levels of science and technology policymaking. Central to these debates are the expansion of markets—biotechnology companies in particular—into spaces previously occupied largely by state-sponsored or nonprofit research alone, and a shift toward increased property claims that has led many to question to what extent it has become possible to own "life itself" (Jasanoff, chapter 4, this volume; Juma 1989).

The mapping and sequencing of rice genomes provides an interesting set of cases for exploring the development of global governance through intellectual-property rights. The recent effort to map and sequence the rice genome not only illustrates the production of new scientific information, but also the simultaneous constitution of new intellectual-property regimes that do not (always) reflect current legal notions of property rights. I analyze these debates through interviews with genome scientists, news-media accounts, scientific-journal articles, and online exchanges among the different groups working on the genome projects. These debates make visible not only the dynamics of property debates and the emergence of new forms of property rights, but also the fact that such rights are negotiated all the way from upstream science to downstream agricultural development, raising questions of ownership and accountability across the entire spectrum.

Hybrid Properties

Debates concerning the usefulness and viability of intellectual-property rights usually center on an all-or-nothing proposition: either private rights can be claimed, in which case a specific owner, usually one person or entity, is designated, or ownership accrues to the public at large, in which case the information becomes a good inheritable by everyone—though who exactly that "everyone" is remains an open question. But classifications that rest on only two possibilities are of little use when the property rights in question are not clearly all of one kind or the other. The binary does not "logically exhaust all the possible solutions" (Carol Rose 1986; also Aoki 1998). Instead, aspects of information can be both public and private, free and owned, and thus are too complex to be fitted into the simple binary of private or public.

The emergence of intermediate or hybrid categories that do not conform to the well-established binaries of scientific objects (nature-culture) and legal reasoning (public-private) has become a focus in the fields of anthropology, critical legal theory, and science and technology studies.[2] The properties that emerge in the rice-genome projects do not easily fall into the private or public domain (also often called a commons) as traditionally defined by law. Rather, they are a mixture of properties, multilayered in conception, availability, control, and reach. These "bundles of rights" that range between fully protected private property and a completely accessible public domain, I term *hybrid properties*. At one end of the spectrum sit purely private property rights, as produced and defined in the legal system; at the other end is the public domain, where there are no owners and no formal protections. In the middle lies a range of hybrid properties: sets of freedoms and restrictions that are constructed in an ad hoc fashion in rice-genome research, in conjunction with the production of genome data.

The critical legal theorist Carol Rose (1986) argues that there are particular types of property—"inherently public property," in her term—that are neither individually owned nor controlled by the state. Rose focuses on landed property such as waterways and roads, but her more relevant point for the genomics case is that some types of property arise and become stabilized through custom instead of through law. And, as the science and technology studies scholar Stephen Hilgartner (2004) importantly reminds us, we do not have to look to national or international governmental institutions to see the emergence of property regimes.

"Hybrid properties" in genomic information represent a set of social classifications that have developed alongside the production of new scientific knowledge. In sharp contrast to traditional conceptions of the relations between science and law, where knowledge generation is seen as separate from rule making, my account suggests that representations of the genome come into being with tacit property regimes attached to them. Thus, property rights for biotechnology are emerging, not only in formal, top-down institutional processes, such as patent suits, regulatory frameworks, or multilateral treaties and trade agreements, but also at diverse sites in the everyday practices of genome research.

Creative Commons

A simple way to illustrate hybrid properties is through an example from copyright law and practice. In 2001 the intellectual-property lawyer, professor, and activist Lawrence Lessig, along with a group of collaborators at Duke, Harvard, the Massachusetts Institute of Technology, and Villanova, developed a nonprofit institution called Creative Commons (see "All Hail Creative Commons" 2002). Creative Commons (CC) works within copyright law to ensure both free and open access to informational public goods and the ability to make ownership claims to creative works through the intellectual-property system. To better understand what Creative Commons sets out to do, I examine both what a *commons* is and what *creative* aspects are protected in this new regime.

Commons

A commons is generally seen as the opposite of private property, an open space that belongs to all rather than a restricted space from which others can be excluded through socially instituted and legally defined rights and protections belonging to an individual or a group. Intellectual-property rights grant restricted access by law to those who make new knowledge; these rights of exclusion are thought to benefit society by striking a balance between increased incentives to innovate and greater market efficiency. That balance is effected in two main ways: First, through different types of property rights. The logic behind patents, for example, is to release the trade-secret status regarding information about an invention to the public so that others can build on the work, while allowing the inventor to charge royalties on the use of their invention, thus creating the incentive to innovate. Second, limits on the length of time property rights can be held—seventy years after the author's death for copyright, twenty years after an application is granted for patents—are meant to ensure that all creative works are eventually available for public use without restrictions. The argument goes that society benefits from innovations, but innovations occur only when the investment of labor and capital produces social utility.

Not every novel thing or thought, however, is subject to intellectual-property claims. If a private-property right means total control, or fully restricted access to those unauthorized, then the alternative to, or the "outside" of, the intellectual-property system is the public domain or commons with no ownership and no restrictions.[3] The terms *public domain* and *com-*

mons, often used synonymously, are poorly defined and contested in legal theory and practice (Hess and Ostrom 2003). Existing formulations occupy a counterposition to, and are meaningful only when juxtaposed against, private property. *Public domain*, or *commons*, often refers to whatever is unprotected by private intellectual-property rights, "either as a whole or in a particular context, and is 'free' for all to use" (Boyle 2003, 30).

Commons, as used by intellectual-property analysts, is a term popularized by Garrett Hardin's article "The Tragedy of the Commons" (1968). Hardin imagined the commons as an open pasture where herdsmen compete for grazing space for their cattle. A commons for him was a rivalrous resource, meaning that several or many individuals or groups compete for the use of that resource, which moreover is not infinitely renewable. The tragedy occurs because everybody gains from using the resource and nobody has an incentive to restrict his or her use of it. Consequently, the resource is overused and fails to "replenish" itself. Typical examples include fish stocks in the open ocean, or timber forests. Private-property rights or government ownership on behalf of the public are the usual solution to the commons problem, but some commons are not amenable to easy parceling or are not beneficially "owned" by any one individual or group. The political scientist Elinor Ostrom, and others working within her school of public-choice theory, have argued that common-pool resources can be organized and managed by the likely users, creating public goods instead of negative externalities.[4]

The "tragedy" of the commons takes a different form in the case of information. The problem of overuse is averted because intellectual products are nonrivalrous; their use by one person does not exclude or otherwise interfere with others' use of that same information. Images, music, and text are examples of nonrivalrous goods. But it is more costly to exclude people from using information than from using more tangible commodities such as fish or timber. In the face of high costs and nonrivalrous goods, property rights are thought to be a necessary incentive to promote innovation (for example, Landes and Posner 2003).

Others, however, have argued just the opposite: that many informational goods *must* be freely available without restrictions because they provide the materials for future innovation (for example, Boyle 2003; Lessig 2001). Moreover, proponents of public access say, these goods will still be produced even without the property right. Evidence that information can freely circulate and result in innovative products can be found in the open-source software movement and to a lesser extent in open-access publication debates (Bollier 2002; Vaidhyanathan 2003 [2001]).[5] In both cases incentives to innovation can be extra-economic, with prestige and accreditation as two of the values that may function in this manner (see Strathern and Hirsch 2004).

Both copyright, which automatically assigns rights, and patents create an additional dilemma, for there is no way for future users of a work to know whether, and to what extent, the rights holder wishes to enforce their property claims.[6] While the public domain or commons might be thought of as the counterpart or even precursor to private property, in current practice "public" entitlements must often be carved out of a space that is first designated "private."[7] Intellectual-property laws as currently conceived shift the claim to ownership from a state in which new rights must be asserted to one in which the claims are in place at the moment of "creation" and can be defended against unauthorized appropriation. Ultimately, intellectual-property laws shift the terrain of property rights so that controlled access takes precedence over free use—with the result that access must be carved out of a realm of controlling ownership instead of the other way around.

Creativity in Creative Commons

Creative Commons is a nonprofit organization that works within intellectual-property laws, but outside legal and legislative institutions, to create a formalized system of public property that complements rather than competes with copyright laws (Lessig 2004). While the developers of cc see value in the current intellectual-property system, they also persuasively argue that greater flexibility would make the innovation system function better. Creative Commons offers creators a chance to specify how they want others to use and build on their work, indicating what is free and what is subject to control. The "spectrum of rights" graphic on the cc website locates Creative Commons squarely between full copyright protection and the public domain (see 1). However, the potential for mixing and matching the desire of a "creator" to allow free use by others and still retain some formal protections is much more flexible in the middle zone. The website explains.

> Too often the debate over creative control tends to the extremes. At one pole is a vision of total control—a world in which every last use of a work is regulated and in which "all rights reserved" (and then some) is the norm. At the other end is a vision of anarchy—a world in which creators enjoy a wide range of freedom but are left vulnerable to exploitation. Balance, compromise, and moderation—once the driving forces of

1. Spectrum of property rights. From the Creative
Commons website, http://www.creativecommons.org.

a copyright system that valued innovation and protection equally—have
become endangered species.

> Creative Commons is working to revive them. We use private rights
> to create public goods: creative works set free for certain uses. (Creative
> Commons 2010, accessed 22 June 2011)

More specifically, cc provides four formal mechanisms (licenses) for demonstrating to others how a new work can be used: attribution, noncommercial, no derivative works, and "share alike." A cc license could include any or
all of the specified conditions that help to make known the explicit wishes of
the copyright holder, given that intellectual-property rights in this area grant
protections against reproduction and sale, import and export of a work, the
creation of derivative works, public performance or display of the work,
and the assignment of these rights to others.[8] Without cc, the copyright
holder can exercise any or all of these rights, while others are prohibited
from doing so unless they obtain consent from the copyright holder. There
are many potential combinations for cc licenses. For example, creators of
music, movies, images, and text can obtain a license that requires attribution to a work, but stipulates no payment and allows derivative works. Or
they can specify that future uses of their work are subject to noncommercial-
ization and that those who use it must not change it in any way (no derivative works).

A "science commons," adhering to the same principles of "some rights
reserved" as for copyright, was introduced by the cc developers in early
2005. Science Commons (sc) focuses on three frequent areas of contestation: scientific publication, material-transfer licenses, and databases. The
debate over scientific publication centers on how to make scholarly work accessible to larger numbers of people, while balancing that goal against the
high costs of production and publication. Science Commons aims to make
copyright negotiation easier in scientific publishing through standardized
licenses for individual works and archiving projects.

Supplementing intellectual-property rules are material-transfer agreements (MTAS), which are contracts between two parties regarding the use

of a tangible research material, usually biological (such as genes and model animals), though chemical and even software transfers are effected through MTAS. Different MTAS cover exchanges between universities or other publicly funded research centers, between industry and a university, or between any combination of these and other research centers, such as hospitals. MTAS can range from extremely restrictive (both the original materials and all derivatives from those materials are subject to intellectual-property claims) to fairly relaxed (if patent claims are not in play, a standard agreement typically stipulates that the materials must be used for nonprofit or teaching purposes only and cannot be distributed without the written consent of the provider). Clearly, even in the least restrictive of situations, MTAS have property claims attached to them, and simply sharing materials without such an agreement is risky business for anything that has commercial potential. Science Commons is working to standardize MTAS, though some standardization is already in place—particularly through the National Institutes of Health (NIH)—and to expedite the acceptance of MTAS through negotiations with universities, industry, and other research centers. MTAS may stipulate the conditions under which patents will be enforced, whether sharing is allowed (or not), and who owns what once the materials exchange hands.

Finally, databases raise a third set of issues for property in scientific knowledge. Though they can be protected by copyright if they demonstrate creativity or originality in their arrangement, most databases of scientific information are not protected by intellectual-property rights in the United States or under international law. The United Kingdom enacted a database law in 1998 that covers any database involving substantial financial investment (Data Protection Act 1998). In addition to what the Science Commons developers see as a further encroachment by intellectual-property laws into previously unregulated domains, they also worry that in current practice, databases fall into the all-or-nothing category of either wholly confidential or publicly available information, in the latter case with no rights reserved by the compiler.

The developers of Creative Commons and Science Commons recognize that social systems can be tools that complement the law to produce desired ends (Kelty 2004a; Kelty 2004b). Both the CC and SC projects rely on private voluntary agreements with the holders and users of property rights, and enforcement comes more from social norms than from legal coercion. While this approach serves a functional goal of expanding a narrow realm of property conceptualization and practice, the dynamic and situated character of property-rights regimes is not captured by either regime (Kelty 2004a).

Both cc and sc open up multiple property arrangements, but they do not reflect the iterative nature of hybrid properties as they are allocated in practice—in a simultaneous, nonlinear process of making and defending rights. In fact, most critical accounts of intellectual property begin at the end, so to speak, arguing that existing rule regimes are too restrictive, not restrictive enough, dealing with the wrong problems, not targeted at the right people, or some combination of these.[9] The "hard question" for these analysts is deciding what can be covered by intellectual-property rights and what should be part of the commons (Boyle 2003). In rice genomics, the cultural norms of science conflict with the terms of intellectual-property laws as suggested by the sc model, but these competing ideals have generated reorderings of both systems. Scientific and other norms get mixed together with those of intellectual-property rights to produce a simultaneous development of new knowledge and new property claims.

Property in Rice Genomes

Creative Commons and, to a lesser extent, Science Commons illustrate how a formally recognized regime for hybrid properties can mediate and facilitate the production and use of new knowledge. One might suspect that these are not the only arenas in which hybrid properties arise, and indeed exactly that is happening around the development of rice genomes. I now turn to that story, picking up where Science Commons has left off, as a direct route into understanding how technoscientific imaginations shape the way genomic properties, both public and private, are being constructed and deployed. Property rights for biotechnology are emerging in everyday practices at diverse sites of genomic research. Compared with Creative Commons, the hybrids one finds in these sites of practice are more subtle and complicated, and they introduce further questions about the way ownership is made visible and thus opened to contestation.

Genome mapping was once a time-consuming and difficult enterprise. Until the early 1990s, a graduate student might have written her entire dissertation on the mapping of a single chromosome. The advent of faster automated-sequencing technologies radically changed the terrain of genomics.[10] Now, the sequence of an entire species can be accomplished in only a couple of years. The last ten years have seen the completion of genome maps for fruit flies, model plants, mice, rice, and human beings among other organisms.[11] Four different sequencing efforts for rice were initiated by early 2000, two of which were publicly funded while two others were based in private firms. To date, all four groups have produced drafts of the rice genome, and one of the public projects has produced complete sequences for all twelve rice chromosomes.

Along with the rapidly growing genome data, property rights are accumulating and being redefined in the work of genome researchers. Debates over private property and the public domain are critically important not only in the area of copyright, but also with regard to patents and trade secrets, and to the fate of databases and access to scientific information.[12] Five types of rights were asserted or negotiated for each rice genome project: trade secrets, patents, database release, data "sharing," and publication. Each site of joint constitution—of genomic information and property rights—can be explored along four lines of variation: who has access, to what, with what restrictions, and subject to whose definition of those restrictions.

The institutional and epistemological orderings and reorderings that occur in the creation of rice genomes and property rights correspond to rice in its multiple guises: as scientific information, as a model cereal, as a major food staple, as a cultural icon, and so forth. The hybrid properties that emerge in the rice-genome projects are built on particular representations of genomes and of the "publics" that are imagined to benefit from genome research. These representations arise in social contexts, through the process of producing, disseminating, and controlling new information. How genomes and their imagined benefiting publics are positioned relies on the cultural, political, and historical contexts in which both are situated.

Biotechnology Companies: Monsanto and Syngenta

Two corporations produced drafts of the rice genome using two different techniques, constituting two different genomes and different but similar hybrid properties. Monsanto, the U.S. biotechnology giant, began its sequencing effort in 1998, funding Leroy Hood's research team at the University of Washington to produce a draft sequence using a "traditional" approach of contiguous mapping in clone libraries.[13] This technique relies on mapping the genome from end to end and expects to meet a very stringent error rate. Their efforts identified approximately 95 percent of the genes, but the map they produced did not contain enough information to count as a complete sequence (Bennetzen 2002).

Syngenta also began its sequencing project in the late 1990s. Though the firm is based in Switzerland, the genome project was run out of the Torrey

Mesa Research Institute (TMRI) in Southern California, in collaboration with Myriad Genetics, a U.S.-based biopharmaceutical company, and the Clemson University Genomics Institute (CUGI), which is funded by Novartis. Syngenta used a new sequencing method, the so-called whole genome shotgun strategy. This technique breaks the genome into small DNA segments, sequences those pieces many times over to reduce gaps and errors, and then realigns the now-sequenced fragments in order using automated supercomputers. A draft was completed in early 2001, identifying 99 percent of the genes at 99.8 percent accuracy (International Rice Genome Sequencing Project 2008; Butler and Pockley 2000).

Both companies present themselves as market-driven enterprises, but also as benevolent stewards of the public interest (see Grefe and Linsky 1995). Monsanto's homepage asks the viewer to "imagine" a world in which small-scale farms in Africa produce abundant cotton, young Asian children do not go blind from vitamin-A deficiency, and a host of environmental problems such as soil erosion, energy consumption, and pesticide use find innovative agricultural solutions (Monsanto 2010). Similarly, Syngenta provides an overview of its dedication to social responsibility through sustainable agriculture, stakeholder engagement, and product stewardship (Syngenta 2011). The websites of both companies present visions of abundance in the face of scarcity and relief for those in need.

A quick look behind each homepage, however, shows the drive behind the vision. Research is focused predominantly on corn and soybeans, and products are targeted toward these crops. There are no commercial rice products and no ongoing rice-genome research projects. The message is clear to the critical reader: although rice itself is no longer a major biotech investment for Monsanto or Syngenta, its genetic structure is similar to that of more profitable cereals such as wheat and maize (Monsanto 2010; Syngenta 2011). As in many scientific research efforts, potential profitability and research investment go hand in hand. Nearly twenty years of independently initiated projects to map the rice genome in China, Japan, and the United States passed with little success, until the early 1990s, when researchers began to realize that rice could facilitate knowledge about more profitable cereals (Normile and Pennisi 2002). This tension between profit-making and philanthropy permeates the corporate process of sequencing rice and deciding how the resulting information is used and held.

A complex set of hybrid property rights was produced through the corporate genome-sequencing projects. First, private information was made more public in at least two ways: through data-sharing agreements and publication. Monsanto announced the completion of their genome draft in 2000, but the results were not published and the data were not placed in a public database. However, Monsanto negotiated with another sequencing effort, the International Rice Genome Sequencing Project (IRGSP), to provide that public consortium with Monsanto's data to facilitate their efforts. This decision released the trade-secret status of the information and at the time was a largely unprecedented collaboration between a private company and public research effort (Bennetzen 2002; Pennisi 2000). A similar agreement was reached between Syngenta and the IRGSP in 2002. Publication represents another layer of public information that was carved out of private property. Syngenta published its genome results in *Science* shortly before it agreed to share its data with the IRGSP. Publication provides additional layers of access to new information, with dissemination to a wider audience than the IRGSP.

Property rights have advanced into realms that were unimaginable even fifteen years ago.[14] Patents can be issued for a gene itself, for plants transformed with the gene, and for the seed and progeny of the patented plants (Barton and Berger 2001). The distinction between "nature" and "innovation" has shifted many times, notably through the U.S. Plant Patent Act of 1930, which authorized patenting of asexually reproduced plants, and the Plant Variety Protection Act of 1970, which extended protection to sexually reproduced plant varieties. In 1980, the U.S. Supreme Court, in *Diamond v. Chakrabarty*, approved patent protection for living organisms, on the ground that the Patent Act covered "anything under the sun that is made by man." That decision was broad enough to include biological organisms, traits, and genes. While patents may be sought for genes that are "isolated and purified" from the chromosome, they may not extend to anything "as it exists in nature." John Doll (2001), former director of biotechnology for the U.S. Patent and Trademark Office, estimates that since 1980 more than 20,000 patents have been granted for genes or gene sequences and that another 25,000 are waiting in the queue.

While no genome in its entirety has been patented, as of 2004 the U.S. Department of Energy cites the number of genome-related patents filed at more than three million. Seventy-four percent of agriculture patents are held by private companies, 40 percent of which are concentrated in just five companies (University of California 2003); Monsanto and Syngenta are two of the top three agriculturally related patent-holders worldwide. With the passage of the Bayh-Dole Act, research conducted at universities and other federally funded institutions has also become patentable. Holding enough key patents on specific genes might amount to ownership of the entire genome if the "portfolio" of patents includes enough of the most important genes.

Creating public knowledge requires information release. Accessing that information, however, is another matter. For example, the companies' agreements to share raw genome data with the IRGSP stipulated that the consortium could not designate in their public databases which data were produced by the IRGSP and which were donated by the companies. Since companies can still patent genes and gene sequences, and readily admit that they did not release their genome data until they applied for important patents, determining what information mined from the IRGSP database may be subject to patent protection is difficult, if not nearly impossible.

Additionally, trade-secret release might be considered analogous to placing information in the public domain (Kennedy 2002), but at the time of the *Science* publication, a spokesperson for Syngenta suggested something altogether different.

> Our data is publicly available. To the IRGSP or any other investigators around the world. It's just not in the public domain. Think of it like a book or a movie. It's available to you, you can get the book, you can watch the movie; but it isn't in the public domain, you've got to go pay for it. Somebody owns it, and provides access to it. But we're not charging people for access to it for non-commercial uses. So to academics and so forth it's available without charge. But what we require is that if a commercial invention is made from the collaboration, that Syngenta has an option to consider a license for it. (Walgate 2001)

When Syngenta published its results in *Science*, the company did not deposit the DNA sequences on which the publication was based in GenBank or any other publicly accessible database. Such deposits are important to understanding the publication results and are crucial for replication. In order to use the publication results, without the aid of deposits, interested parties must go directly to the company to see the genetic sequences and must pay for their use for commercial purposes. Access through publication in this case is contingent: commercial interests are subject to rules different from academic research, which must deposit the materials or information underlying a publication in order to appear in *Science*. Ultimately, what may seem most public, such as data-sharing agreements, databases, and publications, can still be kept private through patents and licensing agreements.

An International Genome: IRGSP

The International Rice Genome Sequencing Project, a consortium of scientists dedicated to sequencing the rice genome, began the first mapping initiative in 1997.[15] Representing a new form of collaborative effort, IRGSP consists of scientists from publicly funded universities and research institutes in ten different countries: Japan, the United States, Brazil, China, France, India, Korea, Taiwan, Thailand, and the United Kingdom. From its inception, the collaborative group has been committed to completing the rice genome instead of producing only a draft. Using the same approach of contiguous mapping in clone libraries employed by Monsanto, research groups in each of the ten different countries agreed to sequence particular chromosome segments (Bennetzen 2002).[16] Following the Bermuda Principles concerning human-genome data and intellectual property, the IRGSP seeks 99.9 percent accuracy in its genome map, which means no more than one error in every ten thousand base pairs sequenced, despite the slow and costly process.[17]

IRGSP represents all of the values of rice, and thus the import of rice genomes for all of the purposes of rice research and cultivation. For example, united by a genome, the research groups participating in the consortium portray themselves as a global research body dedicated to a "global" object—rice. Simultaneously, however, national identities are also mapped into the imagined genome: the globally produced genome is chromosomally country-specific. For example, Taiwanese researchers sequenced chromosome 4, while another team in France worked on chromosome 12. Many of the chromosome-sequencing choices were based on particular national interests regarding the functions expressed on a particular chromosome. Robin Buell, the lead investigator for the rice genome project at the Institute for Genomic Research (TIGR), explained that this genome project is different from others in that it assigns different social, cultural, and economic meanings to rice and thus to the sequencing effort, among different national projects.[18] For example, she suggested that TIGR and other U.S.-based consortium members view rice as a model cereal for understanding the genomes of grains such as maize and barley—much as the biotechnology companies do. Thailand, however, received funding from the Royal Thai government because of the cultural importance of rice and the possibility of catching up in science by participating with more technologically and scientifically advanced countries in the international effort.

Environmental representations of rice are also wrapped up with those of rice as a marketable good. A newsletter from the IRGSP website notes that many consumer concerns are directly tied to environmental concerns: "On the one hand, with a larger and more affluent population there will be greater demands for higher production and better quality rice. On the other hand, the same constraints mean that there will be less land, water, and labor to produce the crop. In short, there will be great demands on biotechnology to improve rice production" (Burr 1997). The idea that genomics will lead to new agricultural products that can overcome ecological stressors better than, for example, their Green Revolution counterparts produces the environment as a projected beneficiary to the rice-genome projects as well.

Through its self-identification as an international consortium working on an international genome, the IRGSP calls for international access to its genome data. From its inception, the IRGSP decided that its mapping endeavor would benefit from full disclosure to anybody and everybody who wanted access. Again following the Bermuda Principles, the IRGSP concluded that all genome-sequencing data should be "freely available and in the public domain" to encourage research and ensure societal benefit (Burr and Sasaki 2002). Consortium participants must agree to "share materials, including libraries, and to the timely release to public databases of physical mapping information and annotated DNA sequences" (ibid.). The IRGSP guidelines also stress that individualistic, competitive efforts to map the rice genome waste both effort and precious resources.

Thus counter to the fundamental reasoning for private intellectual-property rights, the IRGSP members argue that unrestricted *public* access is important and necessary for scientific inventiveness and flourishing. Consortium participation rests on collaboration and timely placement in public repositories. New mapping results are first shared with the entire network of participating laboratories, through their own data-banking system, which is not publicly available outside the consortium, and then with the scientific community at large, through placement in the DNA Databank of Japan (DDBJ), the database of the European Molecular Biology Organization (EMBO), and the NIH-led GenBank in the United States.[19]

Constructing its own database, with its own rules and protocols, the IRGSP instituted a new method for collecting, storing, and sharing genome data. Anyone with Internet access can download the sequence data via any of the three public databases (DDBJ, EMBO, or GenBank) although there is no guarantee that the data can be used without restriction because individual genes and gene sequences may be patented.[20] The IRGSP's requirement that consortium data must be deposited in a publicly accessible genome bank was an attempt to situate information among a set of actors with a set of rules and practices that defined the boundaries of genome-related property (Hilgartner 2004; also Jasanoff 2004).

But the multiple notions of public at play in the IRGSP project do not necessarily match up with the publics that can access the genome data, and the layers of ownership possible for genome data means that while the genome level may be relatively accessible, controls on access can be maintained at other levels. For example, while IRGSP members release their data to "public" databases, the IRGSP's own database is not accessible outside the consortium. Moreover, it is unclear whether consortium members have taken patents on individual genes. A newsletter on the IRGSP website, dating back to the inception of the consortium, notes that while withholding sequence information for patenting is incompatible with consortium rules, there is no rule that says patents could not be obtained "downstream of data generation and release" (Burr 1999).

Additionally, any information in the three public databases (e.g., GenBank) may be patented as well. While the information can be used for research purposes, there is always the possibility that the information is subject to intellectual-property protection at the gene level or via reach-through agreements, and the database user is responsible for determining whether the information is patented.[21] The slippage is not unimportant. It indicates a multivalent and shifting conception of what can or should be public, and a lack of legal definition to stabilize the terrain. Thus, a public consortium of scientists, determined to build a publicly accessible repository for the rice genome, has successfully defined boundaries around what should and should not be considered proprietary, and for whom, both inside and outside the consortium.

Many have argued that the genome map *should* be in the public domain, but that genes can still be patented. In a sense, companies and the consortium have negotiated and defined an upper limit of patentability for genetic information: maps are for everyone, but individual genes that may be isolated from the map are potentially patentable and thus commodifiable—a far cry from "free" or "unrestrictive." Lawrence Lessig (2001), writing about the Internet, eloquently argues that while the information contained therein is described as "free," the code that delivers that information is based on an "architecture of control" that will close off access to new thought and free expression. The same can be said for genomic information, for which access is granted "freely" at the level of the genome, but control is exercised at the level of the gene.

A National Genome: Beijing Genomics Institute

Whereas IRGSP is a global research project working on a global genome, the fourth effort to sequence rice that I will look at is national. In early 2000 the Beijing Genomics Institute (BGI), a publicly funded research group in China, began sequencing rice. BGI used the same whole-genome shotgun strategy as Syngenta. BGI's sequence was produced very rapidly—in only two years—and while it identified only 92 percent of the genes, the sequence accuracy is approximately 99.9 percent for more than 90 percent of the sequence produced thus far (Bennetzen 2002).

The sequencing method was not the only difference between IRGSP's project and BGI's. BGI was also interested in expanding the *kind* of data available by sequencing a different rice variety. The sequencing efforts initiated by Monsanto, Syngenta, and the IRGSP all focused on the Nipponbare cultivar (*japonica* subspecies). BGI chose instead to sequence *indica*, which is the subspecies consumed by most people in the world, including the Chinese; one source places production of *indica* at 80 percent relative to other varieties (Smith and Dilday 2003). While both *indica* and *japonica* varieties are believed to have originated in China, *indica*, with long grains that do not stick together when cooked, is preferred by most rice-consuming nations (Latham 1998).

Explaining the decision, an introductory statement on the BGI website states that the other sequencing projects used the subspecies *japonica* (Nipponbare) as "target materials" even though *indica* is "*dominantly* planted in Asia and other regions in the world, and has provided the *unique* template for the *unique* hybrid rice strain that has greatly contributed to solving the food supply problem in China" (Rice Information System 2003, emphasis added). Commenting on why BGI chose to sequence *indica* instead of *japonica*, the program director said, "There was a feeling that China should sequence its *own* rice" (quoted in Normile and Pennisi 2002, 35, emphasis added).

Thus a genome map that is distinctly Chinese but also Asian is bound up in the way BGI represents its goals. Moreover, BGI's success counts in another way: "China is almost the only developing country in the genomics race, and the only one to make its mark. It has all the usual disadvantages, lack of finance, of supporting technologies and scientists. Whereas the United States and Japan have about seventy researchers and engineers per ten thousand inhabitants, China has only six. Despite that, she has come out well ahead of the west" (Mae-Wan Ho 2001). This text positions China as a developing country, an underdog against better funded and better staffed developed-country projects. At the same time that it places the United States and Japan in the developed-country category, BGI also situates itself as pointedly non-Western. As Kuo argues in this volume (chapter 9), the state remains important in shaping conceptions of life and culture at a global level. While Chinese participants in the IRGSP imagine themselves as part of a global genome effort with chromosome-specific national identities, BGI starts out at the level of the nation in its self-representation.

On its website, BGI explains that the institute initiated its sequencing project to accelerate and "broaden the scope" of genome research, with a commitment to placing all of its sequencing information in the public domain and to eschewing patents generally (Beijing Genomics Institute 2001). Like IRGSP, BGI implemented a data-sharing program with DDBJ, EMBO, and GenBank, but it also made its genome information available to the public via its own online database.[22] On a spectrum of availability, BGI's genome project is perhaps the closest to the fully public domain: a new *kind* of information was made available (that about the *indica* variety); trade secret was not claimed and patents are frowned on; the results were published in many forums, making the information accessible to a range of potential users; the materials and other information underlying the publication were placed in an assortment of databases; and BGI is committed to sharing its results to the widest extent possible.

Publication: *Science*

The fourth site at which to explore the making of genomes and property is in the journal *Science*. Unlike consortiums, companies, and countries, publications do not do their own genome-sequencing work, but they use the work of others to constitute implicit property claims. To map the development of hybrid property in a publication, it is useful to examine the debates that occurred when *Science* published two drafts of the rice genome, in early 2002—one by BGI and the other by Syngenta.

Scientific journals are, of course, institutions with their own identities and discourses. In the case of rice genomics, journals also render questions about new kinds of property more visible than they previously were, and in this way more urgent. In April 2002 *Science* published two rice-genome drafts. In the pages of the magazine, in newspapers and other print media, and on the websites of rice-genome researchers, the drafts were roundly applauded as important developments. The rice genomes were called the

"Rosetta Stone" of cereals, which opens up the possibility of understanding the language of more commercially valuable grains (Normile and Pennisi 2002, 32); the term "Rosetta Stone" was also code for "cereal of the world's poor," and suggested that the rice genome offered potential for unlocking the secrets to ending poverty and hunger (Bennetzen 2002, 60; Cantrell and Reeves 2002, 53).

The publication was also surrounded by controversy. Even before the rice genome was officially announced, a group of twenty concerned scientists, including two Nobel Laureates, wrote a letter to the editor of *Science* voicing dissatisfaction with the decision to allow a private company to publish its results without conforming to what many *Science* readers considered to be the norms of scientific practice. Enacting the Mertonian norm of communism, which holds that new knowledge should be openly available to advance discovery and innovation, they argued that publication should entail free and unrestricted access to the data used to produce the article (Amos 2002; Declan Butler 2002). *Science*, along with other major scientific journals, had long followed the practice of requiring the deposit of genome data in Gen-Bank or a similar databank as a corollary to publication. The letter emphasized that allowing Syngenta to publish its research without this step was a "very serious threat" to genomics research and a potential threat to using genomic data for humanitarian purposes (Gaia 2002; Marshall 2002). The authors suggested that the involvement of market interests in the production and dissemination of scientific information was infringing on cherished (if utopian) ideals of scientific practice.

The editorial in the 5 April issue answered these charges. As in a similar controversy at *Science* involving the publication of the human genome by Celera Genomics, the editor Donald Kennedy justified his decision as follows: "*Science* normally requires that nucleotide sequence data reported in its papers be deposited in GenBank. . . . On rare occasions, however, we make an exception and allow the data to reside elsewhere as long as public access is ensured. We did that with the historic publication of the human genome sequence by Celera, copies of which are still freely available with the sole restriction that it cannot be redistributed" (2002, 13). Effectively, Kennedy adjudicated the construction of a two-tiered property regime for genome information. Weighing the arguments for community standards and unrestricted access to proprietary information against making an exception that would allow greater access to some otherwise restricted information, and reasoning that *accessibility* to data is different from the "place in which it is deposited," Kennedy determined that an exception was acceptable in this case, as it had been for the human genome (ibid.). However, data availability in this case meant an agreement allowing the company to provide access to its information through its own website or via CD-ROM instead of placing the data in GenBank, as would normally have been required (Torrey Mesa Research Institute 2004; Marshall 2002).

This was not an inconsequential decision. While publication releases information about research results, it is generally accepted that results and descriptions of the process are not sufficient in themselves to permit reproduction of the results in most genome work. Let me offer a clarifying analogy: publishing sequencing results but not depositing the information in a database is like producing a telephone directory that tells the reader how many instances there are of each last name and what regions of the city they live in, but does not provide specific names, telephone numbers, addresses, and so forth. Thus, by not placing the information in a database, Syngenta effectively held onto what is considered the most important part of the information necessary for use, namely, the equivalent of the connection between specific names and specific addresses.

Also, while publication released the trade-secret protection on Syngenta's procedures and other sequencing information, it did not prevent the company from withholding important annotations which provide critical information about the usefulness of particular genes or gene sequences. The scientists who wrote to *Science* worried that access to and control over information was at stake; hybrid property is not just about who has rights, but also about how rights to access are defined among multiple users in the face of changing institutional circumstances in which increasing amounts of basic research findings are becoming proprietary (Kennedy 2002).

Though Syngenta eventually agreed to wider release of its raw sequencing data, that is, allowing the IRGSP to have complete access, the sequencing data must still be kept confidential from researchers outside the IRGSP and are only available for noncommercial research purposes, and access is still limited (Normile 2002).[23] Moreover, patents on individual genes are still possible—a major restriction on information that Donald Kennedy asserted is in the public domain. Finally, it is important to point out that this public information is in any case never fully unrestricted or free. Though published by a not-for-profit organization, *Science* is in many respects a market-driven enterprise as well as an academic one. It costs money to subscribe to the magazine or access its online issues, and scientists do not have open access to this magazine in all parts of the world.[24]

Hybrid Property Regimes

The debate in *Science* highlights the importance of rice as a focus of agricultural biotechnology. Indeed many of the reasons that led to the four sequencing efforts were also noted in the pages of *Science*. The following extract from an editorial provides one example.

> Rice, arguably the most commonly consumed food source in the world, is eaten by 3 billion people daily. Rice crops account for about 10% of the arable land mass. The rice genome is estimated at 500 million base pairs, with some 40,000 genes over 12 chromosomes. Agricultural developers around the world are anxious to identify those genes that could improve rice yields on marginal soils and provide greater disease and pest resistance, shorter maturation times, and a wider range of tolerable climates. Rice is important not only in its own right but also as a model for improving other grain crops because wheat, rye, barley, maize, sorghum, millet, and rice have similar genetic maps. The International Rice Research Institute estimates that by 2020, 4 billion people—more than half the population then will depend on rice as a food. (Bloom 2000, 973)

Here rice is represented at once as a model cereal, a market good, a staple crop, and an environmental stressor. Genomics, correspondingly, is portrayed as the answer to a variety of goals. In the process, hybrid properties that were negotiated and delineated by the four rice-genome projects and in *Science* reflect the multiple publics imagined as using and eventually benefitting from the research. Table 1 illustrates the hybrid properties that arose through rice genomics. Each horizontal column represents an ownership claim, from the most restrictive and most legally well-established at the top (trade secret and patents), to those that are newer or less restrictive (databases, data sharing, and publication). The vertical columns represent the bundles of hybrid properties in the rice-genome projects. The column entitled "private property" illustrates those claims along each of the five ownership dimensions that might comprise the strongest set of private-property claims. Of course, trade secret is the linchpin of this regime, as without it, some information is necessarily in the public domain. Patents do release some information, but at the gene level, rather than at the genome level, therefore allowing a genome to be held under trade secret while patents are taken out on commercially important genes or sequences. The far right column entitled "public domain (commons)" illustrates the strongest possible

TABLE 1 A spectrum of rights: Hybrid properties

		Private property	Monsanto	Syngenta	IRGSP	BGI	Public domain (commons)
Established	Trade secret	Upheld	Released	Released	No	No	No
	Patents	Yes	Yes, with multilevel enforcement	Yes, with multilevel enforcement	Contingent, downstream only	No	No
	Database release	No	Not through public databases: accessible on company websites	Not through public databases: accessible on company websites	Yes, but also has a private database	Yes	Yes
Contested	Sharing	No	Yes with IRGSP, but commercial use not allowed	Yes with IRGSP, but commercial use not allowed	Yes, but unclear re annotations	Yes, but unclear re annotations	Yes, including annotations
	Publication	No	No	Yes, in Science	Yes, online	Yes, in Science and online	Yes, in journals, online, etc.

set of public information along the five dimensions I have tracked in this chapter. In the middle, from strongest private regimes to strongest public regimes, I have outlined the mix of properties recognized in the four rice-genome sequencing projects.

These sets of hybrid properties are imagined to most immediately benefit publics that include participating scientists and other interested researchers. However, access is contingent at many levels. Across all four projects it depends on Internet access at a minimum, but also on the resources and knowledge to utilize the data. Availability, therefore, does not equal access. As Rosemary Coombe evocatively observes, "Rather than a planet evenly comforted by a blanket of warm information flows, we have instead created a loosely-woven net that is attached at very few points—from which many are left out in the cold" (1996, 243–44). There still remains a major gap between the skilled and unskilled, between those who have the resources to gain access to information and those who do not. The organization of power may have spread out to more dispersed locations, but it is still concentrated. Tracing hybrid properties raises questions about ownership and relationships, access to scientific information, and responsibility for where that information travels, as well as the very social connections along the chain of knowledge production.

Conclusion

The publication of the rice genomes followed five years of mapping and sequencing efforts by four independent sequencing projects. While the publications in *Science* made this research and the property debates surrounding it public for the first time, this was simply the culmination of long-standing and ongoing negotiations over the kinds of property rights that would emerge with the new scientific information. From the time the IRGSP began its mapping effort, in 1997, followed by Monsanto in 1998, new conceptions of property rights arose alongside the generation of genomic data. Debates over this information have occurred and continue to occur at different levels of social organization—within scientific practice, at the national level, and in international arenas.

National interests, corporate social responsibility, scientific norms, and profitability all affect the way new genome research is transformed into property. Creative Commons demonstrates that categorical distinctions between private property and the public domain (or commons) provide a structure for argument and in some cases become codified in law. Rice genomes, and the property claims negotiated around them, link the scientists who make and use the new information for basic research to scientists involved in experimentation on rice and other cereals. Private companies and scientific publications also assert their organizational objectives—defining and defending how and where they will deploy rice genomes and for whose benefit. Hybrid bundles of property rights for genomic data result from and reflect tacit understandings of how these competing pressures and interests should be resolved. As such, hybrid property regimes, finally, are important modes of governance through which varied values and interests come together, creating practices that will inform the way we eventually use the information and understand the meanings of genomes and of rice.

Notes

1. Dennis Normile and Elizabeth Pennisi, "Rice: Boiled Down to Bare Essentials," *Science* 296 (2002), 32–36.

2. See for example Jasanoff 2004; Hayden 2003; Coombe 1998; Haraway 1997; Haraway 1991; Latour 1993.

3. For further analysis of the terms *public domain* and *commons*, see Boyle 2003; Hess and Ostrom 2003; Litman 1990.

4. See, for example, Dietz, Ostrom, and Stern 2007; Hess and Ostrom 2003; Ostrom, Burger, Field, Norgaard, and Policansky 1999.

5. "Open code" software was created to allow programmers the ability to view, amend, and distribute copies of software code (LINUX is the archetypal example). It comes with a general public license (GPL), which stipulates that derivative code must retain all the rights claimed for the original code, that is, it must be made available for others to view, amend, and distribute. For more information about the debate concerning open-access publications, see "Access to the Literature" 2004.

6. The Berne Convention of 1886 established that copyright protection accrues from the moment an original work is created; the creator does not need to formally register or apply for copyright protection for protection to take effect. Minimal qualifications of originality apply: a few words or a diagram may qualify. These protections have been standardized internationally through the World Intellectual Property Organization of the United Nations. All World Trade Organization members are subject to the Berne Convention through the Agreement on Trade-Related Aspects of Intellectual Property Rights.

7. This argument does not necessarily apply to that regarding property rights in land, which may conform more to a line of reasoning that considers a commons to be the predecessor to private property.

8. There is one exception: "no derivative works" and "share alike" licenses cannot be used together because "share alike" licenses only apply to derivative works.

9. Bollier 2002; Barton and Berger 2001; Rai and Eisenberg 2003; Boyle 1996; Buttel and Belsky 1987.

10. Particularly the introduction, in 1986, of the automated DNA sequencer, developed by Leroy Hood.

11. Generically, a genome is all of the genetic, or hereditary, information contained in an organism. A genome is comprised of chromosomes, which are found in the nucleus of a cell. Together, the chromosomes contain a full set of the DNA base pairs, necessary to construct a genome; the total number of base pairs determines its size. In the case of rice, there are twelve chromosomes and approximately 430 million bases. Rice is small relative to other grains; corn has 3 billion bases and wheat 16 billion. But the synteny between the different cereals (i.e., the fact that they have the same genes in the same order) makes rice a particularly attractive model plant for scientists. Many discussions of the rice genome include the size of the human genome, which has 3.2 billion base pairs, as a point of reference.

12. Hilgartner 2004; Rai and Eisenberg 2003; Haraway 1997; and Boyle 1996.

13. A draft sequence is one in which the order of base pairs has been assembled four or five times, ensuring accuracy and assisting in the eventual ordering of the base pairs into their correct sequence in the chromosome. In a draft, only approximate chromosomal locations are known for each ten thousand base pair fragment. A finished sequence, while not truly "complete," because there are still gaps and errors, is considered a "high-quality reference" with only one error per ten thousand base pairs. This is typically thought to require anywhere from eight to twelve base pair assemblies and precise chromosomal locations (see U.S. Department of Energy 2008).

14. See Boyle 2003; Hayden 2003; Rai and Eisenberg 2003; Mackenzie, Keating, and Cambrosio 1990.

15. The members of this international consortium are from Asia, Europe, and North and South America. Japan is participating through the Rice Genome Research Program, a joint collaboration of the National Institute of Agrobiological Sciences and the Institute of the Society for Techno-innovation of Agriculture, Forestry and Fisheries. In the United States participants include the Institute for Genome Research; Cold Spring Harbor Laboratory; Clemson University Genomics Institute; Washington University; the University of Arizona, Arizona Genomics Institute; Rutgers University Plant Genome Initiative; and the University of Wisconsin, Rice Genome Project. The National Center for Gene Research of the Chinese Academy of Sciences is doing sequence work for China, and the Academia Sinica Plant Genome Center is working in Taiwan. The remaining participants include France (Genoscope), Korea (the Korea Rice Genome Research Program), India (the Indian Initiative for Rice Genome Sequencing), Thailand (the National Center for Genetic Engineering and Biotechnology), Brazil (the Brazilian Rice Genome Initiative), and the United Kingdom (the John Innes Center). See International Rice Genome Sequencing Project 2008.

16. Their sequencing responsibilities included: Japan, chromosomes 1, 2, and 6–9 (approximately half of the genome); the United States, 3, 10, and 11; Brazil, part

of 9; China, 4; France, 12 and part of 11; India, part of 11; Korea, parts of 1 and 9; Taiwan, 5; Thailand, part of 9; and the United Kingdom, part of 2.

17. The Bermuda Principles originated at a meeting in 1996 sponsored by the Wellcome Trust at which a group of organizations working on human-genome research agreed on two main imperatives: first, genome information should be freely available and in the public domain; and second, genomic information should be released as rapidly as possible. The Bermuda Principles were upheld in a second meeting, held in Florida in 2003, and extended to other large-scale genomics projects, specifically the mouse genome.

18. Robin Buell, personal communication and interview, 20–21 April 2005.

19. GenBank is an international collaboration between the European Institute of Bioinformatics in the United Kingdom, the NIH's National Center for Biotechnology Information in the United States, and the DNA Data Bank of Japan in Mishima. It is a genetic-sequence database providing an annotated collection of all publicly available DNA sequences (National Center for Biotechnology Information 2008).

20. IRGSP's genomic data can be obtained at the Rice Genome Research Program website, http://rgp.dna.affrc.go.jp.

21. A reach-through agreement stipulates that royalties must be paid on end products that are not covered by the original patent because they include some element or involve some process that *is* subject to patent protections.

22. Genomic data can be found on the BGI website, *Rice GD: Genome Database of Chinese Super Hybrid Rice*, at http://rice.genomics.org.cn/rice/index2.jsp.

23. Researchers can only search up to 15,000 base pairs of the sequence at a time, and download only 100,000 base pairs of sequence each week. More can be freely downloaded if the researchers' institutions sign an agreement with Syngenta stating that the data will not be used for commercial purposes (see Torrey Mesa Research Institute 2004), http://www.sciencemag.org/content/suppl/2002/04/04/296.5565.92 .DC1/Goffweb2.pdf, "Accessing Torrey Mesa Research Institute (TMRI) Rice Genome Data."

24. The not-for-profit organization, the Science and Development Network (http://www.scidev.net), does post the content of selected articles from *Science* and *Nature* without subscription or usage fees, though the full content of these journals is not available through the site.

6

TRAVIS TANNER

MARX IN NEW ZEALAND

"This story makes no sense," exclaimed the graduate student. "Where is it going? It's like a labyrinth with no end in sight. I feel lost. Had Patricia Grace the language skills of Toni Morrison, her novel would have been better." These comments, made by a fellow graduate student in a course on indigenous literature, stunned me. How could someone be so naïve, I thought to myself? The insinuation that Patricia Grace, an indigenous writer from New Zealand, did not measure up to the "great" writers of the Western canon deeply offended me. I objected, as I often do to these remarks, "Better how?" The graduate student replied, "You know, less 'real.' Like Toni Morrison. Where's the literariness in Patricia Grace's novel?"

For that graduate student, native "literature" is really just glorified ethnography. According to this view, good literature lacks the strong cultural inscriptions found in indigenous texts. I understood the graduate student to mean that fiction (itself a nonpure category) is better than ethnographically informed writing (another nonpure category) because it steers clear of cultural and historical realities. Reminding the class of the proverbial death of the author, the graduate student argued for freedom of textual interpretation, which was being denied to the "Western" readers in our class (another nonpure category). But to me this was ethnocentrism masquerading as a critique of ideology. "Don't get me wrong," the graduate student continued. "Patricia Grace's novel is very good—from an anthropological perspective. But as literature it is not very effective." The student was serious, and had classroom support. "What is this resistance to indigenous literature?," I asked some friends after class. My friends thought for a moment and replied, "Indigenous literature is labeled inferior by detractors of ethnic literature. The task of the critic is to show how these texts are *just as good* as any other." I did not feel any better.

The naïve graduate student and my well-intentioned friends share a complicity that can be instructive for a politically minded reading practice of indigenous literature that wants to be neither congratulatory nor accusatory. Attempts to measure indigenous literature by the standards and values of Western literature facilitate the expansion of global capitalism under the cover of human betterment (bourgeois human rights, the financialization of the globe, the commodification of culture, various kinds of development practices, and so on). The idea that indigenous literature can be "just as good" as Western literature accomplishes this complicity in the name of so-called equality. Discrimination need not be overt to be dominant. In fact, it can be, and often is, more oppressive when the rhetorical signs point the other way. Thus, we must be vigilant in dredging neoliberal discourse that purports to be redemptive, to ensure that it is not deployed in the name of equality. We miss this complicity if we overlook how indigenous texts are devalued.

In response to this complicity, we can resist the reduction of indigenous texts to a common denominator of value by reading the textual elements that resist the cultural abstractions that frame so-called good literature from the "West."[1] Attention to voice, narrative structure, and culturally informed modes of communication challenges the norms of good literature in manifold ways, offering critical insight into the past of indigenous-white encounters, and opening up ways of reading not shadowed by globalized Western aesthetics. Resisting the complicity of valuing indigenous literature according to Western criteria is a political act of extreme importance; it teaches us to be culturally attuned readers who can negotiate shifting grounds without losing our convictions. This balancing act informs my readings of politically charged indigenous texts in this uncertain world.

Using these two exchanges—with the grad student and with my friends—as a grid to orient my chapter, I explore a nonequivalent mode of reading in Patricia Grace's novel *Baby No-Eyes*. To trace this nonequivalence, I read a Western theorist (Marx) through Grace to detect the causal chains of straight storytelling that buttress literature for the West before showing how indigenous modes of storytelling can reshape the parameters of meaning-making. The parameters I work within are law and temporality in the context of capitalism as a means to legitimate exploitation. Because I do not want to be complicit in the same way as my interlocutors, I will show how Grace's novel resists these parameters while working within them; that is, I trace the deformations of meaning-making with and against these parameters. While law and temporality under capitalism are my main focal points, I want to talk specifically about how Western law and temporality structure history and knowledge in modernity.[2] This requires an analysis of the formal dimensions of the story of capitalism. How these structures frame our world and, with some textual resistance, can be deframed is my readerly project.

Two cases of dispossession serve as the background for Patricia Grace's novel: the ongoing struggle for land rights in New Zealand, and an incident of genetic theft that occurred, in 1991, in Wellington Public Hospital in New Zealand. Grace juxtaposes the two cases to suggest that dispossession continues today in old and new forms. Land and bodily dispossession parallel one another in that both are mapped and mined for their precious properties: natural resources, labor, DNA, and organs. By folding into each other both land and bodily dispossession, the novel emphasizes that Maori subjectivity extends beyond the Western ego and how subjection can occur at the environmental and biological (read spiritual) levels. The connection is crucial because it situates Maori subjectivity within Maori cosmology. The destruction of one spells the destruction of the other.

Marx makes a similar observation in the first volume of *Capital*. In the section titled "So-Called Primitive Accumulation," Marx claims that capitalism began when the dispossession of public lands forced the peasantry to sell their labor for survival. But Marx does more than describe a new mode of production being ushered in by the dissolution of older modes of production; he also notes the cultural destruction wrought by dispossession. In a passage on the expropriation of agricultural production following the enclosures of the commons in England, he says, "By the nineteenth century, the very *memory of the connection* between the agricultural labourer and communal property had, of course, vanished" (Marx 1976 [1867], 889, emphasis added). In these lines we glimpse the loss of identity hastened by dispossession. If land is more than a commodity to be exploited for profit—if it has social value connecting people to time and place—then we can begin to appreciate the cultural void capitalism creates.[3] The connection that links subjects to space is radically severed. But this is more than a literal severance; it is also symbolic, affecting the psychic dimension of cultural life. Whether or not these memories are accurate is less relevant than how a people's conceptual universe is affected by capital, for whenever people are removed from spaces that shape their sense of self, the psyche suffers. Dispossession is not just a material phenomenon. The symbolic-psychic component is equally important, particularly in the case of indigenous peoples who have sacred ties to space that extend beyond the physical to the metaphysical. Of course, Marx was talking about the loss of identity that comes when the private European laborer is severed from the land. Nonetheless, there is a line connecting his thoughts on value and dispossession with the biotechnological predicament Grace critiques in her novel. This line illustrates the ongoing phenomenon of dispossession by accumulation.

In Graces's novel, the character Te Paania wakes from a coma following a deadly car crash that killed her husband and her unborn baby. On waking, she also learns that experiments were performed on the fetus by white doctors without her or the family's consent in the matter: the baby's eyes were removed and tested by doctors, reflecting hopes that, in the words of one Maori character, "remote communities['] . . . genes may have something different to offer" science, nations, and pharmaceutical companies (280–81). The novel's Maori protagonists ardently reject this type of research on the grounds that commodification destroys identity. As one character, Mahaki, notes, their "genetic bits are about to become some scientist's big discovery. They're after endangered species . . . Up for grabs, up for patenting, up for sale, but no proper processes . . ." (187, ellipses in original).

There is historical cause for Mahaki's alarm. In the 1990s population geneticists recruited by universities and various government-funded projects began to map human genetic diversity as part of the race to map the human genome. The project, known as the Human Genome Diversity Project (HGDP), received worldwide opposition from indigenous activists, who protested that it was yet another form of colonialism, this time targeting indigenous biology. The term *biocolonialism* became a catchword used by activists in defense of the cultural values they saw the project destroying. The project was eventually terminated, in 1996, in the face of growing international political pressure brought by indigenous groups and their constituents.[4]

While Maori cosmology is cited in the novel as a reason for rejecting DNA research—to quote Te Paania, "Genes are the ancestors within us" (280)—I believe a subtler point can be made. Mahaki does not dismiss science altogether, but argues that "no proper processes" are being implemented to guard against exploitation. Te Paania elaborates on this view later in the novel, when she speaks before a group of scientists.

Stop targeting remote communities just because their genes may have something to offer. At least wait until there've been *proper codes* of ethical practices and legal confinements established, *proper processes* for consents to be obtained—processes acknowledging whole community and intergenerational ownership, processes free from extortion and pretext, processes that positively acknowledge the right to say no—of people who may be opposed to their genealogy being interfered with; who don't like the idea of their life patterns being taken and owned by someone else; who don't want the essence of themselves being altered or disposed of, or transferred into plants or animals or other humans. Stop pretending that indigenous people will benefit from this research. (280, emphasis added)

The political message is unambiguous in these lines: tampering with Maori DNA (*moti*, or life essence) would do irreparable harm to the people's sense of self and community. Cultural knowledge and history could not be tracked if the community was divorced from its moti. Something like a biological division of labor would set in, reducing humans to the bare life Giorgio Agamben (1998) saw operating in the Holocaust camps. Life would cease to have meaning if knowledge and history became readable for rational information alone. This suggests that the flattening of temporality in the name of DNA as a transcendental signifier would, in the making of a general bioequivalence, greatly limit how people see themselves in the universe. With fewer mysteries to be explored, human faith and reason, in all their fallibility, would become ever more susceptible to the influence of an ever more powerful matrix of knowledge and power. Human sociality would come to depend on biology as never before (see Rabinow 1999). Freedom would become bioinstrumentalized.

While we should heed these warnings and take strong political positions alongside indigenous groups fighting for their cultural freedom, we also do indigenous communities a tremendous disservice if we relegate their histories to a pre-genomic history. This disservice comes to light if we take account of the flexibility *and* pervasiveness of global capitalism. Situations like the HGDP, or Grace's fictional situation, are cases where the rhetorical signs point the other way, for even the most seemingly anticapitalist values can be appropriated in the capitalistic juggernaut. Proof in point: culture sells, especially the exotic kind. Yesterday's Manichean us-them politics buckles under this type of logic, which is why we cannot simply read Grace and HGDP opponents as antiscience advocates. Some no doubt are, but that position does not reflect the entirety of Grace's message. The distinction between antiscience advocates and HGDP opponents turns on what Mahaki and Te Paania mean by "processes" of knowledge. In terms of legal and temporal parameters, "processes" are the ways in which meaning is made structurally. If Grace is not rejecting science per se, she is rejecting the ways in which Western science tells its good, straight story. Thus, the novel prompts us to rethink our hard-and-fast politics along formal grounds. The processes of meaning-making can be tracked by attending to the novel's staging of conflict between Western and indigenous knowledge systems in its temporal and narrative dimensions, which will lead us to the messiness of genealogy the novel endorses.

So what might "proper processes" look like for indigenous thinkers like Grace? The ellipses that Grace inserts after Mahaki's comment on proper processes are ambiguous. Because no answer is given as to what counts as proper processes, the reader is left to wonder. Te Paania's statement is equally vague. No answer is given. It is clear that both Mahaki and Te Paania disagree (as do I) with DNA research for political reasons, but their rhetoric slips and reopens the debate. What would count as good science?, they seem to want to know. What does that practice look like? Perhaps Marx knows.

In "So-Called Primitive Accumulation," Marx critiques the bourgeois's "idyllic methods of primitive accumulation" by showing that, contrary to the absurd prehistory capitalism tells as its origin story, the history of expropriation is "written in the annals of mankind in letters of blood and fire": colonialism, brutal child labor, slavery, land theft, and murder (1976 [1867] 874–75). Marx likens this ridiculous story to the story of Christian original sin that separated the haves from the have-nots.

> This primitive accumulation plays approximately the same role in political economy as original sin does in theology. Adam bit the apple, and thereupon sin fell on the human race. Its origin is supposed to be explained when it is told as an anecdote about the past. Long, long ago there were two sorts of people; one, the diligent, intelligent, and above all frugal elite; the other, lazy rascals, spending their substance, and more, in riotous living. The legend of theological original sin tells us certainly how man came to be condemned to eat his bread in the sweat of his brow; but the history of economic original sin reveals to us that there are people to whom this is by no means essential. (ibid. 873)

Against this theological original sin, Marx's writes the unofficial history of brutality at the heart of capitalism. His text demystifies the prehistory of primitive accumulation by exposing its ruthless endeavors: "The spoliation of the Church's property, the fraudulent alienation of the state domains, the theft of common lands, the usurpation of feudal and clan property and its transformation into modern private property under circumstances of ruthless terrorism, all these things were just so many idyllic methods of primitive accumulation" (ibid. 895). Marx's rewriting of primitive accumulation serves as a model for how to rethink colonial histories. The task it sets for the cultural critic is to read against the grain of official histories for the gaps and silences that bear witness to the oppressed. Many have been inspired to undertake similar projects (the Subaltern Studies Collective, various postcolonial scholars, cultural theorists, and so on) to powerful effects. My interest in Marx, however, is slightly different. What I find intriguing is his methodology for how to write our counterhistories. As a universal history of "mankind," capitalism can tell only a straight story: the Christian original-sin story that Marx critiques. In this sense, the genealogy of capitalism (the haves and the have-nots) eschews complexity for simplicity. In this historical scheme, past, present, and future are diachronically aligned in time as a means to justify class, gender, and race discrimination. Circularity and temporal entanglements—the contingencies of life—are ruled out from the beginning. Capitalism's history, in short, is straight . . . and good. No imagination.

Capitalism needs this straightness to organize itself in terms of generalized equivalence: money, but also the phallus and law. As universal signifiers, these markers make equivalence possible. DNA is quickly finding its place as a transcendental signifier alongside these others. Their logic is similar: to be universal, they must circulate according to abstract principles. Under their reign, men and women can be compared, as can shoes and houses or genes and land. Singularity is effaced by the formal properties of these signifiers. Marx gestures to the abstract form of capitalism in various places throughout *Capital*, and analogies can be tracked in Grace's novel. For the purpose of this chapter, however, I want to track how law and time code the propriety of DNA research. If science itself is not the culprit, as our reading of "processes" in Grace and Marx suggests, then there must be another dimension of capital that engenders physical and metaphysical domination. This additional dimension is what I call the secret law of value.

Social forms like law are formal sets of codes that order, consciously and unconsciously, our moods and desires. They mediate life in a universal way by making everyone structurally "equal." Typically, commentators of Marx have shown how labor is the cause of this equalization process that leads to real (meaning abstract) subsumption. Recall the famous passage in *Capital*:

"With the disappearance of the useful [qualitative] character of the products of labour, the useful characters of the kinds of labour embodied in them also disappears; this in turn entails the disappearance of the different concrete *forms [Formen]* of labour. They can no longer be distinguished, but are all together reduced to the same kind of labour, human labour in the abstract" (Marx 1976 [1867], 128, emphasis added). Advancements in technology and machinery mark the distinction between formal and real subsumption of labor, and it is only natural to place biotechnologies on the side of real subsumption. According to real subsumption, value is determined by the socially necessary labor time needed to produce things in comparison to other producers. The amount of surplus value to be generated from labor is indirectly related to newer forms of technology. As technology develops, labor becomes less physically demanding, but no less exploitative. That is a secret of Marx's *labor* theory of value, but not the only one.

Everything in Marx's theory points to labor (formal subsumption) as the fundamental form of social inequality. Historically this has been the case as physical bodies and energies were needed to industrialize the planet. But as we entered a neoliberal era, bodies and their capacities were brought more and more in alignment with capital's tendencies to exploit humans. Collapsing the old division between producers and consumers previously consolidated under practices of formal subsumption has been the project of neoliberal capital in search of new bodies to exploit. Yet with fewer and fewer sources to exploit and with growing global dissent against the forces of globalization, as we are now seeing in the fallout of the economic crisis of 2008, the instruments of capital have become truly perverse, and it is now commonplace for people to work against their best interests *in the process* of resisting the forces that tyrannize us. This is what Marx called the "real subsumption" of capitalism. Where once capitalism identified an external source to exploit for profit, when these sources are exhausted the only thing left for capital to do is to incorporate these antithetical elements and make them work against themselves. On the face of it, real subsumption of labor resembles theories of ideology that seek to explain how people are deceived about what's best for them. Ideology critique assumes a class divide between producers and consumers along which knowledge takes shape. Traditional Marxism sought to identify this divide and awaken the working classes to a new consciousness in the hopes of forging different social futures. But what is the course of action when neoliberalism has collapsed this divide? How do we even develop a new consciousness when concepts like "equality" and "freedom" have been corroded by capital? To address these questions is to think about the ways in which our symbolic universe has been coopted by capitalism to the point of making it hard to speak other than in its name.

We arrive at symbolic domination as the cause of exploitation by a gap in Marx's own thinking. In *Capital*, sandwiched between chapter 1, "The Commodity," and chapter 3, "Money, or the Circulation of Commodities," is a curious chapter on "The Process of Exchange." In this chapter Marx reveals the hidden mechanism of exchange and commodity circulation.

> In order that these objects may enter into relation with each other as commodities, their guardians must place themselves in relation to one another as persons whose will resides in those objects, and must behave in such a way that each does not appropriate the commodity of the other, and alienate his own, except through an act to which both parties *consent*. The guardians must therefore recognize each other as owners of private property. *This juridical relation, whose form is the contract, whether as part of a developed legal system or not, is a relation between two wills which mirrors the economic relation.* (Marx 1976 [1867], 178, emphasis added)

These lines call into question the *labor* theory of value on symbolic grounds. What appeared to be the "disappearance of the different concrete *forms* of labour" is in fact a product of a "juridical relation." While it seems in this passage that law comes to work on behalf of the owners of the means of production to ensure their unfair advantages over workers, which is true, we should also take note of how power is refashioned in this historical shift from the commodity form to the money form. More than an additional method by which to secure surplus value from laborers, symbolic domination like the kind we see in laws that purport to speak on behalf of humanity become, in perverse fashion, the inroads to a more entrenched subjection. This is evidenced in how we enter into these relations with consent and "free" of external force. While exploitation seems to be lacking in this scenario, in reality it has become dematerialized but no less destructive in the shift from commodities to abstract signifiers like money. It is in the spirit of skepticism toward this transition that I reevaluate the intentions of genomic science in indigenous communities.

Far from resulting in a nonexploitative relationship, however, consent in fact obscures the horrors of capitalism. This is the mystery of symbolic power. Marx is never explicit about symbolic power in *Capital*, but we can glean its importance from various passages on time and discipline scattered throughout the book. Étienne Balibar draws us deeper into this connection when he notes, "We would today call the analysis of symbolic structure . . . [a] double language 'spoken' by the world of commodities: the language of equivalence and measurement, given formal expression in the monetary sign, and the language of obligation and contract, formally expressed in law" (1995b, 71). And later: "The structure common to economic and juridical (and moral) fetishism is *generalized equivalence*, which abstractly and equally subjects individuals to the form of a circulation (circulation of values, circulation of obligations). It supposes a *code* or *measure*—both materialized and idealized—before which 'particularity,' individual need, must yield" (ibid. 72, emphasis in original). In the place of "individual need" rises a general, social necessity governing everything from time, labor power, states of being and consciousness, and social relations. Balibar attributes this generality to a juridical fetishism, shifting the terrain of value from labor to law. Thus, there is a "symbolic structure," in addition to a "world of commodities," that determines value. Appropriating Balibar a bit, we might wonder how the codes of law affect consciousness and culture? After all, Balibar notes that these codes are more than material; they are also "idealized" at the group level, displacing individual identity. This conjecture moves us toward a theory of subjectivity in Marx based on the symbolic structures of capitalism. It suggests that the one common characteristic of social life necessary under capitalism is its abstraction (see Postone 1993). Time, history, morality, culture, and labor—all forms of life are rendered abstract in the name of "generalized equivalence."

If we were performing a deconstructive reading of capitalism, a rhetorical analysis could perhaps suffice as our critique up to this point. The excavation work we have performed to unearth the duplicitous nature of capitalistic "equality" and "freedom" would be performed. But it would not move us beyond the crucial negative critique that is deconstruction's bread-and-butter. This is its downside: alternatives are hard to come by. In an effort to move toward a normative position of some sort, we must realize that there is more to rhetoric than its power of signification. Something comes through the symbolic order that is more than symbolic. This something is super-symbolic in that it has residues of the symbolic and is something more. I cannot help but think of the Maori moti as this something more that shapes (forms, *Formen*) our world.

Our rereading of Marx's theory of value has uncovered the symbolic structures that undergird capitalism's history. In this reading, symbolic domination is the key to the abstractions produced by transcendental signifiers like money, DNA, or the law. There is *something more* about the objective character of money or DNA that creates subjection under capital. This *something more* is fetishism, and it imprints our psychic worlds in consequential ways. The question we must ask ourselves is: how do we resist the powers of fetishism? If power functions symbolically at the level of abstract signs, where do we locate freedom? I propose that new signs are necessary to resist the allure of capitalism. These new signs are not radically new in the sense that they exist apart from our world or outside it. I firmly reject, as do Marx and Grace, the notion that resistance comes from outside capitalism in the name of "culture." There is no Archimedean point beyond this world from which to launch our critique. To make such a claim supports yet another position of transcendence that threatens to engulf us in abstraction. However painful it may be, capitalism can only be resisted internally.

Think about the narrative of *Capital*: why does Marx begin with the commodity and end with primitive accumulation? The chronology seems backward. Interestingly, Marx tells a counterstory of capitalism from within capitalism, beginning his analysis with the predominant concepts and categories defining capitalism, like commodities and surplus value, only to arrive somewhere else: at a critique of its gaps and silences, founded on the exploitation and colonization of labor, land, and social relations. Something similar happens in Grace's novel as she twists language and novelistic conventions to account for Maori subjectivity, knowledge, and history.

Grace's novel is composed of a prologue, thirty-seven chapters, and an epilogue, presented in alternating voices. At the beginning of each chapter is a nautilus-like design called a *koru* to remind the reader of the spiraling narrative form running throughout the novel. The koru represents open-endedness and is an integral part of the Maori worldview. It describes how the Maori understand time, history, and narrative. *Whaikōrero*, or speechmaking, follows a similar pattern. In a recent essay on *Baby No-Eyes*, the critic Michelle Keown discusses how *whaikōrero* or speechmaking structures the novel: "*Baby No-Eyes* features a polyphonic narrative structure which approximates the patterns of Maori *whaikōrero* or speech-making, in which different orators take turns to offer individual perspectives on a topic of discussion" (2005, 152). Like the polyphonic voices Mikhail Bakhtin reads in Dostoyevsky (Bakhtin 1984, especially chapter 1), or the mosaic voices Michael Fischer finds in autobiographies by scientists (Fischer 2003, especially chapter 6), Grace's alternating voices disrupt the otherwise abstract narratives conditioned by capitalism to provide "individual perspectives" on shared events. Voices, time, and narration weave in and out of focus in the novel, overturning the straight-stories that govern capitalism's conceptual universe. No wonder the graduate student was confused by and uncomfortable with Grace's storytelling practice. Accustomed to a hegemonic linear narrative, the student had a hard time seeing the storied alternatives that exist in other narrative modes within capitalism. I want to stress this last point about the *alternative narratives within capitalism*, because there is no outside of capitalism. A more sophisticated argument is required if we want to avoid accusations of nihilism and cultural relativism that stalk the either-or spectrum. Specifically, the critical perspective we gain of capitalism is afforded by literature like Grace's novel because in it we see how to disrupt the symbolic domination that crystallizes in the narrative of capital, which has also become the narrative of indigenous lives.

While Grace's polyphonic approach to narrative may be a cultural practice, I choose to read her as a participant in modernity, rather than position her on a parallel track alongside modernity. To relegate ethnic literature to its own separate plane of existence, apart from modernity, risks being exclusionary in the worst possible ways and refuses to acknowledge a more plural present in favor of a more monological one. This is precisely what Marx and Grace reject.

If Grace's story is enmeshed in the sordid history of capitalism and colonialism, she will seek to de-range and reconfigure the narrative and temporal structures of capitalism to allow other voices to be heard. One such voice belongs to Gran Kura, Te Paania's grandmother, who counsels her family after years of being "good" and "obedient" in missionary schools and society to be irreverent and resistant. We hear in Gran Kura's name the echoes of koru weaving throughout the generational voices speaking in the novel. Gran Kura's voice passes through Te Paania as she learns how to tell "improper" stories—the ones disallowed by colonial society.

> We knew we'd been attacked but were not equipped to fight the outstretched arm or the insinuations about being proper. I didn't know then that a curse was a matter of potent ill-wishing, and that if we were not to die from it we needed to turn speakings back on those who spoke them in order to make them void. . . . Even if I did it artlessly and without dignity, it was an attempt at dignity, a rejection of the idea of us not being proper people with ordinary hopes and a normal desire to learn and be part of the ordinary world. (89–90)

Similarly, Te Paania embodies Gran Kura's defiance when she resists learning "proper" cooking habits (89) and rebukes a boss who refuses her adequate pay (106). Stooping to indignant acts of cursing and artless speech, these characters defy the type of subjects they are made out to be in the world of capital's embrace. Doubly subject as native and woman, Te Paania taps into the channels of resistance when she professes, in an anti-Bartlebyian tone, that there are "no proper processes" to hear the complaints of indigenous peoples. What we hear in her defiant speech is not, crucially, a denunciation of genomic science as such, but a plea for the "proper processes" that are lacking. That she has to make her case in this form of speech is less a reflection of her personal opinion about science than the result of symbolic domination that has made it hard to tell different stories. Because capitalism doesn't typically take notice of other voices, Te Paania and her grandmother make their case for social inclusion in a rhetoric of impropriety. This form of speech isn't dismissive but radical in its insistence on being heard when no one is listening. It is my contention that such forms of speech attempt to punch holes in symbolic domination in politically powerful ways.

These resistances are also cultural in that they represent a method of refusing identification conferred by centers of power: racism, sexism, and pharmaceutical capital. Te Paania's resistance alters the narrative and temporal grids of capitalistic intelligibility by sending these messages of exploitation back to the sender. I repeat what Te Paania notes while reflecting back on her days in school, "I didn't know then that a curse was a matter of potent ill-wishing, and that if we were not to die from it we needed to turn speakings back on those who spoke them in order to make them void. [Later,] I wasn't taken by surprise again. No one was able to shut me down from that day on" (89–90). Te Paania's resistance is internal to the system in that she never withdraws from it. Like Gran Kura, who refuses to speak English after years of repressing her native tongue, Te Paania redirects the curse back onto the colonial world responsible for its production.

A similar redirection occurs later in the novel when the community rejects an offer made by the New Zealand government to buy back land originally stolen from the people. The conflict leads to a massive demonstration and eventual occupation of the sacred territory under negotiation. This act of resistance confuses the white community leaders and government officials, who are convinced of their own altruism. Why don't the people want their land back, they muse? The problem for the native community is one of representation. Land cannot be represented symbolically in document signatures or dollar bills. The same goes with DNA. Moti cannot be represented in a genetic map or a drug. Translation doesn't work that way for the Maori. Yet this is exactly what law attempts to do: translate, in a generalized manner, land and bodies into universal codes of signification (health and wealth). What the white officials and bureaucrats deem a miscommunication between them and the community is actually a failure of imagination on their own part.

The novel unambiguously associates law with this failure of imagination. Landed dispossession and bodily dispossession are equated by the power of legal doctrine. For example, "They [the scientists] were allowed to [steal the baby's eyes for DNA research] because they were allowed to. Law allowed them. Power allowed them. We had no right to say no, or yes, because we weren't people. Baby wasn't a baby, wasn't the family's baby. Baby was a body, and legally belonged to the coroner" (188). We can compare this with Mahaki's thoughts on land theft later in the novel: "They'd [the Maori community] tried to get it across that it was laws, not people, that were the enemy, that it was justice at stake; or it was fear inside people that was the enemy, not the people themselves" (214). The compositional structure of law cannot do justice to culture. Its reductive form works against the Maori and other indigenous groups by effacing the rich nuances of tradition and knowledge like *whaikōrero* told in more elaborate storytelling practices. As a vehicle for knowledge and cultural transmission, storytelling connects indigenous people to time and place in complex ways. This complexity is lost when cultural ways of knowing the world are translated into Western categories. This is precisely what happened in the failed case made by the government men to the community members in the novel, and the failure of the HGDP's effort to map genetic difference in the 1990s.[5]

As a mode of resisting Western categories that reduce Maori culture to static signifiers, Grace's novel disrupts any easy translation process. I recall another student in the indigenous course saying, "You never know where you are in Grace's novel. The answers come later, and this is very frustrating." The student was responding to the following passage.

> There's a way the older people have of telling a story, a way where the beginning is not the beginning, the end is not the end. It starts from a centre and moves away from there in such widening circles that you don't know how you will finally arrive at the point of understanding, which becomes itself another core, a new centre. You can only trust these tellers as they start you on a blindfold journey with a handful of words which they have seemingly clutched from nowhere: there was a hei pounamu,

a green moth, a suitcase, a birdnosed man, Rebecca who was mother, a man who was a ghost, a woman good at making dresses, a teapot with a dent by its nose. (28)

The student is correct to say that this passage does not explicate the action being described. We don't find out until much later what the line "there was a hei pounamu, a green moth, a suitcase, a birdnosed man, Rebecca who was mother, a man who was a ghost, a woman good at making dresses, a teapot with a dent by its nose" means. Rather than containing everything we need to know, this passage refuses to let us identify with Grace or her characters. The importance of this gesture should not be missed, for it expounds a theory of culture that is emerging, as opposed to one that already exists in a contained set of assumptions and beliefs. While reading, the reader is woven into the text-ile of the story as listeners. As listeners, we have an ethical obligation to reformat the ways we process knowledge in accordance with the Maori method of storytelling, whaikōrero. Content is less important than our recognition of culturally specific knowledge practices. The benefit of a formal versus a content-centered epistemology like the ones presented in Grace's novel is that it doesn't exclude singular attitudes and beliefs about the world. There is no contradiction, for example, in combining religious ideas or sentiments. Native Americans have long practiced Christianity alongside traditional healing ceremonies without trouble. This is why indigenous people could conceivably be in favor of DNA research, as Te Paania suggests in her speech at the end of *Baby No-Eyes*. The dilemma isn't over *what* knowledge is produced, but *how* knowledge is constructed, although some will surely disagree with the construction and the product. Be that as it may, Grace's unbounded "processes" aren't prescriptive in their design (koru). Rather, they present another grid of intelligibility—one constantly in the making between speakers and their complex identities.

Whatever union may form between science and culture, the stories that will be told will necessarily be "improper," twisted, and nonequivalent. As we envision what these stories might look like, Grace's novel might serve as a model for how to write and read "improperly." Coupled with a meandering plot that jumps effortlessly between past, present, and future, Grace's narrative is a dynamic performance of *twisted genealogies*. Kinship in the novel isn't defined in terms of origin or place, but in terms of entanglements. Against the organicist notion of indigeneity some uphold in the name of cultural rights and sovereignty, Grace offers something much different. Te Paania and her son, Tawera, live with Mahaki and Dave, a homosexual couple, despite the jeers they get from Shane (Te Paania's first husband) and others. Gran Kura, Te Paania's grandmother, learns that her birth mother is actually her aunt, who had been approached by her parents, according to the "old ways" (162), when they couldn't conceive. And then there is the spectral relationship linking each character with the ghost of the baby who died in the car wreck. More than a metaphor, the baby's ghost reminds the reader of the multiple dimensions of time and understanding running throughout the novel. These genealogies, together with the reader's role as listener, combine and disrupt one another in unpredictable ways. It is the nature of Maori stories to run deep and shallow.

To account for a Maori subjectivity that extends beyond the self to include the environment, extended kin, and oral histories, Grace composes a text of layers, folds, and disruptions that functions at once as a political act of cultural determination and as a strategy to resist the very forms she is working within: the temporal structures of capitalism. If her mode of storytelling cannot extricate itself from the dominance of abstract time and history, it can at least re-form these categories in new and inventive ways.

These dynamic stories are simultaneously a critique of capitalism's abstract historical and narratological forms, and an attempt to make something more of them. As critique, they show the formal limitations of capitalism by exceeding the very parameters within which they are situated. They emphatically resist the inscription of monological knowledge in the twisted intricacies that buck straight storytelling. This is why we cannot say definitively whether indigenous cosmologies like the one Grace presents are anti-science. If we take Mahaki's earlier elliptical comment seriously, the "proper processes" that might bring together science and culture are yet to be decided—they are emerging.

Perhaps to tell such stories, we will have to learn to appreciate different value systems, where voices and histories morph into one another and clash in a multitude of noise, mystery, and information. Rather than seeking to control these unknowns, maybe we can somehow foster them without fear or anxiety. Telling and reading these twisted stories surely will not be easy, but in them resides our collective futures. They require a new art of storytelling and of listening, to resist the flat ones that threaten to reduce the rich knowledges and histories at play in the world.

Notes

1. By putting quotation marks around "good" and "Western" I do not want to suggest that these constructions have no reality effects. They do, and if I leave off the quotes in the rest of the chapter, I do so fully aware of their ideological baggage.

2. I include the contemporary moment in my understanding of modernity.

3. Later Marxists like Antonio Gramsci have talked at length about culture under capitalism, but the kernel of the discussion can be found in Marx. After all, his definition of capitalism is entirely social: "Capital is not a thing, but a social relation between persons which is mediated through things" (Marx 1976 [1867], 932).

4. It should be noted that although the HGDP was terminated, the spirit of this research lives on in new projects, most recently in a joint effort conceived by IBM, the Waitts Foundation, and the National Geographic Society. The moniker for this bio-avatar is the Genographic Project, http://genographic.nationalgeographic.com. See also Reardon 2004; Amani and Coombe 2005; Cindy Hamilton 2001; and the websites of the Indigenous Peoples Council on Biocolonialism (http://www.ipcb.org) and the Action Group on Erosion, Technology and Concentration (http://www.etcgroup .org), formerly Rural Advancement Foundation International.

5. In her book *Race to the Finish* (2004), Jenny Reardon argues that the HGDP project failed because scientists and indigenous communities targeted for DNA research "failed" to work co-productively across technical and cultural idioms. Against this view I am arguing that the failure of the project, like the rejection of DNA science and the struggle for land rights in the novel, is due to the representational aspect of land and the body captured by the transcendental signifiers money and DNA.

7

KRISTIN PETERSON

AIDS POLICIES FOR MARKETS AND WARRIORS

Dispossession, Capital, and Pharmaceuticals in Nigeria

Most of the literature on globalization that theorizes flexible capital, flows (media, migration, technology), global cities, cosmopolitanism, and local-global relationships proceeds from an analysis of finance and manufacturing capital.[1] Such paradigms account for accumulation, speed, and the migratory patterns of both people and technology via capital circulating among cybernetic and physical spaces. As one imagines the enormity of capital movement, what is said of the spaces and places that are emptied out, from which these voluminous forms of capital are originally extracted? As it is widely recognized that the African continent continues to provide raw material in the form of oil, minerals, and cash crops to the rest of the world in crumbling and nonreproducible ways, can there be an analysis of an emptied-out space as the left-behind effect of such movement? Can there be an accounting of this space that is connected but defies overlap with other spaces in the transnational realm, one that cannot always imagine how raw material and capital are transformed and consumed beyond its boundaries, yet one that is not parochial in the estimation of its own loss?

Dispossession and Its Organizational Strategies

When it comes to theorizing Africa's relationship to globalization, there is remarkably little said other than that Africa is simply marginalized in the global political economy.[2] However, Africa is being rigorously "reinscribed" in the world via trade, development, and economic policies, which suggests an importance greater than simple marginalization. How African states comply with the World Trade Organization, for example, will largely determine the role and activities of trade and global governance in ways that are yet to be imagined, and in ways that are alarmingly on the horizon, such as the slow and rigorous wiping away of the generic drug industry via legal measures found in numerous free-trade agreements.

This chapter assesses a form of "lively capital" that begins with the following assumptions: wealth accumulation as described by analyses of speculative and manufacturing capital, global cities, and so on cannot solely account for the contours or performance of global capitalism and Africa's relationship to it. Rather, Africa is an imperative and integral part of current processes of globalization that include the continent's cultural and economic representations, the building of new capital markets, and the redirected efforts of foreign aid that are increasingly being tied to global securitization. Instead of thinking about globalization as a unitary capitalism, I am more interested in theories of capital that may better capture and complicate Africa in the world, as more than one capitalism is at play and at stake here.[3] In these paradigms, therefore, more attention needs to be paid to *wealth extraction* and *dispossession*, whereby the emptied-out material space is generated by both extractive industries and overlapping configurations of policymaking and capital mobility.[4] This dispossessed space provides the ground in which emergent and competing kinds of capital, as well as social and institutional exchange, find their roots and growth. In this particular instance, they manifest as varying forms of pharmaceutical capital, whose circulation and existence are tied to oil, debt, and military economies.[5]

By describing the institution of pharmacy (defined as the discipline of drug dispensation and composition) and, to a lesser extent, drug manufacturing in Nigeria as exemplaries of emptied-out space, I am not referring to *terra nullius*, which would imply sheer absence in a colonial imaginary, the modern ghost of which is invoked by the pharmaceutical industry as "lack of infrastructure" and used as a prime reason to refuse adequate drug price reductions. Nor am I privileging the colonial state as a robust entity that extended its drug-distribution efforts beyond the citizen to the subject, a task begging the attention of the postcolonial state. Rather, I am referring to two means of dispossession: The first is the structural adjustment program (SAP) of 1986 of the International Monetary Fund (IMF), which was strongly tied to a rise in militarism in Nigeria and to a protracted prodemocracy movement. The SAP initiated a massive emptying-out of existing health institutions and pharmacy, and it disabled drug manufacturing (via currency devaluation, wage decreases, state privatization, and dismantling, etc).[6] New therapeutic institutions emerged, replacing dying institutions in a process in which professional and patient agencies, strategies, and subjectivities came into being, literally enveloping other ones. The second is a more refined form of dispossession that attempts to dismantle the generic drug industry's market viability through two routes: trade-related intellectual-property law and specific AIDS treatment policies, both of which emphasize and privilege proprietary transnational drug companies and the circulation of their products.

By exploring the near-death of an industry and the subsequent rise of neoliberal health policies, I show how wealth extraction and other forms of dispossession are preconditions for generating contradictory *imperatives of capital* as they relate to old battles over the social contract, but are largely being reterritorialized in these scenarios (Ferguson 2005). Marx described two competing forms of capital, one that perpetuates further production and the other that perpetuates further circulation (see Marx 1977 [1852]; Marx 1959). More recently, David Harvey's important insights on capital mobility, described as "accumulation by dispossession" (2003, 137–82), rethink the importance of primitive accumulation since 1973, a process that— in contrast to the formulations of Rosa Luxemburg (1968) and Marx, yet following Hannah Arendt (1968)—remains an important strategy for capitalist expansion in the twenty-first century.[7] Using Marx and Harvey to frame the larger politics and stakes, I would argue that what we are seeing in these dispossession and reinvestment strategies is the combination of territorial and capital logics, with policy logics that specifically emerge in the context of AIDS.[8]

Here, I refer to the financial interaction among, and capital movement facilitated by, policy organizations as well as financial institutions overseeing the implementation of policy. In being "brought together" by AIDS, policy organizations, the state, corporations, and AIDS activists make *implicit agreements* with each other that generate particular kinds of capital flows. In this context, implicitness represents the crux of policy logics reacting to neoliberal strategies and reform. That is, healthcare systems are increasingly overlooked in favor of policies that address "the gaps" in care, such as prevention and treatment for HIV, rather than comprehensive care. Nonexistent robust health-systems infrastructure is a prerequisite for implementing policy that addresses these infrastructure gaps. This has led to new ideas of health, bodies, and surveillance collated by myriad consenting actors and institutions that generate abstractions and analysis of "the gaps." As a result, the "gaps" in healthcare systems are transformed into the system itself, for which humanitarian and government organizations deploy millions of dollars dedicated to new infrastructure, while health systems are left to wither in neglect; it is a scenario wherein the logics of health and economic crises both presuppose and require each other. "Implicit agreements" thus points to how economic and health abstractions become naturalized as normative social and institutional exchange. The policy-driven capital form is thus simply a question of how capital naturalizes its own mobility and operation.

Until the 2003 implementation of the U.S. President's Emergency Program for HIV/AIDS Relief (PEPFAR) for select states, including Nigeria, most AIDS development agencies had favored HIV-prevention policies over widespread treatment. This means that in Nigeria, for example, HIV education and prevention programs function as an AIDS humanitarian apparatus that provides protectionist measures for the oil-extraction industry. That is, long-term and sustainable AIDS-treatment policies would require, first and foremost, converting oil wealth into funding for treatment for the nearly five million who are HIV-positive and many more who are infected with numerous other infectious diseases.[9] Fundamentally, any attempt toward widespread treatment necessitates reconfiguring the relationship between African states and their corporate partners, between external debt and foreign aid, between African states and their creditors. Because Africa's creditors are also Africa's AIDS donors (the World Bank, Paris Club members, the United States), it is perhaps no coincidence that a very particular biopolitical regime manages both HIV bodies and relationships between the state and humanitarian, and international financial institutions.

In these interdependent contexts, the state's own role is to facilitate, orchestrate, and permit "accumulation by dispossession to occur without sparking a general collapse" (Harvey 2004, 115)—constituting the general gist of an IMF structural-adjustment program. While Janet Roitman (2005), Achille Mbembe (2000), Jean-François Bayart (1997), and Sean Brotherton (2008) have all demonstrated how the state in Africa does act in its own interests, turning dispossession into new kinds of accumulation, I would furthermore suggest that the state's technological capacities are shifting into other technological priorities that inscribe new patterns of capital flow and formation.[10] In such cases, the state and capital are not always at odds with one another; it is not always the case that the state erodes while capital flourishes. Certainly for Nigeria, the primary source of accumulation is not based on wage labor, but rather on government contracts and oil-rent politics, through which accumulation is fundamentally channeled via the state, so that the state and capital actually rearticulate each other.[11] If anything, stringent and lax laws can coexist in the same space to enable and disenfranchise certain capital manifestations, where at once the state's own interests are both fulfilled and erased. This is a contradiction that largely emerges in the aftermath of a state privatization as well as within Nigeria's current efforts to be more squarely inserted into global markets.

In the rest of this chapter I describe the shrinkage of pharmacy and decline of drug manufacturing since the IMF structural adjustment, and the lack of quality drugs that has resulted. I also describe the excess of counterfeits, illegal drug markets, self-medication, and drug-labeling problems as produced materiality and practice in a "post"-IMF space. I then examine how two treatment policies (one in Nigeria, one in the United States) map onto these spatial environments. I pay special attention to the merging of security and health discourses in the implementation of these treatment policies. "Household security," in the context of AIDS, loses ground to "national security," giving rise to new policy logics that bypass the aftermath of healthcare infrastructure dispossessions. Ultimately, I argue, the 1986 IMF SAP in Nigeria constitutes a particular historical moment that emptied out Nigerian and other African pharmacies while inaugurating new protectionist measures for the global circulations of pharmaceuticals, where new markets and security cultures thrive.

Frantz Fanon wrote, "It was not the organization of production but the persistence and organization of oppression which formed the primary social basis for revolutionary activity" (1966, 88, cited in Robinson 2001, 134).[12] In following Fanon, and in viewing *dispossession* as a form of *oppression*, I argue that dispossession is the primary organizational strategy that generates such protectionist measures, which alter health and medical practices, including drug production and consumption. The very drive of this dispossession are the implicit agreements among institutions that enable capital to thickly accumulate in ways that contradict the interests of public health. It may be counterintuitive to imagine that state and other forms of dispossession negate capital accumulation and wealth.[13] Indeed, dispossession ultimately curtails incentives for foreign direct investment when state services like electricity no longer function properly and social conflicts carry on amid scarce resources. But dispossession actually serves as a productive contradiction in a Marxist sense. The AIDS policy is a pivoting anchor that enables the subsidizing of new drug markets and, particularly for Nigeria, keeps the flow of oil wealth sustained in ways that continually reproduce national and international elites. In the process, the state both consumes and negates its own interests; and the population must negotiate a therapeutic economy and knowledge that edges on a dangerous medical pluralism.

Emptying-Out of Pharmacy, Dismantling Drug Manufacturing

In the mid-1980s the pharmaceutical-manufacturing industry comprised over fifty manufacturing firms that produced generics for malaria and other pertinent endemic parasites and diseases. By 1996, ten years after structural-adjustment implementation, nearly two-thirds of the industry bottomed out. Two IMF austerity measures (among others) impacted this decline. The first, a high tax placed on imported raw materials, was viewed as a step toward increasing local production of raw materials and justified by the IMF as addressing the need to wipe out "nonessential" state imports. The second was the devaluation of the currency, which cut earning power in half across the country within the first month of structural-adjustment implementation (and led to further declines after that).

The decline in earning power had two effects: 1) it eliminated purchasing power for local manufacturers, who could no longer invest or reinvest in raw-material production (as of today, 100 percent of all raw materials for drug manufacture are imported to Nigeria at extraordinary costs, undercutting the IMF's original claims that its impetus was to improve self-sustainability); 2) the purchasing power of the consumer was also devastated, whereby those who sought generic Nigerian drugs were left to either pay more or seek alternatives such as traditional medicines or practitioners who claimed to have a cure for AIDS. Traditional medicine has always comprised part of the therapeutic economy. It is estimated that 70 percent of the population seeks out traditional healers (due to cost and familiarity) as primary healthcare providers, which matches most estimates that 70 percent of the population lives on less than one U.S. dollar per day (Maiwada 2004).

Both private capital flight and the debt repayment are derived almost exclusively from oil wealth, and both represent primary forms of wealth extraction.[14] State privatization, trade liberalization, the removal of petroleum sub-sidies, and the devaluation of the naira led to decreased earnings, and food prices nearly quadrupled. Nigeria faced increased black-market expansion, heightened poverty, increased crime, food riots, and worker strikes. Primary healthcare services collapsed, which, again, impeded the IMF's stated goal of building self-reliance, as the fund envisioned total cost recovery from patients who could not afford even basic food commodities (Salako 1997). In Nigeria capital investments and recurrent payments, such as salaries and essential drugs, and facilities maintenance were suspended (Samba 2004). With the introduction of user fees and the sale of drugs liberalized, the pub-lic consumption of drugs drastically declined. By 1990 the domestic pro-duction of pharmaceuticals had ceased almost entirely throughout Africa; most pharmaceutical and medical-supply industries were pushed into bank-ruptcy, and medical workers fled to the private sector both within and out-side of Africa (ibid.).

The SAP also affected drug-distribution systems, which were already facing great difficulties and challenges. The original drug-dispensation pro-gram was based on a colonial administrative system whereby drugs were transported to central stores and dispensed by government pharmacists. After the Nigerian civil war (1967–70) and the oil boom of the 1970s, there was massive hospital and healthcare expansion. Additionally, overseas manufacturers found the seventy-million-person market to be highly lucra-tive, and started to pack and distribute imported drugs in Nigeria. Compa-nies such as Pfizer, Abbott, Glaxo, Wellcome, and Roche came to Nigeria and manufactured drugs (Ovbiagele 2000). The colonial system of drug dispensation could not meet the needs of an expanding healthcare system, and the government was slow to react. Steeped in postwar reconstruction efforts, the government was unable to reconstitute or expand the regulatory structures quickly enough to forestall the growing chaos of drug distribu-tion (ibid.). With an oil bust producing a severe economic crisis, the govern-ment took a desperate measure, liberalizing drug-import policies via an "im-port license" that allowed nonpharmacists to freely import and sell drugs at huge profits. As a result, massive quantities of fake drugs entered the coun-try, and military and civilian counterparts together assumed control of drug markets. It was not until 1990 that the National Drug Policy was executed, which gave rise to the National Agency for Drug Administration and Control (NAFDAC), Nigeria's drug regulatory agency, which remains to this day highly underfunded.

By 2005 Nigeria had accumulated a total debt of 36 billion USD, at which point the country threatened to repudiate what it deemed an illegitimate accumulation of debt by former corrupt military leaders who did not pay as scheduled; late-payment fines and arrears amounted to billions, even though the principal had been paid off at least three times. By 2001, with the end of military rule, annual payments to foreign creditors amounted to $2 billion, while only $300 million was allocated toward the entire national healthcare budget, designated for 120 million people. The health budget of 2001 was 1.9 percent of the total national budget (or U.S.$7 per capita) (World Health Organization Statistical Information System 2001).[15]

In the same year, *nearly half* of all drugs in circulation were found to be counterfeit or substandard, and such drugs are mostly found in thousands of drug markets across the country (R. B. Taylor et al. 2001). The 2001 statistic on fake and substandard drugs may be declining, as Dora Akunyili, the former director of Nigeria's drug regulatory agency, NAFDAC, started a campaign to confiscate and burn fake drugs in great media and public displays. As a result, she and many NAFDAC workers were attacked in markets; several car bombs were detonated; various assassination attempts were made on their lives; and the NAFDAC headquarters were burned down in 2003. Market sellers in open drug markets have been rarely prosecuted in the past despite good laws on the books, and in using their own union protection, they are fighting the prospect of unemployment. Whether or not Akunyili's efforts were effective, the public attention brought to fake and counterfeit drugs marks a shift in consciousness with regard to the presence of fake drugs in the country.

In contrast to the numerous illegal drug markets, very few pharmacies exist, and over 90 percent of those pharmacies are in the urban areas, 30 percent of which are concentrated in the city of Lagos. Out of the thirty-six states, only six have more than a hundred pharmacies and five states have fewer than fifteen, which are intended to serve potentially millions of people, given a population of 130 million (Pharmacists Council of Nigeria 2000).[16] There are nearly four times the number of registered pharmacists as there are pharmacy premises, perhaps pointing to the problems of gaining start-up capital, unstable electric supplies, and unemployment. Moreover, doctors dispense drugs themselves and are not eager to hand over dispensing responsibilities to pharmacists—a state of things to which many regulatory officials, some of whom spoke to me, have resigned themselves.

As the profession has declined, pharmaceutical practice has itself changed. Even in the urban areas where pharmacies are accessible, how pharmacists dispense drugs is highly mystified for many patients, the names and dosages of drugs prescribed, for example, rarely being labeled (O. Taylor 1998). Indeed, to omit such information is a common practice and even the policy of many hospitals and pharmacies. When I was in Owerri, in the eastern part of the country, with Mary, a nurse and AIDS activist, we visited her family in a nearby village and found that her mother-in-law was ill. After taking her to the doctor, we walked with her to the hospital pharmacy, where she picked up her prescribed medication. A sign posted next to the pharmacy window, in both English and Igbo, encouraged patients to ask questions about the drugs they were receiving. Mary's mother-in-law received her drugs in a plastic bag, which, marked in pen, stated how many pills she should take per day. Neither the name nor dosage of the drug was listed. Mary went back to the pharmacy and asked them to label it "correctly." The conversation escalated in the street, with Mary yelling at two hospital administrators about their nonlabeling policy and a couple dozen patients who had gathered around to listen. The administrators calmly told Mary that they could not label prescriptions because too many patients self-medicate, which Mary countered with, "And what happens when your patients have adverse side effects or allergic reactions to prescribed drugs? How will any medical worker ever know what was prescribed when the patient has no idea? And what if the patient dies? Then what?" To these questions, there was no response.

Indeed, concern about self-medication with controlled drugs—which is, after all, the most common method of treatment—is the most common justification pharmacists give for nonlabeling. But conversations I had with pharmacists indicate that something else is at work: the desire to keep drug knowledge circulating only among pharmacists. Many articulated what seemed like a mantra of using self-medication as an excuse for non-labeling. Very few wanted to explore the notion of assisting a very large self-medicating population through labeling.[17] The profession of pharmacy, long held in esteem, has been increasingly devalued, due to the inhospitable climate of drug distribution and difficulties in competing against "illegitimate" businesses. Perhaps nonlabeling acts as a reconfiguration of expertise wherein making certain knowledge secret confers a sense of authority on a profession struggling to regain legitimacy or status. Indeed, one of the newsletters of the Pharmacists' Council of Nigeria stated at the 1999 annual meetings that a newly institutionalized honorific, "Pharm," would precede the honorifics Mr. and Mrs. The use of this honorific is now common practice among pharmacists. Together, the mystification of drug knowledge and the implementation of the new titles establish a sense of control over professional loss.

Of course, the control of knowledge does not preclude patients from seeking the information they want. Patients often obtain their knowledge of drugs from market sellers, most of whom are not trained as pharmacists. I have walked through the Lagos drug markets, where the conditions for storing drugs are generally not ideal, and found both controlled and uncontrolled substances. I have watched buyers give sellers lists of symptoms, for which the seller proceeds to find the appropriate medications in his supply. Moreover, hospitals and clinics also get their supplies from these markets, and pick-up trucks with hospital logos regularly pull up to restock their supplies. Not only do patients prefer to buy in markets, but so do physicians. In Lagos, a report estimated that 58 percent of all physicians identify drug markets as their vendors of choice, because of availability and ease, despite the fact that substandard and counterfeit drugs are to be found in some of these markets (Pharmaceuticals Manufacturing Group, Manufacturing Association of Nigeria, 2001).

Many people I interviewed preferred to buy their drugs in the markets, to save money by avoiding the additional costs generated by seeing a doctor for treatment and care. Drug availability extends even to roadside "hawkers" who sell medications of all sorts in traffic jams and along busy and commercial roads, literally appearing and disappearing with the traffic itself. Sellers can also be found on public transport. While riding on buses, I encountered traveling drug salesmen who took turns wooing the crowds by offering candy and lame jokes about gender relations before launching into pitches about the efficacy of their goods. The attitude of the crowds on such bus rides often evolved from boredom and annoyance into enthusiasm and consumer passion. Pain pills, antibiotics, acne busters, and aloe vera were offered for sale at various times, and I myself bought some imported Indian Neem toothpaste.

Manthia Diawara has argued that "West African markets provide a serious challenge to the scheme of globalization and structural adjustment fostered by the World Bank and other multinational corporations that are vying to recolonize Africa. . . . [W]hat makes these traditional schemes of globalization special is the structural continuity they maintain with contemporary markets in opposition to the forms and structures of modernism that the nation-states have put in place in West Africa since the 1960s" (1998, 114–15). Indeed, economies are largely controlled and determined inside the markets (stationary and mobile), not by the banks. Diawara rightly claims that this poses a challenge to financial institutions that see the nation-state as the only legitimate vehicle to conduct business (ibid. 116). Diawara describes a typical scenario: to cope with the struggling economy, state officials depend on markets, where currency can be exchanged at higher rates than what banks offer, where low civil-service wages can be enhanced by bribes, and where forms of emergency cash may even be provided to a strapped government (ibid.). This is a system of recycling indebtedness that actually helps to stabilize a financial crisis and is tied to a conglomeration of state practices where notions of the public and private are blurred (ibid. 117).

The extent of counterfeit, fake, and substandard drugs located in the markets and on the streets actually represents a remarkable contradiction. On the one hand, the dispossession of the Nigerian pharmaceutical industry's capacity to, at best, carry out good manufacturing practices freed up space for new, mostly imported drug products to take root—drugs that can bypass regulatory organs of the state. During the 1990s these drugs especially competed with global company products that had distribution outlets in the country. In the process of an IMF-generated dispossession that would lay the ground for new proprietary pharmaceutical capital, counterfeit drugs, which made upward of 80 percent of the national drug market, became the thorn in the proprietary distribution agenda. In this sense, the drug markets provided the very challenge to state privatization and neoliberal reform that Diawara claimed.[18] For Diawara, "As postmodern reality defines historicity and ethics through consumption, those who do not consume are left to die outside of history and without human dignity. The traditional markets are the only places where Africans of all ethnic origins and classes, from the country and the city, meet and assert their humanity and historicity through consumption" (1998, 120–21). On the other hand, there are a great number of counterfeits and fakes that can create severe side effects and injury, which puts the "right to consume" in jeopardy and thus the role of both the state and market in question. The "right to consume" needs to be situated in the context of the "right to produce" and the "right to regulate"—a messy configuration. As these rights of the individual and the nation-state surface as out of joint, these forms of medical pluralism continue to thrive.

At the very same time that fake and counterfeit drugs were substantially outcompeting the global proprietary drugs in Nigeria (and counterfeits of all sorts were doing the same in other parts of the world), demands for increased intellectual-property protection were being issued by private industry and Western governments via the World Trade Organization. While NAFDAC has made moves to break down drug markets in the interest of both protecting proprietary drug businesses (something explicitly emphasized to me) and public health, and this certainly does protect the intellectual-property rights of the global proprietary industry, it is also giving an unexpected boost to the local generic industry, whose products were also largely copied during the 1990s. In the context of this rebirth, a different form of dispossession is creeping in, one that poses threats to the local manufacturing industry. This clearing is making room for U.S.-subsidized proprietary antiretroviral drugs via the politics of trade-related intellectual-property law as well as AIDS treatment policies. But this more refined emptying-out may actually destabilize the very new capital forms that are just beginning to be cultivated in this dispossessed drug landscape.

AIDS Treatment Policies and New Capital Imperatives

In Nigeria there are two major antiretroviral (ARV) treatment policies. One is the Nigerian government's, which uses only generic drugs produced in India by Ranbaxy and Cipla for the 20,000 enrolled patients (out of four million HIV-positive people, 400,000 of whom are estimated to be in immediate need of ARVs). The second uses both generic and proprietary drugs that are being supplied by PEPFAR, the largest international health initiative ever to target a single disease. In Nigeria, PEPFAR subsidizes and distributes U.S. proprietary drugs to over 100,000 patients.[19] Over $20 million for fiscal years 2004–6 were allocated to PEPFAR in Nigeria. There is no long-term plan of sustainability for either PEPFAR or the government's program. The initial PEPFAR plan was to allocate drugs and treatment for five years, which was extended to an additional ten years by the Bush administration. In the meantime, the Nigerian government plan is slowly yielding ground to PEPFAR, which is less expensive for patients.

Both programs involve different treatment regimens, are differently subsidized, and are highly politicized. As a Nigerian government supplier, Ranbaxy cornered the generic market in the early 2000s, and their drug prices actually exceeded the cost of similar generics. There have been other tensions: while at the 2004 Nigerian National AIDS conference, I witnessed a confrontation between Ranbaxy representatives and Nigerian AIDS activists over the fact that the company was selling ARVs at its booth without requiring prescriptions. There have also been numerous incidents that indicate that Ranbaxy does not want the more extreme problems of drug distribution in Nigeria to become public knowledge, particularly around the issue of counterfeits. The company must contend with a popular Nigerian opinion, held since SAP, that the majority of counterfeits are made and exported from India (generics and fakes can be often confused for each other, but fakes are often referred to as "India" drugs). Moreover, the Nigerian government has been accused of poor distribution and operations, the worst of which was a two-month-long drug shortage in 2002, due to bureaucratic complications, and with the implementation of PEPFAR, the government is slowly surrendering ground to the new U.S. operations.[20]

PEPFAR constitutes the largest roll out of public-health funding in history. It follows a particular policy logic initiated by Bill Clinton, who declared HIV/AIDS a national security threat in the mid-1990s. Since then, the rationales of health and security policies have increasingly merged in several different international arenas. For example, the emerging U.S. Africa Command (AFRICOM) intends to integrate staff structures from the U.S. Department of State, the U.S. Agency for International Development, and humanitarian organizations into existing military structures to conduct tasks ranging from managing AIDS to sharing intelligence. Indeed, the Office of the U.S. Global AIDS Coordinator, which heads PEPFAR, has been moved out of U.S. offices managing traditional development work and now answers to the U.S. Department of State under the secretary of state. Furthermore, the U.S. Department of Defense has a large role to play in the PEPFAR countries, including Nigeria, which rank high among U.S. security concerns. In Nigeria, not only does the Department of Defense have well-endowed AIDS projects, but so too does the Henry M. Jackson Foundation, a philanthropic organization dedicated to subsidizing what it calls "military medicine," which largely constructs infectious disease as part and parcel of security discourse.[21]

Stefan Elbe (2005) shows how these events reflect the ways a range of actors (international organizations, governments, and NGOs) are cast in the name of the survival of communities, economics, militaries, and governments. Key here is how such mobilization is enrolled by the language of security and emergency, which as Alan Ingram describes "takes HIV/AIDS out of the sphere of 'normal politics' and creates obligations to respond in ways that are adequate to the new salience of the problem" (2007, 516). As a result, policy becomes less directed toward civil society and more directed to security and intelligence (Elbe 2006; Ingram 2007). Vinh-Kim Nguyen sums up much of these new rationales and deployments through his term *experimentality*, which he describes as exercising "a new form of legitimate domination through highly mobile, disaggregated and mutable governmentalities. The latter are biological and political technologies for constituting populations and transforming subjectivities in a focused manner around a particular predicament of government. These predicaments are framed in humanitarian terms and call for urgent measures designed to save lives and prevent suffering, which is understood as an immediate and embodied (or even biological) phenomenon" (2007, 1).

Aside from the general merging of health, development, and security organization and rationales, Nigeria occupies a particular place in these new activities as it is a country that is viewed by the United States both as strategic to peacekeeping operations in Africa and, perhaps more important, as key to security efforts related to oil supplies throughout the Bight of Benin. Oil is crucial because it constitutes a significant chunk of the country's income, and several authors have demonstrated how security and oil are abstracted, where private forms of health development erase the politics and violence of extraction.[22] In order for the rationales and linkages of partnership development and security to actually take hold and play out, HIV prevention and treatment policies must be rationalized as normative in precisely the same way as oil extraction and security paradigms (Zalik 2004). Translating these paradigms into policy requires a particular form of management: indeed, PEPFAR under the Bush and Obama administrations mirrors the ways in which the Iraq and Afghanistan wars are managed—largely contracted and outsourced to private partners. The indirect result is that as West African security concerns and expansion appear to be continually facilitated not only by anti-terrorist efforts and the search for steady oil supplies, but also off the back of subsidized pharmaceutical products that rely on AIDS treatment policies for their mobility and consumption.

One of the early predecessors to the Bush administration's PEPFAR initiative was a lesser-known program that may mark the beginning of multilateral AIDS policy networks: the United Nations (UN) Accelerated Access Initiative (AAI) of 2000, a joint initiative among the Joint United Nations Programme on HIV/AIDS (UNAIDS), the World Health Organization (WHO), and the proprietary pharmaceutical industry that utilized public-relations firms to bilaterally negotiate the reduction of high and out-of-reach drug prices in Africa.[23] In exchange, stringent intellectual-property laws were conceptualized, proposed, and often implemented for African states in a manner that favored and protected multinational pharmaceutical companies' business practices in Africa. After a coalition of drug companies withdrew a well-known suit against South Africa, in 2001, claiming that its 1997 Medicines Act violated World Trade Organization (WTO) regulations on compulsory licensing and parallel importation—legislation that South Africa never acted on—companies taking part in the AAI began to heavily recruit many African countries to negotiate bilateral confidential agreements with the apparent aim of wiping out the generic drug industry. Two years into the program, ACT UP Paris reported that UNAIDS and the WHO, which orchestrated the negotiations, never provided technical assistance to participatory countries in protecting intellectual-property law or creating guidelines on relations between countries and companies. The WHO and UNAIDS forfeited power and follow up, which empowered the companies to take advantage of the lack of UN oversight (ACT UP Paris 2002). The Health Gap Coalition declared, "UNAIDS drug access policies are currently being structured, by and large, in response to big pharma's displeasure" (2000).

While these negotiations were hailed as some of the best and only options to access treatment, even though only 0.1 percent more people were put on treatment, other issues were crucially erased (ACT UP Paris 2002). At the end of this program, drug prices were not heavily slashed, but the AAI served as one of many now existing gateways for the proprietary pharmaceutical industry to outcompete generics by making policy that eradicated the generic industry. Such policies and actions have shaped the compilation of future drug markets, not simply in Africa, but throughout the world.

In addition to bilateral intellectual-property negotiations, the Trade Related Intellectual Property (TRIPS) Agreement of the WTO, to which Nigeria is a signatory, gives proprietary pharmaceutical companies exclusive twenty-year manufacturing, pricing, and distribution rights on their drug patents. The U.S. government's Agency for International Development (USAID) funds the Commercial Law Development Program (CLDP), an initiative of the U.S. Department of Commerce to "assist" Nigeria in complying with TRIPS. The CLDP sponsored several meetings jointly with the Nigerian Intellectual Property Law Association between 2000 and 2004. At these meetings, there were many panels and instructions on how to comply with the TRIPS/WTO geared around how Nigeria can "be on the right side of globalization." Consistently, the discourse, without any explanation, was that the stronger a country's intellectual-property law was, the more economically viable and powerful it would become in the global economy. Compared to the vast numbers of intellectual-property lawyers in the United States and European patent offices who have access to worldwide databases that can easily determine if an invention is new or discern an intellectual-property violation, the Nigerian patent office awards a patent if the two-page application form is filled out correctly. Given such technological and expertise disparities, Nigeria can hardly be expected to compete internationally or instantly instantiate power in the global economy.

The United States submitted its own drafts of a new Nigerian intellectual-property law to the Nigerian government, in 2002. I acquired these drafts, which clearly showed that the United States desires a law that favors U.S. businesses while it wipes out all legal provisions to import less expensive generic drugs. At the "final" drafting meeting, AIDS activists (with technical support from international actors like Médecins Sans Frontières [MSF] and Ralph Nader's Consumer Project on Technology) muscled their way into the meeting to demand the inclusion of "healthcare safeguards," which were incorporated into the draft. This was perceived as a great victory.

However, less than a year later, the CLDP returned to Nigeria, apparently (at least according to rumor) under the instructions of the U.S. Patent and Trademark Office, which had determined that Nigeria's new intellectual-property law draft did not satisfy U.S. preferences. A secret meeting took place, without activists' or government health officials' knowledge. But the meeting became known when its "successful conclusion" was announced on national television.

I acquired the latest intellectual-property draft, which may or may not be the official document, as, historically speaking, multiple drafts have circulated among Nigerian and U.S. officials, who have generated confusion over the "real" document. However, lawyers at MSF in Europe analyzed the draft in my possession and found that it included "data exclusivity" measures that, in short, effectively reduce the generic drug industry's capacity to quickly manufacture generics coming off patent. Such provisions already carried out in other free-trade agreements allow proprietary drug companies to keep data confidential. Such an act actually undermines the original intent of a patent that exchanged inventive data for short-term exclusive marketing. Moreover, it may be a strategy that slowly whittles away the public domain. That is, without the data in hand, a generic company is prevented from developing the technological design to engineer a generic product, a delay that can essentially extend the life of a patent.

USAID simultaneously funds a great number of local AIDS NGOS to carry out prevention and education programs. To some AIDS activists, there is the appearance of a USAID policy contradiction, which supports AIDS activism yet also works to severely curtail drug access. But there may not in fact be a contradiction, as prevention and education campaigns are located in the realm of individual empowerment and responsibility, drawing attention away from the legal structures that generate obstacles to pharmaceutical flows. AIDS activists and NGOS have objected to the relationship between the Nigerian and U.S. governments. But this relationship demonstrates a conflict that the state itself has with multilateral organizations. That is, the state opposes U.S. and European stances on treatment access at global trading negotiations, but at the same time attempts to meet the pressure to comply quietly behind the doors of federal ministries. This represents an increasingly common strategy utilized by the United States, whereby it capitalizes on the lack of communication between ministries, and between ministries and Nigeria's Geneva representatives; and bilateral and regional (trade or otherwise) agreements become the alternative avenue and means for compliance when global negotiations continually fail. Yet what exactly does it mean for Nigeria to buy generic drugs for its own national antiretroviral program while at the same time it cooperates with the U.S. government to legally wipe out generic drug access? Such an action will effectively make its own antiretroviral program illegal. Nigeria still has not complied with TRIPS, and it is not clear when or if it will happen.

Conclusion

The "implicit agreements" made among multilateral institutions fundamentally drive a political economy that relies on dispossession as its primary organizational strategy. Unlike the massive state and economic adjustments made under the IMF that literally teeter economies on the edge of collapse, a sustained dispossession is very particular and targeted; it takes place amid already chaotic economic and social environments and therefore must operate in more delicate ways that do not threaten the existing thresholds of disintegration. The most particular example is found in struggles over intellectual-property designs, which do not necessarily destroy entire economies, but target specific industries that are viewed as especially competitive; and the massive introduction of free ARV drugs that will not be sustained over time will also have a similar impact on this industry.

While dispossession on a large scale produces sustained clearings for new transnational capital to take root, there is something far more productive in the margins, as Janet Roitman would put it. But at the same time, dispossessed capital and its new formations and mobilities have to rely on other orchestrations of individual and institutional struggles and consent that should suggest just how we might rethink political economy. Rationales for public health and imperatives of capital, and humanitarianism efforts and security cultures, combine and conflict, but ultimately show that the AIDS crisis itself is the greatest thorn in any country's national neoliberal agenda, or perhaps its greatest opportunity.

Notes

1. Some of the literature on the global and globalization that informs my thinking follows. On the global city, see Sassen 2001. On cosmopolitanism, see Cheah and Robbins 1998; Derrida 2001. On globalization, transnationalism, and global networks, see Grewal and Kaplan 1997; Appadurai 1991; Appadurai 2001; Jameson and Miyoshi 1998. On neoliberalism, see Comaroff and Comaroff 2000. On finance markets, see Lee and LiPuma 2002.

2. Notable recent exceptions include Ferguson 2006; Mbembe 2001; Cooper 2002; Moore 2005. James Ferguson (2006) in particular has referred to Africa as the "inconvenient case" that runs against contemporary works of globalization, which describe unencumbered transnational flows; particularly provocative is the analysis by Ferguson (2005) of the way that capital does not flow but "hops" through privatized spaces, especially of oil extraction.

3. I acknowledge Kaushik Sunder Rajan for helping me think through this argument.

4. I acknowledge Don Moore who spoke to me about how "emptying-out" as a form of dispossession is generated by political technologies that contribute to creative destruction accompanied by consequential material realities.

5. Few of the globalization paradigms in circulation today entirely apprehend the complicated ways that African states are positioned in these scenarios. That is, global cities, finance markets, and widespread manufacturing bases do not exist on the continent in congruent ways found elsewhere in the world; and local-global relationships do not adequately capture the nature of the state where kinship and private and public forms of power commingle (Ake 1996).

6. Contrary to widespread popular dissent, the IMF negotiated a SAP with then head of state General Ibrahim Babangida shortly after he took over by coup d'état. This event marks the beginning stages of a protracted prodemocracy movement that coexisted with a heightened culture of militarism characterized by the quest for wealth and violence. With rising poverty and lack of security, a culture of militarism lingers now into the current civilian era and importantly informs retrospective debates and discourses about IMF fallouts such as governance, and the politics of healthcare, pharmacy, and drug manufacturing.

7. Julia Elyachar (2005), Janet Roitman (2005), and Donald Moore (2005) examine the micropolitics of dispossession and the various forms of reinvestment and political stakes in Egypt, the Chad Basin, and Zimbabwe, respectively.

8. George Caffentzis (1995) made an almost identical argument about structural adjustment and dispossession in Africa, where he traces its dramatic impact on social transformation.

9. According to official statistics, oil wealth makes up over 46 percent of Nigeria's gross domestic product and accounts for 85 percent of the country's foreign exchange (World Bank Report 2004). The rumor on the street in Lagos is that 40 percent of Nigeria's crude oil goes missing each year (Apter 2005).

10. Brotherton, however, refers to the Cuban state.

11. For more recent excellent analyses on oil in Nigeria, see Watts 2001; Watts 2004; Apter 2005; Naanen 2004; Okonta 2004.

12. I first found an analysis of this quote in Avery Gordon's introduction to Cedric Robinson's *The Anthropology of Marxism* (2001), which is reprinted in Avery Gordon's *Keeping Good Time* (2004).

13. James Ferguson (2005), however, points to how these scenarios become possible when capital hops, rather than flows, into privatized sectors, like oil extraction.

14. A. P. Robertson estimates that "capital flight from Nigeria alone vary from $50 billion, to 135–150 billion, to 3,000 billion British pounds" (n.d. 5).

15. The Nigeria debt-cancellation deal of 2006 included forgiving all but the U.S. $12 billion owed to Paris Club members. This deal included adhering to the controversial IMF Support Policy Instrument, which is extended to countries that do not need IMF loans but that nonetheless seek IMF endorsement signaling an easier release of funds from multilateral donors and banks. This mechanism extends older forms of legitimacy-making in the context of debt-worthiness (not necessarily credit-worthiness). In return, countries must adhere to the usual privatization schemes geared toward foreign direct investment. Nigeria agreed to completely privatize the national energy sector and reorganize its banking sector, among others, in exchange for the deal.

16. In fact, one pharmacist who has worked in the national inspectorate estimated to me that 95 percent of all registered pharmacies would not actually pass inspection; they manage to get their licenses in exchange for money to "look the other way," as he put it.

17. A survey by A. O. Bright and O. Taylor (1999) showed that irrespective of socioeconomic stratum, self-medication is very high, recording 75 percent as the lowest figure. It concluded that if pharmacists refuse to assist the self-medicating population, then morbidity, mortality, iaotrogenicity, and other adverse effects will be on the increase. Their statistics on reasons for self-medication are worth reporting here: cheapness, 32.9 percent; effectiveness, 71.35 percent; sure relief or cure, 85.23 percent; time saving, 66.19 percent; respondents with clear knowledge of pharmacists, 88.79 percent; respondents who would ask for pharmacist intervention, 52.49 percent; respondents who would see doctors first if ill, 9 percent; respondents who feel they do not need a pharmacist, 36.11 percent.

18. Julia Elyachar (2005) shows how in the aftermath of structural adjustment the poor—their networks and social practices—have been incorporated into the rhetoric and logics of free market expansion by the World Bank and other institutions that use an explicit neoliberal vision (and not other alternatives) to implement such policies and practices.

19. There are several other international bilateral treatment programs that are also part of this mix.

20. It should be noted that at least one indigenous Nigerian drug-manufacturing firm was poised to begin producing anti-HIV medication. The subsidization of both

U.S. and Indian products has essentially eliminated the prospects of Nigerian manufacturing at this time.

21. The other top two PEPFAR countries that get Department of Defense funding are Tanzania and Kenya, perhaps matching antiterrorist efforts since the 1998 U.S. embassy bombings in those countries.

22. Apter 2005; Okonta and Douglass 2003; Watts 2001; Zalik 2004; Ingram 2007.

23. The pharmaceutical industry was represented by five companies: Boehringer Ingelheim, Bristol-Myers Squibb, GlaxoWellcome, Merck, and Hoffmann-La Roche.

PART THREE

GLOBAL KNOWLEDGE FORMATIONS

8

ANDREW LAKOFF

DIAGNOSTIC LIQUIDITY

Mental Illness and the Global Trade in DNA

"Information is, at the end of the day, the coin of the genomics realm."

ANTONIO REGALADO, "INVENTING THE PHARMACOGENOMICS BUSINESS"

Discussions of globalization processes often describe an increasingly rapid flow of information, capital, and human bodies across national borders in the wake of technological innovation and political-economic transformation. As a number analysts have noted, such increasing global circulation operates in relation to regulatory techniques and governmental strategies — at local, national, and transnational levels — that both encourage and constrain these flows (Brenner 1999; Sassen 2000).[1] Examples of such techniques include intellectual-property regimes, immigration policies, and environmental standards. The negotiation of institutionalized regimes of coordination or harmonization — the linking of places through the creation of commensurable standards — is often necessary to make such circulation possible. By the same token, technical and regulatory regimes can also block the movement of goods or persons, as in the case of barriers to the sale of genetically modified foods or ethical codes concerning the sale of human organs.[2] Recent studies of the creation and enactment of standards regimes provide helpful tools for the analysis of the micropractices involved in creating zones of potential circulation (Bowker and Star 1999; Espeland and

Stevens 1998; Alder 1998). This chapter combines an ethnographic description of the process of transnational standards coordination with an analysis of the macropolitical contexts in which such commensuration practices unfold. It follows a particular set of transnational flows, involving human DNA, biomedical knowledge, and capital, whose direction is intimately related to the relative presence of regulatory and technical regimes within different national spaces. The goal of the chapter is to indicate the salience of epistemic practices for understanding the generation of value in an economy of biological information.

Specifically, I follow an attempt by a French biotechnology company to find and patent genes linked to psychiatric illness among a group of Argentine mental patients. This genomics research was significant in its institutional form, as well as in its potential implications for the reconfiguration of knowledge about mental illness. As an alliance between biotechnology and psychiatry across continents, and between public and private institutions, it represented a new type of assemblage oriented toward the understanding and regulation of human behavior. The central problem it raised—both practical and epistemological—concerned the *potential universality* of genomic knowledge about mental disorder. The success of the company's gene-hunting effort hinged on the global validity of a set of diagnostic standards that, it was hoped, would make possible the commensuration of divergent illness experience into a common classificatory scheme. Such commensuration, in turn, would enable psychiatric illness to be represented as genomic information, and would thus make the illness experience of Argentine patients convertible with that of patients in other parts of the world. How such experience was rendered liquid—that is, able to circulate and to potentially attain value as information—is the focus of this chapter. I show that in the case of mental illness, the effort to generate a space in which information flows seamlessly between biomedicine and the market is challenged by the difficulty of knowing just what a psychiatric disorder *is*. The extraction of valuable knowledge from patients' DNA relies on the development of diagnostic standards whose validity and extendibility remains in question.

Diagnostic Liquidity

In June 1997 the French genomics firm Genset announced a collaboration with the psychopathology department of a public hospital in Buenos Aires to collect and map the DNA of patients suffering from bipolar disorder. The genes or markers linked to susceptibility to bipolar disorder, if found, were to be patented by Genset as part of its strategy to enter into partnerships with major pharmaceutical companies for the development of new diagnostic and therapeutic technologies.

The process of gathering large amounts of data about the prevalence of illness in populations has historically been linked to state-based public-health initiatives: in order to gauge and improve the health of the population, national and multilateral health agencies have, in collaboration with public-health experts, sought to accumulate knowledge about the spatial incidence of disease. Recent genomics research in places like China, Iceland, Russia, and Argentina is distinctive in that it is often conducted by private database firms in collaboration with local clinics. The case I describe here—in which the actual collection of DNA was carried out by local clinicians working in public hospitals, under contract to a genomics database firm—is not atypical of such arrangements—though it was more informal and less visible than, for example, deCODE's work in Iceland. This pattern of collaboration is conditioned on recent economic and technoscientific developments: on the one hand, the emergence of health as a significant global marketplace, and on the other hand, the rapid development of DNA-sequencing technology and bioinformatics in the wake of the Human Genome Project. For this reason, highly capitalized biotech firms have become interested in the possibility of attaining valuable genetic information through research on specific local populations. In this emergent space of exchange between industry and the life sciences, the role of government remains salient: the health marketplace as a target of technoscientific innovation is structured by the legal forms that ensure that biological information can attain value—that is, intellectual-property regimes.

It should be emphasized that this strand of genomics research targets health consumers in the advanced industrialized countries. The most valuable information in the health marketplace pertains to specific kinds of populations: North Americans and Europeans at risk of chronic illness, whose insurance will pay for the extended use of patented medications.[3] In this context, the type of DNA collection and analysis I describe here seeks to demarcate specific illness populations that are simultaneously potential market segments. As this case illustrates, in poorer parts of the world patients serve as potential sources of valuable knowledge rather than as target markets, and they are often easier to access due to relaxed regulatory controls.

The Genset bipolar study was one of a number of transnational projects in the late 1990s involving newly minted genomics database firms based in the United States or Western Europe and health clinics in other parts of the world that were contracted to provide supplies of DNA from sample populations.[4] The sense was that there were hidden riches buried in the genomes of these clinically diagnosed patient populations.[5] As the Human Genome Project progressed, what the legal theorist James Boyle called "an intellectual land grab" (1996, 9) began as genomics database start-ups competed to find and patent genes or genetic markers linked to common, complex disorders.[6] Like other claims to property rights in genetic material, the Genset study was, as Sheila Jasanoff writes, "a project in mining nature for extractable entities that can freely circulate" (chapter 4, this volume).[7] While the value of such genes was still a matter of speculation, genomics database companies were confident that patented sequence information would prove a marketable resource in the burgeoning health marketplace. In Argentina, Genset sought to secure a supply of blood samples from an ethnically diverse patient population whose genetic background was similar to that of European and North American target markets, but without certain of the regulatory and legal complications that characterized such work in the North.

At stake in the process of gathering, analyzing, and developing proprietary knowledge from patients' DNA samples was the relation between truth and value in the global biomedical economy. At the scientific level, the translation from genetic material to significant information depended on the validity of the diagnostic criteria used in gathering sample populations— criteria that in the case of psychiatric disorders had emerged in local and contingent circumstances. The economic value of such information, meanwhile, hinged on an intellectual-property regime that granted monopoly rights to genomic innovation and on a market—or an imagined future market—that structured demand for such information. Transnational epidemiology, in turn, made it possible to locate that market and estimate its size.

A key question arose in Genset's research that focused attention on the classificatory devices to be used in gathering the sample population: to what extent could these diagnostic criteria be claimed to measure the same thing across different spaces? How to know, for instance, whether a case of bipolar disorder in the United States was the same "thing" as a case of bipolar disorder in Argentina? Recent work in the social studies of science and medicine has investigated the processes through which the apparently universal validity of biomedical knowledge is materially and discursively forged via the standardization of practice across multiple domains.[8] This work indicates that the spread of standardized protocols does not necessarily produce equivalent practices in diverse sites (Mol and Law 1994). Thus, for example, Timothy Choy demonstrates the variable ways in which a "standard" technique for measuring air pollution is implemented across structurally differentiated global space (Choy, chapter 3 in this volume). In the case of the Genset study, what must be examined is the complex process of commensuration that was necessary for subjects with diverse histories and experiences of illness to both recognize themselves as having bipolar disorder and to be so classified by doctors. At the same time, the difficulties faced in conducting the study illustrate local epistemic and political challenges to producing such equivalence.

To analyze the process of forging consistent illness populations so that Argentine patients' DNA could enter into circulation, I borrow the term *liquidity* from the field of finance. Bruce Carruthers and Arthur Stinchcombe (1999) analyze liquidity in futures markets as an example of the production of standardized value—the creation of generalized knowledge about value out of idiosyncratic personal knowledge. They argue that producing equivalence out of specific entities involves both social regulation and political negotiation. Standardization is a social and cognitive achievement: buyers, market makers, and sellers have to share the conviction that "equivalent" commodities are really the same. Turning an illiquid asset into a more liquid one is a process of reduction and standardization of complexity.

To be transferable—liquid—an asset must lose its particularity and locality. Classificatory technologies work to simplify, stratify, and standardize such assets. Thus—to use Carruthers's and Stinchcombe's example— a distinctive house becomes a liquid asset only when there are agreed-on conventions for evaluating it in comparison with other houses. Similarly, William Cronon (1991) has shown how wheat was made into a liquid commodity in nineteenth-century Chicago through the invention of a set of technical standards for classifying the characteristics of specific bushels of wheat in terms of more general quality grades that made it unnecessary for buyers to inspect each bushel purchased.[9] Individualized evaluations of quality were thus shifted into collectively sanctioned criteria, enabling bushels of wheat to be abstracted and circulated as currency. In order to successfully implement such a system, the existence and legitimacy of a governing body that regulates the practice of measurement is crucial.

It is possible to consider the circulation of bipolar patients' DNA in terms of this process of abstraction through technical classification: the patients' illnesses assumed potential informational significance—and therefore, value—only insofar as their specific life trajectories could be brought into a shared space of regulated measurement. That is, their illnesses had to be made "liquid." From the vantage of genomics research, one should not need to know about the specific life trajectory of the person from whom DNA has been extracted in order to evaluate the significance of the information it bears. Diagnosis is the convention that produces such equivalence. In the case of bipolar disorder, what might seem like an implausible association then becomes natural: a young woman who has attempted suicide in Buenos Aires is brought into potential relationship with a middle-aged man in Chicago who goes bankrupt through risky business ventures. They are both members of a group of previously distinctive individuals who now share a diagnosis.[10] The emergent group is alternately an epidemiological population, a market segment, and a community of self-identity.

Thus, while "liquidity" is typically understood in terms of finance, here techniques of classification enable biomedical knowledge to be assimilated to the domain of market exchange, shaping a hybrid commercial-epistemic milieu. In biomedicine, forging such a space of liquidity requires consistent classificatory practice among doctors—a problem that remains fraught in psychiatry, especially in Argentina. In what follows I describe how doctors in Buenos Aires performed classificatory work with psychiatric patients in order to render their illnesses liquid—that is, abstract and therefore exchangeable. This process involved the temporary extension of both a technical and an ethical standards regime. The setting of the DNA collection in Argentina revealed not only the reliance of technoscientific objects, such as bipolar genes, on such regimes, but also the limits to their extension.

Circulatory Networks

The bipolar study crystallized through a contingent set of associations and opportunities—though one structured by the processes of biological value creation. In 1997 Daniel Mendelson, an unemployed Argentine molecular biologist, was making a living by supplying genetic material from human organ tissue to Genset, a French biotech company that was building a cDNA library—a compilation of expressed human genes for use in detecting significant genetic information. Mendelson's work was a bit grisly. He would call up contacts who worked in forensic-pathology laboratories in Buenos Aires hospitals, and ask them to send over healthy tissue from newly dead cadavers. Genset wanted various organs for its collection: kidneys, hearts, even brains. Once the tissue was sent over to him, Mendelson would process it in a lab he had rented at the Campomar Institute, a well-known biological-research center near the Parque Centenario in Buenos Aires. He had been trained there before leaving to do postdoctoral work at the Pasteur Institute in Paris with his wife, Marta Blumenfeld, also a molecular biologist. Now she was vice president of genomics at Genset, and he was struggling to establish a beachhead back home in Buenos Aires.

Mendelson had a new idea: they could expand their business of providing genetic material by obtaining DNA samples from patients with mental disorders. Genset was looking for populations of patients who had been diagnosed with schizophrenia and bipolar disorder for its gene-discovery program in complex diseases. An old friend of Mendelson's and Blumenfeld's from school now worked as a psychiatrist at a general hospital in Buenos Aires where patients could be recruited. After some back-and-forth negotiation, the details were worked out: Genset would give a hundred thousand dollars to Hospital Romero for structural improvements, and in exchange, doctors there would provide blood samples from two hundred patients diagnosed with bipolar disorder, types I and II.[11]

Genset was in a hurry to acquire such material. As a genomics database company, its strategy depended on finding and patenting genes linked to susceptibility to common, complex diseases. With its growing patent portfolio and proprietary genomic-search technologies in hand, Genset sought partnerships with large pharmaceutical firms to develop new diagnostic and therapeutic applications. It had recently formed strategic alliances with Abbott Pharmaceuticals, a leader in the diagnostics market as well as the maker of the leading medication for bipolar disorder, and with Janssen Pharmaceuticals, producers of the antipsychotic drug Risperdal. Pharmaceutical-industry strategists expected the next series of significant discoveries of drugs for mental disorder to come out of the Human Genome Project; closer on the horizon was the prospect of diagnostic tests linked either to disease-susceptibility or medication-response. In order to have commercial rights to such products, Genset had to beat a number of competitors, in both the academic realm and the private sphere, to the relevant genomic loci. The alliance with Abbott was an early signal that major players in the pharmaceutical industry saw genomics as an important strategic arena. Given the possibility of royalties on a range of products, it seemed in the late 1990s—a moment of intense speculation in the life sciences, both conceptual and financial—that genomic information had potentially exponential value. As one biotech analyst wrote of the collaboration, "The Genset-Abbott deal is clearly geared toward creating a resource that the pair can sell again and again" (Regalado 1999, 45).

The value of such resources relied not only on the construction of a market in genomic information, but also on the prospect that something scientifically significant would be found—which was by no means a foregone conclusion. Despite decades of academic research and a string of false alarms, no genomic loci had yet been confirmed to be linked to any of the major psychiatric disorders. According to Mendelson, it had only recently become possible to hunt seriously for such genes. First, developments in molecular biology and information technology now allowed genome-wide searches for disorders characterized by complex genetic and environmental interactions. Genset's proprietary SNP (single nucleotide polymorphism) map provided dense markers to guide its researchers through the immense human genome, giving it an edge over academic and private-sphere competitors.[12] And second, it was now possible to forge coherent populations of clinically diagnosed patients: standardized criteria for diagnosing bipolar disorder had been spelled out in 1980 with the publication of the third edition of the diagnostic manual of the American Psychiatric Association and had evolved in subsequent editions.[13]

Mendelson explained the process of looking for single nucleotide polymorphisms—natural variations in the genome—associated with bipolar disorder: if Genset could find a corresponding variation in multiple patients, it was likely that a susceptibility locus would be near, or statistically associated with, that variation. It was not a new or original scientific idea, he admitted, but it was one that was, practically speaking, daunting, requiring massive gene-sequencing and bioinformatics capabilities. Five years before it would have been technically unimaginable.

Once Mendelson and Blumenfeld made arrangements with Genset on the one hand, and with the hospital on the other, there was some delay in getting the DNA collection going. First, the Buenos Aires city government blocked the project on the grounds that it violated a law against trafficking in blood. After the concerned parties convinced the city's legal office that DNA was distinct from blood, and therefore saleable, another problem emerged: according to city regulations, a public hospital could not be paid by a private company for its services. This regulation was eventually circumvented, with the help of contacts in the municipal government, by changing the wording of the contract from "payment" to "voluntary donation." By the time such regulatory hurdles had been taken care of, six months had passed.

And then, when the study finally began, doctors at the hospital faced an unexpected problem: they could not find enough bipolar patients. It turned out that bipolar disorder was rarely diagnosed in Argentina. The North American diagnostic system in which it was recognized had not permeated the Argentine mental-health world, nor had "bipolar identity" spread to raise awareness of the condition among potential patients (Jamison 1997). Without such techniques of classification in place, the extraction and exchange of DNA could not begin. Doctors in the men's ward at Hospital Romero remained in need of donors even after recruiting at a nascent self-help group for patients with bipolar disorder, and were forced to make announcements in the newspapers asking for volunteers. In July 1998 a number of articles appeared in the city's major dailies describing the symptoms of bipolar disorder and promoting Romero's study.[14] These articles were in part geared to inform the public about what bipolar disorder was, given the absence of general knowledge of the condition. The publicity campaign turned out to be quite successful in drawing volunteers to the hospital, and by late September, psychiatrists in the men's ward were almost two-thirds of the way through their assignment to compile two hundred samples. I was able to observe some of the collection process.

Collection (I)

Hospital Romero is located in a working-class neighborhood in the southern part of Buenos Aires. The hospital does not seem a likely place to be linked to cutting-edge genomic research. Built in the 1930s, much of Romero is visibly crumbling, testimony to the current conditions of public-health infrastructure in Argentina. The psychopathology service is in especially poor shape—Genset's promised donation would go a long way, it turns out. On a Tuesday morning in September, a diverse group lingers around wooden benches in the entryway, all waiting to be attended: patients and family members, pharmaceutical-company representatives (known as *valijas*, or "briefcases," because of the satchels full of samples and promotional literature they carry around), and various cats who have wandered in from the hospital grounds. Through a swinging door, I enter the men's wing, passing a dozen old cubicles, where a few patients lie on sagging cots, on the way to the examination rooms. Some of the other patients are playing cards, or listening to the radio. The floors are of once white, broken tile; the smell of ammonia is in the air.

A woman in her fifties, led by her daughter, is shown into a small room, bare except for a few chairs. They have traveled to Romero from a town about an hour away, having seen an article in *La Nación* on the study of bipolar disorder being conducted there. After a preliminary phone interview, they were invited for an examination at the hospital. The mother and daughter do not seem particularly interested in the details of the gene study. They have come not so much to give blood as to ask for help: a diagnosis, a drug, a competent doctor. Gustavo Rechtman, a staff psychiatrist in his thirties, interviews them for about five minutes. He is formal and to the point, asking first whether the woman has had any depressions. Yes, she answers, looking to her daughter for reassurance. Very serious ones, adds the daughter—with suicidal thoughts. And are these sometimes followed by euphorias? She nods. Has she used any medications? She has taken antidepressants in the past, and lithium—though, Rechtman notes, perhaps at too small a dosage. Her weight indicates that there might be a thyroid condition. Rechtman gives his diagnosis: bipolar disorder, type II—with hypomania. He mentions Fundación Bipolares de Argentina (FUBIPA), the support group for bipolar patients and their families that helped publicize the study, but discourages the woman from seeking further treatment at Romero: it is very busy here, he says, and besides, this is a men's ward.[15] Instead, he will write a note to the doctor at her health clinic telling him of the diagnosis.

Rechtman then explains the scientific research to them: a French laboratory is doing a genetic study in order to eventually create a treatment, to see if the genes of patients are different from normal genes. It will have no direct benefit for her. Is she willing to participate? A glance at her daughter. Sure. A form is filled out: age, gender, marital status, occupation, ethnicity, financial status, familial antecedents, medication history. Then she is brought to a larger room, where test-tubes sit on the table, some already filled with blood. A male nurse, after considerable difficulty, finds a vein. A notebook is annotated, a code number put on the test tube. While the blood is drawn, the woman is handed a consent form, which she glances at briefly before signing. The blood will then travel the same route as the organ tissue before it—DNA will be extracted at Campomar and sent by special courier on to the Genset research campus at Evry, outside of Paris.

The transferability of genetic material depended not only on Genset's technical capacities to derive information from the patient's blood, but also on the extension of an ethical-legal regime that sanctioned the technique: norms and regulations surrounding the circulation of genetic material between public institutions and private companies, and across national boundaries. The consent form legally detached the DNA from the patient. Drawn up by psychiatrists at Romero, it did not mention the possibility that the extracted genetic material might be patented. In a context where biomedical research was relatively rare and doctors retained significant authority, the consent form was not a well-recognized device, and therefore was something of a hollow ritual designed to meet the demands of the North Atlantic ethical sphere—it would be a part of the protocol that the firm would include along with any scientific achievements in a patent application or publication. What the patient received at the hospital was not a payment, but a diagnosis and a referral.

In general, the circumstances of the study did not especially concern observers I spoke with in Buenos Aires. Only a foreign company, some commented—and certainly not the Argentine state—could possibly do such advanced scientific work here. As for the role of the private sphere, given Argentina's recent history of state violence and political corruption there was little sense that private companies were less trustworthy than the state. And compared to some scandalous medical experiments that had recently been publicized, this one seemed fairly innocuous, involving only the taking of blood, and might lead to scientific advance.[16] Meanwhile, there was little worry over the political implications of finding genes linked to mental illness or discussion of a return of eugenics—although, especially from members of the city's large corps of psychoanalysts, there was considerable skepticism as to whether anything significant would be found. Nor was the question of whether genes should be patentable much broached, except insofar as transnational bioethics discourse was beginning to be imported by a small circle of legal scholars.[17] Both anxieties and promises around the Human Genome Project, so prevalent in the North, had not yet arrived in Argentina.[18]

For some Argentine scientists, publicity around the Genset study provided an opportunity to encourage more local attention to such issues. Mariano Levin, a molecular biologist who had worked in France with Genset's scientific director, Daniel Cohen, suggested that Argentina was an appealing place for the study precisely because of its lack of regulations on genetic research and patenting, not to mention that it was a good bargain for Genset. Cohen is a "marchand de tapis," he remarked, a rug merchant. For what was pocket change in the field of genomics, Genset would receive samples of diagnosed patients from a population whose ethnic origins were similar to those of target drug and diagnostic markets in Europe and North America. As Blumenfeld said, the city's "outbred population," predominantly of Italian, Spanish, and Jewish descent, was one reason, along with its large supply of well-trained psy-professionals, that Genset chose to work in Buenos Aires.

Of several articles that appeared in the Argentine media concerning the gene study, only one, in the short-lived progressive weekly *Siglo xx*, was critical. This piece was accompanied by pictures of multiple Barbie and Ken dolls, and a table, translated from *Mother Jones*, showing multinational pharmaceutical companies' claims to patented genes. The article began with a joking reference to an Argentine penchant for melancholia: "In Canada they study the gene for obesity. In Chile and in Tristan da Cunha, that of asthma. In Iceland, that of alcoholism. In Gabon, that of HIV. In the international partitioning of the body by the Human Genome Initiative . . . the French private company Genset chose Argentina to investigate the genetic roots of manic depression, as if this illness were an innate characteristic of the national being" (Goobar 1998, 66).

The Genset study was used, in the article, as an opening for a discussion of the potential abuses of transnational genomics research. "It's a huge business straddling the frontier between medicine and biopiracy," said a geneticist who wished to remain anonymous. Why did Genset bother to go to Argentina to look for the genes? Mariano Levin was quoted in the article: "In this country there are no laws on genetic research and patenting, which diminishes the risks and costs if something goes badly, and increases the benefits if the research is successful" (ibid.). Levin's argument was not that such research should not be conducted in the country, but rather that Argentina needed to adopt and implement new forms of regulation—and ideally, to develop its own biotechnology research sector—in order to avoid being exploited by multinational firms seeking inexpensive genetic resources.

Alejandro Noailles, the director of Romero's psychopathology ward, suspected that the peripheral status of Argentine clinicians made the country an especially good place for Genset to do the study. This is a private company, he emphasized in our first meeting, with a purely cost-benefit logic, and it is relatively inexpensive for them to do the study in Argentina. But even more, they won't have to share patent rights with those who do the work of collecting the samples: if the company were to do the study in Europe or the United States, he surmised, they might have to split the proceeds with the clinicians.

The key legal device making illness-susceptibility genes potentially valuable was the agreement that well-characterized genes could be registered as intellectual property, which had been supported, though not without controversy, by European and United States patent offices since a landmark 1980 Supreme Court decision allowing living organisms to be patented (Rabinow 1995; Jasanoff 1995). Patents guarantee an exclusive license to commercialize discoveries for a limited time—normally twenty years. The question of what kind of DNA sequence information was sufficient to grant patent rights was a matter of some contention. In 1998 the director of bio-technology examination at the U.S. Patent and Trademark Office gave a pro-visional answer: "For DNA to be patentable, it must be novel and non-obvious in light of structurally related DNA or RNA information taught in nonpatent literature or suggested by prior patents" (Doll 1998, 690). After an initial stage of broad acceptance of patent claims on new genetic information, the tendency by the late 1990s was toward a narrower vision of patentability— an insistence that the biological function and potential uses of the sequence be well demonstrated. Patent or no, the eventual value of such information was uncertain, as genomics-based products remained far on the horizon.[19]

Genset's research strategy of opportunistically seeking genetically heterogeneous patient populations was distinct from that of some other genomics companies, such as decode, which sought to leverage the ethnic homogeneity, detailed genealogical records, and comprehensive clinical data available on the Icelandic population for its potential informational value.[20] Genset's research also provoked a far more muted response from the public than decode's work in Iceland: while decode's project led to a na-tional referendum and a spirited transnational debate on its ethical implica-tions, research like Genset's remained mostly within the background noise of the 1990s biotech boom. An exception was an article in the *Guardian*, which noted that gene patenting was far from an exclusively North Ameri-can phenomenon: "European firms have become some of the most enthusi-astic stakers of claims on human DNA. Patent applications on no fewer than 36,083 genes and DNA sequences—28.5% of the total claimed so far—have been filed by a single French firm, Genset. Andre Pernet, Genset's chief ex-ecutive officer, said: 'It's going to be a race. The whole genome will have been patented two years from now, if it hasn't been done already'" (Meek 2000).

Genset had fashioned itself as a company specializing in disorders of the central nervous system—specifically bipolar disorder and schizophrenia. As its founder, Pascal Brandys, said, "I believe that the brain is the next frontier, not just in genomics but in biotechnology as a whole" (*Genetic Engineering News* 2000). Given the increasing size of the central nervous system (CNS) market, genes linked to mental illness that might provide new targets for drug innovation or lead to diagnostic technologies were potentially quite lucrative. Worldwide drug sales for CNS disorders were $30 billion in 1999, and CNS was the fastest growing product sector in the United States pharmaceutical market; by 2000 CNS disorders had overtaken gastroenterological illness as the second largest market segment, after cardiovascular conditions.[21] A venture capitalist noted the increasing interest in the CNS market, invoking the familiar image of the land rush: "Every doctor knows that the brain is the final frontier of medicine, but VCs [venture capitalists] are just now starting to sniff opportunity. There'll be a lot of opportunities to play this sector because there are just so many problems that fall under the heading CNS" (Herrera 2001). Such opportunities ranged from Alzheimer's disease to attention deficit disorder, anxiety, and schizophrenia. One question that was crucial to the eventual success of such ventures was whether illness populations as they had been classified according to the diagnostic standards of the American Psychiatric Association could be delineated at the genetic level.

Diagnostic Infrastructure

Noailles had recently returned from a visit to Genset's high-tech laboratory near Paris, stocked with millions of dollars worth of gene-sequencing machines and high-speed computers. There a committee of European psychiatrists had gone over the research protocol for the study with Romero's staff to ensure consistent diagnostic practice. It was hoped that such standardized diagnostic protocols would mediate between the subjective interpretation of the clinician and the impersonal evidence of the gene. Genset's protocol presumed that for the purposes of gathering consistent populations, psychiatric disorders were not inherently different from other common illnesses with complex inheritance patterns, like osteoporosis or diabetes. From this vantage, the process of making illness liquid should have been a relatively straightforward process, at least at the level of diagnosis. However, as Genset's experience in Argentina proved, the ecology of expertise and the dynamics of patient identity in psychiatric disorders are considerably distinct from more stabilized areas of biomedicine.

Genset's collection process was based on a more general assumption, in cosmopolitan psychiatry, of the existence of an undifferentiated global epidemiological space. The World Health Organization (WHO) estimated that 2.5 percent of the world's population between the ages of 15 and 44 suffered from bipolar disorder (World Health Organization 2001).[22] If this were the case, where were the Argentine bipolar patients? Why was it so difficult for Romero's doctors to come up with two hundred samples? Like the who, Genset's research protocol presumed that bipolar disorder was a coherent and stable entity with universal properties. But as a number of science studies scholars have argued, the existence of a given technoscientific object—here, bipolar disorder—is contingent on its network of production and stabilization.[23] An individual experience of suffering becomes a case of a generalized psychiatric disorder only in an institutional setting in which the disorder can be recognized, through the use of specific concepts and techniques that format the complexities of individual experience into a generalized convention.

Beginning with the publication of the third edition of the *Diagnostic and Statistical Manual of Mental Disorders* (DSM-III), published by the American Psychiatric Association in 1980, a diagnostic infrastructure came to underpin diverse phenomena in U.S. psychiatry, ranging from drug development and regulation, to third-party reimbursement, clinical research, and patient self-identity. The goal of these classificatory standards was reliability: if the same person went to two different treatment centers, he or she should receive the same diagnosis and treatment in each place. Such standards made it possible to forge comparable populations for research and to measure the relative efficacy of specific intervention techniques. The Research Diagnostic Criteria (RDC), forerunner to DSM-III, was developed in the 1970s for just this reason—the need to have a standard gauge so that researchers could meaningfully measure response to given medications across populations. The RDC was addressed to the problem of the low reliability of diagnostic procedures, which hampered large-scale, comparative research in psychiatry. As its creators wrote, "A major purpose of the RDC is to enable investigators to select relatively homogeneous groups of subjects who meet specified diagnostic criteria" (Spitzer, Endicott, and Robins 1978, 774).

The connection between the RDC and the DSM-III is significant in that it shows how government regulations on market entry for pharmaceuticals — the FDA's clinical-trial requirements—eventually played a key role in transforming psychiatric epistemology, structuring the need for diagnostic standards that led to the DSM revolution. Once enacted, these conventions then proved useful across a number of arenas of administration and practice— for insurance administration, transnational epidemiology, patient self-identification, and the re-biologization of psychiatry as a clinical-research enterprise.[24] Diagnostic standardization in psychiatry thus made mental illness potentially transferable between the domains of industry, government, and biomedicine.

However, psychiatrists trained to see patients in terms of an individual life course are often resistant to the imposition of systems of standardized diagnosis.[25] Such classification presumes a distinctive model of illness, which Charles Rosenberg (2002) has described as the "specificity model" characteristic of modern biomedicine. According to this model, which first came to prominence in the mid-nineteenth century, illnesses are understood as stable entities that exist outside of their embodiment in particular individuals and that can be explained in terms of specific causal mechanisms located within the sufferer's body. Disease specificity is a tool of administrative management: it makes it possible to mandate professional practice through the institution of protocols; to engage in large-scale epidemiological studies; to rationalize health practice more generally (Timmermans and Berg 2003). At the intersection of individual suffering and bureaucratic administration, the technology of standardized nosology helps to "make experience machine readable," as Rosenberg writes (2002, 250).

While the DSM met the demand for consistent diagnostic practice across diverse sites, the question remained whether the forging of such populations was based on valid—rather than simply reliable—criteria of inclusion.[26] Standardized psychiatric measures are founded on contingent agreements on behavior rating scales among experts rather than on pathophysiological measures. This is where psychiatric genomics research such as Genset's faced a conundrum. On the one hand, this research required that codified diagnostic standards be in place. At the same time, it sought to eventually remake these standards by producing a new technology of measurement, one that would transcend the subjective evaluation of the clinician: the gene-based diagnostic tool.

The problem of how to definitively recognize a given illness phenotype remained critical to psychiatric genomics research, leading to professional reflection on the process of mutual adjustment between the surface and substrate of mental disorder. In a review of "psychiatry in the postgenomic era" in 2002, two leading experts focused specifically on this challenge—as a conceptual as well as a practical problem.

> There will be critical conceptual difficulties and none are more important than readdressing the phenotypes of mental disorders. The ability of genomic tools to find the appropriate disease-related gene(s) is limited by the "quality" or homogeneity of the phenotypic sample. . . . There will be a somewhat circular process of understanding phenotype as we gain a better understanding of genotype; this, in turn, will affect our under-

standing of phenotype. All of this circularity may seem unsettling and unsatisfying to philosophical purists and it is difficult to see any way out of a process of constant adjustment. However, in the meantime, it is critical that we collect broad and thoughtful phenotypic information and not be handcuffed by diagnostic criterion sets that have reliability as their strong suit but were never meant to represent valid diagnostic entities. (Kopnisky and Hyman 2002)

Thus experts were at once using the agreed-on definitions of illness phenotypes such as bipolar disorder and assuming that these definitions were provisional and would be superseded by advances in genomics. Indeed, the psychiatrists who were gathering blood samples at Romero were skeptical that the diagnostic protocol given to them by Genset would be sufficient to find a gene. In our discussions they remarked that several different forms of the illness were being conflated in the study's protocol. A journalistic account of the study characterized this concern: "For the Argentine psychiatrists, this classification could be insufficient. As a matter of fact, they admit, other classificatory schemes point to the existence of up to six types of presentation of the illness, which for a long time was considered a psychosis and now is characterized as an affective disorder" (Navarra 1998, section 6, page 4).

Genotype and Phenotype

How did the modern form of "bipolar disorder" come into being in the first place? It is an especially intriguing category of illness because it seems to exist on both sides of certain key boundaries of mental disorder—the boundary between affective and thought disorder, or in psychoanalytic epistemology, between neurosis and psychosis. Moreover, its increasing visibility over the past two decades relates to the rise of pharmaceutical treatment in psychiatry.

From the early twentieth century until the introduction of psychopharmaceuticals, in the 1950s and 1960s, the "functional psychoses" such as manic-depression and schizophrenia were seen as chronic conditions requiring life-long institutionalization. With the introduction of psychotropic medication and then regulatory demands for randomized clinical trials, a drug market in antipsychotics and mood stabilizers was created and populations for clinical research were delineated.[27] Following confirmation of the effectiveness of lithium in the 1960s, bipolar disorder became a rare success story within psychiatry, able to be managed if not cured.[28] Despite the disorder's relatively privileged place in the field, the boundaries of the disorder as well as its origins and its defining symptoms remained at issue up through the 1990s.

According to the American Psychiatric Association's DSM-IV (1994) — which guided Genset's protocol—bipolar disorder was characterized by fluctuations in mood, from states of manic excitement to periods of abject depression. The presence of affective disorders within the patient's family was also a diagnostic clue. There were at least two types of bipolar disorder: type I was "classic" manic-depression, characterized by severe shifts in mood between florid mania and depression; type II included cases where severe depression is punctuated not by full-blown mania, but by mild euphoria, "hypomania" (American Psychiatric Association 1994). The condition had to be diagnosed longitudinally, since in its synchronic state it could be difficult to differentiate the manic phase of bipolar disorder from the delusional symptoms of schizophrenia, or at the other extreme, from the melancholia of major depression.

But it was uncertain whether bipolar disorder was truly distinguishable from schizophrenia or depression, as the ambiguous status of "schizoaffective disorder" suggested. Genetic and neurological studies continued to confound researchers trying to establish consistent means of differentiation. Estimates of its prevalence in the population ranged from 0.5 percent to 5 percent, depending on the criteria of inclusion used.[29] Some psychiatrists argued that there was a "psychotic continuum" from bipolar disorder to schizophrenia, from predominately affective traits to thought disorder.[30] Meanwhile, expert advocates of the diagnosis claimed that many actual bipolar patients had been incorrectly diagnosed with unipolar depression and given antidepressants, which could set off a manic episode (Akiskal 1996). Such proposals would radically expand the bipolar population. Geneticists struggled to define the disorder's boundaries in order to gather consistent populations for research: "There is growing agreement that in addition to BPI [bipolar illness], MDI [manic-depressive illness] encompasses several mood disorders related phenomenologically and genetically to BPI. These include bipolar disorder type II . . . some cases of major depressive disorder without manic symptoms . . . and some cases of schizoaffective disorder (in which symptoms of psychosis persist in the apparent absence of the mood disorder). The MDI phenotype may include other, milder manic-depressive spectrum disorders such as minor depression, hypomania without major depression, dysthymia, and cyclothymia, but this is less certain" (MacKinnon, Jamison, and DePaulo 1997, 356).

Would finding susceptibility genes once and for all pin down the *thingness* of the disorder? In academic studies of the genetics of bipolar disorder, the late 1990s were a time of frustration. While twin and family studies had indicated heritable susceptibility since the 1930s, hopes that the advent of techniques for gene identification in molecular biology would quickly make it possible to find the biological mechanisms involved were disappointed. After a period of excitement in the 1980s, as various reports of loci for linked genes appeared, a decade later the glow had receded after repeated failures to replicate such studies (Leboyer et al. 1998). Experts gave dour assessments of the state of the field: "In no field has the difficulty [of finding genes linked to complex disease] been more frustrating than in the field of psychiatric genetics. Manic depression (bipolar illness) provides a typical case in point," wrote two Stanford geneticists (Risch and Botstein 1996, 351). By 2001 newly reported findings of a susceptibility locus on chromosome 10 were greeted warily by researchers (Bradbury 2001, 1596).

There were a number of possible suspects for the mixed results: "The failure to identify BP-I loci definitively, by standard loci approaches, probably reflects uncertainty regarding mode of inheritance, high phenocopy rates, difficulty in demarcation of distinct phenotypes, and presumed genetic heterogeneity," wrote a team at University of California, San Francisco (Escamilla et al. 1999).[31] In other words, these researchers thought that conceptual difficulties around defining the phenotype for diagnostic purposes posed an insuperable technical challenge. The Stanford researchers, in contrast, argued that no dominant gene had been found because of the biological complexity of the inheritance mechanism.[32] Surveying the state of the field, some geneticists posed a worrisome question about the diagnostic entity they were looking at: "The question remains: do our modern definitions of clinical syndromes (presently considered as phenotypes) accurately reflect underlying genetic substrates (genotypes)?" (Leboyer et al. 1998). In other words, for the purposes of genetic studies, was there really such a thing as bipolar disorder?

The phenotype question created a paradox for these studies: on the one hand, genetic research promised to resolve such problems by making clear the underlying biological processes: "Currently the major problem is the unknown biological validity of current psychiatric classifications and it is worth bearing in mind that advances in molecular genetics are likely to be instrumental in providing the first robust validation of our diagnostic schemata" (Craddock and Jones 1999, 586). In order for such validation to occur, researchers had to know what they were working with. Yet they lacked objective tools to do so: "In the absence of a clear understanding of the biology of psychiatric illnesses the most appropriate boundaries between bipolar disorder and other mood and psychotic disorders remain unclear" (Craddock and Jones 1999, 586). Genetic studies might even turn out to undermine the notion of a clear distinction between these disorders: "One of the exciting developments has been the emergence of overlapping linkage regions for schizophrenia and affective disorder, derived from studies on independently ascertained pedigrees. These results raise the possibility of the existence of shared genes for schizophrenia and affective disorder, and the possibility that these genes contribute to the molecular basis of functional psychoses" (Wildenauer et al. 1999, 108).

The unfulfilled promise of genetics led psychiatry back to an old curse: the problem of how to stabilize its objects—that is, how to ensure that its illnesses were "real" things, whose contours could be recognized and agreed-on by diverse experts. Despite the discipline's adoption of neuroscientific models, and ongoing genetic and neuroimaging research into mental disorders, the question of the relation of psychiatry to biomedicine remained: to what extent could psychiatric conditions be considered equivalent to "somatic" illnesses? The effort to achieve such equivalence was one rationale for the re-biologization of U.S. psychiatry beginning in the 1980s (Lakoff 2000; Luhrmann 2000). Difficulties in confirming genetic linkage challenged the legitimacy of psychiatric knowledge, and the very existence of its objects. A leading researcher expressed frustration at the place of psychiatry in genetics research: "[Psychiatric geneticists] continue to face an obstacle that does not hinder their colleagues who investigate non-psychiatric diseases; psychiatric phenotypes, as currently defined, are based entirely on clinical history and often on subjective reports rather than directly observed behaviors. . . . In no other branch of medicine have investigators (and practitioners) been called on to demonstrate time and again that the diseases they study really are diseases" (Gelernter 1995, 1762, 1766).

Epistemic Milieu

This problem was especially palpable in Buenos Aires, as doctors struggled to locate patients who had been diagnosed with bipolar disorder. The dearth of bipolar subjects in Argentina was due not to a cultural difference in the expression of pathology or to the country's genetic heritage, but to a different set of conceptions and practices, within its professional milieu, of the salient forms of disorder and the tasks of expertise.[33] The nosological revolution in North American psychiatry—the shift to DSM-III and its successors beginning in 1980—had not extended to the Southern Cone. In Argentina, the DSM faced professional resistance on both epistemological and political grounds. The pervasive presence of psychodynamic models among psy-professionals led to an emphasis on the unique clinical encounter between doctor and patient, and a suspicion of diagnostic categories that purported to generalize across cases. Meanwhile, there was political opposition to the incursion of such standards on the grounds that they were being imposed in the interest of managed care and pharmaceutical industry interests. Many Argentine psychiatrists associated the use of the DSM with neoliberalism, the privatization of state industries, and the dismantling of the welfare state.

A number of absences also made resistance to standardization more feasible: in contrast to the North American situation, the Argentine psychiatric profession was not structured by a demand to forge populations for epidemiological or neuroscientific research. Disciplinary prestige did not come from producing scientific articles in transnational journals, and professional training did not include an emphasis on standardized diagnostic classifications. Further, insurance reimbursement systems did not require the use of "evidence-based" protocols in diagnostic and intervention decisions. Thus, while the Argentine patient population had been made available for genomic research in ethicolegal terms by Genset's contract with Hospital Romero and the consent form, it had not been rendered equivalent in epistemological terms.

Across the hallway from where the genetic study was being conducted, the women's ward of Romero's psychopathology service achieved the surprising feat of practicing Lacanian analysis within a public hospital that served a predominantly poor and socially marginal population.[34] A number of times women who had received a bipolar diagnosis and then given blood samples for the Genset study in the men's ward were later hospitalized across the way, during psychotic episodes. Such patients' claims to be bipolar were mostly disregarded by the physician-analysts there, who saw such self-diagnosis as a form of resistance to subjective exploration in psychoanalytic terms, and considered "bipolar disorder" to be a condition that owed much of its existence to the promotional efforts of the pharmaceutical industry.[35] As they saw it, their task was to penetrate beneath these generalizing categories to understand the distinctive life history and process of subject-formation of the patient.

Meanwhile there remained the question of how the patients themselves understood their condition. Given the prevalence of psychoanalysis in Argentina, and the absence of the kind of patient self-help movements that have transformed the North American milieu, it was not necessarily a receptive site for the inculcation of "bipolar identity." The problem for Genset was at one level a technical one: how to find a pool of patients that would prove amenable to genomic research. But insofar as psychiatric diagnosis also names a subjective mode, the question involved self-identity as well. Bipolar self-identity—which emerged in the United States as part of a burgeoning self-help apparatus—was not widespread in Argentina.[36] To what extent did subjects who entered Romero come to see their own life trajectories in terms of an illness characterized by extreme mood swings that had a biological underpinning? An unusual case during the sample collection illustrated some of the complex interactions between patient self-understanding and professional diagnosis that characterizes psychiatric conditions.

Collection (II)

On a Thursday morning in the men's ward, more potential DNA donors have come for their appointments. In one examination, a young woman does most of the talking, rapidly and in disjointed bursts. She is a psychoanalyst, she explains, and so she does not believe in genetic explanations. But a patient of hers, a friend—who had read about the study in the paper— told her that she had certain characteristics that seemed like they could be "bipolar," so she decided to come, just in case, out of curiosity. She does not want to give her name: professionally, she says, it would be bad for her reputation if it were known that she had come to find out about her genetic makeup. It soon becomes apparent that the woman thinks that there is already a genetic test available for bipolar disorder, and she has come to Romero to take it. She is not sure whether she really wants to know, or even if it would be possible to know such a thing through a blood test. When Rechtman finally makes it clear that in fact there is not yet a genetic test, but the hospital is collecting samples in the hopes of finding such genes, she begins to protest the very premise of the study.

"How can you possibly know a person's diagnosis if you haven't been treating them?," she demands. She cuts off Rechtman's response, explaining that in psychoanalysis, you have to establish a transferential relationship with the patient in order to see the psychic structure. The whole operation seems rather suspicious to her. She eyes the anthropologist: what is he writing down? Rechtman tries to calm her, explaining the rationale for diagnosis: "There are certain signs of the disorder—for instance, what was it that your friend noticed?" The woman lists a few symptoms: insomnia, cocaine use, depressions, an eating disorder. "My analyst says that I'm an obsessive," she explains.

"But the psychoanalytic clinic has its limits," she muses. "Perhaps if there were something physical?" They debate further, back and forth, and the discussion becomes acrimonious. Finally, Rechtman tries to close off the examination: "I wouldn't include you in the study, because it's not clear what you have." "But what else could it be?," she asks, now almost wanting to be convinced. "Maybe it's what your analyst says, obsessive neurosis," he suggests skeptically. "But I suspect that it is bipolar disorder." She muses for a moment, then poses another question: "What does Prozac have to do with all this?" Rechtman throws up his hands. At last, they reach a labored conclusion, agreeing to disagree. Her DNA will not be among the samples sent by courier to Paris. She has rescued her professional pride, and declined to shift her identity. Educated, middle class, and *porteña*, she retains her model of mental distress.

Despite her protestations, the woman's presence at the hospital indicated a certain urge to shift her conception of herself, to try new explanations and interventions. Because the experience of psychiatric disorder interacts with the characteristics of the disorder, its diagnosis is a moving target.[37] Psychiatry, in part because it depends on patients' subjective reports of their symptoms, has had a difficult time shifting the disorders under its purview into stable things in the world. The search for genes related to mental illness is, among other things, an attempt to turn mental disorders into more durable entities. Yet it seems that the discovery of loci of genetic susceptibility would not necessarily make such illnesses less complex. As the interaction above illustrates, knowledge of susceptibility is likely to add a layer of complexity to patient self-understanding and to provide a set of new possibilities of intervention, rather than to reduce mental illness to purely organic determination.[38]

Local Conditions

The historian Ken Alder writes, "understanding the process by which artifacts come to transcend the local conditions in which they are conceived and produced should be one of the central tasks facing any satisfactory approach to technology" (1998, 501). The DSM emerged from a specific conjuncture within North American psychiatry in the 1970s, and spread to other sites—both administrative and scientific—because of its ability to make behavioral pathology transferable across domains. The DSM was not just an isolated set of technical innovations within psychiatry. Its eventual widespread use in professional milieus (and resulting controversies from such use) had to do with its ability to serve a diverse set of needs: for drug development given regulatory guidelines; for insurance protocols based on "evidence-based medicine"; for the re-professionalization of psychiatry as a biomedical science.

Technical protocols such as diagnostic standards structure the production of a space of liquidity: they mediate between the domains of science, industry, and health administration. Such technologies are part of an infrastructure, both material and conceptual, that enables goods, knowledge, and capital to flow across administrative and epistemic boundaries. They link social needs such as health to profit-seeking ventures and to scientific communities. The use of such technologies in material practices such as professional standardization and DNA collection underlies the abstraction of a global biomedical information economy. In this sense they point to the "conditioned contingency" (Rabinow 2004) of phenomena such as the Genset study and the value of bipolar genes. There was no inevitable telos—no underlying logic of capital—directing the creation of value from Argentine patients' genes. At the same time, there were "structural" reasons for the form that the assemblage took: it was no coincidence that patients from Buenos Aires were providing material for the creation of products destined for consumer markets in the North.

The case also points to limits to the capacity of standards to transcend the local. The setting in Argentina indicates that the extension of a diagnostic infrastructure does not occur uniformly across space but rather through networks, and must be supported or imposed by institutional and regulatory demands. The shift in North American and Western European psychiatry from "clinical" to "administrative" norms had not taken hold there by the late 1990s, despite efforts to privatize parts of health management along North American lines. The advance of the DSM was an element in a health apparatus oriented toward bureaucratic management that had not suffused the Argentine milieu. Nor was there a significant patient-activist movement shaping collective action around the recognition and legitimacy of specific disorders. And a professional culture whose epistemological forms were incommensurable with the DSM was entrenched. For these reasons, individual

clinicians retained considerable autonomy in terms of diagnostic and therapeutic practices.

The difficulty of finding bipolar patients in Buenos Aires pointed to the
halting extension not only of diagnostic standards, but also of modes of
self-identification around illness labels such as bipolar disorder. In order
to be a viable diagnostic entity, the disorder needed an epistemic niche in
which it could take root and thrive. Bipolarity came into being temporarily
in the men's ward of Hospital Romero, but only through the imperative to
find a sufficient sample of patients for the Genset study. In turn, it disap-
peared when patients traveled to the women's ward. Patients' illnesses were
rendered liquid without permanently transforming patient-identity, since a
diagnostic infrastructure for managing health in terms of specific subpopu-
lations was not in place. Thus, while information may be "the coin of the
genomics realm," the extraction and circulation of such information is not a
simple matter. In the case of mental illness, the value of genomic information depends on the stabilization of the very thing it claims to represent—
the disorder itself.

Notes

Thanks to Kaushik Sunder Rajan and to the participants in the "Lively Capital" workshop for their generous comments and suggestions. Earlier versions of this chapter
have appeared in *Theory and Society* 34 (2005) and in Andrew Lakoff, *Pharmaceutical Reason: Knowledge and Value in Global Psychiatry* (Cambridge University Press,
2006).

1. For the description of an emergent anthropology of global technoscientific and
administrative forms, see Collier and Ong 2004.

2. For the case of the organ trade, see Cohen 2004. For the case of Europe as a
"technological zone," see Barry 2001.

3. For an analysis of the logics of value underlying the identification of "unhealthy"
populations in need of chronic medication in rich countries, see Dumit (chapter 1 in
this volume).

4. A project conducted in rural China by Millennium Pharmaceuticals in collaboration with Harvard University and seeking genes linked to asthma provoked a scandal after an investigative report appeared in the *Washington Post* (Pomfret and Nelson
2000).

5. "Mining" was a common metaphor for the search for potential riches hidden in
the human genome. As an article about a Genset research collaboration in China put
it in 1996, quoting Genset's president: "China's population is a gold mine of genetic
information. The country's rural populations have remained relatively static this century, so each region has a unique blend of genes and diseases. This makes it much

easier to trace hereditary diseases back to defective genes, which are unusually abundant where the disease is prevalent. 'You can treat regional local populations almost like single families,' says Brandys" (Coghlan 1996, 4).

6. The kind of mapping Genset was engaged in—which sought to find markers of genetic variation (single nucleotide polymorphisms, or SNPS)—did not presume a *causal* relation between a given DNA sequence and the presence or absence of disease; rather, it hypothesized that certain markers of variation could be correlated to greater *susceptibility* to that disease.

7. Important recent anthropological analyses of the conditions of possibility for genomic value include Hayden 2003 and Sunder Rajan 2005.

8. Examples include studies of organ donation, evidence-based protocols, and government-funded biomedical research. See Hogle 1995; Timmermans and Berg 1997; Epstein 2007.

9. Michel Callon (1998) describes the process whereby objects are "disentangled" from their immediate surroundings and made calculable as one of "enframing."

10. Genset would eventually check the validity of its findings of a "psychosis gene" among Quebecois and Russian populations against the sequence information extracted from its Argentine bipolar samples (Blumenfeld et al. 2002).

11. A note on the names used in this chapter: I have used pseudonyms both for the hospital and for the doctors where the research was carried out, in order to protect the privacy of my informants there. Because the Genset research is public knowledge, I have identified the firm and its employees by name.

12. This strategic edge proved temporary: I interviewed Mendelson before the announcement, in 1999, of the "SNP Consortium" by a group of major pharmaceutical companies, academic centers, and the Wellcome Trust, a collaboration designed to undercut the strategic position of database firms like Genset. By 2000, Genset was forced to remake itself as a drug-development company given the unproven profitability of genomics database firms.

13. Its precursor, manic-depression, was first named by Emil Kraepelin around 1900. See Kraepelin 1904.

14. As an article in *La Nación* put it: "Currently, the hospital needs more sporadic patients to complete the sample that is awaited in France" (Navarra 1998).

15. Unlike patient groups in the United States, FUBIPA and similar groups are a relatively marginal phenomenon in Argentina and are typically run by local experts in the disorder rather than by patients and family members.

16. One highly publicized example was the Wistar Institute's field trial, in 1986, of a recombinant rabies vaccine in cattle outside of Buenos Aires, in which no Argentine authorities were informed of the experiment (Dixon 1988).

17. It appeared, for instance, through the UNESCO Bioethics initiative, represented in Argentina by the legal scholar Salvador Bergel, who opposed the licensing of genetic material (Bergel 1998).

18. To the extent that imagery of a dystopian genetic future entered the popular imagination, it was via the film *Gattaca*, rather than through newspaper editorials by vigilant watchmen. This can be contrasted with the deCODE case in Iceland (Palsson

and Rabinow 1999). On the other hand, Argentina was one of the first countries to ban cloning in the aftermath of Dolly, a result of the power of the Argentine Catholic Church to define the boundaries of reproduction.

19. Indeed, a group of major pharmaceutical companies, in partnership with the Wellcome Trust, was able to circumvent the biotech effort to patent and license SNPS—markers of human genetic variation—by forming a consortium, in 1999, to make such markers publicly available, significantly hindering the business strategy of companies like Genset. Database companies were then forced to shift into drug development, an even more treacherous and uncertain field.

20. For contrasting analyses of the deCODE project, see Palsson and Rabinow 1999 and M. Fortun 2001. Another difference was that the Argentine subjects of Genset's study were not prospective consumers of the technologies under development, whereas the deCODE project guaranteed Icelanders access to Hoffman-LaRoche products developed from the research.

21. See the IMS Health website, http://www.imshealth.com, accessed August 2000.

22. Typical estimates in the bipolar-genetics literature were around 1 percent. But some experts thought it was as high as 5 percent. Much depended on the criteria of inclusion, and the means of distinguishing bipolar disorder from overlapping syndromes such as schizophrenia, unipolar depression, and attention deficit disorder.

23. For a description of how certain mental disorders come to thrive in specific political, cultural, and professional niches, see Hacking 1998.

24. In this sense the DSM can be considered a "biomedical platform," as Peter Keating and Alberto Cambrosio (2000) describe it, operating to connect clinical conventions with biological conventions, individuals with populations.

25. For the distinction between the clinical and the administrative as different modes of justification in medical work, see Dodier 1998.

26. For a lucid analysis of questions of reliability and validity in psychiatric diagnosis, see Allan Young 1996.

27. For the story of the emergence of the "specificity" model in psychopharmacology in relation to regulatory demands, see Healy 2002. For the history of FDA regulations and the demand for proof of efficacy and safety in clinical trials, see Marks 1997.

28. Though discovered in 1949, lithium was not widely adopted until its effectiveness was confirmed in the early 1970s—in part because it was not a proprietary compound and so there was little marketing incentive for conducting the requisite clinical trials, but also due to the lack of interest in biological treatment of manic-depression among psychodynamic psychiatrists, then predominant in U.S. psychiatry.

29. A key difference is in whether both bipolar type I and type II are included. In the Romero study both types were included (Kessler et al. 1997; Angst 1998).

30. For example, the psychiatrist Timothy Crow (1986) wrote, "The psychoses constitute a genetic continuum rather than two unrelated diatheses." By the same token, there was some question as to what sort of entity "psychosis" was. Did the

term cover a specific set of syndromes, or simply give a new name to madness? The meaning of psychosis had shifted from its original distinction from neurosis; at first, the term had indicated mental disorders not linked to organic lesions, whereas the neuroses were associated with the nerves themselves. It then settled into indicating serious mental pathology, usually marked by delusion or hallucination.

31. In another paper, the same group placed blame on the uncertainty of the relation between phenotype and genotype on the seemingly multiple ways the "underlying disease" expressed itself: "Genetic studies of psychiatric disorders in humans have been inconclusive owing to the difficulty in defining phenotypes and underlying disease heterogeneity" (McInnes et al. 1999, 290).

32. "We believe the explanation lies elsewhere [than genetic heterogeneity], namely that the genetic mechanism underlying the disease in these families is more complicated than postulated, leading to a reduction in [statistical] power" (Risch and Botstein 1996, 352).

33. Lawrence Cohen (1998) describes a similar problem on his arrival to study old age in India: an apparent lack of patients with senile dementia.

34. For the historical context of psychoanalysis in Argentina, see Plotkin 2000 and Vezzetti 1996.

35. I describe these dynamics in more detail in Lakoff 2006.

36. For an analysis of the "cultural life" of bipolar disorder both as a diagnosis and as a mode of self-understanding in the contemporary United States, see Martin 2007.

37. Ian Hacking (1999) argues that psychiatric identity is an example of the type of classification he calls "interactive kinds." As opposed to "indifferent kinds," like trees, these are classifications that interact with the thing being classified.

38. Paul Rabinow provides some guideposts for thinking about the way in which genetic identity might interact with new forms of political rationality. He identifies groups whose affiliation is based on a common disorder or genetic risk, and who influence health policy and scientific research, as emerging signs of such biosociality. "Such groups," he writes, "will have medical specialists, laboratories, narratives, and a heavy panoply of pastoral keepers to help them experience, share, intervene, and 'understand' their fate" (Rabinow 1996, 102).

9

WEN-HUA KUO

TRANSFORMING STATES IN THE ERA

OF GLOBAL PHARMACEUTICALS

Visioning Clinical Research in Japan, Taiwan, and Singapore

If we follow only one value, . . . no options can be chosen other than an "ultimate solution." . . . In other words, choosing an ultimate solution implies the confirmation of a
particular value and viewpoint.

YOICHIRO MURAKAMI, *ANZENGAKU* [ON SECURITY]

Over the past two decades pharmaceuticals have become more standardized and globalized, and since the 1990s an increasing number of prescription drugs have been developed and marketed. This trend has turned global
pharmaceuticals into an emergent subject of research consisting of analysis
of lively interfaces between life and capital, science and society, on levels
from the individual to the global.

   In this area, two perspectives hold special significance. One concerns the
standards and regulatory techniques that facilitate the spread of globalization. In *Global Pharmaceuticals: Ethics, Markets, Practices*, Adriana Petryna,
Andrew Lakoff, and Arthur Kleinman show that standards and regulations
are used not only to make pharmaceuticals more scientific and reliable, but
also to give pharmaceuticals a cross-cultural and thus indisputable quality
(Petryna, Lakoff, and Kleinman 2006). As Yoichiro Murakami notes, any
"ultimate solution" for all variations and discrepancies is inevitably not free

from a particular viewpoint that channels all interests in a certain direction. In connection with this, Andrew Lakoff presents in this volume (chapter 8) an ethnographic account that traces how the "liquidity" of an illness (a term borrowed from the field of finance) is achieved through rapid flow and circulation of genetic information and capital across national boundaries. This account reiterates the warning by Petryna, Lakoff, and Kleinman to anthropologists: "As standards travel, their social and economic embeddedness is revealed" (2006, 12).

The second perspective on global pharmaceuticals concerns clinical trials. Certainly, the pharmaceutical industry has its reasons for running clinical trials globally: nowadays more trials are required for approval and more subjects are required for each trial; moreover, the subjects must be recruited in a more timely fashion. Nonetheless, what is at stake in the demand for larger pools of human subjects is how body and life are conceived in the pharmaceutical industry—how can the latter turn these bodies, which are distinct individuals embedded deeply in various social and environmental contexts, into seemingly interchangeable clinical-research subjects? How can it be assumed that pharmaceuticals that are tested on a narrow range of subjects will be "available" for all bodies?

This necessitates a notion of biopolitics beyond Michel Foucault's (1997), and requires new frames in order to be comprehensible.[1] For example, in the chapter "Globalizing Human Subjects Research" (2006), Petryna observes the move of clinical-research trials to regions considered ethically lax and thus competitive. On the other hand, looking at how "ideal treatment responders" are defined and located, Lakoff (2006) questions the seemingly perfect triad of illness, pharmaceuticals, and target populations. Both Petryna and Lakoff call attention to how life and body should be reconceptualized in the presence of global pharmaceuticals.

Complementing these two perspectives on global pharmaceuticals, the present chapter discusses the regulatory regime, focusing not on how people may be affected by regulations, but on how the world of pharmaceuticals itself formulates operational standards.[2] Echoing Sheila Jasanoff (chapter 4 in this volume), who addresses the judicial aspects of how life and nature are constituted for public concerns, my approach here is institutional and transnational. The field for this chapter is the International Conference for Harmonisation (ICH), a series of meetings initiated by the United States, the European Union, and Japan with the aim of establishing uniform standards for proprietary drugs.[3] In particular, I will analyze a controversial topic discussed at the ICH, concerning the acceptability of foreign clinical data, and will look at how Japan, Taiwan, and Singapore have responded.[4]

This topic arouses discussion, especially in Asia, because of how differences among populations may be negotiated. In theory, this discussion has to do with the question of whether any easy equation can be made between the test population and the target population for a drug. The existing literature often refers to this debate in terms of racial differences, which in the U.S. context means the differences between whites and nonwhite minorities.[5] Although the configuration of racial categorization deserves serious study, as debated at the ICH, I focus on the state, exploring the intricate processes of negotiation and self-government that occur when states deal with global pharmaceuticals. While it is generally regarded from the corporate viewpoint as a "non-tariff barrier," race in this chapter is taken rather as a reference that triggers "frictions," to echo Anna Lowenhaupt Tsing's notion of how the local deals with the universal (2004). This approach lays bare the emerging characteristics and visions of Japan, Taiwan, and Singapore as they negotiate the requirements for imported pharmaceuticals, while addressing the overarching need to not compromise the health of their populations.

Before looking into the East Asian encounter with the ICH, I would like to clarify what I am taking "global" and "the state" to mean here. The former concerns the qualification of international conferences as an ethnographic site, for which I offer three justifications. First, the conference is itself a discursive site at which participants present their current attitudes on certain issues or even open new topics. Unlike extant regulations and published papers, these cutting-edge ideas are fresh and have the potential to change or to influence one another. The second reason relates to the conference's interactive function. As an occasion for the interchange of opinions, the conference involves the seemingly contradictory functions of comparing different opinions, on the one hand, and working together, on the other. It is a complex, highly dynamic zone, where various actors are compelled to trade information, persuade one another, and exchange visions.[6] Thus the conference is an arena where both controversies and conventions are expected.[7] The third reason can be found in the conference's accumulative and periodic nature: participants at a conference act on the information they receive, and their actions will be recounted at future meetings. Through publication of proceedings and presentation slides (now often online), the information is disseminated in a lasting way, spreading and generating new information. The conference is a living archive, and participants rely on this kind of reference material in order to reconstruct memories which themselves represent the nature of the conferences.

Following closely the performance of Asian states within and outside the global site of the ICH allows us to appreciate their characteristics. Certainly, some bureaucracies and institutions do constitute the state as an acting entity. Nonetheless, this chapter is not intended as a simple return to the "bring the state back in" approach (see Evans, Rueschemeyer, and Skocpol 1985), which would discuss the mechanisms by which each state formulates policies on pharmaceutical regulatory trends.[8] Rather, this chapter aims to capture how, seen from a global viewpoint, one state interacts with others. As Ernest Gellner (1983) reminds us, the state is a political shell in which culture can be shared and nationalism crafted. The aim of the inquiry at hand is thus not to define what the state is or should be; rather, it is to distinguish one "political shell" from another in the face of globalization.[9]

The political shell analogy fits well with our discussion of pharmaceutical regulations. Although the ICH aims to establish universal standards for drug approval, agreement is not necessarily easy to achieve, even if each state recognizes the importance of regulations and is capable of making them. While they may look alike, regulations can be distinct from one state to another, depending on the political culture in which they are derived and nurtured. A study by Sheila Jasanoff (2005a) of the United Kingdom, Germany, and the United States is a salient example. Jasanoff advances two arguments concerning these states' attitudes toward biotechnology. First, the policies concerning life sciences have become a more or less self-conscious project of nation-building at a critical juncture in world history. Second, political culture does matter to contemporary politics. The present chapter follows this trajectory, supplementing it with cases drawn from Asia. Taking advantage of the rich archive compiled by the ICH, I carry out an interpretative ethnography of the real-time behavior of Asian states in regard to globalization.

Indeed, it would be worthwhile undertaking an investigation into the state in the era of global pharmaceuticals at the complex interfaces between the Western and the non-Western, the global and the local, business and science. I am not just interested in a simple explanation of how globalization is sweeping over the non-Western world. As Michael Hardt and Antonio Negri suggest (2000), globality "should not be understood in terms of cultural, political, or economic *homogenization*. Globalization, like localization, should be understood instead as a *regime* of the production of identity and difference, or really homogenization and heterogenization" (2000, 45). I am interested in how "state" and "race" were first challenged by science just as the ICH was about to merge regulatory territories, and then referenced by the state strategies and visions that are currently being developed.

Looking into the Secret Box of Regulations:
Understanding the ICH

The industry likes to portray itself as producing pharmaceuticals as highly regulated commodities.[10] Since the 1960s major pharmaceutical markets have established their own regulations primarily in response to emerging drug disasters such as that surrounding thalidomide, and as a result the regulations have developed into a huge, complicated system.[11] Scholars at the Center for Drug Development, Tufts University, estimate that in the United States it takes an average of eight to twelve years to get a product to market and costs millions of dollars.[12]

Increasingly rigorous requirements have lengthened the premarketing period for new products. However, this is not the biggest obstacle the industry has had to face in the past two decades. Like other sectors, the pharmaceutical industry has long sought a global market, and this desire is becoming more intense. Extremely high standards have protected monopolies by raising the barriers facing new market entrants. But those who can afford entry need to recover costs as soon as possible for a simple reason—the effective marketing period of approved drugs has shortened and so has the period for which drugs are protected by patent.

What makes this situation even more difficult is that regulations created in the different major pharmaceutical markets may look similar, but are not simply interchangeable, as each must respond to specific requirements and concerns. A seemingly trivial but crucial difference that regulators like to cite is the "room temperature" criterion used in tests to assess the stability of a drug. Before the introduction of universal guidelines, temperature controls in such studies were set by the sponsor and were appropriate to the locality. This earlier system necessitated new stability data for registrations in different regions, even if the climatic difference between the original location of study and the location where the drug went to market was no more than one or two degrees Celsius.[13] Such additional regulations are undoubtedly an impediment to the industry, as it is impractical to meet all such demanding standards, and this delays still further the launch of products to market. At the same time, regulatory agencies may have to respond to public-health crises that the delay may cause among patients who cannot afford to wait.

From the perspective of standardization, the ICH is a unique project serving as a communication platform for the regulator and the regulated, thus presenting what seems to be the perfect way to accelerate the global accessibility of the latest cures—each party understands that universal standards cannot be achieved without the involvement of the other. Even so, unlike the World Health Organization (WHO), which specializes in health issues under the conventional political scheme of the United Nations, the ICH is selective and technology-oriented.[14] It invites only a handful of members, whose pharmaceutical innovations and markets represent over 85 percent of the world total, and claims to work only on practical issues concerning the quality, safety, and efficacy of pharmaceuticals.[15]

The dynamics between industry and regulators, and among different regulatory bodies, may be observed in the way in which the ICH tries to achieve consensus. The "ICH-process," a five-step procedure, has been drawn up to ensure that each guideline is properly discussed and implemented in all ICH regions. Proposals for new harmonization must be brought to the steering committee to initiate an ICH action. If accepted, a proposal is assigned to an expert working group, which advises on the technical aspects of harmonization topics (step 1). When a primary guideline is drafted, it must first be distributed to and achieve consensus among all the invited experts (step 2). When the draft is complete, it is brought back to each region for feedback on related topics (step 3). Each guideline must be agreed on by all experts and each ICH region before being submitted to the steering committee, where the guideline is confirmed (step 4). The final step, which makes the ICH unique, is a follow-up mechanism applied to determine whether a guideline is adopted by local regulatory agencies within six months of its release (step 5). It takes an average of twelve to eighteen months for a guideline to be implemented through this five-step process.[16]

This may strike a chord with readers who are familiar with Sheila Jasanoff's notion of "republics of science," according to which democracy "is not a singular form of life but a common human urge to self-rule that finds expression in many different institutional and cultural arrangements" (2005a:290). Indeed, acknowledging that science cannot be immune from politics (or, as Jasanoff would suggest, "scientific cultures are at one and the same time political cultures"), the ICH's intention in creating this diplomatic harmonization procedure is that diverse regulatory practices be respected in the making of universal standards. Nonetheless, in accordance with the wishes of all participants, the ICH has proven to be not a hurdle but a catalyst for the standardization of standards. From its first meeting, held in 1991, to its sixth (ICH 6) in 2003, the ICH has set up fifty-four guidelines in the categories of quality, safety, efficacy, and multidisciplinary issues.[17] The industry appreciates this. For instance, Stuart R. Walker, of the Center for Medical Research International, comments, "As a result of this initiative, the drug regulatory process has become smoother, quicker and less burdensome" (Nutley 2000, 9). Praise like this has also come from regulators, for whom the ICH has been a resounding success by making the world of pharmaceuticals "flat." It seems to have achieved both phases of a complicated commercial and scientific mission.[18]

The ICH is not satisfied to stay within its initial three regions, but has furthered the attempt to spread its new standard to the rest of the world. Based on a resolution made at the ICH 4 in 1997, the ICH has begun to expand, seeking to implement as many guidelines as possible in non-ICH regions (ICH 1997, revised in 2000). In 1999 the ICH Global Cooperation Group (GCG) was formed to serve as a bridge to other countries affected by these guidelines. Although the GCG's mission may be to raise the quality of clinical research in these countries to global standards, it functions passively to ensure that the direction of regulatory flow from the ICH to non-ICH regions is irreversible. The primary objective of the group, it is claimed, is "to act as a resource for the understanding, and even acceptance, of many of the guidelines" (Nutley 2000, 10).[19] Although the markets outside of the ICH regions are too tiny to be incorporated into the original plan, the ICH has decided to extend the margins of its guidelines and incorporate these places. By standardizing standards, it continues to work toward the goal of a single global market and health community. For the ICH, the ultimate solution is already here; it is merely a question of when it will be implemented.

One Problem, Three Answers: Asian States and the ICH

How would the ICH affect Asian states? A conventional account is provided by the anthropologist Kalman Applbaum, who, tracing the introduction of a new group of antidepressant SSRIs (selective serotonin reuptake inhibitors) to Japan, shows how medical knowledge and capital work together in a top-down, West-initiated approach (Applbaum 2006). Applbaum portrays a hierarchical, dichotomous world, with typical American pharmaceutical companies that zealously sell their products to Asia, on the one side, and a silent Japan that is unwittingly "educated" about diseases and accepts its treatment, on the other. Mediating between the two, the ICH is cast as a mere instrument through which this capitalist wish is fulfilled.[20]

The globalization process that Applbaum presents is fascinating, but one-sided, not only omitting mention of local reactions, but also skipping the technical yet subtle debate at the ICH, and this weakens the account. I shall apply a "slow-motion" approach to trace how Japan, Taiwan, and Singapore (in that order) encountered the ICH, step by step, in various situations at different times. The ICH decision "to eliminate redundant trials" was universally applicable, yet each state responded differently.

Japan was the first of the states to come across the ICH. Unlike the European Union and the United States, where most global pharmaceutical companies are located, Japan was invited to the ICH not because of its ability to carry out pharmaceutical research and development, but because of its huge market, ranked second in the world, and its tough regulatory requirements.[21] Before 1986 a company that sought to sell drug products in Japan nearly always had to repeat the clinical trials required in Japan, using Japanese subjects. Even after a notification was enacted allowing acceptance of foreign data "in principle," it was very rare for concessions to be granted.[22] The industry was frustrated and complained that Japan was practicing pharmaceutical protectionism.

Japan did not let this accusation lie, but brought up at the ICH the issue of race in relation to drug behavior. Osamu Doi, then representative of Japan's Ministry of Health and Welfare, thought it crucial for the acceptability of foreign clinical data and argued that a consensus on the issue should be reached at a global scientific occasion like the ICH. On his insistence, it was agreed that population differences, such as sex and age, would be discussed.[23] At the time, nobody imagined that this topic, later called "E5," would become one of the most difficult ever discussed by the ICH. Six years of discussion produced only a vague guideline. Its vagueness, however, gave Taiwan the chance to speak for itself.

It is not necessary to review exhaustively the discussion of how differences among populations were dealt with by the expert working group (for a detailed science and technology studies analysis, see Kuo 2008). In brief, Western nations stressed the fundamental unity of humankind and claimed that further clinical trials should be pursued only if it could be determined that real differences, unique to Asians, exist. Japan, on the other hand, taking for granted the genetic and cultural distinctness of its population, insisted that no trials should be foregone unless the *similarity* between the Japanese and other ethnic groups, such as whites, could be proved.[24] Thus after an agreement was reached on waivers for pharmacokinetics studies on the basis of a study comparing individual variations and interethnic differences, proposals were submitted in two divergent directions. While experts from the European Union and the United States asked for more waivers in late-phase clinical trials, which are more expensive and time-consuming, Japan, in order to discover possible differences, expected a trials system with equal representation of different ethnic groups; that is, whites, blacks, and Asians (in this instance, Japanese).[25] Neither of these proposals pleased all parties, and there was a deadlock.

It was Roger L. Williams, an expert from the FDA, who saved the discussion by bringing in the concept of the bridging study. Technically, "bridging" means undertaking extra studies to generate the necessary information for extrapolation of late-phase clinical data to the population of an untested region. In practice, however, it was a diplomatic compromise intended to release tension by leaving part of the procedure ambiguous and open. From the Western viewpoint, the bridging study was a way to test whether existing data could be extrapolated to a local region where the product was to be marketed, and was only to be applied if the product was suspected of being ethnically sensitive. From Japan's viewpoint, however, bridging studies were considered a way to allow additional studies designed especially for Japan, formatted as full studies but using fewer Japanese samples. Because the ICH respects local agencies' judgment as to whether a drug may have ethnic-related effects, Japan could require each producer to provide additional data.[26] Although confusion remained, the guideline was agreed on in 1997 and implemented a year later.

Taiwan followed the E5 issue at the ICH3, but did not get actively involved until implementation, for two reasons. First, despite its considerable record as a buyer of pharmaceuticals, Taiwan's market is much smaller than Japan's.[27] Like other markets of its size, Taiwan always feels under great pressure to bargain for more local trials before granting approval, notwithstanding global industry regulations on new drug approvals.[28] Second, and more general, is Taiwan's political situation. In spite of its economic vitality and connectedness, Taiwan has not been allowed to join international governmental organizations, including medical ones. Although some Taiwanese experts have seen the need to form a network for regulatory science in Asia, it is hard to realize this without a specific focus.[29]

The E5 guideline gave Taiwan a concrete topic on which to speak. In contrast to Japan's ambiguous attitude toward the guideline, the Taiwanese government immediately announced that it wanted to be the first non-ICH state in Asia to adopt the E5 guideline, including the controversial sections.[30] The Center for Drug Evaluation (CDE), an FDA-like institute, was established in

Taiwan in July 1998 to offer high-quality in-house reviews of new drug applications. In effect, it took responsibility for the implementation of ICH guidelines and the handling of all international affairs relating to pharmaceutical regulatory science.

What made the CDE famous were its evaluations of differences among populations. On the one hand, it recognized differences between Asians and Caucasians; but it did not insist that clinical data produced in relation to one population could not be applied to another after careful evaluation. Using a genetic survey of Asian populations, the CDE required bridging studies only when the application was considered ethnically sensitive, and welcomed Asian data from outside Taiwan.[31] By 2003 local trials had been deemed necessary for bridging studies only for fifteen out of sixty-two applications, and there were convincing reasons for bridging in each case (Lin, Chern, and Chu 2003).

The CDE's aggressive approach attracted international attention and led it to head a forum on bridging studies at the Asia-Pacific Economic Cooperation (APEC), the only influential international forum where Taiwan is recognized as an independent political entity.[32] Starting in 2000, the APEC network provided Taiwan with a way into the ICH, where it was invited, as the APEC representative, to the satellite meetings of ICH5 and ICH6. At both of these meetings, Taiwan was characterized as an exemplary non-ICH country that had engineered a situation enabling the industry to make available the latest medicines while protecting the health of its people. In fact, this win-win situation has other implications for Taiwan. Unlike Japan, which engaged only reluctantly in the globalization process, Taiwan, which has been long isolated from the world, grabbed the chance to be heard and made the best use of it.

Compared to these two states, Singapore was rather behind in following ICH developments. Although in 1995 APEC proposed that its headquarters for the Coordinating Centre for Good Clinical Practice (GCP) be based in Singapore, as an aid to its burgeoning biomedical-research industry, in drugs regulation Singapore was behind Japan and Taiwan. It did not renew its regulatory system until 1998, when a center for drug evaluation was established, involving the collaboration of the National Science and Technology Board, the Ministry of Health, and Singapore General Hospital.[33] The system developed slowly before being incorporated in 2001 into the Health Sciences Authority, a new institute derived from the existing regulatory section of the Ministry of Health. Singapore missed not only the E5 debate, in which Japan was closely involved during the early 1990s, but also the chance to form a professional and independent regulatory institute, as Taiwan did in the late 1990s.

However, this did not prevent Singapore from attempting to catch up with other Asian states. Where APEC served as a gateway for Taiwan's globalization, the Association of Southeast Asian Nations (ASEAN) served as Singapore's platform.[34] In order to create a single pharmaceutical market, ASEAN countries hope to harmonize regulations for pharmaceutical registration on the basis of ICH guidelines and through mutual recognition among regulatory agencies.[35] With the initiative of the ASEAN Consultative Committee for Standards and Quality, a Pharmaceutical Product Working Group was formed in 1999 to establish common technical requirements and develop quality guidelines for product registration.[36]

Even so, Singapore's role in ASEAN is ambiguous. Although Singapore shares with other ASEAN states the will to form a single pharmaceutical market, their developmental statuses are different. Some states in this region continue to rely heavily on high-quality generic drugs. But Singapore has made much progress in healthcare, and its pharmaceutical market, though small, is able to utilize cutting-edge medications. Singapore has also begun to involve itself in ASEAN activities and regularly shows up at the APEC network. But although some global companies have set up their Asian subsidiaries in Singapore, Singapore remains unsure of its strategic position in the global network of pharmaceutical regulation.[37]

Crafting Genomic Race, Bridging States, Going Global: Three Post-E5 Positions

Let us continue our analysis of the agendas that Japan, Taiwan, and Singapore developed to cope with the ICH in the new millennium. In retrospect, these positions might be seen as responses to the E5 guideline and the concept of the bridging study.[38] The previous section introduced the three states in the chronological order in which they encountered the ICH. But in this section, the three states will be presented in a comparative frame: their agendas are related but distinct, each reflecting a particular concern of the regulatory environment and vision, thus interacting with and negotiating commercial concerns in specific ways.[39]

Bridging studies were not the solution that Japan expected. Using two analogies, the cartoon below expresses how the Japanese authority conceived of itself in the bridging-studies scheme (see 1). Japanese clinical data are portrayed as either a tiny ant on a huge elephant (foreign data), or as a baby

1. Echoes from the past: Japan's perception of the E5 Guideline.

turtle (nascent bridging data) on the back of its mother (existing foreign data). From this perspective, Japanese clinical trials are considered inferior to those of the world leaders. Japan has to wait until other countries finish their trials, and its contribution to the world, compared to multinational clinical trials, is extremely small. Believing itself a world power, Japan views the "bridging study" as a design that discriminates against the Japanese and thus needs to be corrected.

The Japanese vision of clinical trials is clearly about ensuring the involvement of the Japanese at each stage of the trials. If we apply the conventional dichotomy of race versus ethnicity (i.e., biology/heredity versus culture/society), it may be hard to understand what Japan means by "the Japanese." But my analysis of the E5 debate at the ICH reveals that in the globalization scheme Japan's existence is at stake (Kuo 2008). Japan used two regulatory practices to achieve its vision. The negative one, as the industry soon realized after implementing the E5 guideline, was to reject so-called retrospective bridging studies—studies that aim only to generate Japanese clinical data to complement the application package. By October 2003, when the ICH6 was held, only a handful of cases which claimed to apply the concept of bridging had been accepted.[40]

Meanwhile, as a positive strategy, there appeared an agenda called "global drug development," proposed by the Japanese regulatory authority as an alternative way of doing global clinical trials. In relation to the various populations among which the tested product may be applied, trials using a global-drug-development scheme have to fulfill two requirements: sufficient numbers of Japanese test subjects and early involvement in the design of clinical trials. It was in this context that the idea of the "genomic" race was put forward, for Japan had to provide a scientific basis for what it means by "Japanese subject" in this clinical-research scheme.

Only by understanding this can we understand the aims of the Advanced Life Science Information System (ALIS) and pharmacogenetics — two scientific endeavors urged by the Japanese government. Both are necessary for supporting the concept of genomic race to justify clinical trials. Sponsored by the Japan Science and Technology Agency, ALIS is an integrated, informational gateway that englobes all projects concerning genomic research by and about the Japanese.[41] Consisting of several databanks and open to the public, ALIS is intended to make the Japanese accessible and "visible." For instance, one of the projects that ALIS links to is the database of Japanese Single Nucleotide Polymorphism (JSNP), which has identified and collated up to 150,000 SNPs from the Japanese population to construct a dataset in order to probe the relationship between polymorphisms and common diseases or reactions to drugs.[42] Obviously, this is the new definition of "the Japanese" that the Japanese regulatory authority favors for global drug development. As an official of the Ministry of Health, Labor, and Welfare (MHLW) revealed, "From now on, the intrinsic factors of racial difference can be replaced by the genome" (*Yakujinippo* [pharmaceutical news] 2001).

Even so, this new definition requires a theoretical tool in order to work in clinical trials. To serve this need, a statistical method called "genomic statistics" was developed by Masahiro Takeuchi at Kitasato University and later adopted by the MHLW to explain how differences among populations should be dealt with to make better global trials.[43] At the APEC symposium in 2003 on statistical methodology for evaluation of bridging studies, Takeuchi presented the main concept of genomic statistics as follows: in order to avoid statistical biases in clinical research due to inappropriate population selection, genomics should be applied to identify the right groups for testing. In other words, the test populations are chosen by their genomic characteristics, or "molecular profiling."[44]

At first glance, the above methods have nothing to do with the Japanese in particular. Indeed, pharmacogenomics is not a "Japanese" science, nor does genomics belong exclusively to Japan. However, Joan Fujimura's study of Japan's Human Genome Project has suggested that culture should be considered a set of particular practices existing at particular times and places; it is "both a heuristic device for discussing local and global actions and movements and a concept that is being continuously produced through actions and discourses about these actions" (Fujimura 2000, 84). Echoing this point, the present chapter further emphasizes how these universal sciences find their strategic uses in Japan's vision of global trials. The reason is simple: genomics is so extremely expensive that few states can afford it. When these sciences are introduced to create a higher standard for global clinical research, Japan, which is willing to spend as much money as necessary to prove its distinctness, will participate more in these trials, or at least have the opportunity to represent all Asian populations. This is Japan's way of living with globalization. In fact, during my interview with Masahiro Takeuchi, I reminded him that in his explanation of global clinical trials, he had said "Japanese" when he should have said "Asians." He did not deny it: "Well, yes. But do you think it will make any difference?"[45]

While Japan has chosen to focus on the concept of the Japanese, Taiwan has tried to prove the existence of the Taiwanese state. Taiwan's regulatory authority, the CDE, successfully made itself visible at the ICH by way of the E5 policy separating race from the state—Taiwan recognizes the differences between Asians and other populations, but does not insist on Taiwanese data as a requirement for every drug proposed for sale in Taiwan. However, this policy advantage is losing importance, as the paradigm is shifting to global drug development. In response, three statistical methods were presented in 2003 at the APEC symposium on statistical methodology for evaluation of bridging studies.[46] Despite their scientific merits, these methods also have policy implications: they provided the CDE with tools to "save" bridging studies in the scheme of global clinical trials.

The first is the group sequential method, which proposes a practical approach that includes patients from new regions, such as Asian states, as part of the recruitment for the whole study for submission to the "original" region (Europe and the United States in most cases). In this sense, the bridging study is considered a substudy of a global trial. In order to ensure the consistency of the study protocol, special sample sizes and designs are required. In the same scenario, the second method can be called a "weighted-discounted" approach. It is derived from the traditional Z-test method and is based on the argument that information already obtained from the original study will have a huge effect on the partition of sample space in the bridging study. Thus, the results of the latter must be weighted according to the region in which it is conducted. I call the third statistical method "multicentered-hierarchical." Recognizing that each state market in the Asia-Pacific region is too small to bargain for a full clinical trial, this method suggests a hierarchical operational structure that groups the centers recruited in a transnational clinical trial. In order to establish reasonable measures for all regions in which the product would be marketed, this method insists that every region should have a representative center and that a "state effect" should be attached.

Like pharmacogenomics, these methods appear abstract scientific elaborations; they seem to have nothing to do with any state in particular. However, as with genomics and Japan, there are policy assumptions implicit in these methods, and despite differences in methodology and statistical techniques, they share the same goals. They aim not only to prove that bridging studies are still workable in a multinational situation, but also to emphasize the importance of "regional differences," which do not appear in the E5 guideline, and to ask for the inclusion of subjects from every state where the product is to be marketed. Only from this perspective can we understand Taiwan's strategy for embracing globalization. Taiwan is attempting to demonstrate its existence by building a statistical network through bridging studies. Not having been considered a state for over thirty years, Taiwan appreciates the ways of extending the use of bridging studies, which allow every state to make a "fair" contribution to global clinical trials. Only thus can Taiwan's functionality as a state be preserved.

Singapore lagged behind Japan and Taiwan in the E5 debate. The ethnicity question was even more complicated for Singapore than for the other two states. It seemed impossible to use a scientific tool to deal with the ethnic complexity of Singapore's population.[47] However, Singapore chose simply to ignore the differences altogether, so that it could "skip" the dispute over bridging studies.[48] Identifying its state as a rising star in Asia's booming biobusiness and a node in the global network of clinical research, Singapore's Centre for Drug Evaluation arrived at a clear policy decision regarding the E5 guideline and bridging studies: it made the best use of the former while ignoring the latter.[49] As the center's director John Lim commented, "If we are global, there will be no need for bridges."[50] Singapore claimed to be able to provide the best sites for clinical trials of Asian people, but did not seek to apply the results to its own nationals.

This policy reflects Singapore's vision of making itself a global state by effacing regional differences. Singapore is not in a position to be a primary reviewer of drugs, so it hopes instead to be the first East Asian country with access to the latest drugs marketed in the most advanced countries.

This standpoint can be discerned in the pharmaceuticals reviewing system introduced by the Centre for Drug Evaluation, which aims to streamline the global flow of pharmaceuticals through Singapore (Wong and Lim 2003). For example, "verification" evaluation is the quickest route for new drugs that have been granted marketing approval by the benchmark regulatory agencies recognized by the Health Sciences Authority.[51] This process was designed to accept results from those authorities in order to shorten the time between primary review and marketability. To this end, Singapore has drawn up a dossier format and review requirements comparable to those of the ICH, following the leading regulatory authorities. In addition, its English-language environment enables Singapore to maintain ties with its former colonizer and its former colonies outside Asia.

The Southeast Asian pharmaceutical network consists of both external links to the West via Singapore and internal connections within the region itself, and Singapore is well aware of this. Over the past five years Singapore has been enhancing its regional connections through ASEAN. In 2003 Singapore was appointed to chair the ASEAN Implementation Working Group on ASEAN common technical documents (ACTD), leading the realization of ACTD implementation with Malaysia and Thailand (which is currently the APEC representative for the ICH GCG).[52]

One cannot say that Singapore is the only nation making this enhancement of regional connections happen; however, considering the strategic position of its global industries, Singapore will certainly benefit from this homogenous regional market of pharmaceuticals. It has deregulated all requirements, including those concerning race-related effects, in order to have access to the most advanced products as soon as the West. In addition, it is engaged in regional networks, such as the APEC and ASEAN, to make "pathways" through which these products can be sold in other Asian states. Unlike Japan's vision of a nation-state made up of a Japanese race, or Taiwan's struggle to gain recognition as a state, Singapore sees its advantage in global networking. The network exists and therefore so does the state.[53]

Monitoring Transforming States in the Global-Genomic World

What is the future of the pharmaceutical industry as it marches into Asia? Some reports, such as *Scrip 100* (2006), portray the situation much like a typical market-research report and dwell on identification of the right markets and penetration of their trade barriers. For example, Ian Schofield, a

Scrip 100 reporter, forecasts that China and India, two rising economies in Asia, are potential markets as well as growing threats. Japan is still criticized by the industry for the tardiness of its regulatory practice and the difficulties in conducting acceptable clinical trials, but seems to have been let off the hook by the industry, which is more interested in making sales than in settling scores (Schofeld 2006). For such reports, national difference seems to be out of the question. Japan's changing attitude toward the acceptability of foreign clinical data is noted, which the industry considers an "improvement," and explained as a response to emerging business rivals in the region, like Singapore and Korea. This story of pharmaceuticals and globalization fails to take account of any cultural or social aspect of the state.

Yet if one uses the example of the regulatory standardization of pharmaceuticals, it is clear that no two states behave alike in the face of global capitalism. Only at the lively interfaces where the state meets the global can we identify their distinct characteristics. This chapter may be read as a rough-and-ready sketch of how Asian states resist the mighty wave of globalization, a kind of story that may be found anywhere. However, that is not my goal. Echoing the notion of Michel Foucault (2001:170–71) of the epistemic changes which allow only some problems and not others to come to the fore, I attempt to trace a particular process in the dawn of the genomic era, where pharmaceuticals and related clinical research have to be global and universal. This specific moment problematizes the previously invisible issues concerning the state and population differences as issues in the world of pharmaceutical regulation. What is at stake is perhaps why state and race, which are inextricable from the making of nation-states, merge at the frontier of global clinical research.

Michael Fischer's notion of anthropological voice is useful in this context (Fischer 2003). The challenge for anthropological voice, Fischer argues, should be considered with a keen understanding of the modern world and of the role that anthropology can play in it. The challenge in renewing the notions of the ethnographic and anthropological voice "is not the disappearance of difference, of different cultures, or of ways of organizing society any more than it is not the disappearance of class, capital, unethical exchange, power, or gender relations. On the contrary, the challenge is that the interactions of various kinds of cultures becoming more complex and differentiated at the same time as new forms of globalization and modernization are bringing all parts of the earth into greater, uneven, polycentric interaction." Considering "the aspiration for cross-culturally comparative, socially grounded, linguistically and culturally attentive perspectives," the challenge for anthropology is "to develop translation and mediation tools for helping make visible the differences of interests, access, power, needs, desire, and philosophical perspective" (Fischer 2003, 3).

What I hope to have achieved in this chapter is consonant with the above challenge. I have explored on the national level what anthropological voices we can appreciate when the world consciously becomes global and genomic. I have argued elsewhere (Kuo 2008) that different conceptualizations of race and ethnicity cannot easily be understood unless encountered in the arena of clinical research. In this chapter, my focus has been on the state, an artificially organized yet strategically lively actor, which, together with other states, composes the world politicoeconomic infrastructure.

Only with the above intention in mind can we understand that when race became an issue at the ICH, it did not reflect simply nationalism. This chapter has shown the two functions of race in this story. First, race is itself a socially contextualized topic which is fluid and always questionable. Any scientific attempt to clarify it creates confusion and reveals the cultural and social assumptions behind it.[54] Second, as an issue for discussion at the ICH, race is a point of reference by which we monitor state actions. In other words, it is a "lens" that allows one to observe and appreciate the "deep play" of Asian states, a notion inspired by Clifford Geertz's famous article (1973) on this global platform. Thus, the examples of Japan, Taiwan, and Singapore may not be used simply to "fill in" a conceptual gap between individuals and the world, as political scientists may claim. For us, they are anthropological voices, and always in transformation.

Let me end this chapter by summarizing these voices. Japan seems to be the only state that holds strongly to the concept of a coherent Japanese nation. Yet this idea may not faithfully reflect either the composition of the Japanese population or what Japanese nationals think of themselves. Rather, it presents a vision of the Japanese state clashing with globalization. According to this vision, the Japanese race and the Japanese state are two sides of the same coin. As we can see from Japan's reaction at the ICH, this vision understands race more as a collectivity than as a question of purity. Race is the basis of Japan's claim to uniqueness, and the state is the subject that insists on the category and benefits from it.[55]

Taiwan has other concerns about statehood. Unlike "normal" states, Taiwan has been politically isolated for some time and is always eager to prove itself a good "citizen" in the global village. As seen in the ICH discussions, although bridging studies presume a "West-center, East-peripheral" worldview, this presumption is not a problem for Taiwan. By the same logic, the

E5 topic gave Taiwan a bridge to the world. While Japan pushes the genomic view for global clinical trials, Taiwan survives through bridging studies by making statistical bridges to other regions. Although these methods have not yet been accepted and implemented as policies by other Asian states, Taiwan's CDE is ready to promote this vision as long as it makes Taiwan's voice heard in the global arena.[56]

A shining star in biotechnology, Singapore is recognized as a competitive hotspot in the network of global business.[57] Very few may remember the small city-state's cultural complexity, and its government tends to ignore this fact whenever it is a barrier to business. This willful ignorance is how the Singaporean state is attempting to survive globalization. Unlike Taiwan, which recognizes the ethnic distinctness of Asians and has taken advantage of this in its bridging studies, the Singaporean state does not see any benefit in such a strategy.[58] Singapore's vision is thus to boost and to be at the hub of the pharmaceutical sector in Southeast Asia.

Although the state and its transformative nature form the main concern of this chapter, I do not intend to reject other concerns that link the local to the global within the topic of the body and pharmaceuticals. As Lakoff and Jasanoff have argued in this volume (chapters 8 and 4), the role of the state remains salient in the space of exchange between life and business: by creating regulatory regimes, states ensure the economic value of pharmaceutical innovations. All these factors distract the state from its ideal and naïve aim of "protecting its people's health."

The present text situates itself within a literature that attempts to understand global pharmaceuticals through various research schemes and regulations. Further analysis of the state's role in global schemes is carried out by looking at its transformations through the transformations of its regulatory systems. Hardt and Negri have pointed out that the world that capital is faced with is not really "smooth" but "defined by new and complex regimes of differentiation and homogenization, deterritorialization and reterritorialization" (2000, xiii). Echoing this argument, the present chapter has shown that each of the states in question has its own concerns, which are not exclusively related to either health or capitalism. One is led to the conclusion that it is too early to declare the demise of the state in the name of globalization. Instead, the world is being referenced and fundamentally changed in terms of state and race.

The introduction of this new world perspective inevitably refreshes our understanding of the state as an entity which is "co-produced" with other states. As an open conclusion, let us return to Ernest Gellner's observation on the state in the modern world. Viewing two ethnographic maps before and after the age of nationalism and looking at a political map of the modern world, he observes, "There is little shading; neat flat surfaces are clearly separated from each other. . . . We see an overwhelming part of political authority has been concentrated in the hands of one kind of institution, a reasonably large and well-centralized state" (Gellner 1983, 139–40). In concert with this reflection, the present chapter argues that every state deserves ethnography. Gellner is right. In the era of global pharmaceuticals, the state still matters.

Notes

1. One theoretical elaboration of how capital functions in the era of genomics and clinical research is Kaushik Sunder Rajan's *Biocapital* (2006).

2. A study that does look at the effects of regulation is Joseph Dumit's "Drugs for Life" (2002), which argues that people are "destined to become ill" when the criteria for normality broaden.

3. The full name is the International Conference on Harmonisation of Technical Requirements for Registration of Pharmaceuticals for Human Use. For a concise introduction to the ICH, see Nutley 2000.

4. The topic was titled "Ethnic Factors in the Acceptability of Foreign Clinical Data." This topic was also named "E5," the fifth topic under the category of efficacy at the ICH.

5. For example, there has been a heated debate about Bidil, the first drug approved by the FDA to treat heart failure among "self-identified" African Americans, with regard to its impacts on the racial politics in the United States. Numerous forums, such as websites, discussion boards, and conferences, have been formed, among which the conferences organized, since 2005, by the Massachusetts Institute of Technology's Center for the Study of Diversity are typical in discussing race along with the social contexts it locates and the policy implications it carries. Jonathan Kahn, one of the regular presenters at these conferences, has studied Bidil intensively and wrote a paper, "Harmonizing Race" (2006), that compares the ICH E5 guideline to the FDA's regulations on ethnic difference. Additionally, Steven Epstein, who wrote *Inclusion* (2007), has also addressed the ICH in reference to the ethnic politics of the United States on clinical trials.

6. With the notion of actors, I refer to Bruno Latour's works on the making of social actions. See Latour 1987.

7. Compare Peter Galison's notion of the "trading zone" in *Image and Logic* (1997).

8. Although still understudied, the state has been investigated by anthropologists from various perspectives. For instance, Michael Fischer's *Iran* (1980) attributes the Shiite denial of the legitimacy of the Pahlavi regime to tensions located in the context of dualistic dynamics involving the global and the national, capitalism and Islam, and

modernity and tradition. Tracing the introduction of modern agriculture to India, Akhil Gupta's *Postcolonial Developments* (1998) criticizes the discourse of underdevelopment by showing how being drawn into global political economy has become a dilemma for the Indian state. James Scott's *Seeing like a State* (1998) looks at the state by analyzing tensions between state authorities and various "unstable" individuals in the failures of state projects throughout history. Emily Martin's work on the state (1999), by contrast, views the evolved modern state as a flexible organism capable of responding to requests.

9. In order to study the problem of the nation-state in the global scene, the political sociologist Horng-luen Wang (2000) proposes an institutional approach. He defines a nation-state as a political and cultural product derived by demarcating a territory within the networks of the global. According to Wang, when dealing with a modern nation-state, one can draw no clear line between culture and politics, or between nationalist reality and pure nationalism. Instead, the nation-state is an institutionalized form of life whose existence relies on the operation and context of a given institution.

10. A typical discourse can be seen in a publication of the Pharmaceutical Research and Manufacturers of America, *Pharmaceutical Industry Profile 2006* (PhRMA 2006).

11. To be exact, this growth includes the expansion of the pharmaceutical industry and the FDA. In the mid-1970s statistics became involved, as clinical trials became highly complicated and difficult to manage. Both regulators and industry hired more experts and statisticians for clinical trials, and their efforts made clinical trials data management highly technical and abstract that was impenetrable to outsiders. For further description of the dynamics that developed between the regulated and the regulator, see Kuo 2005, chapter 2, 100–109.

12. There have been debates over the actual cost of innovating each new drug. The figures given by the PhRMA (and challenged by critics such as Marcia Angell, James Love, and Sidney M. Wolfe) show that the actual cost per innovated drug increased from $259 million in 1990 to $302 million in 1995. According to a 2001 study by the Tufts Center for the Study of Drug Development, this has lately increased to $802 million. The most-cited study evaluating the average cost of bringing a new chemical entity to the market is DiMasi et al. 1991. For PhRMA's reasoning on this cost, see PhRMA 2000. For the main criticisms of the figures released by PhRMA, see Public Citizen 2001.

13. This does not consider the problem of humidity, which was not specified before any universal standard was established.

14. The WHO founded the International Conference of Drug Regulatory Authorities (ICDRA) in 1980 to allow the drug regulatory authorities of WHO member states to meet and discuss ways to strengthen collaboration and exchange information. Before the inauguration of the ICH, the ICDRA was a means of aiding the WHO and regulatory authorities in their efforts to improve the safety, efficacy, and quality of medicines. Even so, as a purely administrative meeting of regulators, ICDRA did not achieve as much as it aimed to. In fact, in the opening speech of the third ICH con-

ference, in 1995, Hiroshi Nakajima, then director general of the WHO, urged that the achievement of the ICH should be shared by all WHO member states, not just ICH regions.

15. The ICH has six voting members, including regulatory agencies (the FDA, the Japanese Ministry of Health and Welfare, and the European Community) and industry representatives (PhRMA, the Japan Pharmaceuticals Manufacturing Association, and the European Federation of Pharmaceutical Industries and Associations). Some organizations, such as the WHO, Health Canada, and the European Free Trade Association, are nonvoting observers. The International Federation of Pharmaceutical Manufacturing Associations, which is also a nonvoting member, functions as secretariat.

16. For an updated and detailed description of the ICH process, see "The ICH Process for Harmonisation of Guidelines," available at the ICH website, http://www.ich .org.

17. According to the ICH website (last visited on 18 July 2011), the number of finalized guidelines has increased to sixty. The number of guidelines in the categories of quality, safety, efficacy, and multidisciplinary issues is 24, 13, 18, and 5, respectively. I shall refer to ICH conferences using the letters "ICH" plus a number denoting chronology; for example, the sixth ICH conference will be indicated by "ICH6." Although the seventh conference—originally scheduled for 2007—was canceled, the steering committee and expert working group meetings were held in Brussels in May and in Yokohama in October.

18. This is a remark often heard in my interviews with regulatory agencies and industry representatives. For them, there seem to exist no conflicts between public health and the sale of drugs as all regulatory boundaries are dissolved. Even so, some of the ICH's achievements have been assessed critically by scholars such as John Abraham and Tim Reed (2002), who criticize the ICH for working to accommodate the industry's desire to loosen requirements. In another essay, John Abraham (2002) even casts the ICH in the longer trend of the development of pharmaceutical regulations, in which the industry plays an aggressive role to making drug approvals easier and faster.

19. This statement was later modified: "To promote a mutual understanding of regional harmonisation initiatives in order to facilitate the harmonisation process related to ICH guidelines regionally and globally, and to facilitate the capacity of drug regulatory authorities and industry to utilize them" (ICH website). Nevertheless, the hierarchal structure imposing ICH standards on non ICH regions remains the same.

20. For this, Applbaum cites exactly the ICH's E5 topic, claiming that it facilitates the move to accept foreign clinical data, and thus achieves the "most significant for opening the door [of the Japanese market] to global activity in the industry" (2006, 91).

21. According to Visiongain's report *Japanese Pharmaceutical Market, 2006–2011*, in 2006 Japan's pharmaceutical market generated sales totaling $60 million, approximately 11 percent of the world market.

22. Notification no. 660, "Handling of the Data of Clinical Studies for Pharma-

ceuticals etc. Conducted in Foreign Countries," announced by the director general of the Pharmaceutical Affairs Bureau, 29 June 1985.

23. Personal conversation with Osamu Doi in Tokyo, 7 August 2006.

24. Japan uses the term *minzoku* to summarize its conceptualization of bodies and cultural distinctions. This is the term used in the Japanese version of the E5 guideline (also in the official Japanese version), and is key to understanding this debate. In brief, the dichotomy between biological concepts and social-cultural concepts offers little insight into the idea of *minzoku*, which relates more to the idea of "nation" or, in a practical sense, to that of "nation-state," which blends ideas of race, ethnicity, and the political institutions that manifest them.

25. In addition to the pharmacokinetic study, which belongs to phase 1 of clinical trials and uses a small group of healthy volunteers to assess the basic behavior of a drug, phase 2 and 3 studies must be completed before approval can be granted to a new drug. Performed on a larger group of healthy people and patients, phase 2 studies are designed to assess how well the drug works and at what dosages. Phase 3 studies are more expensive than phase 2 studies; most are multicentered trials involving large patient groups, of three hundred to five thousand or more, depending on how the disease works, and are intended as definitive assessments of the drug's efficacy.

26. Another ICH example—similar to the E5, but less controversial—is the E10 guideline (titled "Choice of Control Group in Clinical Trials"). According to this guideline, the use of placebos should not be universal, but depend on the research design and population. This issue was brought up for discussion at the ICH in part because comparative clinical trials were the dominant method of gathering data, yet the use of placebos was considered unethical in Japan. Recognizing this regulatory difference, guideline E10 leaves it to local regulatory authorities to judge the acceptability of clinical data.

27. According to IMS Health, Taiwan is the twentieth-biggest individual market in the world, consuming U.S. $2.2 billion in pharmaceuticals in 2003. On the global pharmaceutical market map (in U.S. dollars), it trails Japan ($52.4 billion), Korea ($3.9 billion), Australia ($3.1 billion), India ($3.4 billion), and China ($4.0 billion) in the area known as the Africa-Asia-Australia region; http://www.imshealth.com/portal/site/ims, accessed 11 August 2003.

28. The development of clinical-trial regulation in Taiwan has a complicated history which deserves a serious study of its own. In essence, it is not determined solely by the government but also by the PhRMA and the U.S. trade representative. After 1989, through U.S.-Taiwan trade negotiations, high-quality (and costly) clinical trials were first imposed, at the request of the industry, to block local competitors. This bar was elevated even further before the ICH guidelines were introduced. However, as pointed out by Oliver Yoa-Pu Hu, former director general of Taiwan's Bureau of Pharmaceutical Affairs, it was also the PhRMA that, after 1994, asked Taiwan to remove clinical-trial requirements because they were found to be no longer necessary. By claiming that the existing trials were not "scientific" enough and thus formed an

"inappropriate non-tariff barrier," the PhRMA accused the Taiwanese government of erecting trade barriers in its annual reports to the U.S. trade representative, asking for higher pressure to suppress these local trials (from a personal conversation with Hu in Taipei, 20 October 2003). For selected annual reports the PhRMA prepared for U.S. trade representative on the estimation of trade barriers during the years Dr. Hu mentioned, see PhRMA, *Trade Estimate Report on Foreign Trade Barriers* (NTE). Available at http://www.cptech.org/ip/health/phrma. Last accessed on 18 July 2011.

29. In 1995 Taiwanese experts and medical advocates founded the ICH-Taiwan, a strategic, mission-oriented committee under the auspices of the Department of Health. Of the three working groups, one is designated to promote Taiwan as an operational center for clinical trials. For more details on ICH-related initiatives in Taiwan, see Chen 1998.

30. Another competitor in this region, Korea, revised its Good Clinical Practice regulation in accordance with ICH guidelines and in 2001 introduced the concept of the bridging study.

31. Lin et al. 2001. It is the first survey widely covering populations in East Asia and the first to genetically locate the so-called Taiwanese on a global map. It became, and still is, the core scientific study supporting the CDE's bridging study policy.

32. The full name of this network is the APEC Network of Pharmaceutical Regulatory Science-APEC Joint Research Project on Bridging Study.

33. The most notable regulation in this revision is the Singapore GCP guidelines, which had not been updated since their establishment in 1978. As Taiwan and Korea had done for their GCP guidelines, Singapore revised its GCP by adopting the ICH E6 guideline (Ministry of Health, Singapore, "Singapore Guideline for Good Clinical Practice (SGGCP)," available at www.gcphelpdesk.com/index.php/knowledge-base/item/download/6, downloaded on 18 July 2011). For the development of Singapore's regulatory environment in the 1990s, see Fong 1998.

34. The ten member countries are (in alphabetical order): Brunei Darussalam, Cambodia, Indonesia, Laos, Malaysia, Myanmar, the Philippines, Singapore, Thailand, and Vietnam.

35. Indonesia's National Agency of Drug and Food Control, Malaysia's Drug Control Authority, the Philippines' Bureau of Food and Drug, Thailand's Food and Drug Administration, and Singapore Health Sciences Authority, Centre for Pharmaceutical Administration, and Centre for Drug Evaluation represent the most progressive agencies in this region and play active roles in the harmonization process.

36. Before this initiative, pharmaceutical review processes varied. Different documents were required for registration, and the average time needed for registration ranged from seventeen weeks to eighteen months. Although progressing slowly, this project is still ongoing, as reported in the ICH-GCG meetings (the latest of which took place in Brussels in November 2008). For an evolutionary account of clinical trials in Southeast Asia and harmonization initiatives, see Ellick Wong 2003.

37. Starting with Eli Lilly's the Lilly-NUS Centre for Clinical Pharmacology in 1997, leading pharmaceutical companies, such as AstraZeneca, Bristol-Myers Squibb, GlaxoSmithKline, Merck KGaA, MSD, Novartis, Novo Nordisk, Pfizer, Sanofi-

Aventis, Lundbeck, Schering AG, and Schering-Plough, have set up their biomedical research centers in Singapore.

38. It may be noticed that it is not only the Asian states discussed in this chapter that need to readjust their standards of clinical trials to take into account ethnic difference. Referring to the U.S. context, Jonathan Kahn's "Harmonizing Race" (2006) compares the different racial categorizations in FDA regulations and the E5 guideline, pointing out that although the FDA has adopted the ICH categorization in its regulations, it still advises product sponsors to use the original categorization, which includes Hispanic or Latino as an independent category, in its data collection, even outside the United States.

39. For more information on the comparative approach, see Sheila Jasanoff's *Designs on Nature* (2005), which contains a good discussion of why comparison is necessary to understand regulations in the transnational context.

40. At the meeting of the APEC Network of Pharmaceutical Regulatory Science in 2003, Elaine Esber reported complaints from the industry such as "E5 has resulted in a request for more studies, rather than less"; it is "a convenient excuse for requiring a local registration study and calling it a bridging study"; "requests are for data, country by country, not as a region"; "there are ulterior motives for requesting that studies be done, e.g., protect local industries"; "E5 is being implemented as a trade barrier"; "most companies are doing studies just to not get into an argument"; "governments are not being flexible"; and many others. For the successful cases using bridging studies, see Uyama et al. 2005.

41. These multi-institutional projects include Human Organized Whole Genome Database, Japanese Single Nucleotide Polymorphisms, Human Genome Sequencing, Eukaryotic Comparative Genome Browser, Structural Initial Data Library of Amino Acid Residues, SNP Database Network in Japan, and the Japanese node of the International HapMap Project. Each project involves about three to five research institutes from Japanese industry and academies.

42. For a brief history and discussion of the JSNP project, see Hirakawa et al. 2002.

43. Personal conversation with Kazuhiko Mori in Tokyo, 20 September 2007.

44. Takeuchi cites a study in the *New England Journal of Medicine* (NEJM) in which patients of diffuse large-B-cell lymphoma are divided into three subgroups according to their genetic profiles; each subgroup had a distinct t survival rate after treatment. In other words, there is a correlation between genetic expressions and cancers, as well as between genetic expressions and clinical outcomes. See Rosenwald et al. 2002.

45. Interviewed at Kitasato University, Tokyo, 23 August 2004.

46. Connected with leading universities, industry, and regulatory agencies, Taiwan has a strong pool of Taiwanese experts in biostatistics, especially in the field of clinical trial design. This also makes this symposium significant in terms of global clinical trials. The following methods are presented in the symposium held. For their presentations, see National Health Research Institutes 2003.

47. Unlike Japan, which tends to vaunt the "homogeneity" of its population,

or Taiwan, where Han Chinese are predominant, Singapore, whose population is a mere 3.44 million, consists of people of at least three origins—the Chinese, the Malay, and Indians, along with others.

48. The decision not to study ethnic effects in pharmaceuticals can be also seen in another state on the Asian side of the Asia-Pacific region, Australia, whose population is over 30 percent Asian.

49. As a part of the Health Sciences Authority, the Centre for Drug Evaluation merged with Centre for Pharmaceutical Administration to found the Centre for Drug Administration in 2004.

50. Cited from John Lim's PowerPoint presentation, titled "Procedure of Consultation and Evaluation," delivered at the 2001 APEC Network of Pharmaceutical Regulatory Science meeting, Taipei, 25–26 May. This presentation is not in the public domain.

51. These agencies are the FDA, the Medicines and Healthcare Products Regulatory Agency, the Therapeutic Goods Administration (Australia), the European Agency for the Evaluation of Medicinal Products, and Health Canada.

52. It is expected that there will be complete implementation of common documents for pharmaceutical applications and twinning-system programs allowing information exchange for specific areas within ASEAN states.

53. Although beyond this chapter's scope, as a meditative footnote on social theory I would like to note that the difference among Japan, Taiwan, and Singapore is not just one of national variation; it reflects fundamental concepts of how the state forms a vision of its future. I would like to refer the case of Japan and Taiwan to Benedict Anderson's *Imagined Communities* (1983), according to which visions are somehow formed on the basis of the infrastructure of people living in the land. In other words, visions are formed from within the state. However, this comment may not be entirely applicable to Singapore, which is too tiny to claim state value without considering influences from outside. This characteristic becomes more obvious when researchers of globalization start working not only on individual states, but on multiple states and their interaction. In addition to the present chapter's discussion of the development of pharmaceutical regulations, Aihwa Ong and Charis Thompson have recently completed work on global trends in biotechnology and stem-cell research. Both discuss Singapore in their problematic.

54. A recent failed attempt to "fix" race is the Human Genome Diversity Project, an international project that seeks to map the diversity and unity of human species by DNA sampling. For a critical analysis of how this scientific enterprise was initiated and collapsed, see Reardon 2004.

55. This view can be found in the recent resolution by the Pharmaceuticals and Medical Devices Agency (PMDA) to initiate a research agenda for global drug development based on the diversity of populations. For more detail, see the PowerPoint reports in "PMDA Challenges for Global Drug Development" 2007.

56. In a study that I am currently conducting, I follow up on what the CDE proposes regarding the bridging of clinical data among Asian states. In accordance with the vision presented in this chapter, this project has the following preliminary result:

moving beyond making statistical bridges, Taiwan has been seeking to construct a regional network of regulatory science. This network should function not only as a scientific one that integrates clinical data produced from Asian states, but also as an administrative platform, where each state respects the presence of other states by mutual recognition.

57. Readers may be reminded here of a study by Aihwa Ong (2000) of the changing nature of citizenship and governance as seen in Southeast Asia. Ong argues that the developmental state encouraged spatial fragmentation of citizenship and governance in order to cope with the demands of the global economy. Although Ong is also working on different aspects of how states function, she considers the Singaporean state as a mere collection of functionaries, whereas I regard it as an agency or gate that controls the flow of capital. However, we both agree on and capture the characteristic flexibility that Singapore uses to survive. For a recent evaluation of the state endeavor of Singapore's biopharmaceutical performance, especially the Biopolis Science Park and the science and technology plan for 2010, see Holden and Demeritt 2008.

58. This standpoint was clearly stated by John Lim, currently the chief executive officer of the Health Sciences Authority, at the Symposium of APEC network on Pharmaceutical Regulatory Science (Tokyo, 12–13 October 2006). Even so, a state's vision or voice does not necessarily reflect that of its people. For example, Kerry Holden and David Demeritt question the ICH guidelines introduced by the Singaporean government, asking "why international guidelines, such as the ICH-GCP, appear neutral in the face of multi-racial, cultural, and ethnic societies; or what these guidelines do for science and how they improve it" (Holden and Demeritt 2008, 80).

KIM FORTUN

BIOPOLITICS AND THE INFORMATING

OF ENVIRONMENTALISM

Consider Texas City, in Galveston County, Texas, in the middle of the biggest petrochemical corridor in the United States. Texas City is home to a huge Union Carbide plant, which I have kept an eye on for a long time as a way to continue my engagement with the Bhopal disaster, my focus in earlier research. Texas City was the site of a catastrophic industrial disaster in 1947, when a freighter loaded with ammonium nitrate blew up, igniting a chain reaction that ripped through the chemical plants that surrounded the city. Over 500 people were killed; thousands were injured; over 3,000 homes were destroyed.[1] In 1987, Texas City was the site of what union workers call an "almost Bhopal." A contract worker at Marathon Oil dropped a compressor on a tank of hydrofluoric acid: 1,000 people were injured; 3,000 people were evacuated. The local economy still revolves around chemicals. There are so many point sources that it can seem impossible to know where to start an effort to reduce local pollution (Fortun 2001).[2]

By typing in the zip code for Texas City at Scorecard.org, in 2004, I got to a webpage titled "About Your Community," Galveston County.[3] There were sections on air, waste, land, and water, and also on environmental justice and on "setting environmental priorities." At the bottom of the page, I was encouraged to "explore the maps" to see how air pollution in Galveston

County compared with other communities, and to locate polluters and see how close they were to my home and workplace. In the section on air, I was told that, based on the most current data of the Environmental Protection Agency (EPA), Galveston County was among the dirtiest 10 percent of all counties in the United States in terms of noncancer hazards from hazardous air pollutants (HAPS). I was also told that 250,158 people in Galveston County faced a cancer risk more than 100 times the goal set by the Clean Air Act—84 percent from mobile sources (i.e., cars and trucks), 14 percent from point sources (i.e., large industry), and 2.5 percent from area sources (small businesses; see 1).

I then clicked on "What's Your Risk?" This page told me that the average inhabitant of Galveston County had an added cancer risk of 790 per million, and that the highest contributor to cancer risk was diesel emissions. I could click through to more information on diesel emissions, or instead click through to more information on "Cancer Risks" or "Noncancer Hazards." I could also click through to more information on "Caveats." Here I could read why exposure estimates from 1996 might not accurately predict current exposures; why exposure modeling is uncertain; why uncertainties increase with focus on small geographic areas or individual sources; why health-risk assessment also involves important uncertainties; and about caveats applicable to specific chemicals or emissions sources.

Backing up to the page on "Cancer Risks and Noncancer Hazards," I realized that I could also have clicked on two specific chemicals: diesel emissions, the HAP with the highest contribution to cancer risk in Galveston County, or acrolein, the HAP with the highest contribution to noncancer hazards. Clicking on either of these chemicals took me to a full chemical profile that included information on human health hazards, on how the chemical ranked as a hazard compared to other chemicals, on who used the chemical, and on where it was released across the United States. The profiles also included information on regulatory coverage of the profiled chemical, on basic tests done (or not done) to identify hazards associated with the chemical, and on the data that was missing yet needed to make a safety assessment. At the bottom of the profile was a list of links to numerous other sources that had more information on the chemical profiled, collected from states, the U.S. EPA, and international sources. As I scrolled through the page on diesel emissions, a pop-up window blinked at me to type in my zip code again, in order to take action. I could send a prewritten email to the EPA or a similar letter to the Texas governor, Rick Perry.

- **Cancer Risks and Noncancer Hazards in GALVESTON County**

Cancer Risks from Hazardous Air Pollutants:

Average individual's added cancer risk: 790 per 1,000,000

Population in areas where cancer risk exceeds 10^{-3}: 54,145

Population in areas where cancer risk exceeds 10^{-4}: 250,158

HAP with the highest contribution to cancer risk: DIESEL EMISSIONS

Noncancer Hazards from Hazardous Air Pollutants:

Average individual's cumulative hazard index: 3.5

Population in areas where hazard index exceeds 1: 246,305

HAP with the highest contribution to noncancer hazards: ACROLEIN

[top]

- **Sources Contributing to Health Risks from Hazardous Air Pollutants**

	Contribution to Added Cancer Risk	Contribution to Cumulative Hazard Index
Area source	3%	9%
Mobile source	84%	37%
Point source	14%	55%

See source categories contributing to chemical-specific risks
See the 1996 emissions data EPA used to estimate exposures for this area

[top]

- **What We Don't Know About HAPs in GALVESTON County**

5 hazardous air pollutants in GALVESTON County lack the risk assessment values required for safety assessment
32 hazardous air pollutants in GALVESTON County lack the exposure estimates required for safety assessment

1. Cancer risks and noncancerous hazards in Galveston. Screenshot from http://www .scorecard.org, 2004.

- Distribution of Environmental Burdens in GALVESTON County

DISTRIBUTION OF BURDENS BY RACE/ETHNICITY

Releases of Toxic Chemicals	(indicator of chemical releases)	Ratio
People of Color	████████████████████ 1200000	1.52
Whites	██████████████ 790000	

Cancer Risks from Hazardous Air Pollutants	(added risk per 1,000,000)	Ratio
People of Color	████████████████ 470	1.09
Whites	███████████████ 430	

Superfund Sites	(sites per square mile)	Ratio
People of Color	████████████████ .18	1.64
Whites	██████████ .11	

Facilities Emitting Criteria Air Pollutants	(facilities per square mile)	Ratio
People of Color	███████████████ .6	1.11
Whites	██████████████ .54	

DISTRIBUTION OF BURDENS BY INCOME

Releases of Toxic Chemicals	(indicator of chemical releases)	Ratio
Low Income Families	████████████████████ 1500000	2.00
High Income Families	██████████ 750000	

Cancer Risks from Hazardous Air Pollutants	(added risk per 1,000,000)	Ratio
Low Income Families	████████████████ 450	1.02
High Income Families	███████████████ 440	

Superfund Sites	(sites per square mile)	Ratio
Low Income Families	███████████████ .16	1.14
High Income Families	█████████████ .14	

Facilities Emitting Criteria Air Pollutants	(facilities per square mile)	Ratio
Low Income Families	████████████████ .64	1.16
High Income Families	██████████████ .55	

2. Environmental burdens, Galveston.

Back on the page titled "About Your Community" for Galveston County, I could click through to learn "Who Is Polluting Your Community" and "What Are the Major Pollutants." By clicking on the first of these, I got to a list of companies that were required to report their emissions to the U.S. EPA's Toxic Release Inventory (TRI), a pollution database established by law in 1986 and the first federal database that Congress stipulated be accessible to the public in a computer-readable format. Union Carbide's big Texas City plant was fourth on the list, releasing 1,101,343 pounds. The worst polluter was Sterling Chemicals, which emitted 8,812,611 pounds. I could also pull up a list of TRI companies ranked across the state of Texas. Here, Union Carbide's big facility in Texas City was ranked 58th. The worst polluter, in terms of total pounds released, was a BASF plant in Freeport, which released 21,492,909 pounds.[4]

Instead of following up on BASF, I returned to an "Environmental Release Report" focused specifically on Union Carbide's Texas City plant. On this page, I clicked to view a pop-up map that showed TRI facilities in Galveston County. Clicking on any one of the icons that dot the map pulled up a company name and a graph that showed trends in environmental releases between 1988 and 1999. The map allowed me to zoom way in, or way out, but it did not provide street names. Since I was familiar with the area, I still got a sense of how the ten TRI facilities in the area were clustered. I was also able to view a line graph that showed how the Texas City Union Carbide facility ranked among all TRI facilities in the United States for major chemical release and waste generation—it was in the 90th percentile. I could also see that the top-ranked ozone-depleting chemical at the facility was chlorodifluoromethane, and that the top-ranked cancer risk came from releases of benzene while the top-ranked noncancer risks came from vinyl chloride. Next I viewed a list of pollution releases at this facility sorted by health effect. The list told me that 184,513 pounds of recognized carcinogens, and 180,600 of suspected carcinogens, were released into the air in 1999. The list also told me the pounds of air releases that were recognized and suspected blood toxicants, developmental toxicants, endocrine toxicants, immunotoxicants, kidney toxicants, gastrointestinal toxicants, muscular toxicants, neurotoxicants, reproductive toxicants, respiratory toxicants, and skin or sense organ toxicants.

Scorecard told me how many pounds of toxics were released in a given year by a given facility. I was also able to learn about probable risk, body system by body system, based on a hazard-ranking system that compared all chemicals to benzene, a known carcinogen, to indicate cancer potential, or toluene, a developmental toxic, to indicate noncancer risk. The ranking system provided users with relatively stable reference points for thinking about an otherwise confusing array of health risks.

Yet another section on Scorecard's report on Union Carbide's Texas City plant was "What We Don't Know about Chemical Safety and Harm." Clicking there opened up yet another storyline—a story about the information that was *not* available for this or any other facility, even for high-volume chemicals.

Informationalism in Practice

A functional change in a sign-system is a violent event. . . . Yet if the space for change (necessarily also an addition) had not been there in the prior function of the sign-system, the crisis could not have made the change happen. The change in signification-function supplements the previous function. "The movement of signification adds something . . . but this addition . . . comes to perform a vicarious function, to supplement a lack on the part of the signified." The Subaltern Studies collective scrupulously annotates this double movement.

[What one gets] is a theory of change as the site of displacement of function between sign-systems.

GAYATRI CHAKRAVORTY SPIVAK, "SUBALTERN STUDIES"

The information experience that I have just described took place at Scorecard.org, a website that was supported by a relational database that contained profiles of over 6,800 chemicals. The website integrated local pollution information for the United States with information on health risks and with information on relevant environmental regulations. It allowed users to produce customized reports and encouraged communication with the U.S. EPA or with a polluting company.

Environmental Defense launched Scorecard in 1998, saying that its purpose was to make the status of the environment as easy to check on as the local weather. Shortly after, *Chemical Week* described the website as the "Internet Bomb" because of its potential effect on the reputations of chemical companies. Greenpeace has referred to Scorecard as the gold standard of environmental information systems because it provided opportunities for movement from information to collaborative action, and because it was partly built on open-source software, which Greenpeace says operates according to the same tenets as radical environmentalism.

I describe Scorecard here to draw out how environmental information systems are an important site of biopolitics today, indexing what I think of as the "informating" of environmentalism.[5] With the term *informating* I want to draw out how information technology and culture animate change at multiple scales, sometimes provoking critical changes in sign systems. Such shifts enable articulation previously impossible or unrecognizable, displacing what I call "discursive gaps." It is, of course, widely acknowledged that "informationalism" is a key dimension of global order today.[6] One of my broad goals is to map how the environmental field is a site of and influence on this order. I also want to make a specific argument about how informationalism works. In addition to driving and undergirding policy and law, organizational priorities, and the practice of various social actors, informationalism is routing desire and shaping subjectivity, configuring what people want to do and what they think is both possible and obligatory. Informationalism sets up, for example, how policymakers conceive of efficacy, how scientists think of civic commitments, how citizens conceive of knowledge and rights, and how people experience health and illness. Its effects are cultural, as well as political and economic.

It also can be said that the extraordinary productivity of information-alism emerges from the level of practice and from the particular type of labor that informationalism engenders. Informationalism, through its material grounding in informatics, facilitates information production, flow, and processing. Of particular critical interest are moments of processing, when "information" is put into the boxes and categorization schemes that informatics depends on. It is a stage of necessary reduction, a stage when substantive differences are massaged away. Moments of processing are also moments of play. Informatics provide the means to archive, order, visualize, and reorder data, of different types, in large quantities. The capacity to order and reorder—quickly and without great expense—is of critical importance. Through informatics, differences are worked out, powerful relationships are established, or not, and it becomes possible to say some things but not others. This makes informatics an important site of cultural production and ethical action today.

In allowing for and encouraging reorderings, informatics operate with what can be called a logic of supplementarity (following Derrida's conception), enabling substitutions and additions that have the potential to reconsider what a system can say and do, encouraging displacements and realignments. Informatics thus destabilize established systems by design. Though the context always matters, informatics can be conceived as a material cultural form that is valenced in particular ways. Through the facilitation of constant reordering and revisualization of one's "object" of concern, informatics tends to push fields in which they operate into iterative rather than into reproductive modes.[7]

I am driven toward a generalizing argument here in part to provoke reconsideration of the now common argument that digitization is a disenchanting, reductive project that compels what Donna Haraway calls "the god trick" by promising total knowledge. This argument is not incorrect. But neither does it account for all that is going on through informatics. I want to tell the other side of the story. A story about the ways people are investing in informatics, often in intensely creative ways. A story about the critical potential of informatics as a mode of production. A story about how informatics are allowing for new ways of sorting and arranging differences— between social groups, between the normal and the pathological, between the acceptable and the unacceptable, between knowledge and overwhelming complexity. A story about work within and around discursive gaps.

Discursive gaps are gaps in what discourses can say or even recognize.

They are what people can't get their heads and tongues around. They operate through disavowal and ignorance.[8] Important cultural work is done in efforts to displace discursive gaps, and environmental information systems can provide critical tools. Designers and users of environmental information systems work in NGOs, in government labs, and at home computers, as professionals, parents, citizens, and journalists, connecting to data resources and other users in ways unimaginable even a decade ago. Bit by bit, together, they are changing what counts as an environmental problem. Their work produces what Gayatri Spivak calls "discursive displacements": shifts in sign systems brought about through movement within the system, which brings previously latent signification to the surface.

Environmental politics, particularly toxic politics, includes people, creatures, and issues that are difficult to assimilate into established narratives about truth, responsibility, and health. Extraordinarily expert environmental information systems, somewhat ironically, draw these people and issues into visibility. They make "the environment" accessible to understanding and governance, configuring—quite literally—what counts, and what does not.

Bit by Bit

Greenpeace and Worst-Case Scenarios

"Information strategies" for dealing with environmental risk became the explicit focus of law in the United States, in 1986, through passage of the Community Right-to-Know Act, Title III of the Superfund Amendments and Reauthorization Act (SARA). Widely regarded as the primary legislative response to the Bhopal disaster in the United States, the act mandated a range of initiatives to support emergency planning and public access to information (Hadden 1994). High-risk facilities, for example, had to provide the information needed by local rescue personnel to plan emergency evacuations.[9] By the time amendments to the Clean Air Act were passed, in 1990, this had evolved into a mandate for "worst-case scenarios" for 66,000 high-risk facilities around the United States. A worst-case scenario shows the radius within which people will die if there is a massive toxic release from a plant without adequate evacuation, as happened in Bhopal. Although worst-case scenarios were supposed to be ready for distribution by June 1999, in August 1999 President Bill Clinton signed the Chemical Safety Information, Site Security and Fuels Regulatory Act, which blocked Internet posting of information about any facility's "offsite consequence analysis"—worst-case scenarios.[10] The chemical industry argued that this legislation was needed to prevent dangerous information from falling into the hands of terrorists. Information thus became the hazard.

Greenpeace took up the work of publicizing worst-case scenarios nonetheless, insisting that "chemical security" depended on it. Only if dangers were publicly known, Greenpeace argued, would initiative be taken to secure plant premises and to substitute high-risk chemicals and processes with safer alternatives. According to a Greenpeace editorial in the *New York Times* in September 2004, a study conducted by the Army Surgeon General after 9/11 found that up to 2.4 million people could be killed or wounded by a terrorist attack on a single chemical plant. The EPA followed through with the study, identifying 123 chemical plants where an accident or attack could threaten more than a million people, and 7,605 plants that threatened more than a thousand people. The EPA also determined that it could use the Clean Air Act to compel chemical plants to improve security. Responsibility for chemical security was nonetheless given to the Department of Homeland Security, which did not have the power to enforce security measures and thus had to rely on voluntary efforts. According to Greenpeace, the Department of Homeland Security also "tried to reduce the threat of catastrophic attack with the stroke of a pen, . . . announcing that the number of plants that threatened more than 1,000 people was actually only 4,391, and the number that endangered more than a million people was not 123 but two" (Hind and Halperin 2004).

"Chemical security" has become justification for withdrawing environmental risk information from the public domain. Environmentalists have pushed back with a double logic, arguing both that the chemical industry is a weak link in the U.S. homeland security program *and* that the risk of a terrorist organization taking advantage of worst-case scenarios does not outweigh the hazard of an uninformed public. Nevertheless, the Chemical Safety Site Security and Fuels Regulatory Act made it illegal to electronically publish detailed worst-case scenarios. Most significant, high-risk components of a plant—such as a storage tank of ammonia—cannot be identified, which not only keeps "terrorists" from locating them, but also thwarts citizens hoping to formulate and implement specific risk-reduction plans. The potential for local environmental action has been undercut. Greenpeace has skirted the restrictions by moving to another scale, with visualizations that are effective without the outlawed level of detail. Other kinds of information

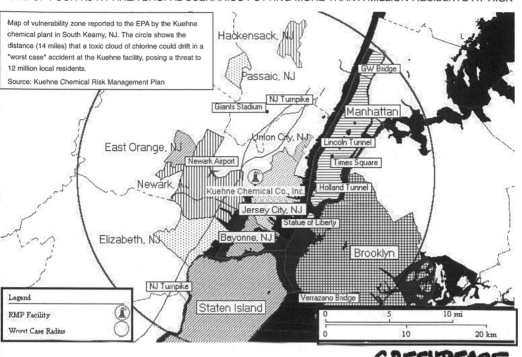

3. New Jersey–New York Bhopal scenarios. Courtesy of Greenpeace,
http://greenpeaceusa.org.

substitutes for the information that the law has taken offline. In mapping
a potential worst-case scenario at the Kuehne Chemical Company in South
Kearney, New Jersey, for example, Greenpeace draws out the population
density of the area and many well-known landmarks—Times Square, the
Statue of Liberty, Newark Airport, Giants Stadium—that would be affected
by a worst-case release (see 3). In another visualization, Greenpeace overlays
worst-case scenarios for forty chemical plants between Baton Rouge and
New Orleans. While plant-specific information is not included, the cumulative warning is powerful (see 4).

Greenpeace does not deny the significance of "chemical security." Information about potential catastrophic environmental risk is recognized as a
charged resource. Rather than skirting the complexity, Greenpeace has engaged it, creatively using digital-mapping capabilities to make catastrophic
risk potential visible, without providing the kind of detail that could contribute to sabotage at the local level. Environmental hazards remain on-

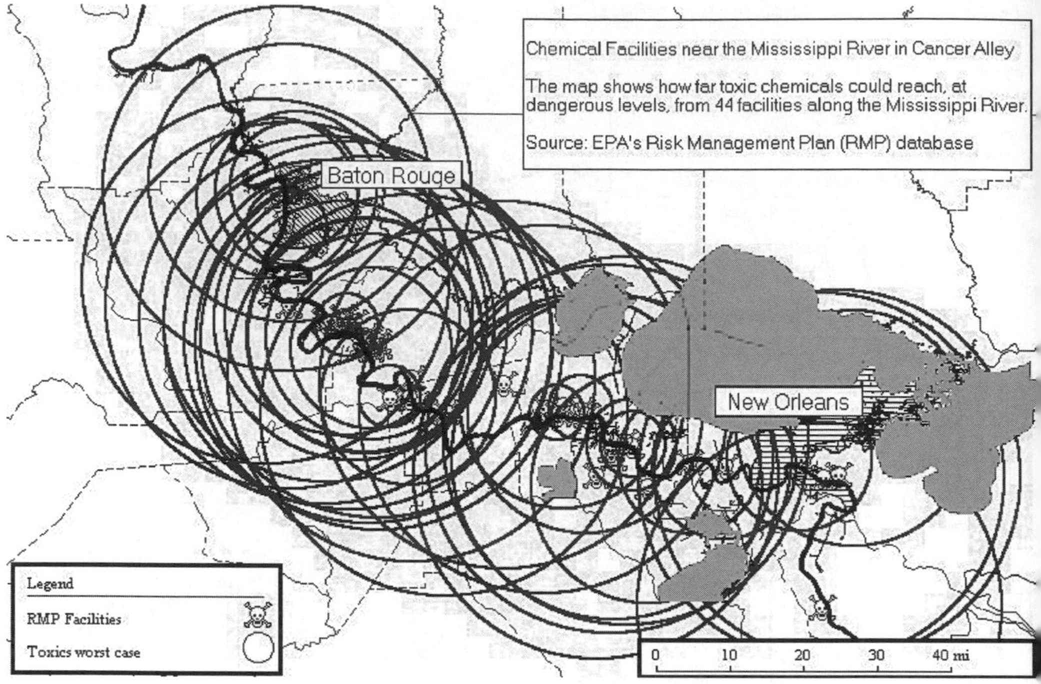
Chemical Facilities near the Mississippi River in Cancer Alley
The map shows how far toxic chemicals could reach, at dangerous levels, from 44 facilities along the Mississippi River.
Source: EPA's Risk Management Plan (RMP) database
Baton Rouge
New Orleans
Legend
RMP Facilities
Toxics worst case
0
10
20
30
40 mi

4. Bhopal scenarios on the Mississippi. Courtesy of Greenpeace, http://greenpeaceusa.org.

screen—and are highlighted as being of particular concern in a world riveted by terrorist threats—while crude reifications of homeland security are disrupted.[11]

We-Acting against Asthma

Consider the work of West Harlem Environmental Action (We Act) on asthma. Asthma incidence in the United States has increased dramatically and unevenly in recent years. Poor and minority children get sick and die from asthma much more frequently than wealthier, white children (see 5). Access to care has long been acknowledged as part of the problem. Indoor pollutants—mold, rodent and cockroach feces, dust, second-hand smoke, all often found in low-income housing—have also been recognized as significant. The force of the outdoor environment—air pollution—is still coming into view, however, and the development of informatics is a key part of the story.

Connecting outdoor air pollution to asthma incidence is enabled by digital mapping, modeling, and visualization tools—tools that draw previously

| ⬥EPA | United States
Environmental Protection Agency | Indoor Environments Division
Office of Air and Radiation (6609J) | EPA 402-F-04-019
May 2006 |

ASTHMA FACTS

Asthma is a rapidly growing public health problem. According to the Centers for Disease Control and Prevention:

- 20 million people, including 6.1 million children, have asthma.

  - Asthma prevelance is higher among families with lower incomes.

- 12 million people report having an asthma attack in the past year.

- Asthma accounts for more than 14 million outpatient clinic visits, and nearly 2 million emergency department visits each year.

- African Americans continue to have higher rates of asthma emergency department visits, hospitalizations, and deaths than do Caucasians:

  - The rate of emergency department visits is 380% higher.
  - The hospitalization rate is 225% higher.
  - The asthma death rate is 200% higher.

5. EPA asthma facts.

invisible connections to the surface. For example, polluting industrial facilities can be seen in proximity to low-income and minority neighborhoods. Connecting particular exposures to particular health outcomes—a notorious challenge—also becomes more manageable, if far from straightforward. Asthma incidents can be connected to ozone levels, for example, or asthma-related hospitalization can be connected to traffic flows and to the uneven distribution of these flows in different neighborhoods.

We Act, an environmental-justice group, has done important, cutting-edge work of this sort. Using data available from the U.S. Census, the U.S. Department of Transportation, the N.Y. State Department of Health, and the N.Y. State Metropolitan Transportation Authority—in collaboration with researchers at Cornell, Columbia, and Mount Sinai Hospital—they have simulated real-time asthma hospitalization rates for neighborhoods around Manhattan (see figures 6–7). The picture presented is not "real"—it is based on data from previous years—but its real-time effect creates a sense of urgency while making clear significantly uneven distributions of disease incidence.

In another set of visualizations, We Act connects asthma hospitalization rates to proximity to diesel-fume-producing facilities, which are linked to

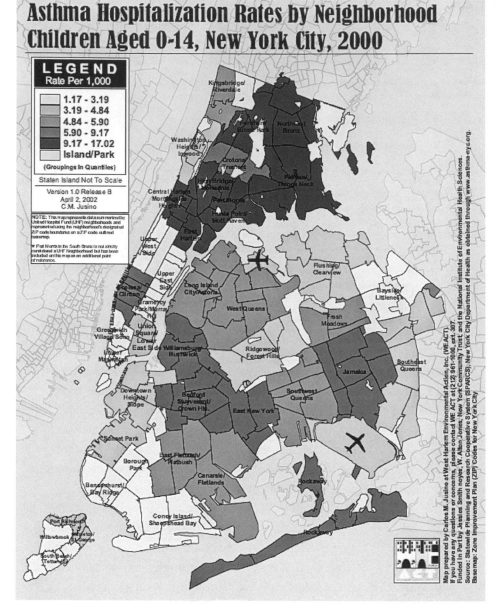

Asthma Hospitalization Rates by Neighborhood
Children Aged 0-14, New York City, 2000

LEGEND
Rate Per 1,000

- 1.17 – 3.19
- 3.19 – 4.84
- 4.84 – 5.90
- 5.90 – 9.17
- 9.17 – 17.02
- Island/Park

(Groupings in Quantiles)

Staten Island Not To Scale

Version 1.0 Release B
April 2, 2002
C.M. Jusino

6. Hospital admissions for asthma.

demographic data to show how asthma risks disproportionately burden minority communities (see 8). With a series of snapshots of the same problem, from different angles, asthma *becomes visible* as an environmental-justice problem. We Act has reordered existing data to draw previously invisible relationships, old problems, and possible remedies into visibility.[12]

Polluting Facilities & Asthma Hospitalizations
For Children 0-4 Years Old
in Manhattan in 1996

LEGEND

Asthma Hospitalization Rates
for Children 0-4 Years Old by
Manhattan ZIP Codes (Rate Per 10,000)
- 0 to 32
- 33 to 75
- 76 to 110
- 111 to 240
- 241 to 508

▲ DOT Diesel Truck Depot
⬠ Port Authority Bus Terminal
● MTA Bus Depot
☐ Marine Waste Transfer Station
▨ Sewage Treatment Plant
▨ Train Yards
∿ Major Highways
✦ 96th Street Demarcation

96th Street

Northern Manhattan Facilities

MapID	Facility Name
1	Kingsbridge MTA Bus Depot
2	MTA Train Yards
3	Department of Transportation / Division of Highways Diesel Truck Depot
4	George Washington Bridge Port Authority Bus Terminal
5	Mother Clara Hale MTA Bus Depot (Scheduled to Expand)
6	North River Sewage Treatment Plant / Riverbank State Park
7	135th Street Marine Waste Transfer Station
8	Manhattanville MTA Bus Depot
9	Amsterdam MTA Bus Depot
10	126th Street MTA Bus Depot
11	Wards Island Sewage Treatment Plant
12	100th Street Bus Depot (Scheduled to Expand)

Southern Manhattan Facilities

MapID	Facility Name
13	91st Street Marine Waste Transfer Station
14	59th Street Marine Waste Transfer Station
15	41st Street MTA Bus Depot
16	42nd Street Port Authority Bus Terminal
17	Hudson MTA Bus Depot (Scheduled to Close)

7. Polluting facilities and child hospitalization.

Scorecard Redux

Before Scorecard, the task of gathering data on pollution in a particular area, or related to a particular health risk, was tedious and frustrating. Bill Pease, the designer of Scorecard, learned about this in his first few months at Environmental Defense, in 1995. As its senior environmental health scientist, he was swamped with requests from grassroots groups needing help obtaining and interpreting information about toxics in their community. Pease needed a way to save people the time required to go from government office

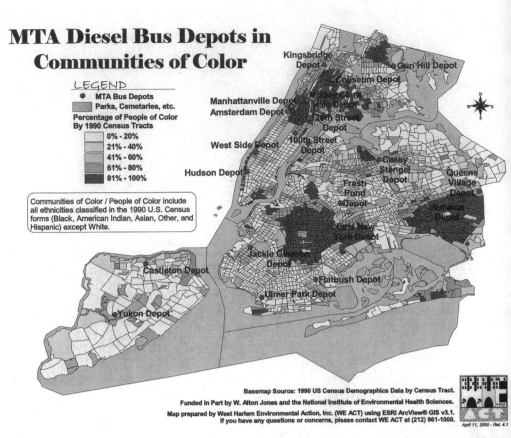

MTA Diesel Bus Depots in
Communities of Color

LEGEND

- ● MTA Bus Depots
- ▨ Parks, Cemeteries, etc.

Percentage of People of Color
By 1990 Census Tracts

- 0% - 20%
- 21% - 40%
- 41% - 60%
- 61% - 80%
- 81% - 100%

Communities of Color / People of Color include
all ethnicities classified in the 1990 U.S. Census
forms (Black, American Indian, Asian, Other, and
Hispanic) except White.

Kingsbridge Depot
Gun Hill Depot
Coliseum Depot
Mother Clara Hale Depot
Manhattanville Depot
Amsterdam Depot
126th Street Depot
West Side Depot
100th Street Depot
Casey Stengel Depot
Hudson Depot
Fresh Pond Depot
Queens Village Depot
Jamaica Depot
East New York Depot
Jackie Gleason Depot
Castleton Depot
Flatbush Depot
Ulmer Park Depot
Yukon Depot

Basemap Source: 1990 US Census Demographics Data by Census Tract.
Funded in Part by W. Alton Jones and the National Institute of Environmental Health Sciences.
Map prepared by West Harlem Environmental Action, Inc. (WE ACT) using ESRI ArcView® GIS v3.1.
If you have any questions or concerns, please contact WE ACT at (212) 961-1000.

April 11, 2000 - Rel. 4.1

8. Bus depots in communities of color.

to government office, to the library to the polluting facility in search of information that often wasn't available without argument or delay. He also needed to provide grassroots groups with tools for interpreting the data they collected. His solution was to build an internal database and to hire a team of environmental scientists and database consultants. Their plan, until they consulted with the MIT computer scientist Phillip Greenspun, was to build a standalone program that could be downloaded, or distributed on CD-ROM. Greenspun convinced him to go the way of the Web.[13]

Over a billion pages can potentially be produced in Scorecard. The result, for the users, is both exhilarating and overwhelming. Consider the possibilities for environmental action produced by Scorecard in Texas City. One could decide to work at the county level, knowing that mobile sources and diesel emissions pose the greatest health hazards. Or one could work at the facility level, focusing on Union Carbide's Texas City plant, trying to reduce benzene releases, for example—since benzene is the chemical released from this facility that poses the highest cancer threat. As part of one's campaign, one could link up with other communities at high risk from benzene releases and point out to the company and the press that Union Carbide's Texas City plant is the fourth worst polluter in Galveston County and is in the 90th percentile of worst-polluting companies in the United States. Another strategy would be to focus specifically on ozone-depleting chemicals, targeting chlorodifluoromethane at Carbide's Texas City plant, while targeting other chemicals that are major contributors to ozone depletion released from other facilities in the area, and perhaps around the country.

Any of these efforts would be riveted by uncertainties. Scorecard provided the information base and encouraged one to act on it, at the same time highlighting how little is known about chemical toxicity, and how data in the Toxic Release Inventory is incomplete, out-of-date, and often inaccurate. This double gesture—providing information as the basis for action, while qualifying the validity of the information—is characteristic of environmental information systems. Often, they produce working knowledge, which is nevertheless claimed to be imperfect knowledge.

The goal of Scorecard was not to reassure the user, but to interconnect her—with different types of information, with the regulatory process, with people in both similar and different locales, with ways of visualizing and spatializing phenomena that were usually represented in abstract, impersonal terms. Getting "straight to the point" was not the goal. Instead, users were encouraged to wander through different kinds of information, much of it flagged as uncertain, visualizing comparisons, piecing together a picture of reality to work with.[14] High levels of information literacy were required, and cultivated.

Scorecard allowed users to zoom in to the local and out to the national, clicking through graphs that provided snapshots of pollution dispersion, and through to chemical profiles that characterized pollution hazards. The experience of Scorecard could be dizzying. But Scorecard took on some of the most recalcitrant problems within environmental politics—the need to deal with too little, as well as too much, information; the need to deal with contested scientific findings and intractable uncertainty about long-term effects; the need to think locally, as well as comparatively and globally.

Informating Environmental Ethics and Politics

Understanding and governance of "the environment" has always required extraordinarily ambitious information collection and processing.[15] Consider, for example, Richard Groves's account of the specimen processing involved in the development of India's famous botanical classification systems, and the role of the printed book in the uptake of these classification systems in Europe in the sixteenth century (Grove 1995). Consider, too, Groves's description of how understanding of tropical deforestation emerged from networks of scientific societies throughout the colonial world (Grove 1995). Since the Second World War, the ever-growing quantity and diversity of industrial chemicals in use have made the information collection and processing needed to keep track of "the environment" even more challenging. Philosophical discourses about the environment often miss this, however, remaining caught in frustratingly rigid oppositions between nature and technology, between lay and professional knowledge, between truth and uncertainty, and anthologies compiled for teaching environmental studies tend to underwrite this exclusion, despite dramatic growth in the amount of environmental information produced and circulated in recent years, despite the important though often undramatic role that information politics has come to play within environmental politics.

Environmental information systems need to be recognized as significant and contested sites of political action, because they are sites where conventional ways of thinking about the environment are being reconfigured. Quite literally. Setting up a comparison or connecting bits of information previously unrelated performs cultural work. So do click-throughs. Zooming in and out, learning to consider the implications of scale involves what Antonio Gramsci termed "elaboration," the labor of working out common sense. This kind of labor can't be reproductive. It involves a play of signs and systems that is always unsettling.

The stakes are high. In the environmental field, technology and culture shape what is and is not perceptible in particularly powerful ways. Many environmental issues—toxics, climate change, biodiversity—are simply unseeable without technical prosthetics. Technical prosthetics also enable complex operations of difference. The environmental field, like so many fields, is riveted by such differences and hasn't yet learned to deal with them.[16] Environmental health problems, for example, may be "caused" by industrial pollutants, vehicle emissions, class, race, and individual susceptibility. Such multiple determinism is not well dealt with in law, regulation, scientific practice, or the public imagination. Environmental information systems like Scorecard have begun to shift the grounds. They are a site where common sense is being reconfigured.[17] Knowledge is being made, circulated, legitimated, and internalized in new ways. Complex orderings of reality are constantly emerging and continually being displaced. Biotruth becomes a moving object. Biopolitics are being reconstituted.[18]

The informating of the environmental field deserves critical attention, as does the work of informating writ large. To be informated is to be beset by possibilities for constant reordering and revisualization. When fields of practice are informated, previous latent signification often comes to the surface; discursive gaps—spaces where established analytic and explanatory language fail, spaces where hegemony comes to crisis—can be displaced.[19] Fields of practice that have been informated are thus sites "of the displacement of function between sign-systems" (Spivak 1987, 198). They are a place where change happens, sites of transaction between the past, present, and future.

Notes

1. In the spring of 2010 the *Houston Chronicle* posted recently discovered photos of the Texas City disaster. See "Texas City Blast of 1947" 2010.

2. Texas City suffered yet another disaster on 23 March 2005, when a British Petroleum (BP) refinery there exploded, killing 15 workers and injuring more than 170 others. A follow-up report by the U.S. Chemical Safety and Hazard Investigation Board documented a litany of plant design and safety problems behind the disaster. On 30 October 2009 the U.S. Occupational Safety and Health Administration (OSHA) imposed an $87 million fine, the largest in OSHA's history. BP challenged the fine. "Remember the 15," a website to commemorate workers killed in the disaster, provides a timeline of related events, at http://www.rememberthe15.com.

3. The information experience described in the forthcoming description took place in 2004; details reflect this. On 1 November 2005 Environmental Defense transferred responsibility for and ownership of Scorecard to another nonprofit organization, Green Media Toolshed (GMT). GMT stopped updating Scorecard in 2006. Meanwhile the EPA developed better tools of its own to encourage use of environmental-risk information. See, for example, a review by Timothy Barzyk and others at EPA's National Exposure Research Laboratory (Barzyk, Conlon, Hammond, Zartarian, and Schultz 2009). The EPA's online resource most similar to Scorecard is at the "My Environment" link at http://epa.gov. The United States National Library of Medicine's ToxMap site also works rather like Scorecard, allowing users to visualize TRI data (as well as Superfund data) along with census data, health data, and so on. See http://toxmap.nlm.nih.gov.

4. I have also watched BASF over the years, since a labor lockout at a BASF plant in Geismer, Louisiana, helped mobilize the first major labor-environment coalition in the mid-1980s. To draw the public into the issues, the Oil, Chemical and Atomic Workers union ran a billboard suggesting the high stakes, asking if Geismer could become "Bhopal on the Bayou."

5. Important recent work on environmental information systems examines their uptake to address biodiversity, climate change, and a range of other issues (Bowker 2001; Edwards 2010; Edwards 1999; Sieber 1997; Sarewitz, Pielke, and Byerly 2000). Erich Schienke (2006), Lane DeNicola (2007), and Alex Sokolof (2006), doctoral students at Rensselaer, wrote dissertations on environmental information systems in China (regarding the development of geographic information systems in panda conservation and air-pollution management), in India (regarding the development of remote-sensing expertise for environmental applications), and at Greenpeace International (regarding the development of information infrastructure in a globalizing NGO). These dissertations significantly advanced my understanding of environmental information systems.

6. Manuel Castells contrasts informationalism to industrialism, locating its origins in the 1970s and in technologies like the microprocessor, optical fiber, transmission control protocol (TCP), Internet protocol (IP), and so on. The productivity of informationalism, according to Castells, "lies in the technology of knowledge generation, information processing and symbolic communication" (2000, 17). Castells also explicates informationalism as a global ordering force, noting that the "multimedia world will be populated by two essentially distinct populations: the interacting and the interacted, meaning those who are able to select their multidirectional circuits of communication, and those who are provided with a restricted number of prepackaged choices. And who is what will be largely determined by class, race, gender and country" (ibid. 371). Alberto Melucci argues that "in the contemporary context, we can define domination as a form of dependent participation in the information flow, as the deprivation of control over the construction of meaning" (1996, 182).

7. Elsewhere, Mike Fortun and I have developed the idea that environmental information systems (particularly microarrays and databases in the emerging field of toxicogenomics) can be "experimental systems" of the sort theorized by the biologist and historian of biology Hans-Jörg Rheinberger (Fortun and Fortun 2005). Rheinberger explains that "an experimental system in which a scientific object gathers contours and becomes stabilized, at the same time must open windows for the emergence of unprecedented events. While becoming stabilized in a certain respect, it must be destabilized in another. For arriving at new 'results,' the system must be destabilized—and without a previously stabilized system there will no 'results.' Stabilization and destabilization imply each other. If a system becomes too rigid, it is no longer a machine for making the future; it becomes a testing device, in the sense of producing standards or replicas. It loses its function as a research tool" (1998, 291).

8. Focusing on discursive gaps is particularly relevant in the anthropology of technoscience today because of the pace of change across technoscientific fields, partly driven by impressive developments in information processing and sharing capabilities. Scientists and technologists themselves now routinely comment on how established concepts and methods have become somewhat exhausted. They need new ways of handling data and of judging its significance, and new ways of collaborating across disciplinary boundaries. They are socially and intellectually situated to recognize needs and possibilities for new types of knowledge, produced and vali-

dated in new ways. They are subjects-in-doubt, struggling for language, defying predictable cultural patterns. This is the case in much work around toxics, for example.

9. Another key component of the Community Right-to-Know Act (1986) was the TRI, which provided the base data for Scorecard.org. The TRI contains data reported by a range of (but not all) industrial and federal facilities that produce or handle listed toxic chemicals. The goal of the TRI was to allow the EPA as well as citizens to track and evaluate routine, legal emissions.

10. The EPA provides an overview and links to the full text of the act at its website, http://www.epa.gov.

11. Greenpeace's portfolio "Worst Case Scenario Chemical Disaster Maps," last updated 5 May 2005, is available at its website, http://www.greenpeace.org. Also see "Greenpeace 'Dossier' on Chemical Security," last updated 22 July 2009, available at the same website. Also see Orum 2008; Ember 2007a; Ember 2007b.

12. In 2005 coverage by the *New York Times* of a program in Harlem that increases access to healthcare for children with asthma noted that rates of asthma in Harlem were "found to be more than five times the national average, with 31.4 percent of children found to be sick" (Santora 2005). It was also noted that "experts can provide no specific explanation for either the dramatic rise in asthma generally or why it is so prevalent in poorer communities like central Harlem" (ibid.). Researchers at Columbia's Center for Children's Environmental Health have conducted a number of studies in West Harlem, attending to a range of possible asthma drivers and triggers, including air pollution (Patel and Miller 2009; Gilliland et al. 2005; Kinney, Chillrud, Ramstrom, Ross, and Spengler 2002).

13. Greenspun's website, http://philip.greenspun.com, explains that the software behind Scorecard emerged from the work of the Scalable Systems for Online Communities research group at MIT, which Greenspun founded and then spun out into ArsDigita, built by him into a profitable ($20 million in revenue) open-source enterprise software company.

14. The argument that discursion—a meandering through material—produces an outcome valued differently than linear movement toward a stable conclusion is now often made with regard to the value of hypertext, as in the assertion by George Landow (1992) that there has been a "convergence" between critical theory and technology. This reasoning, however, is not new. In his introduction to a collection of personal essays, for example, Philip Lopate (1995) describes how the essay's "unmethodological method" has been utilized across time, from Montaigne and Bacon, through the Frankfurt School, and in different cultures. Theodor Adorno, for example, is said to have seen rich, subversive possibilities in the "anti-systemic" properties of the essay—in the way an essay wanders through information and thought, rather than working within dominant frameworks of thinking and straight through to a conclusion. For Adorno (1991), the essay was a technology for thinking outside the grand philosophical systems of his time. Environmental information systems, in my view, have a critically similar potential.

15. I cast my critique here against "environmental ethics" to encourage anthropologists working on environmental issues to follow the lead of anthropologists who

have developed very important, empirically grounded critiques of bioethics (Cohen 1999; Das 2002; Michael Fortun 2008; Rabinow 1996; Rapp 2000b). While bioethics is much more professionally codified and funded than environmental ethics, both set many terms of debate and standards of judgment.

16. Recall, for example, the "intersectional sensibility" advocated by the critical race theorist Kimberlé Crenshaw (1990, 1991). Crenshaw criticizes identity politics for asking people to be *either* raced, woman, or queer, and for ignoring intragroup differences—which makes it difficult to deal with domestic violence in black communities, for example, and limits the standing that women, in particular, have before the law. Intersectional sensibilities involve recognition of multiplicity: the simultaneous examination of race, ethnicity, sex, class, national origin, sexual orientation, and so on. Toxicologists would call this "cumulative effect" and recognize that both scientific and legal-regulatory worlds have great difficulty dealing with it.

17. Note that for Gramsci common sense is never immobile. Common sense is always continually transforming itself, leveraging available discursive resources. Common sense is an open system, so to speak. Its porosity is what makes it an important site of struggle.

18. Note Kaushik Sunder Rajan's interesting use of *constitutionalism* within biopolitics (Sunder Rajan 2002).

19. One can also think in terms of discursive risks. If discursive gaps are where hegemonic language fails, the discursive risk is that hegemonic constructs will, nonetheless, be imposed on these gaps, causing a misrecognition of what is going on that effectively quells the possibility of a different kind of future. Discursive risks threaten to make the future a reproduction of the present. In the environmental field today, such risks have truly tragic dimensions. I draw here on Derrida's opposition between the future as it can be known and calculated now, and "l'avenir," that which will come, unexpectedly perhaps with the force of revolution (Derrida 1990). This relates to my interest in Rheinberger's notion of "experimental systems" that, through their technical operation, open up such a future.

PART FOUR

PROMISSORY EXPERIMENTS AND

EMERGENT FORMS OF LIFE

11

MIKE FORTUN

GENOMICS SCANDALS AND OTHER

VOLATILITIES OF PROMISING

Promises and promisings suffuse the life sciences today. But what is it that is marked, gestured toward, or accomplished by these words, *promise* and *promising*, in the multiple scientific, legal, political, cultural, and other contexts in which they are written or uttered? How many ways of promising are there on the plateaus of genomics? I will not count them all, but offer only a few ethnographic waypoints.

Linkage analysis and positional cloning have had a remarkable track record in leading to the identification of the genes for many mendelian diseases, all within the time span of the past two decades. Several of these genes account for an uncommon subset of generally more common disorders such as breast cancer (BRCA-1 and -2), colon cancer (familial adenomatous polyposis [FAP] and hereditary non-polyposis colorectal cancer [HNPCC]), Alzheimer's disease (ß-amyloid precursor protein [APP] and presinilin-1 and -2) and diabetes (maturity-onset diabetes of youth [MODY]-1, -2, and -3). These successes have generated a strong sense of optimism in the genetics community that the same approach holds great promise for identifying genes for a range of common, familial disorders, including those without clear mendelian inheritance patterns. But so far the promise has largely been unfulfilled, as numerous such diseases have proven refractive to positional cloning.
NEIL RISCH, POPULATION GENETICIST (RISCH 2000, 850)

As I've said in the past, [Genentech] has a promising pipeline and room to grow based on its cancer and cardiovascular drugs. . . . The potential for IDEC [Pharmaceuticals] will get better as the company's promising compound Zevalin comes to market. . . . With that in mind, the outlook for the remainder of 2000 is quite promising.

"JUDGMENT DAY IS HERE FOR BIOTECHS," WORLDLYINVESTOR.COM,

20 JULY 2000 (NADINEWONG 2000)

Genomics is real. In the end, I am sure its promise will materialize.

JURGEN DREWS, FORMER PRESIDENT FOR GLOBAL RESEARCH AT

HOFFMANN-LA ROCHE (QUOTED IN LICKING ET AL. 2000)

"Biotech valuations in Germany are not built on anything fundamental," says Michael Sistenich, a fund manager at DWS. He says chief executives promise more than they can deliver to build up their valuations.

"TAKEOVERS CURE FOR GERMAN BIOTECHS," *FINANCIAL TIMES*,

14 JUNE 2000 (FIRN 2000)

You can only pay so much for promises.

"THE REASON TO AVOID BIOTECH STOCKS," MOTLEY FOOL,

5 APRIL 2001 (MCCAFFERY 2001)

DeCODE promises that Icelanders will get any drugs or diagnostics based on their genes for free during the patent period—a promise [Jorunn] Eyfjord calls "a joke. . . . How many drugs do you think are going to be developed, and how many people will really benefit from that?"

"PHYSICIANS WARY OF SCHEME TO POOL ICELANDERS' GENETIC DATA,"

***SCIENCE*, 1998 (ENSERINK 1998)**

A promise was made somewhere.

MEMBER OF ALTHINGI (ICELANDIC PARLIAMENT), INTERVIEW WITH

THE AUTHOR, REGARDING HOW ICELAND'S HEALTH SECTOR DATABASE

ACT (1998) CAME TO PASS

These ethno-epigraphs span genres, contexts, nations, and domains of activity in their multiple instantiations of promising's multiplicity. They cover the entire territory of the "lively capital" of genomics: a complex landscape comprising fissured zones of biotechnological research, economic predictions and bets parlayed in what was perhaps the most speculative economy we've witnessed, political deal-making, and personal oath-taking. Each zone is in turn layered and imbricated with the others, begging for some hyper–Geographical Information System interface to help us visualize the spectral superimpositions.[1]

But this chapter will have to do in the meantime. The statements quoted above make it clear that molecules are promising, the sciences of those molecules are promising, outlooks are promising, markets are promising, and, yes, people in particular contexts are promising. The differences—swearings, speculatings, vowings, hopings, anticipatings, sheer happenings—are far from trivial. Indeed, in the path that I follow below, it is the trivialities of promising—its mundane instantiations in specific situations—that become essential, and essential to map in careful detail. The specificities of promising that I diagram are drawn from my fieldwork on deCODE Genetics—a field which, I hasten to point out, is not limited to the territory of Iceland, where deCODE has its corporate headquarters and research facilities. My fieldwork, conducted mostly from 1998 to 2003 but in some ways still ongoing, occurred in the multiple territories of genomics gestured toward in the ethno-epigraphs: the densely intermeshed worlds of politics, finance, and emergent genomic technoscience in which deCODE Genetics is one particularly volatile "hot spot."[2]

I offer a brief sketch of deCODE's financial history, drawn in part from my book *Promising Genomics: Iceland and deCODE Genetics in a World of Speculation*. The book is structured like a genome, consisting of twenty-three *chs*—chapters, chromosomes, chiasma—twenty-three pairs of twisted doubles, double binds, intertwined differences that cannot be resolved or settled, but remain ever volatile. The book is structured like a genome because both are structured like writing: material marks riddled with chiasma, promising much more than their signs are said to "code for." Each volatile couplet is a site where it is necessary to play, to muddle through, to speculate, or above all, to promise.

Acts of promising are perhaps the performative speech acts par excellence. Donald MacKenzie, Karin Knorr-Cetina, Michel Callon, and others in science studies have recently become interested in performativity pertaining to financial markets, but the genealogy by which I approach and understand promising diverges from theirs.[3] While we all acknowledge the importance of J. L. Austin (1962) as an early analyst of promises and other performatives, my genealogy swerves more toward the "Continental" tradition represented by figures like Jacques Derrida, Avital Ronell, and Shoshana Felman. MacKenzie, Knorr-Cetina, Callon, and company tend to be interested in performativity in a rather restricted domain: economists' equations, financial instruments like derivatives, electronic stock-trading systems, and the like. Derrida, Ronell, and Felman encourage us to trace the dissemination of promises and other performatives into a far more general economy.

In *How to Do Things with Words* (1962), Austin diagrammed not the *distinction* between constative and performative statements, but their zones of entanglement and indiscernibilities. In the process, Austin loosened the coupling between the intentional subject and the performative utterance.

> Thus "I promise to . . ." obliges me—puts on record my spiritual assumption of a spiritual shackle.
>
> It is gratifying to observe in this very example how excess of profundity, or rather solemnity, at once paves the way for immorality. For one who says "promising is not merely a matter of uttering words! It is an inward and spiritual act!" is apt to appear as a solid moralist standing out against a generation of superficial theorizers: we see him as he sees himself, surveying the invisible depths of ethical space, with all the distinction of a specialist in the *sui generis*. Yet he provides Hippolytus with a let-out, the bigamist with an excuse for his "I do" and the welsher with a defence for his "I bet." Accuracy and morality alike are on the side of the plain saying that *our word is our deed*. (Austin 1962, 9–10)

Over numerous writings, Derrida articulated how promising is disseminated throughout language, rather than being the exclusive quality or effect of a particular subset of speech acts, the ones Austin called "commissives." Derrida thus dislocates promising not only from any intentional subject, but from any localizability in a particular set of speech acts. He dislocates the location of promising entirely, making promising occur as a kind of general feature of language—although that's not the most promising way of putting it.

> Each time I open my mouth, each time I speak or write, I *promise*. Whether I like it or not: here, the fatal precipitation of the promise must be dissociated from the values of the will, intention, or meaning-to-say that are reasonably attached to it. The performative of this promise is not one speech act among others. It is implied by any other performative, and this promise heralds the uniqueness of a language to come. (Derrida 1998, 67)

> An immanent structure of promise or desire, an expectation without a horizon of expectation, informs all speech. As soon as I speak, before even formulating a promise, an expectation, or a desire *as such*, and when I still do not know what will happen to me or what awaits me at the end of a sentence, neither *who* nor *what* awaits whom or what, I am within this promise or this threat—which, from then on, gathers the language

together, the promised or threatened language, promising all the way to the point of threatening and *vice versa*, thus gathered together in its very dissemination. (Ibid. 21–22)

Derrida's analysis, if trusted, would indicate several things for future analyses of promising like this one. First, promising occurs everywhere in language, which is never sufficiently present to itself to fully guarantee all its workings or capacities. Language runs on credit, if you like—which is not to say, as you surely know, that bills don't fall due. Second, the fact that promising occurs everywhere in language means it would also be at work, or in play, within the analytic language which produces such a statement as "promising occurs everywhere in language." *Et voilà*: promising turns threatening—and just as quickly turns back again. The promise of language is its threat, its poison its gift.

The stage is set for scandal, of an unusual if all-too-common kind. Shoshana Felman's *The Scandal of the Speaking Body* (2002) carries on in the Austinian and Derridean legacy to detail how promising is a material, embodied, and necessary feature of any speaking—that is, writing. Promises and similar performatives represent an impropriety that inhabits everything that is proper—a genome, a science, an economy, an oath, an ethnography. The most solemn and profound promise is simultaneously quotidian and singular, inhabiting the field of practice rather than some theoretically generalizeable plane.

> With Austin as with psychoanalysis, the irreducible triviality of the idiosyncratic is that of a *practice* of the singular.
>
> Of a practice, that is, of what belongs to the order of *doing*. For unlike saying, doing is always trivial: it is that which, by definition, cannot be generalized. . . . Thus true History, belonging to the order of acts or of practice, is always—however grandiose it may be—made up of trivialities.
>
> The same is true of writing. (Felman 2002, 78)

Analyzing promising—or coming to better appreciate its forcefulness—is therefore a job for history, ethnography, or other practices that find the trivial important.

This "irreducible triviality" amounts to a "radical negativity" that, according to Felman, links promising indissolubly to scandal, or at least to the possibility of scandal. The reasons for this are complex, as are the linkages themselves, and it's important to plumb some of this complexity to keep from falling into a moralism in which "scandalous" is simply equated with "sinful" or "shameful."

> Radical negativity . . . belongs neither to *negation*, nor to *opposition*, nor to *correction* ("normalization"), nor to *contradiction* (of positive and negative, normal and abnormal, 'serious' and 'nonserious,' 'clarity' and 'obscurity')—it belongs precisely to *scandal*: to the scandal of their nonopposition. (Ibid. 101–4)

> The scandal, in other words, is always in a certain way the scandal of the promise of love, the scandal of the *untenable*, that is, still and always, the scandal—Don Juanian in the extreme—of the promising animal, incapable of keeping his promise, incapable of not making it, powerless both to fulfill the commitment and to avoid *committing* himself—to avoid playing beyond his means, playing, indeed, the devil: the scandal of the speaking body, which in failing itself and others makes an act of that failure, and makes history. (Ibid. 111)

Continuing briefly in this highly philosophical and thus nontrivial vein, I would return to the promises of genomics and say: genomics can't avoid playing beyond its means. The genomicist, like Don Juan—or a genome, or this chapter—can't help but overcommit. That this scandalous fact opens the way for shameful events should not become cause for an analysis and politics based on resentment: *those seductive genomicists need to be restrained, made to speak and do only what is proper, what they can guarantee.* . . . But the alternative to this kind of Nietzschean *ressentiment* ("science studies, all too science studies") is not the nihilism of "anything goes," but an exposition of what Felman calls the "scandal of the *outside of the alternative*, of a negativity that *is* neither negative nor positive." These are to be found in the trivial events that "history cannot assimilate" but are "nevertheless the *cornerstone* of History" (ibid. 105, 107, emphasis in original).

That's what I promise myself, anyway, and here is where my questions begin again: since promising structures genomics (and structuring means rendering it volatile, speculative, and potentially scandalous), and since promising structures finance, rendering it volatile, speculative, and potentially scandalous, then what are the historical "trivialities" that structure promising? More specifically: what structured the promises that could be made about genomics as science and as business, generator of future knowledge and generator of future capital, in turn-of-the-millenium U.S. and Icelandic discursive, institutional, and cultural regimes? If genomics as science and business depends on performative utterances and devices, what can possibly function as a performative utterance or device? And like a genome or a book, what kind of environment do they have to be crossed with or folded into if they are to work at all?

Those are the structuring questions of the chiasma chapter chromosomes of my book, and I can only take up a few of them here, reshuffled and re-mixed outside of the context of the book. And in recounting the deCODE Genetics story, I will have to be even more selective, and essentially give only a financial history of deCODE, a narrow slice through what Eve Sedgwick (2003) calls the "periperformative vicinities," that vast expanse of forces at work in the deCODE events, including an expatriate celebrity killer whale; territorial fishing limits and the wars, laws, and new forms of property they provoke; not fish-mongering, but what Skuli Sigurdsson (2001) calls saga-mongering, as Icelandic history is repackaged in mythological wrapping; the prescient novels of Halldor Laxness; the chiasmic double bind that verges on doublespeak, in the system of "presumed consent" that all living and dead Icelanders were said to have given in legislation deemed unconstitutional years later; and last but far from least, promises of the Icelander's genetic homogeneity.

Promises are a kind of forward-looking statement (I only foreshadow this important concept here), and if you are what the U.S. Securities and Exchange Commission (SEC) calls a "sophisticated investor," you will know that you are not supposed to base your investments solely on forward-looking statements—for example, those statements made in the early history of the Human Genome Project to the effect that "the human genome is the Holy Grail of biology." Such statements, especially when uttered by Nobel Laureates such as Walter Gilbert, do have their performative effect, not the least of which was leveraging $3 billion out of the U.S. Congress in the late 1980s to support the kind of technoscientific infrastructure development for genomics for which the private sector showed little interest. So in 1987, when the Nobelist and Biogen founder Gilbert promised venture capitalists and Big Pharma companies that genetic-sequence information was valuable, no one bought it, but a mere ten years later, you could take the promise of genomic information to the bank.

Or at least to venture capitalists. In the late 1990s, when the small handful of genomics companies then existing—Incyte Pharmaceuticals, Human Genome Sciences (HGS), Millennium Pharmaceuticals—were only beginning to ink multimillion dollar agreements with "Big Pharma" corporations, the first financing for deCODE Genetics was $14 million from venture capital firms Polaris Ventures and Atlas Ventures. Shortly thereafter, decode and its hitherto unknown CEO Kári Stefánsson lit up the scientific press and internet finance chat rooms when it announced it had been promised up to $200 million in research and operating funds by Hoffman-LaRoche for genomic research into a number of complex diseases, utilizing the genealogies and genotypes of what was described as the "homogeneous" Icelandic population. Years later, you could learn from SEC filings that Roche never paid anywhere near this promised potential sum, but at the time the only thing that mattered was the simple fact of its circulation as the biggest number ever promised to a genomics company. Not only did decode have no products, but it also had no scientific publications to its name and no real research program or track record; it promised to work toward certain research milestones involving the isolation and characterization of genes and gene products extracted from the DNA extracted from the blood of Icelanders, then analyzed against the extensive genealogies constructed by Icelandic families over decades and centuries, now centralized and computerized. Like many such oaths, this one had to be ritualized, and here the nation-state enters the frame of the story, as the Icelandic prime minister David Oddsson passed the pen between Roche's Jonathan Knowles and decode's Kári Stefánsson as they inked their vows in the futuristic Perlan dome atop Reykjavík's geothermal hot-water tanks.

Nearly simultaneously, legislation was introduced into the Icelandic parliament (Althingi) that would create a Health Sector Database (HSD) containing the medical records of all Icelanders, alive and dead alike, that would be licensed exclusively to one genomics company (nameless, but hardly a mystery) for a research and business effort that combined genealogical, genotypic, and phenotypic analysis. "A promise was made somewhere," in the words of one Althingi member, that bound private and state fortunes: decode drafted the legislation, the "charismatic" Stefánsson lobbied for it while simultaneously building the decode infrastructure and brand, and Prime Minister Oddson led the conservative-centrist coalition that, after months of political wrangling, media sideshows, and social upheaval, managed to pass the HSD legislation late in 1998. Some promise, somewhere, was kept, and a year or so later Stefánsson prepared for decode's initial public offering on NASDAQ.

In that spring of 2000 decode still hardly even had a pipeline, let alone anything resembling a promising product like a patented gene or a therapeutic protein. All this time and for years afterward, decode was routinely referred to as "an Icelandic company" by journalists, anthropologists, and other people who might have been expected to know better. Despite, or rather because of being legally incorporated in Delaware, started with capital from U.S. firms, and running entirely on Roche financing—which if it could be assigned a meaningful address, would have to be Switzerland or New Jersey—this performance of Icelandicness was crucial to deCODE's brand, the mark that distinguished it from the rest of the genomics pack like Millennium, Human Genome Sciences, and Celera. In fairness, I should point out that a numerical argument in favor of deCODE's non-Icelandicness only really became available when it filed its registration statement with the U.S. Securities and Exchange Commission, in March 2000, when it was confirmed that Roche had the single largest stake in the company, at 13.4 percent of the shares, followed by three venture-capital firms holding roughly 7, 7, and 6 percent of the shares. When you counted in the individual holdings of the venture capitalists, it was clear that over 50 percent of the company's stock was in non-Icelandic hands.

I have come to value the filings of the Securities and Exchange Commission, and think they are a much underutilized resource for historians. The SEC is far from a perfect institution, and its mechanisms meant to force disclosure of corporate information are also imperfect. For example, if you wanted to know the actual amount of money that Roche promised to pay deCODE for reaching a particular milestone, you would find that that information had been given confidential treatment and was filed separately, inaccessible to investors and the stray ethnographer:

(a) Major Research Programs. Subject to the terms and conditions of this Agreement, for each Major Research Program then in effect on the relevant due date, Roche shall pay to deCODE the following amounts, which shall be due and payable as follows:

<PAGE> 33

[CONFIDENTIAL TREATMENT REQUESTED]

<PAGE> 34

[CONFIDENTIAL TREATMENT REQUESTED]

<PAGE> 35

(b) Minor Research Programs. Subject to the terms and conditions of this Agreement, for each Minor Research Program then in effect on the relevant due date, Roche shall pay to deCODE the following amounts, which shall be due and payable as follows:

[CONFIDENTIAL TREATMENT REQUESTED]

<PAGE> 36

Similarly, the exact financial terms of the agreements between deCODE and its collaborators, like the Icelandic Heart Association, who were channeling patients, blood, and information to deCODE in exchange for payments, were also blacked out.

> COLLABORATION AGREEMENT BETWEEN THE ICELANDIC HEART ASSOCIA-
> TION, HJARTAVERND AND ISLENSK ERFDAGREINING EHF.

> If IE or its parent company, deCODE genetics Inc., succeeds in contract-
> ing for the sale of individual Research Projects or their conclusions or
> sells the results of a Research Project to a third party, HV will receive as
> its share [CONFIDENTIAL TREATMENT REQUESTED] of all payments to IE
> or to deCODE genetics Inc., . . .

> If IE or deCODE genetics Inc. concludes a contract with a third party pur-
> suant to Paragraph 2 above, IE will pay to HV [CONFIDENTIAL TREATMENT
> REQUESTED] on signature of such contract, and thereafter an annual
> amount of [CONFIDENTIAL TREATMENT REQUESTED] during the course
> of the Research Project in question, the total amount never to exceed
> [CONFIDENTIAL TREATMENT REQUESTED]. In the event that the Research
> Project is concluded in a shorter time than five years . . . HV will receive
> on such conclusion the amount which remains unpaid of the [CONFIDEN-
> TIAL TREATMENT REQUESTED] for the Research Project in question. . . .
> This payment is in addition to and independent of the [CONFIDENTIAL
> TREATMENT REQUESTED] payment pursuant to Paragraph 2 of this Chap-
> ter III.

Nevertheless, the SEC forced deCODE to disclose and not simply promise, if it wanted to get investment from eager Americans dreaming that genomics was the next big thing after the internet and the dot-coms. The SEC also forces disclosure of more generic risk information, so Icelanders who had been treated to little but glowing accounts that practically guaranteed future genomic riches for themselves and for their nation could read for the first time in March 2000, often in solid caps supplemented by finer print, that deCODE's "RELIANCE ON THE ICELANDIC POPULATION IN OUR GENE DISCOVERY PROGRAMS AND DATABASE SERVICES MAY LIMIT THE AP-PLICABILITY OF OUR DISCOVERIES TO CERTAIN POPULATIONS," THAT "ETHI-CAL AND PRIVACY CONCERNS MAY LIMIT OUR ABILITY TO DEVELOP AND USE

[OUR DATABASES] AND MAY LEAD TO LITIGATION AGAINST US OR THE ICE-
LANDIC GOVERNMENT"—INDEED, "CERTAIN PARTIES HAVE ANNOUNCED AN
INTENTION TO INSTITUTE LITIGATION TESTING THE CONSTITUTIONALITY
OF THE ACT"—and, in my favorite chiasmic double bind, "OUR COLLABORA-
TORS MIGHT ALSO BE COMPETITORS."

You can argue that this is rather thin gruel for critical analysis or political opposition, and I might agree, but it is way thicker than the press releases that decODE had been issuing until then, which was pretty much all
that investors, Icelandic or otherwise, had to go on. *Promising Genomics* describe how the rules about making such statements about genomic futures
had changed in the mid-1990s. Performative utterances are utterly context-
specific, and the institutional and legal context for making what the SEC
calls "forward-looking statements" and I call promises changed dramatically when the U.S. Congress became dominated by the Republican Party
in 1995. The Private Securities Litigation Reform Act of 1995 was one of the
first pieces of legislation pushed through by Speaker of the House of Representatives Newt Gingrich as part of his "Contract with America." Corporate
CEOs could promise anything—I exaggerate, but I trust that you are what
the SEC calls "sophisticated investors" and that you know I exaggerate, and in
any case I include this cautionary statement about my exaggeration—CEOS
could promise anything, orally or in writing, and not be sued later, as long as
such performative statements were accompanied by other statements (like
SEC filings) that said, in effect, don't rely on that performative statement.
These expanded safe-harbor provisions structured the expanding bubble of
Internet and biotech initial public offerings (IPOs) in the late 1990s.

Genomics companies like decODE Genetics, Celera, Millennium, and
Human Genome Sciences have existed entirely in a historically specific
regulatory framework of corporate disclosure—a framework that sanctions
and encourages the promissory quality of the "forward-looking information"
which these corporations produce and thrive on. What is forward-looking
information? An SEC document in 1994 made this distinction: "Required
disclosure is based on currently known trends, events, and uncertainties
that are reasonably expected to have material effects, such as: a reduction
in the registrant's product prices; erosion in the registrant's market share;
changes in insurance coverage; or the likely non-renewal of a material contract. In contrast, optional forward-looking disclosure involves anticipating
a future trend or event or anticipating a less predictable impact of a known
event, trend or uncertainty" (U.S. Securities and Exchange Commission
1994b, 10).

Since its establishment in 1933 as part of the New Deal, the SEC had prohibited disclosure of forward-looking statements; corporations, it might be said, swore not to promise, but only to disclose. With the market crash of 1929 fresh in its institutional memory, the SEC regarded such forward-looking statements as "inherently unreliable," the kind of speculative activity that had bubbled and burst the economy so recently; in its role as protector of investors and preserver of trust in capital and its markets, the SEC worried that "unsophisticated investors would place undue emphasis on the information in making investment decisions" (ibid. 2). This philosophy prevailed for thirty years.

But in the mid-1960s the SEC formed the Wheat Commission to reconsider these matters, and to speculate once again on the question of speculative statements. Despite the fact that the Wheat Commission's 1969 report found that "most investment decisions are based essentially on estimates of future earnings"—that is, "based on" something that, at least in part, was spectral or speculative—they did not recommend changing disclosure requirements at that time, apparently for the sole reason that that was the way things had been done for thirty years: "Commission stated that its decision not to mandate disclosure of forward-looking statements was based on its desire not to deviate too far from its historical position of prohibiting such disclosure" (ibid.).

In 1972 the SEC held further hearings and solicited more comments, only to announce its "intention" to promulgate rules that, while not *requiring* the disclosure of forward-looking information, would nevertheless *encourage* its voluntary disclosure; the SEC also promised corporations some limited protections from any subsequent antifraud litigation. But it never issued those new rules, citing "opposition from commenters," and simply deferred to the future once again, hoping that the issue would be taken up again by its newly formed Advisory Committee on Corporate Disclosure.

The courts, in the meantime, had begun to issue rulings that bore on similar questions revolving around such notoriously difficult yet ever necessary concepts as "facts," "intentionality," and "prediction." In the case with the almost allegorical title, *Marx v. Computer Sciences Corporation*, the courts first addressed the question of whether "predictions or statements of opinion could ever be considered to be 'facts' which could be said to be false or misleading for purposes of liability under the securities laws" (1974, 507 F.2d 485). The court found that "while predictions could properly be characterized as facts, the failure of a prediction to prove true was not in itself actionable. Instead, the court looked at the factual representations which it found were impliedly made in connection with the prediction; namely that, at the time the prediction was made, it was believed by its proponent and it had a valid basis" (ibid.).[4] Promising, in this conceptualization, was predicated on the purity of some interiorized intent—a common formulation of promising that Austin demonstrated "paves the way for immorality" (1962, 9).

In its "concept release" in 1994 the SEC noted new trends in capitalism that suggested the need for change in disclosure requirements. There was "increasing interest" in both corporations and in the analyst and investment communities for "enhanced disclosure of information that may affect corporate performance but is not readily susceptible of measurement in traditional, quantitative terms. . . . Among such qualitative informational items are workforce training and development, product and process quality and customer satisfaction. . . . Companies are beginning to experiment with voluntary disclosure of the utilization of an intangible asset termed 'intellectual capital,' or employee knowledge. In this connection, another federal agency has urged more corporate disclosure of the use of measures of 'high performance work practices and other nontraditional measures' of corporate performance" (U.S. Securities and Exchange Commission 1994b, 10).[5]

The SEC also included in its "concept release" some suggestions from legal scholars as to how the speculative status of forward-looking statements might be recodified. John Coffee offered a "Bespeaks Caution" doctrine under which "a forward-looking statement would be protected so long as it were properly qualified and accompanied by 'clear and specific' cautionary language that explains in detail sufficient to inform a reasonable person of both the approximate level of risk associated with that statement and the basis therefore" (U.S. Securities and Exchange Commission 1994b, 20). (The "reasonable" if not "sophisticated" investor returns as ideal assessor here again, precisely at a time when the sheer increase in the number of investors into the U.S. securities market might have called for some reconsideration of the security of these characterizations.) In addition, "the suggested safe harbor would not require that the forward-looking statement have a 'reasonable basis' (as under existing Rules 175 and 3b-6) because, according to Coffee, this requirement often raises factual issues that cannot easily be resolved at the pre-trial stage" (ibid. 21). These suggestions—in which the volatile and hence litigation-inviting concepts of "fact," "reason," and "basis" are devolatilized by a second cautionary statement that simply names them as volatile—would in fact become central features of the new disclosure landscape.

The Private Securities Litigation Reform Act (PSLRA) of 1995 was supposed to have multiple effects, the two most vital being to reduce the number of "frivolous lawsuits" brought against corporations by their shareholders and to expand the "safe harbor" for forward-looking statements. The high-tech, infocentric corporations of Silicon Valley were a main force behind the passage of the legislation. As an editorial in the *Washington Post* stated it on the day of the final congressional vote (which would override President Clinton's veto of the bill): "When the price of a company's stock drops sharply, the present law invites suits on the questionable grounds that the company's past expressions of hope for its future misled innocent stockholders. This kind of suit has turned out to be a special danger to new companies, particularly high-technology ventures with volatile stock prices. . . . The bill would protect companies' forecasts as long as they did not omit significant facts."[6] *Money* magazine spoke in the name of protecting "small investors like you," and was more critical of the proposed legislation, characterizing it as "a license to defraud shareholders" by "help[ing] executives get away with lying": "High-tech executives, particularly those in California's Silicon Valley, have lobbied relentlessly for this broad protection. As one congressional source told the Washington bureau chief for *Money*, Teresa Tritch: 'High-tech execs want immunity from liability when they lie.' Keep that point in mind the next time your broker calls pitching some high-tech stock based on the corporation's optimistic prediction."[7]

What the PSLRA of 1995 accomplished is perhaps best described by Pillsbury Madison and Sutro, the law megafirm that occupied a central position in these events as counsel to the Securities Litigation Reform Coalition, a political action committee consisting of close to two hundred corporations and their CEOs.[8] By their own description, Pillsbury Madison "assisted in framing the debate surrounding the bill, aided in drafting the statute, and provided members of Congress and Congressional and White House staffs with 'real world' information about the frivolous lawsuits that have plagued both mature and emerging companies over the past decade."[9] The webpage from which these words are excerpted is a part of their ongoing "framing" work (since the establishment of frames, margins, and reading practices of exclusion and inclusion is a never-ending task), explaining to their clients and others the possible meanings and effects of the new law.

Pillsbury Madison called the safe-harbor provisions of the 1995 legislation "the most important provision" for most companies. It embodies the "belief that the U.S. capital markets will benefit from an increased flow of forward-looking information" (ibid.). The final legislation adopted a strong form of John Coffee's "bespeaks caution" doctrine, making a company which makes forward-looking statements "immune from civil liability if the forward-looking statement is identified as a forward-looking statement. The legislative history accompanying the bill makes clear that it is unnecessary to state explicitly that 'This is a forward-looking statement.' Instead, cautionary words such as 'we estimate' or 'we project' likely will be sufficient" (ibid.).

In sum, the PSLRA authorized statements about the economic future by establishing a "safe harbor" that protected corporations from shareholder lawsuits.[10] "Oral forward-looking statements" (such as things said on IPO "road shows" or teleconferences) and "written forward-looking statements" (such as ebullient press releases) were sanctioned, so long as they were marked as such, and so long as potential investors — "sophisticated" or not — were also directed to other, more "cautionary" texts. Pillsbury Madison again: "The general requirement that a forward-looking statement be accompanied by a listing of 'important factors' can be met by a statement identifying the information as forward-looking along with a further statement clearly conveying the message that actual results may differ materially from the results predicted in the forward-looking statement and referencing a 'readily available written document' that contains cautionary language" (ibid.).

Such documents often turn out to be SEC filings: registration prospectuses, quarterly and annual reports, and the like. But before turning to some examples of such forward-looking and cautionary statements in the genomics world, I will roughly sketch a few additional effects that this legislation seems to have had on the economy more broadly. The full and final effects of the PSLRA, according to most sources that I've seen, remain unclear. Whether it has reduced the number of lawsuits (and whether there were, as certain surveys "suggested," so many of them in the first place), whether it has given CEOs "a license to lie," and whether it has increased market volatility while parting the sophistication-challenged investor from his and her money — all remain relatively uncertain, unknowable, or at least open to debate (see, e.g., Perino 1997). But some signs of this uncertain future are still worth reading and pondering.

In their delectably named article, "Enter Yossarian" (1998), Elliott Weiss and Janet Moser analyze the Catch-22 set up by the PSLRA: plaintiffs in securities-fraud cases need to show that a company knowingly misrepresented matters in a forward-looking statement, but can only do so by being granted discovery, which they can't get since they can't file suit based on forward-looking information now protected by the expanded "safe harbor." This fissure is also plumbed by Douglas Branson, who describes the effect as one of the substitution of "mere" words or ink for "due diligence."

> The PSLRA provision provides that if a statement is "accompanied by meaningful cautionary statements," plaintiffs are denied discovery and the court must dismiss allegations as to the false or misleading nature of the statement when presented with a dispositive motion. (Branson 1998, 517)

> The effect of the PSLRA provision, as with stronger forms of the judge-made bespeaks caution doctrine, is to substitute mere cautionary words for the due diligence traditionally associated with preparation of statements in disclosure documents. Rather than building a due diligence file that, through research, establishes more than plausible, and often alternative, bases for making the statements made in the documents, securities lawyers will plaster forward-looking statements with cautionary warnings. . . . No incentive now exists for hiring the traditional, classical securities lawyer who quarterbacked the research and other effort necessary to do the due diligence exercise. Instead, Congress has placed a premium on buying and using ink by the barrel. (Ibid. 518)

Are there nonvolatile rules for distinguishing a "forward-looking statement" from a historical matter of fact, a promise from a disclosure? The legal scholar Richard Rosen suggests not:

> Of course, it would be wonderful to have a rule that allowed courts and litigators unfailingly to be able to discriminate among the cases and to allow discovery to proceed only in those in which there was some reason to believe that, notwithstanding adequate cautionary language, the company's management really did know that the predictions were not likely to come true. But because no such rule could ever be constructed, any rule will either allow a few dishonest issuers to win motions to dismiss or will impose enormous costs on honest companies and their shareholders. Congress has struck the balance in favor of protecting companies and their shareholders—a balance which strikes me as absolutely correct. (Rosen 1998, 657–58)

One might agree with Rosen insofar as it is indeed impossible to construct "unfailingly" discriminatory rules for calculating fissures, volatilities, and chiasma such as these opened up by forward-looking information. But it would be prudent and even reasonable to question the "balance" that he hyperbolizes as "absolutely correct," since it in fact depends on a far-from-certain calculus of the fissure: how does one know that, in the chasm opened up by forward-looking information, there won't be an "enormous" number of dishonest issuers and only a "few" costs to honest companies? Rosen's unsubstantiated calculation is little more than a political endorsement of the balance that Congress—no, more specificity is in order: that a Republican-dominated, Newt Gingrich–led, futures-market-giddy Congress—indeed "struck": a forceful cut that transfers many of the financial risks and the burdens of proof, discovery, and due diligence from corporations to their presumably "sophisticated shareholders." Like the ones in Iceland.

Let's return to 1999. The Health Sector Database legislation, which presumes the consent of all living *and dead* Icelanders to have their medical records placed in a database to be built and exclusively licensed to deCODE, has been passed by the parliamentary coalition led by David Oddson's Independence Party, in late 1998. The $200 million promise from Roche has been sworn, but is far from being in the bank. (As later SEC filings by deCODE show, Roche only ever paid $76 million on this promise: basically, the operating expenses it had agreed to, plus a very few—certainly nowhere near the number that would have made it possible to reach the $200 million promise—"milestone payments" for work done by deCODE.) The IPO on NASDAQ has not even been mentioned in public. Like all biotech startups with a high "burn rate" (in the neighborhood of $2 million per month), deCODE needed operating capital. Where to get it? On the "gray market."

As in the United States., in the mid- to late 1990s "the securities market changed quite a bit" in Iceland, Finnur Sveinbjörnsson told me in 2001; he was then president of the Icelandic Stock Exchange, or ICEX. Trained as an economist, Sveinbjörnsson had worked for the Central Bank of Iceland early in his career, then for the Ministry of Commerce on issues related to the European Economic Area, and from 1995 to 2000 for the Bankers Association, a lobbyist organization. He continued:

> The securities departments in banks have grown tremendously, mostly because of the abolishing of foreign exchange controls, which meant that there was interest in foreign securities, and the banks responded. Also because there was appetite for corporate and local government bonds, and these entities responded by switching some of their financing away from traditional bank loans to the bond market. Furthermore many investors suddenly realized that there were more possibilities than just

opening a savings account. They could invest in securities, both bonds and stocks, and get much nicer returns than previously. So there was basically an explosion, from both the supply and the demand side in the securities market.[11]

There was also more money to invest.

Many individuals have become rich in this country. The main source of wealth has been the quota system in fishing. There are a number of individuals who have sold their quotas and left this business, so they have gotten out of this business, and suddenly found themselves with considerable amounts of money in their hands. This money has to a large extent gone into the securities market. Furthermore, some of this increased activity has been financed through foreign money—money borrowed by Icelandic banks from overseas and lent to their local customers. Some individuals have borrowed and used the proceeds to speculate in the market.

"There was an unclear, or gray, provision in the securities legislation," Sveinbjörnsson explained. "Apparently the legislation did not prevent professional investors from selling shares to the public without having to fulfill the stringent requirements of a public offering, as long as prospective buyers came to the investors asking for the shares—that is, active promotion and selling was clearly out of bounds but responding passively to demands was within limits."

While the Icelandic financial institutions and professional investors may not have taken out ads in the newspapers, they did talk actively to journalists and played, like their foreign counterparts, a certain part in the securities hype of the late 1990s. Pumped up by these glowing media accounts and the general securities exuberance, hundreds of Icelanders bought shares in deCODE on this "gray market," through these Icelandic financial institutions.

This was a very high number in the eyes of the Icelandic financial authorities, especially for securities so widely acknowledged as risky and requiring long-term financial strength to withstand likely losses. "The problem of the gray market was discussed thoroughly last winter and into the spring [1999–2000]," Sveinbjörnsson continued, "in a committee that the Ministry of Commerce set up with representatives from the market, from the stock exchange, from the ministry itself. This committee reviewed the legislation in our neighboring countries, and in the U.S., and came up with recommendations that were enacted by Althingi in the fall of 2000, thereby closing the legislative loophole that had allowed the gray market to develop." The journalist James Meek, writing for the *Guardian*, helps us with a sketch of those whose money passed through this loophole before it was closed, in what amounts, and with what personal effects.

"I'd never bought stocks before," Henrik Jónsson told Meek. "I watched the TV news carefully and saw the stock prices rising consistently. You had stories in the media telling people how high Decode stocks were rising. My heart is crying" (Meek 2002). Jónsson had bought 5 million krónur worth of deCODE stock at $56 per share in the spring of 2000. He had been disabled in a traffic accident, and the sum represented nearly a quarter of the financial settlement he had fought to obtain from the insurance companies. The shares were priced at $6 when he returned to the National Bank of Iceland, in 2002, to sell the shares they had sold him.

Meek collected numerous such stories when he was in Iceland in the fall of 2002. Perhaps the worst story was that of the anonymous E., who had also bought in at the height of the exuberance in the spring of 2000, plunking down 30 million krónur—about $400,000—over three separate occasions. Some of the money he borrowed from the state-owned National Bank, from which he also bought the shares; in that case, E. mortgaged his house. He agreed to buy more from the Agricultural Bank, and arranged another loan with the Bank of Iceland. A day later, the Bank of Iceland canceled all its loans to buy shares, but the Agricultural Bank would not let E. off the hook. He spent the next month trying to arrange other financing, during which time the share price fell from $64 to $50, but the bank insisted he still pay $64. Meanwhile, the National Bank moved to foreclose on its loan. "I asked them if I could wait at least until the shares rose to $15. The lawyer from the bank said, 'You must know these shares won't go up.' I said, 'You sold me these shares, telling me they would go up to $100.' He said, 'If you're going to blame us you'd better pay up'" (ibid.). There are hundreds of such stories, at least, each one deserving a place in a future archive.

The loophole that was the gray market in Iceland had a structure in which the media played an important part. But for the intricacies of the mechanism, you have to turn back to the fine print of the SEC filings. So let's allow deCODE to disclose the basic elements of the story: "There is no public market for the Preferred Stock although some banking institutions in Iceland have been making a market for privately negotiated transactions among non-U.S. persons in the Series B preferred stock. Our stock transfer records indicate that approximately 10 million shares of Series B preferred stock were transferred during 1999 in approximately 7,000 transactions and approximately 1.1 million shares of Series B preferred stock were transferred during January 2000 in approximately 2,700 transactions. The majority of these transactions had an Icelandic financial institution as one of the counterparties."[12]

To elaborate: in June 1999 deCODE, a good eight months before it would be bound by the U.S. SEC to disclose this to any investors and a full year before its stock went public on NASDAQ, sold 5 million shares of its Series B stock at $7.50 per share to a holding company in Luxembourg, Biotek Invest, apparently set up for this and only this function. At the same time, an "oral agreement"—and while deCODE, like any U.S. corporation, is legally a person, one may wonder what mouth it employs to make these kinds of spoken promises—was made that deCODE would be paid twice that amount if two conditions were met: if the Luxembourg company sold at least half of the shares at $15, and if the price in "the Icelandic market" didn't go below $15 for the six months remaining in 1999. Those two conditions were not only met, but exceeded: 6,000 transactions were made in deCODE stock at prices between $30 and $65 per share on "the Icelandic market"—known as "the gray market" when you're not reporting to the SEC. By comparison, at deCODE's IPO on the NASDAQ, in July 2000, the share price opened at $18, bubbled briefly to $27, and was never again in the vicinity.

Biotek Invest sold the shares to Icelandic banks and securities companies; the Icelandic financial institutions, empowered through an "interpretation" of existing law—interpretation is never far from speculation—sold the shares to Icelanders eager to invest in "an Icelandic company" that they had been told was uniquely positioned to understand and perhaps cure genetic diseases, a story publicized by Icelandic journalists, with the encouragement of the Icelandic financial institutions, in newspaper stories and electronic media that would later be cited by some anthropologists, bioethicists, and assorted others as evidence of a "democratic discussion"; and the financial risks accrued at the bottom of this speculative chain. As part of the promise, written and oral, Biotek Invest got a 7 percent commission, plus interest. By this mechanism, a total of $69,750,000 flowed from the bank accounts of Icelanders to the bank account of deCODE. Biotek Invest was liquidated in December 2000, having existed for barely a year and a half; assets of $5.25 million were transferred to some equally anonymous company, called Damato Enterpises, in Panama City, Panama. This is where the disclosure ends; anything else would be pure speculation.

Eventually deCODE would get its largest capital injection from its IPO on the NASDAQ, for which it had filed its S-1 registration statement in March 2000. In this event, too, promising was at work in yet another way. At deCODE's IPO, on 17 July 2000, some 11 million shares were sold, grossing $198,720,000. Morgan Stanley, the underwriters, took nearly $17 million of this as commission and expenses.

In December 2001 a class-action suit was brought against deCODE Genetics by the law firm of Milberg Weiss. This was hardly unusual at the time; over a thousand such securities-fraud class-action suits were filed in the Southern District Court of New York alone (where the deCODE case was filed) in the wake of the market's collapse that year. Milberg Weiss, a law firm which "dominates its field [securities class-action suits] to a degree that no private firm does in any other kind of litigation," accounted for a large number of these. A co-founder of Milberg Weiss, William Lerach, would later bring a case against Enron. Lerach was among those who had convinced President Clinton that he should veto the Private Securities Litigation Reform Act of 1995, a veto overturned by the Republican-controlled Congress. Indeed, according to one lawyer who lobbied for the PSLRA, "The whole idea behind the law was to destroy Lerach" (Toobin 2002).

The complaint was complex, but centered on the charges that deCODE and Kári Stefánsson, along with the co–lead underwriters of their IPO, Lehman Brothers and Morgan Stanley, had made "materially false and misleading statements" in their SEC prospectus and "engaged in a pattern of conduct to surreptitiously extract inflated commissions" greater than the ones they had disclosed.[13] One mechanism by which this extraction occurred was that the "Underwriter Defendants agreed to allocate deCODE shares to those customers in the Offering in exchange for which the customers agreed to purchase additional deCODE shares in the aftermarket at pre-determined prices."[14]

Between January 1998 and December 2000, 460 "high technology and Internet-related companies" had come up for sale on the public markets; class-action suits against 309 of these, including deCODE, were consolidated for coordination and decision of pretrial motions and discovery, to be conducted by Judge Shira Scheindlin. The consolidated defendants, including Morgan Stanley, asked for dismissal of all the cases.[15]

Judge Scheindlin's 238-page decision attempts to sort out the various complaints against the various underwriter defendants, and decides which complaints will be dismissed, and on what grounds. The decision does not discuss the decode case directly, but it is listed as one of the cases that is dismissed on the basis of "No Allegation of Motive" to violate Rule 10b-5, which prohibits market manipulation through devices such as tie-in agreements. In other words, even if Morgan Stanley and Lehman Brothers did in fact require their customers to buy decode stock after the ipo in a tie-in agreement, because there was no evidence presented that they sold the stock in this period, the claim was dismissed. As Judge Scheindlin put it, "Mere ownership in the absence of profit-taking does not establish a motive that would support a 'strong inference that the defendant acted with the required state of mind.'"[16] Here again is a case in which intentionality, the ethical criterion based on what Austin called "an excess of profundity," in fact "paves the way for immorality"—or at least getting out of court.

But the sec can prevail where courts may not.[17] In a settlement with the sec in January 2005, Morgan Stanley agreed to pay a $40 million civil penalty for "attempting to induce certain customers who received allocations of ipos to place purchase orders for additional shares in the aftermarket." "These cases underscore the Commission's resolve," read the sec's press release, "to ensure the integrity of ipo markets by prohibiting conduct that could artificially stimulate demand or higher prices in the aftermarket—whether or not there is manipulative effect."[18]

Some Morgan Stanley sales representatives referred to this practice, of their customers' buying in the immediate aftermarket, as fulfilling their "promises" or "commitments." In the suit against Morgan Stanley, the sec at one point used the decode case to illustrate how such promises took place: "In a third ipo, a sales representative e-mailed the syndicate manager, 'just for the record . . . despite a less than stellar allocation [a customer] was in buying decode in the aftermarket yesterday. As they said they would. Almost 200m at $28.' The syndicate manager responded, 'i am very aware of their aftermarket. Kudos to them and it will be remembered.'"[19] Here, at the most quotidian level, the sales rep informs his manager about a promise carried through: a customer had fulfilled their oral agreement to buy shares of decode, "as they said they would," even though they had not been treated in a "stellar" manner. One almost admires the informal moral economy of promising at work here, as the manager expresses his familiarity with this gentleman's arrangement, grants his kudos, and hints at the promise to come, as they always do in these kinds of gift exchanges: it will be remembered.

Conclusion

Of course, every promise by its very structure is excessive. . . . Is not every promise—I
do, I will forever and ever—not a little drunk with itself, dipped in some witches' brew
or Dionysian concoction?

AVITAL RONELL, *THE TEST DRIVE*

Promising, and promising genomics in particular, is a matter of excess—the
scandalous excess of quotidian matters of all kinds, from finance to flesh and
everywhere in between. Matter promises, harboring excessive capacities and
thereby offering itself to us and our technoscientific infrastructures for further becomings. Biological matter in particular has the promise folded into
it, and not just spoken about it. An organism like you, me, or a zebrafish exceeds its genetic "code." It exceeds its "nature" or genes, it exceeds its "nur-
ture" or environment (which in turn is always exceeding itself and changing), it exceeds every wonderful technoscientific tool that has ever been
invented and that ever will be invented to handle the truths of it. It's the
"inability" of any given present or any given thing to coincide with itself, or
to identify fully with its opposite partner, or to fully differentiate itself from
its partner, that underwrites the possibility of its becoming something else.
More, more, more: that's why we and zebrafish and every other geno-proteo-
transcripto-culturomic enterprise *develop*. Like the statements we speak, we
ourselves are forward-looking. We promising animals are networks of differential reproduction—evolving biological matter—wired to other networks
of differential reproduction—the experimental systems of the life sciences
that Hans-Jörg Rheinberger has diagrammed for us—webbed into the network of differential reproduction which we speak and which speaks us, language, a language of promising.

Promisings of genomics fan out from what Richard Doyle (2003) calls the
"constitutive ambiguity" that inhabits flesh and finance alike, a real dependency on some "unthinkable and incalculable future" that provokes incessant thoughts, calculations, and other activities that produce the "liveliness"
of bodies and capital. Organisms, illnesses, events, politics, finance, technological development—all of these are matters of excess that happen without
our full understanding or control, as the cumulative, emergent effect of a
multiplicity of forces—*promises*.

Notes

1. Inventing an ethics of promising will be akin to the project of "learning to speak with ghosts" delimited by Derrida in *Specters of Marx* (1994, xix).

2. See Michael Fortun 2008 for a more detailed and expansive map of these territories, from which much of the present text has been extracted.

3. See MacKenzie, Muniesa, and Siu 2008; Knorr-Cetina and Cicourel 1981.

4. The SEC also cited "more extreme positions" taken by some courts, which ruled that "soft," "puffy" statements "upon which no reasonable investor would rely" would not be actionable, unless they were worded in the language of a guarantee.

5. The footnote to this passage refers to two documents to substantiate this historical argument: a Harvard Business School Working Paper, "Improving the Corporate Disclosure Process" (Eccles and Mavrinac 1994), and an article in *Fortune*, "Your Company's Most Important Asset: Intellectual Capital" (1994).

6. This editorial was inserted into the *Congressional Record* (104th Congress, S19147) by advocates of the legislation.

7. In another example of the mass media being targeted to the reading and citational practices of a few readers, this editorial too was included in the *Congressional Record* (104th Congress, S19147).

8. As a side note, Pillsbury Madison (now Pillsbury Winthrop) also did the legal work for Genentech, in the first (and highly volatile) biotech IPO, in 1980. See Stewart 1984.

9. See http://pillsburylaw.com/articles/securities_reform_act_1995.html, accessed 3 March 2000.

10. Protected them in federal courts, at least. One effect of the PSLRA was to create a rush to file suits in various state courts, a problem addressed by the Securities Litigation Uniform Standards Act of 1998, which preempts class-action suits filed in state courts.

11. Interview with Finnur Sveinbjörnsson, January 2001.

12. deCODE Genetics Form 10-12B/A, 6-20-00, 4, available at the U.S. Securities and Exchange Commission website, http://www.sec.gov. Series A preferred stock is the stock owned by the venture capitalists; Series C stock is the stock owned by Roche; the Series B stock sold to the Icelanders came in part from a repurchase of these stocks, and a repurchase of 333,333 shares of Kári Stefánsson's common stock.

13. *Linda Rubin v. deCODE Genetics, Inc, Lehman Brothers, Inc., Morgan Stanley and Co. Incorporated, Kári Stefánsson, and Axel Nielsen*, doc # 01-CC-11219, filed 5 December 2001 in United States District Court, Southern District of New York, 2.

14. Ibid., 8.

15. *Opinion and Order In re: Initial Public Offering Securities Litigation*, Shira A. Scheindlin, U.S.D.J., United States District Court, Southern District of New York, 03-01555, 19 February 2003.

16. Ibid., 165.

17. For a discussion of the SEC's strategies in such cases where legal proceedings

may not be viable, or may simply take too long, see the interview with a former SEC lawyer in Fortun and Fortun 1999.

18. U.S. Securities and Exchange Commission 2005.

19. *SEC v. Morgan Stanley*, No. 1:05 cv 00166, United States District Court, District of Columbia, filed 25 January 2005, available at the U.S. Securities and Exchange Commission website, http://www.sec.gov.

12

CHLOE SILVERMAN

DESPERATE AND RATIONAL

Of Love, Biomedicine, and Experimental Community

1. *The Chelation Kid*. From Tinnell and Taillefer 2007a. Courtesy of Craig A. Taillefer and Robert Tinnell.

How many times have you, now almost nine, sent us back to our lovers' laboratory Dark with dashed expectations, challenging us to let go and try again?

JACK ZIMMERMAN, PREFACE TO JACQUELYN MCCANDLESS,

CHILDREN WITH STARVING BRAINS

A diagnosis of autism is commonly experienced as a devastating blow for the family of an affected child, no matter how deep their concerns may have been prior to the moment of diagnosis. This is not surprising: autism is typically characterized as an incurable and lifelong neurological disorder, strongly heritable and treatable only imperfectly, via behavioral interventions and psychopharmacology. Rates of diagnosis have increased to approximately one in 175 children in the United States, raising concerns of environmental triggers among both parents and clinicians (Shieve et al. 2006). The disorder, now seen as occurring along a spectrum of severity, was first described by the child psychiatrist Leo Kanner in 1943 (Kanner 1943). It is characterized by a triad of deficits, involving significant impairment in communicative, behavioral, and social domains, with onset before the age of three (American Psychiatric Association 2000). Given the dire prognosis offered by most specialists, parents might easily be moved to investigate any promising treatment option, erring in favor of an interpretation of their child as injured rather than congenitally disabled, and of that injury as a chronic but treatable condition, rather than a static and irremediable prenatal event. Even Bruno Bettelheim (1967), notoriously associated with the theory that cold and emotionally distant mothers caused their children to develop autism, believed that a theory of injury offered more hope for both parents and children than a theory of hereditary disability. For many parents, culpability seemed like a small price to pay for the promise that they might be able to help their child.

To speak of treating autism, let alone curing the disorder, is to invite attributions of desperation on the part of parents and charlatanism on the part of practitioners in both professional circles and the popular press (e.g., Levitz 2005; Harris and O'Connor 2005; ScienceDaily 2007). Newspaper articles regularly accept the characterization of parents as willing to try anything to normalize their child's development, and the attitude toward parents' knowledge and ability to critically evaluate treatments is similar in academic studies on the use of alternative treatments in autism (Shartin 2004; Levy et al. 2003; Levy and Hyman 2003). Nevertheless, listservs are alive with discussions of the relative benefits of various detoxification preparations and the minute details of appropriate supplementation relative to body weight, all healthcare practices that take place outside of certified medical knowledge about children with autism. University-based researchers attribute their findings to the insights of parents without formal medical educations, and materials available at parent conferences involve complex explanations of biochemical mechanisms connecting physiological to neurological dysfunction, which parents listen to with undivided attention, and often master.[1] For many parents, "there comes a point when you've just got to experiment," because of the paucity of options available within the realm of conventional treatments (Shartin 2004).

The 2005 revision of Jon Pangborn's and Sidney Baker's popular autism treatment manual opens with the statement "Individuality is key" (2005, 1). The authors suggest that obligations other than those pertaining to conventional medical practice apply to the private decisions regarding care for an affected child: "As you read forward, keep in mind the huge difference between public and private policy, because this distinction will make your job much easier: *to decide what to do next for the child in your care.* The threshold for making a good decision for a single child is tiny compared to the burden of proof for establishing public policy. Public policy is, moreover, disease-oriented, whereas private health policy is oriented almost solely toward protecting and healing the individual" (ibid. 3). Within the community that Pangborn and Baker are addressing, the concepts of recovery and cure are common currency, achieved through what are termed "biomedical" interventions. These encompass a range of conventional and unconventional therapies, united by the premise that autism is a treatable medical condition. These practices involve a fragmenting of the diagnostic category that, if taken seriously, could be quite disruptive to established research programs in autism. Because it is defined behaviorally, autism is a syndrome, a set of related symptoms. It is not a single disease with a known pathophysiological mechanism. Observable autism symptoms may be secondary to a range of other underlying causes, rather than associated with a single, definitive disease process.[2] For parents and practitioners treating autism as a medical condition, the difficulty of producing standard treatment practices validated by clinical trials stems from the fact that "autism," far from representing a "pure culture," is in fact a complex set of overlapping conditions (Herbert 2005b; Happe, Ronald, and Plomin 2006). As Pangborn and Baker suggest, those running carefully constructed research programs or implementing public policies that assume an underlying biological homogeneity in the population of people diagnosed with autism might find such a statement difficult to accept, lacking, as it does, a comprehensive solution tailored to an entire population. Instead, this perspective offers parents a practical avenue through which to act on their conviction that their children are actively suffering as a result of their autism and that they require treatment as much or more than they require benign acceptance of their differences.

In this chapter, I describe the production of biomedical knowledge in one parent-practitioner community and the process of persuasion and affective involvement through which parents and new practitioners come to take part in a set of practices called biomedical interventions for autism. Social relations based on shared experiences of affect alter the possibilities of vision, the formation of trust, and the ways in which the success of therapies are measured in this community. Although I am considering the formation of community as a response to a medical condition, I am less concerned with the shared biological attributes of affected children or the formation of identity around a common diagnosis (as in Rabinow 1996). This community did not arise spontaneously—its growth was orchestrated by a group of concerned parents and practitioners who have been instrumental in promoting alternative treatments for autism.

This community also does not operate somehow "outside" of the economic and commercial frameworks of contemporary biomedicine, or, indeed, apart from forms of expertise or representations of biological systems that are informed by those same commercial interests. Healthcare is serious business, even, or especially, when part of what is on offer is the potential for a child to interact more successfully with their parents and other people around them (see also Haraway, chapter 2 in this volume). Parents describe the work of caring for affected children, including the use of resources, training in medical techniques, and hours of effort, as an act of love, in favor of a description of this work as attentive, knowledgeable labor. This is an ideological choice, akin to what Sharon Hays (1998) has called "intensive motherhood," but pairing the sense that everyday parenting is a special type of expert labor with the urgency of caring for a sick child.[3] By using the language of love parents are at once making a claim for a form of situated knowledge and invoking a terminology of obligation (Haraway 1991). This alliance between the affective, the anecdotal, and the everyday is part of what gives efficacy to biomedical treatments for autism, but it is also what can make them difficult to translate and reproduce in conventional medical frameworks.

The Edge of Reason

Parents of children with autism spectrum disorders are not unique in their experimental bent: chronic disorders or "contested illnesses" which are poorly or incompletely treated within contemporary biomedicine often lead sufferers to seek explanations outside of authoritative accounts of disease causation (P. Brown et al. 2001). Michelle Murphy has described the types of everyday practices of self-care and emphasis on "biochemical individuality" and personal ecology that have informed communities of those who suffer multiple chemical sensitivity (Murphy 2000; also see Kroll-Smith and Floyd 1997). The movement to treat autism as a medical condition through the use of dietary, nutritional, and detoxification regimes is a movement similarly built on experiential knowledge and pragmatic interventions firmly situated within daily life. As such, it depends on parents' knowledge of their particular child's moods and responses, but it also draws on older traditions of medical cookery that have been largely marginalized as "alternative" approaches within conventional biomedicine (Schiebinger 1991). These affectively laden practices of feeding, medicating, or ministering, and the forms of knowledge associated with them, work to persuade members to join the community and reward their commitments. However, they also limit the ability of the community to translate this work into the language of what they call "mainstream" or "conventional" medicine, which prefers labor associated with dispassionate forms of rationality. To the degree that parents and practitioners are successful in reframing autism as a set of treatable medical conditions, they must seek to erase the affective and situated components of their labor, exactly those elements of practice that are most intuitive at the level of their families and communities.

Despite the widespread perception that biomedical treatments involve questionable rationales, these alternatives often lie comfortably within medical epistemology. They are seen by participants as shifts in treatment perspective or renunciations of standard diagnostic frameworks, not as a choice to abandon medical understandings of autism, but to embrace them more completely. However, they do not offer diagnosis-specific treatments of the type that many hope will emerge from genetic and pharmaceutical research. Unconventional treatments are hesitant, experimental, and gradual, based on practices of monitoring and watchfulness. Parents see biomedical interventions as attempts to modify a process of ongoing disruption or damage, not a fixed structure. They reference development, change, and also promise, an idea that is important to biomedicine, most notably genetics and genomics (Sunder Rajan 2003; Haraway 1997). As Mike Fortun has argued (chapter 11 in this volume), it is the "undecidability" of promises that keep investors interested, especially if that investor is a parent with a decided prior commitment to the well-being of their child. However, in the case of autism, parents are invited to monitor progress even as they invest—they may hope for a cure, but they will take note of any improvement. Their commitment is only as strong as their belief that an intervention is working. The price of an unfounded promise is parental mistrust, so while practitioners may work to define the criteria for success in the broadest possible terms, they are also cautious to not promise more than they can potentially deliver. To quote Jon Pangborn and Sidney Baker, "*In other words, every treatment is really a diagnostic trial*," where responses or failures to respond are clues about the underlying condition (2005, 29, emphasis in original). They explain that "in chronic illness, symptoms are very helpful in naming, but seldom in explaining, the exact mechanism of problems that involve a 'snowball' effect in which many symptoms are secondary to the initial disordered mechanism. Chronic illness does, however, present an opportunity for symptoms to indicate progress made as the result of intervention. The key to using symptoms as a guide to monitoring progress is to focus not only on the defining of major complaints, but on the whole pattern" (ibid. 47).

In the process of tracking the effects of interventions, parents must learn to see behavioral symptoms as proxies for underlying metabolic and physiological dysfunctions. While the "major complaints" in autism include the familiar triad of deficits in behavior, communication, and socialization, the "whole pattern" represents an interconnected set of physical symptoms responsive to minor variations in care. In practice, trials, monitoring, and the ongoing process of diagnosis and treatment are carried out by parents. These forms of love are hard work. Because each child is ultimately their own reference point, knowledge about how to carry out biomedical interventions is distributed among parents and clinicians, and expertise is contained within the community as a whole, rather than in reference to an established body of facts about a discrete and defined disease entity.

Domestic scale interactions based on love and care, attachments that are typically seen as opposed to reason and rationality, actively contribute to rational work in this community. Nevertheless, acts carried out in the fulfillment of parental obligation are suspect and not viewed as legitimate sources of knowledge. By dismissing them, we impoverish our language for talking about passionate forms of labor, the ethical ambiguity of commitments to truth, and the functional work of desperation. In other words, when we pay attention only to the desperation of parents and their eagerness to find a cure for their children at any cost, we run the risk of assuming that parents are merely victims of promiscuous hype and of the promise of cures on the part of unreliable hucksters, instead of savvy consumers capable of drawing on a formidable expertise regarding their own children.

Nevertheless, relations of affect are not by definition innocent or harmless. A loved one can become the object of experiment as well as the precious object of devotion, and familial obligations can have tragic as well as productive bodily consequences. An editorial in the *New York Times* argued that by "mythologizing recovery," parents of children with autism are led to feel that they have failed their child if they cannot effect a cure, leading to a perverse logic where even murder can be rationalized as an "act of love" (McGovern 2006). Parents' devotion to their offspring, especially in instances where severe disabilities can make it difficult for parents to understand their children in terms of conventional narratives of kinship, can be used to justify extreme measures (Rapp 2000a; Rapp and Ginsburg 2001). That parents desire a stronger connection with their child is no guarantee that their judgments will always be secure or that their experiments will not be dangerous.

Love runs through this discussion. Parents claim the authority of investment, partiality, and passion as the source of their own "lay expertise" (P. Brown 1992; Epstein 1996). Unlike other communities of lay experts, parents challenge both the mode of production and the products of biomedicine, arguing not only for different treatment priorities, but also for altered frameworks for conceptualizing autism. This is a case study in partial perspectives in the sense in which Donna Haraway has employed the term, meaning that being partial is at once about being situated, embodied, and incomplete, but it is also about being impassioned, not neutral, and unwilling to recognize the claims of neutrality of others as unproblematic (Haraway 1991). Instead of seeing love as a term opposed to objectivity, parents describe love as a process that enables new forms of objectivity, a relationship to medical investigation that occupies the fertile edge of the domain of reason. They apply this passion in the form of intellectual inquiry, collaboration, experiment, and research, to argue against the centrality of neurological difference in autism and for the primacy of molecules and metabolites and the molecular correlates of what we call affect and social reciprocity.

Biomedical treatments for autism remain controversial. Frustrated parents speak out against the treatments after they fail to produce cures for their children (e.g., Laidler 2004), while pediatricians regard even measures as limited as dietary modification as, at best, poorly understood and potentially dangerous (Arnold, Hyman, Mooney, and Kirby 2003; W. Weber and Newmark 2007). Supporters argue that the inconsistent responses of children to treatments is a mark of the heterogeneity of the underlying condition. The issue is further complicated by the fact that treatments occur prior to an established consensus regarding the facts about autism, on rapidly developing children. Perceptions of efficacy are also contextualized, meaning that treatments may work for reasons other than the mechanisms ascribed to them, or they may affect factors not included in conventional outcome measures that focus only on behavioral variables. Arguments over efficacy go to the heart of technologies of knowing, from the randomized controlled trial to unquantifiable forms of intimacy, touch, and scrutiny. Controversies over biomedical interventions in autism address both the epistemological problem of what we can know and the political problem of what we choose to investigate.

The History of a Perspective: Defeat Autism Now! (DAN!) and Biomedical Treatments

In calling attention to a parent movement based on biomedical interventions for autism spectrum disorders, I rely on my experiences over several years with the Defeat Autism Now! (DAN!) project of the Autism Research Institute.[4] At a time when as many as 70 percent of the parents of children diagnosed with autism try some form of "biologically based" treatment as an alternative to conventional approaches, DAN! lies within a broader set of social and communicative networks that share explanations, texts, and practices (Wong and Smith 2006). A number of professionals have multiple affiliations, and DAN! doctors are often present at conferences for Autism One or the National Autism Association, both of which advocate similar therapeutic approaches.[5] The Autism Research Institute is neither the largest nor the best-funded of the parent groups involved in research on the autism spectrum disorders. However, understanding their work is crucial for understanding the ways in which families with autism intervene in medical knowledge production and the complicated ways in which commercial, scientific, therapeutic, and professional interests collide in an organization that seeks to recruit practitioners as well as parents.

DAN! conferences bear a genealogical relationship to both parent advocacy in general and the formation of specific activist groups. They are run by the Autism Research Institute (ARI) in San Diego, founded by Bernard Rimland, a co-founder in 1965 of the National Society for Autistic Children (now the Autism Society of America), which was formed in part to promote Applied Behavior Analysis (ABA), an educational and therapeutic technique developed by O. Ivar Lovaas at UCLA during the 1970s. Premised on a neurological theory of autism opposed to the psychogenic theory then favored, ABA was not yet established as conventional treatment and was regarded by some proponents of psychotherapeutic treatments as a form of conditioning tantamount to abuse (Bettelheim 1967, 111–12). In 1964 Rimland wrote *Infantile Autism* as a challenge to the psychogenic theory of autism promoted by psychologists such as Bruno Bettelheim, drawing together existing research on neurological and genetic factors in autism and proposing a theory of etiology and directions for future research. Rimland's book situated the neurological identity of autism squarely in terms of the case series used by Leo Kanner to characterize autism in 1943, reinterpreting Kanner's characterizations of parental aloofness in terms of a possible set of hereditary factors (an interpretation that persists in contemporary genetics studies).[6] While research into the genetics of autism evolved into a search for associated genes, Rimland turned his attention to the practical problems of treatment and the lack of diagnostic scales that took medical problems into account. He spent years refining instruments to examine factors left out of the definition of autism, producing an "Autism Behavioral Checklist," appended to *Infantile Autism*, that asked questions about regression as well as parent characteristics and the child's eating habits.[7] This instruction in a system of monitoring and care has become part of the practice of the DAN! community.

During the late 1960s, in collaboration with Linus Pauling and based on research in the emerging field of orthomolecular medicine, Rimland began experimenting with megavitamin therapies through his Institute for Child Behavior Research in San Diego (later the Autism Research Institute). He recalled that during the mid-1960s, parents began contacting him regarding the therapy, and that "even though few of the parents were acquainted with each other and each was trying quite a variety of vitamins, the same small group of vitamins was being mentioned again and again. As the number of parent-experimenters grew, it began to include more parents whom I knew personally to be intelligent and reliable people. At that point I contacted a number of doctors in California and on the East Coast who had been experimenting with vitamin therapy. The combined information from the doctors and parents convinced me that I could not, in good conscience, fail to pursue this lead" (Rimland 1976, 200).

Rimland's initial vitamin study recruited parents through the National Society for Autistic Children. The results were suggestive, indicating that when combined with magnesium to prevent deficiencies associated with high doses of B-6, the therapy helped some children (Rimland 1973). At the same time, Rimland was becoming increasingly convinced that the diagnostic entity of "autism" in fact referred to a range of different medical syndromes (Rimland 1971). As early as 1976, Rimland could write, "I am firmly convinced that very little progress may be expected in finding cause and treatment for mental illness in children until the total group of children now loosely called 'autistic,' 'schizophrenic,' 'psychotic,' or 'severely emotionally disturbed' can be subdivided in a scientific way into smaller homogeneous subgroups. . . . I believe the children loosely called 'autistic' or 'schizophrenic' actually represent a dozen or more different diseases or disorders, each with its own cause" (1976, 203).

Rimland was instrumental in convening the first DAN! meeting, in January 1995, with Sidney Baker, a physician with a background in functional medicine, and Jon Pangborn, an industrial biochemist and the parent of a child with autism. At the time, all three were increasingly concerned about rising rates of autism diagnoses, just as all three became more concerned during the 1990s about the possibility that childhood immunizations could trigger autism.[8] As a parent of a child with autism and a scientist (with a doctorate in experimental psychology), Rimland represented the paradigmatic parent-activist, using his set of prior skills in autism research while publicly referencing his own "autistic" tendencies.[9] Other doctors who have attended DAN! meetings over the past decade tell different stories about their involvement with the organization. A number of them, like Rimland, discovered orthomolecular medicine and orthomolecular psychiatry in the 1960s, during or after their formal medical training, while others have backgrounds in functional medicine or nutritional biochemistry, and others associate their use of biomedical treatments with evangelical Christian backgrounds or other religious traditions. Nearly all of them explain their interest in the treatability of autism in similar terms. They became frustrated that the medical conditions experienced by children with autism and reported by their parents were being written off as "comorbidities" by many of their colleagues and were struck, when parents approached them for help, by the physiological variability in the population. Many explain that they initially became aware of biomedical treatments by "listening to parents" and that they have been repeatedly impressed by the responses of children to these treatments and by the nuanced observations of parents.[10] Whether doctors approached treating autism within the context of their naturopathic or integrative medicine practice were recruited by parents or came to the DAN! model through their own search for answers as the parent of a child with autism, most doctors take care to emphasize that while their methods may be drawn from biomedicine, their knowledge of any particular child comes mainly from that child and from his or her parents.

An Experimental Community: Biomedicine as Practice

At the center of the DAN! movement is the DAN! guide, officially *Autism: Effective Biomedical Treatments* (Pangborn and Baker 2005), which has grown from a spiral-bound sheaf of suggestions handed out at the first DAN! meeting to a commercially produced publication that parents are encouraged to share with their child's doctors.[11] The treatment method that the guide outlines requires an explicit though often individualized set of tests and regimes. These proceed stepwise through levels of difficulty and often expense, with more complex treatments requiring monitoring by a medical professional. The first stage is an "elimination diet," which involves removing foods associated with sensitivities in children with autism, which DAN! practitioners associate with both physical and behavioral symptoms. Two popular diets are the gluten-free/casein-free ("gf/cf") diet and the Specific Carbohydrate Diet (SCD).[12] The second stage involves nutritional supplementation, with the aim of repairing the metabolic processes that are understood to be malfunctioning in children with autism, paying specific attention to problems with methylation and oxidative stress.

Chelation, the third stage of treatment, is also the most contentious. Originally developed as a treatment for acute heavy-metal poisoning, it involves the oral administration of molecules that selectively bind to heavy metals, which are then excreted through urine. The process is time-consuming and unpleasant: treatments must take place throughout the day, and a doctor must monitor the child to ensure that heavy metals are being excreted. DAN! practitioners emphasize that chelation only has a positive impact on a subset of children and that it should be carried out with nutritional supports, alongside treatment for intestinal dysbiosis, a yeast overgrowth that can be exacerbated by some chelating agents (Autism Research Institute 2005). The practice of chelation is supported by theories about the toxic effects of mercury preservatives in vaccines (although other environmental exposures to heavy metals are sometimes implicated), lending ideological valences to the choice to begin chelation treatment. Because of the risks involved in the treatment if administered incorrectly and other concerns about medical licensing, only certain doctors will oversee chelation.[13] These are frequently doctors who have a background in functional or environmental medicine and who believe, at some level, in a hypothesis of autism etiology that emphasizes the role of toxic environmental exposures.

All of these steps are supported by parental reports of improved functioning and behavior and the remission or amelioration of medical symptoms such as allergies and gastrointestinal problems. They are also connected, in complex and suggestive ways, to the findings of laboratory and clinical studies of nutrition and metabolism in children with autism, a growing literature produced by academic researchers who have become interested in the anecdotal results reported by DAN! practitioners. This literature proposes that the behavioral symptoms of autism are often caused by altered immune responses to certain foods, oxidative stress, and impaired methylation (which would impede the removal of toxic substances from bodies); potential functional impairment or even genetic vulnerabilities in detoxification enzymes such as methionine synthase or glutathione transferase; neuroinflammation; and overt gastrointestinal pathologies (Jyonouchi, Sun, and Le 2001; James et al. 2004; Vargas et al. 2005; Page 2000). DAN! doctors describe a group of children with bodies that are unable to sustain the daily burden of metabolism in the presence of toxic environmental substances and complex modern foodstuffs. The explicit aim of their collaborative efforts is to produce a metabolic and biochemical model of autism, and to this end, they speak of an "emerging picture" of the biochemistry of autism. Within this framework, designations of biomedical treatments as either "complimentary" or "alternative" are meaningless, because the methods promoted at DAN! conferences have the potential to redefine autism as a medical entity.[14]

Prior to each DAN! conference, presenters and other scientists and doctors gather at a "think tank" to address emerging issues, workshop the talks that they will be giving, and address potential revisions or additions to the DAN! guide. Their exchanges are energetic and occasionally contentious — the practicalities, rationales, and potentials of treatments are argued over in a precise biochemical jargon during sessions of brainstorming that approach the sublime. Many participants in the think tanks are also active members of the DAN! doctors listserv, a list which is used primarily to discuss treatment modalities and therapeutic options. Topics on the list range from attempts to standardize processes like chelation, to speculations from parent-scientists about disease mechanisms, to checks on possible drug or supplement interactions, to requests for help in interpreting confusing laboratory results. Doctors ask for advice on cases that are unfamiliar or recalcitrant to treatment. Some of the most broadly based researchers and acknowledged authorities on the list do not have conventional medical educations.

If doctors focus on the epistemological consequences of framing autism as a treatable medical condition and the practical implications for patient care, parents who attend DAN! conferences describe a process of laborious experimentation and therapeutic trials, characterized by incremental improvement and occasional sudden leaps in functional behaviors. Some of them turn to spreadsheets as a way to visualize the effects of the multiple variables in diet and medication against their child's daily behavior over days, weeks, or months.[15] "Katie," a parent quoted in the recent DAN! treatment guide, says that the most important thing "is daily record keeping": "The advantage of going to all this trouble is that I can look back and say 'Well, treatment A was wonderful for speech, but we also had a lot of night-waking while on it—so what adjustment might I need to make?' So much of what we are doing is trying to restore balance to our children's metabolisms" (Pangborn and Baker 2005, 17).

For both parents and doctors, the DAN! approach is a work-in-progress, a framework for approaching autism as treatable. As such, it is far from proscriptive and is highly responsive to promising new leads or suggestive data. The DAN! community exists primarily for the purpose of circulating knowledge about biomedical treatments, rather than for advocacy, but this does not mean that the circulation of knowledge in the DAN! community is apolitical. Members of the community see the interventions themselves as a politics, involving a choice about how to conceive of autism that upsets established structures of medical and scientific authority. The knowledge that circulates is also broader than "mere facts"; it involves craft knowledge pertaining to treatment choices and treatment preparation, such as tricks for getting children who would prefer to subsist on chicken nuggets to consume a diet consisting of rice, meat, and organic produce, or information about where and how to request and obtain compounded medications for "off-label" uses. Parents solicit and dispense advice of this type on a constellation of listservs, discussion groups, and newsletters that may be only tangentially connected to DAN! meetings and the Autism Research Institute. Although lists like the Autism Biomedical Discussion Group welcome parents who want discussion and feedback on medical interventions, the moderator warns parents in no uncertain terms that they can turn elsewhere for support and commiseration.[16]

Just as the DAN! guide and conferences inform discussions on listservs, parents who have experienced success using biomedical interventions reach out to parents new to the autism diagnosis through memoirs and handbooks, including *Unraveling the Mystery of Autism and Pervasive Developmental Disorder: A Mother's Story of Research and Recovery* (2003 [2000]), Karen Seroussi's account of treating her son with an elimination diet, and Lynn Hamilton's combined memoir and handbook, *Facing Autism: Giving Parents Reasons for Hope and Guidance for Help* (2000), both of which emphasize the importance of parental devotion and determination in the search for information about treatments. Jacquelyn McCandless's book, *Children with Starving Brains: A Medical Treatment Guide for Autism Spectrum Disorder* (2003), details McCandless's own journey from psychiatrist to autism treatment

THE CHELATION KID
STORY - ROBERT TINNELL
ART - CRAIG TAILLEFER
I'M NOT ARGUING WITH YOU. I'M JUST SAYING I DON'T KNOW HOW WE CAN BE SURE THE COD LIVER OIL IS HELPING.
I GUESS WE CAN'T KNOW FOR SURE. AND IT'S NOT LIKE I'M ANTI-SCIENCE OR THE SCIENTIFIC METHOD. BUT WE DON'T HAVE THE LUXURY OF TEN-YEAR DOUBLEBLIND CONTROLLED STUDY.
IF YOU'D HAVE THROWN THE WORD "PLACEBO" IN THERE I WOULD HAVE BEEN REALLY IMPRESSED.
* PRONOUNCED KEY-LAY-SHUN
E-MAIL : INFO@FEASTOFTHESEVENFISHES.COM
EPISODE 36

2. *The Chelation Kid*. From Tinnell and Taillefer 2006a. Courtesy of Craig A. Taillefer and Robert Tinnell.

specialist and DAN! doctor as a result of the diagnosis of her granddaughter, Chelsea.[17] McCandless and her husband, Jack Zimmerman, use the language of love as an explanation for what they see as a reciprocal process of healing that takes the form of diligent experimental medical practice as opposed to the allopathic model of medical treatment: "The children with starving brains challenge our capacity to love—particularly in regard to the qualities of patience and perseverance, devotion beyond the usual parental call of duty and the capacity to think and act creatively 'out of the box.' Often, it is only after a profound challenge to our capacity to love, such as the rearing of an ASD child, that we come to realize how much more there is to discover about loving" (Zimmerman, in McCandless 2003, 194).

Crossing Over: Testimony, Reflexivity, and Recruitment

DAN! conferences are oriented outward, toward recruiting and training new DAN! doctors and parents. Parents may offer to pay the registration fees for their children's doctors as an incentive for them to attend the conference and receive an introduction to the DAN! approach.[18] Attending a conference represents the beginning of a process that involves more than a simple commitment to treating autism using biomedical interventions. It also involves acquiring a specific vocabulary, learning to visually perceive symptoms that may be invisible to those not trained to pay attention to medical conditions in children with autism, becoming skilled in interpreting unconventional laboratory tests and parental reports of idiosyncratic behaviors, tics, or food cravings, and developing the confidence to suggest medical treatments that are regarded as questionable by much of the broader medical community.

Potential DAN! doctors do not see their embrace of biomedical treatments as a choice to move away from reliable medical knowledge. Rather, they experience their choice as an affirmation of medical principles that have been abandoned by allopathic medicine. Within the DAN! community, diets are seen as biomedical interventions, as much or more than the use of antipsychotic medications, because they are understood to address the pathological mechanisms of the disorder, rather than the symptoms alone. The collective sense of parents and physicians that the experimental approach of DAN!, though it may take longer to produce results than psychopharmacology, is a more certain way of addressing the underlying causes of the disorder is reflected in discussions among doctors and in promotional literature. For example, the Autism Research Institute's "Autism is Treatable" campaign focused on using advertising and public-relations strategies to promote the message that, simply put, autism is treatable.[19] DAN! conferences have included panels of parents explaining that most children will respond to some set of interventions. The book produced in conjunction with the campaign detailed tentative success stories, close relationships with advising physicians, and therapeutic optimism based on individualized treatment strategies (Rimland and Edelson 2003).

Conference presentations and treatment recommendations must be translated into practices and reproduced in daily routines if parents are to see results and become committed to the program of interventions. The methods of monitoring and care advocated at DAN! conferences train parents to see their children's symptoms as altered by variations in nutrition, medication, toxin levels, and immune responses. As parents learn to watch their children in new ways, they also learn to attribute changes not to chance or normal processes of development, but to positive effort. Likewise, practitioners using these methods are converted through the experience of seeing children "actually respond" for the first time since they have been treating children with autism; their tangible experiences help confirm their membership in the group.

Modes of interpretation are also built into laboratory tests and the interpretive skills of researchers. DAN! treatments depend on a network of independent laboratories with specialized tests for measuring the baseline and post-treatment levels of various enzymes, nutrients, antibodies, and heavy metals for purposes of diagnosis, treatment selection, and evaluating outcomes.[20] A typical set of tests might check for nutrient levels, both acute and delayed allergies, and markers of heavy-metal toxicity or yeast overgrowth. Doctors who treat autism using conventional approaches regard these tests with some skepticism, describing them as a way of exploiting parents desperate for information or, at best, as tools that are too poorly standardized to be useful. Insurance companies often fail to cover the tests, so parents are compelled to pay out of pocket. DAN! practitioners argue that threshold effects caused by genetic vulnerabilities in conjunction with environmental triggers can manifest in a variety of fashions and that these vulnerabilities and potential treatment strategies become visible through the patterns in lab workups. Under this interpretation, doctors have been quietly characterizing the biochemistry of autism for at least twenty-five years. One researcher commented,

> It is IMPORTANT, if present models are to change, that concrete data about the chemistry of those with autism is made a public record. Professionals with whom I have shared my own daughter's chemical workup have been miffed by what they've seen, because this sort of workup is not usually done; since these sorts of results are unfamiliar they are viewed as only a curiosity when they are seen individually. How can a neurologist become convinced he needs to know about his patient's immune/metabolic status unless the results of a great many of these workups is put to public inspection? How indeed, can we get reimbursement for these sorts of workups by our insurance companies if it is unclear what the data will mean once we get it because there is nothing to compare it to? This all hinges on publication to a wide audience.[21]

At a DAN! conference in 2005, a meeting was organized to introduce "mainstream" practitioners from the local community to the methods of seeing promoted at DAN! conferences, intended as a "bridge" between the two medical frameworks. The practitioners connected to DAN! gave short presentations, oriented toward the concept that individuals with autism have metabolic, immunological, and nutritional variations from typical populations, and that it is possible to intervene and alter these variations to produce behavioral effects, or, as the flyer for the meeting argued, "In the light of these findings, to view autism as a static encephalopathy may mean to overlook potentially valuable treatment opportunities."[22] Each presenter devoted their talk to one metabolic system. During the question period, one participant, an experienced chemist who had not encountered these methods previously, asked if he could use a sheet of butcher paper to write on and, as the presenters watched, he drew a diagram of the ways that the various biochemical pathways implicated in autism formed an interlocking series of reactions. One might, he suggested, intervene at various points in order to produce a response with effects throughout the system—a point that DAN! practitioners regularly make, but one the chemist constructed himself just from hearing the didactic presentations on disease mechanisms, even though these presentations had barely touched on treatment implications.

Via instruction in these methods of interpretation and vision, DAN! practitioners and families focus their efforts on converting qualified practitioners. With increased familiarity, families also work a process of conversion on themselves. Although "conversion" suggests a process based on faith rather than on experimental validation, I do not want to suggest a process of indoctrination or induction into "irrational" nonscientific belief systems, but rather a secular community that is acutely aware of the social context of systems of reason, and the difficulties inherent in translation and proof. DAN! doctors recognize that even if they have presented a rational alternative to conventional medical approaches to autism, the majority of their colleagues will not recognize it as valid because they have not yet learned to visualize autism in this new framework.[23] This understanding comes up frequently in conversation at DAN! meetings because of the reflexivity that is the dubious privilege of a marginal group; individuals whose testimony is continually questioned are more likely to consider the source of their own beliefs. DAN! practitioners struggle with the conflicting demands of maintaining their practices as clinicians, treating children based on detailed parent reports, the need to maintain their credibility and medical certification, and their desire to generate biomedical facts that are legible and credible to regulatory agencies and medical associations.

While the shared practical and affective dimensions of autism treatment, rather than a specific, shared, theory of causation or etiology, help to constitute this emerging community of parents and practitioners, a collective vision of autism as dynamic and treatable does indeed emerge from these practices. This vision is deeply practical. It involves attending to dimensions of the disorder that are physiological as much as neurological, including skin ailments, levels of "stimming" or self-stimulatory behavior and self-injury, exacerbations of gut disturbances or variations in mood, all of which may be mapped to different dietary or other biomedical regimens. This leads to a blurring or reframing of the diagnostic category of autism itself, from one based on behavioral features and an imputed underlying genetics to a (no less real) category based on "measurable factors in the immune, endocrine and metabolic systems," leading to "a diagnosis of biochemical cer-

3. *The Chelation Kid.* From Tinnell and Taillefer 2006b. Courtesy of Craig A. Taillefer and Robert Tinnell.

tainty rather than one build around behavioral things that may change as people respond to a whole host of different sorts of therapy."[24]

While it is possible to seek genetic causes for behavioral symptoms, the search for metabolic biomarkers and the insistence on "biochemical individuality," and that "every child is biochemically unique," points to a deeper reframing that goes beyond seeking interim solutions prior to a final, genetic, answer. Seeking explanations for autism at the level of cellular signaling involves a choice of perspective, potentially away from autism as a monolithic diagnosis and toward autism as systemic disorder with neurological manifestations in which "we do not know how many different kinds of underlying molecular, cellular, metabolic and tissue perturbations may lead to a connectivity disturbance sufficient to produce autism" (Herbert 2005a, 355).[25] This shift in perspective is an option for parents and physicians alike. Physicians welcome it as an intellectually satisfying way to approach the practical problems that they encounter as they try to create treatment programs for children with autism. Parents describe it as a consequence of their work as parents, where attention to individual needs is one qualification for the job.

Visceral Responses: Efficacy, Therapy, and Diagnostic Vision

The work of diagnosing and treating gastrointestinal issues in autism can illustrate the broader concerns related to characterizing the diagnosis, establishing modes of intervention, and developing new techniques for seeing in diagnostic terms. The use of vitamin B6 and magnesium, or subcutaneous injections of methyl-B12, or the elimination diets described in *Special Diets for Special Kids* (Lewis 1998) would provide equally apt examples: each intervention shares similar barriers to legitimacy in the medical community. While parents have long complained about the allergies and gastrointestinal issues experienced by their children, only recently have these issues and the possibility of a new diagnostic category of "autistic enterocolitis" been accepted by anything approaching the broader medical community, and those clinicians who have responded to the claims of parents have been slow to publish their results (e.g., White 2003; Kushak, Winter, Farber, and Buie 2005; Krigsman 2007).

Doctors are often hesitant to associate themselves with discussions about gastrointestinal problems in autism because of the public association between digestive symptoms, the measles-mumps-rubella vaccine, and the controversy that has surrounded scientific articles that have attempted to validate a connection between the two (Wakefield et al. 1998; Murch et al. 2004; Erickson 2005; Richler 2006). In contrast, parents claim on listservs, website testimonials, and at conferences that often the first (and sometimes the only) effect of a strict elimination diet is that their child's chronic diarrhea or constipation was resolved. Gastrointestinal problems have been commonly explained as results, rather than causes, of the neurological disturbance in autism. In order for children with autism to receive treatment for gastrointestinal issues, the problem and the symptoms literally had to acquire diagnostic visibility in children who often cannot communicate the cause of their distress.

A pediatric gastroenterologist was present at the first DAN! meeting that I attended. He seemed uncomfortable among the exhibitors and milling parents, dressed formally in a navy suit. Parents of his patients had requested that he attend, and when I spoke with him, he wanted me to understand that he was not necessarily in support of the broader set of ideas or concerns represented at the conference and was concerned about the claims made by some of the other physicians present. Six months later he was at a meeting on the opposite coast, willingly considering the possibility of gut issues in autism. He allowed that certainly many children with autism had exceptionally difficult gastrointestinal problems that could go unnoticed by doctors focused on their cognitive disabilities. Perhaps, he suggested, stress associated with a child's inability to communicate was the cause of the symptoms that parents reported, possibly as a result of the known neurological connections between the gut and brain (e.g., Gershon 1998). In another six months, he presented preliminary data at a DAN! conference on a set of children diagnosed with autism that he had treated with medications for bowel inflammation, even though he remained skeptical about any distinct gastrointestinal condition associated with autism. One video clip that was screened at this conference showed a child screaming and writhing on the floor, pressing his hands into his stomach. To the audience, the child was clearly expressing agonizing pain due to gut inflammation, and the group reacted as the presenter knew they would, with gasps of sympathy and knowing grimaces. To them, the child's suffering was obvious. To another audience, he would have been merely "tantrumming." Similarly, photographs that parents provided of children angling their bodies over chairs and placing pressure on their abdomens would be seen as "posturing" or forms of stereotyped behavior by an uninitiated audience. Parents knew that they were attempts to attain a degree of comfort and relief.

To the extent that gastrointestinal symptoms in children with autism are acknowledged by conventional practitioners, they are often dismissed as comorbidities or "just the autism." In the world of biomedical interventions, they are reinterpreted as food sensitivities. At my first DAN! meeting I was invited to join a group of young mothers from Louisiana while they talked among themselves, mainly about the bowel problems of their children. One participant leaned over to me and confided that, "If you hang around autistic parents enough, all we talk about is poop." For a community that understands their children's problems as arising at least partially from gut dysfunction, the facts of foul-smelling bowel movements, chronic diarrhea, and the subtle nuances of each become important observational diagnostic tools that are de facto a special province of parents.

Parents share the common currency of experience, but filtered through the conviction that this experience can be rendered legible. Gestures of pain are indeed communicative, and behavioral symptoms can be addressed at the level of malfunctioning physiology, even before the exact nature of the cognitive dysfunction is precisely understood. Technologies of intervention, like the technologies of faith, present tangible results prior to explicit comprehension. Neither parents nor practitioners see their recognition and treatment of gastrointestinal issues in autism as a comprehensive explanation for every case of the disorder, but they do see it as a partial solution to particular cases. Interventions, the success of which are predicated on a parent's willingness to serve as devoted recordkeeper and nurse, are valuable to parents because of their tangible, observed benefits, but they are also valued because they present a rational approach to an otherwise unmanageable experience, that of caring for a child whose moods and bodily functions resist understanding by those who ought to know them the best.

In any practitioner's transition from utilizing conventional treatment modes to more biomedically oriented approaches, such "frameshifts" in vision are especially important. For practitioners considering incorporating DAN! approaches into their practices, a key concern is the threat of liability for using "unproven" treatments, when even recommending elimination diets could lead to investigations by state medical boards. The choice to move away from accepted practice on the strength of individual observations and parental demand is one vital shift. However, the most important transition may be a shift in the relative weighting of the available information for treating apparently neurological conditions. One practitioner who considers a variety of perspectives in treating autism characterized her own experiences this way.

> A key transition in my own understanding of this disorder has been taking careful and thorough medical histories of my autistic and other neurobehavioral patients, and taking their physical complaints seriously. These complaints, once one learns to ask about them, turn out to be so common that it has become impossible for me to ignore them or assume that they are less important than the behavioral features. These children cannot be assumed to have nothing more than brain and behavior problems, since so many of them are also physically ill. A critical question is whether—and if so how—these things are related.[26]

DAN! doctors are by definition converts. While they have conventional medical or nursing degrees, doctors offering therapies outside the consensus treatments contend with identities that are compromised, if not outright "heretical" (Wolpe 1990; Wolpe 1994). Most doctors who are willing to accept the professional costs that come with offering alternative approaches emphasize the importance of listening to parents. Many see themselves as part of a movement with the objective of altering medical practice and with this, the prospects for children with autism. Those doctors who wish to maintain their legitimacy emphasize the specialized or disciplinary aspects of their work as neurologists, gastroenterologists, or immunologists who treat specific problems in children who happen to have autism; those who are less concerned with their institutional status come increasingly to identify themselves as DAN! doctors, first and foremost.

The Heart of the Matter: Desperation, Commitment, and Trust

When I first attended a DAN! conference, I shared the experience of many first-time parent attendees. I was utterly overwhelmed. I thought about this sensation in terms of trust, because this was the determination that I felt called on to make of the presenters and exhibitors. Who could be trusted? Trust and truth claims are the currency of modern science, and have been described in terms of a legacy of rules stipulating economic independence and invulnerability to influence as a prerequisite for scientific "witnessing" (Shapin 1999). "First-time" parents often leave the room during the initial slide lecture in tears, either because they are suddenly faced with the possibility of improving the prospects for a child that had been described as virtually untreatable, or out of sheer confusion at the unmediated outpouring of biochemical terminology and PowerPoint diagrams presented from the podium at the front of the room. I felt equally at a loss, and tried to cope by treating the problem as a task in science studies and asking myself how participants defined evidence and experience, what types of each they valued, and how and when observations and theories stabilized as biomedical facts (Shapin and Schaffer 1985; Latour 1988).

Later, I came to think of the problem in terms of persuasion followed by commitment—in other words, a decision to initiate a trial rather than a relationship of absolute trust on the part of parents. This led me to spend less time working myself into knots of doctors to listen in on their metabolic speculations, and more time chatting with moms in the lobby of the conference hotel, asking them why they decided to try biomedical treatments. One parent, who showed me a photograph of her dreamy-eyed daughter, said that when she first considered biomedical approaches her friends told her that only the desperate would try diets and supplements. She had decided that she was desperate enough when her daughter tried to jump out of the car in the center lane of traffic. Her story was not unusual. Parents grow weary of those that see acts of desperation and rationality as mutually exclusive, especially when it becomes evident to them that few experts recognize the difficulties of their daily lives *or* their ability to critically evaluate treatment options. Their decision to bring the work of medical treatment for their children into their homes reflects a rejection of the despair brought on by a negative prognosis for their child, an embrace of the idea that their encounters with their child can meaningfully shape his or her development,

4. *The Chelation Kid*. From Tinnell and Taillefer 2007b. Courtesy of Craig A. Taillefer and Robert Tinnell.

and a skepticism of the "legitimate complexity" of expert medical knowledge that itself has long been a part of American vernacular medical practices (Starr 1982).

Parents of children with severe genetic disorders are also seen as pushy and driven to distraction by the severity of their children's suffering. However, only the parents of children with autism are suspected of manifesting some kind of *forme fruste* symptoms of the same disorder, themselves. Parents of children with autism are questioned at every level, from their competence as parents (because of their presumed inborn lack of affect, inherited by their children, they are ill-equipped to raise children with affective problems), to their "perseveration" on the effects of chemical exposures or the limitations of available treatment programs. As the parents of disabled children, they are seen as desperate and vulnerable, willing to mortgage their houses to pay for unproven therapies. The image of the desperate parent is so persuasive that it can overshadow the identities of parents who are also researchers or practitioners. Although within the DAN! community, research on one's own child, professional practices, and clinical experimentation can run together, other professionals will go to significant lengths to hide the fact that they are doing research as a result of their child's diagnosis, especially when the research might be associated with a "hysterical" parent belief about vaccines or environmental toxins.

Parents take this suspicion of partiality and turn it on its head, explaining that being part of their child's life means coming to know their child in biomedical terms, if that is the means through which they can help their child. They are acutely aware that their status as caring relatives compromises their claims for objectivity, to the degree that the appearance of objectivity depends on a claim of moral neutrality (Daston and Galison 1992). In contrast, they argue that it is precisely their intimate involvement with their child and their absolute devotion to their child's well-being that allows them to understand how their child responds to treatments in ways that a professional observer with commitments to standard disease models and community standards could never aspire to.

Biomedical Pluralism and Invisible Labor

The combination of accusations that parents are compromised witnesses to their children's physiology and the demands of daily attentiveness associated with biomedical interventions for autism suggests some problems with discussions of contemporary illness-based social groups. Analyses of these groups often assume that communities spontaneously organize around disease categories through either an affinity born of shared experience or the desire for information. They also suggest that the need for advice or information about a condition can potentially be met by attentive medical practitioners, if those practitioners are persuaded to take the condition seriously. Finally, they expect that the work associated with the daily requirements of care for a condition is a "natural" response to being sick, or loving someone who is, and therefore requires no extensive explanation. All of these beliefs are reasonable, but in their evocations of spontaneous or effortless acts, none of them do justice to the knowledge and labor associated with complex, chronic conditions like autism spectrum disorders.

As Bernard Rimland recognized when he first began talking to other parents with affected children, autism only looked like a single condition through the lens of diagnostic instruments designed to view it that way. This story is, in some sense, the reverse image of the story told by Lakoff, where a psychologically complex condition was made to resolve into a biomedically concise characterization (Lakoff, chapter 8 in this volume). For autism, directed efforts to destabilize rather than refine the category, supported by an alliance of functional-medicine practitioners, not-for-profit research institutes, nutritional-supplement manufacturers, maverick researchers, and independent laboratories, encounter entrenched resistance from those whose research programs depend on the stability and reality of autism as a biologically "true" population. The efforts of parents and professionals using biomedical treatments to reframe autism as a chronic illness rather than as a fixed genetic and neurological condition are aided by the absence of genes, diagnostically distinctive patterns of metabolites, or dysmorphologies that consistently mark children with autism as a biologically separate population. The work that constructs autism as a stable population occurs largely at the level of qualitative observations guided by diagnostic questionnaires and checklists (e.g., Lord, Rutter, and LeCouteur 1994; Lord et al. 2000).

A focus on metabolic processes before genetics, environment before "innate biology," and development before hardwiring is a shift in perspective that renders certain processes more or less visible, certain interventions more or less possible, but does not negate any specific approach. DAN! practitioners incorporate genomics, psychopharmacology, and behavioral therapies into their practices. This perspective is replicated in a variety of ways. For parents, it takes hold almost incidentally, as an after-effect of learning to manage the disrupted biochemistry of their children. Practitioners acquire it as a result of their education and background, or their personal experience of the effects of treatment and their desire to help affected children. Doctors and parents refer to their techniques as "biomedical interventions" simply because they understand them to be continuous with techniques drawn from conventional biomedicine, part and parcel of the same system of reason and rationality. Neither parents nor practitioners using biomedical treatments have had much success explaining their approach to skeptical outsiders because of their relentless focus on the individual as opposed to the population and because of their preference for explanations drawn from functional medicine and biochemistry. Only recently have members of the DAN! community made an effort to reach the broader community of researchers and clinicians working on autism by conducting controlled clinical trials and publishing peer-reviewed articles and case reports with well-documented theories of pathological mechanisms (e.g., Adams and Holloway 2004).

In one world, parents sit with their child in a doctor's office, receiving a diagnosis that they have probably already suspected. They are told that autism is a fixed and lifelong neurological condition, which will someday, through the efforts of academic research and family contributions of genetic material, become susceptible to pharmacological interventions or prenatal diagnostic screening. During this visit, they may be given information that will lead them to take part in one of the autism genetics initiatives currently in progress. The parents may leave the office and immediately begin their own research, or perhaps time passes, their child fails to improve with behavioral therapies, and life becomes increasingly unmanageable. Either way, the parents enter another world, seeking advice through handbooks, the Internet, or exchanges with other parents. In this world, parents exchange diet tips and learn to administer vitamin injections, attend conferences, and talk tentatively about improvements and recovery. There are few reports of miracles, but there is a considerable amount of hope. In both worlds, the massive efforts of parents, involving attention, monitoring, care, and observation, are described as acts of love. Both perspectives are firmly part of biomedicine as defined in terms of biological knowledge or commitment to applying that knowledge in the service of biomedical treatments. Their incompatibility in the eyes of so many experts does not necessarily mean that one perspective is irrational or unscientific, but that it is difficult to come to terms with the idea that biomedicine itself can accommodate more than one set of methods or epistemological commitments.

The cognitive and physical effort of making room for those multiple perspectives is part of the affective labor involved in the private, home-bound work of treatment and care, the commitment to an experimental attitude that will yield results only sparingly and without the comfort of expert support. The caring labor involved in treating children medically is not, parents believe, simply treatment carried out by a loving parent, but a kind of care that can only be carried out with love because (in the words of one theorist of disability) "labor unaccompanied by the attitude of care cannot be good care" (Kittay 2001, 560; also Kittay 1999). Our tendency to naturalize caring labor as the only reasonable response of a typically loving parent to a sick child can obscure the varieties of care that are available to parents, and the rational effort involved in making choices amid so much variety. Caring for a child with challenging behaviors or medical complications involves almost endless labor, leading parents to wonder how anyone could suggest that the solution to their financial woes might be to get a "real job," as in Tinnell and Taillefer's comic.

Donna Haraway describes the value that humans attach to companion animals as "encounter value," the worth that individuals place in the connection between them and another living being (Haraway, chapter 2 in this volume). A conventional approach to capitalist investment would assume that people devote resources to someone else—animal or human—because of the rational possibility that they too will benefit in the long run; therefore, the argument proceeds, parents spend themselves into debt exploring therapies for their children in the hope that their children will reward their investments. Any parent knows that such a line of reasoning is nonsensical: parents care for their children out of love and experiment with treatments because they appear to help, even a little bit. It would also be easy to conclude that the resources that parents sink into what are seen as unproven treatments for their children is a symptom of the search for a lost encounter with their children, who (according to self-advocates with autism) are not so much sick as irreducibly other (Sinclair 1994). The parents of "The Chelation Kid" would beg to differ. For them, the central encounter takes place through their ability to see responses to treatments and to work, steadily and unwaveringly, toward recovery. For these parents, faith is not promissory or endlessly deferred (M. Fortun, chapter 11 in this volume), but is found in the experimental practices of care and treatment and in the incremental gains accrued along the way. Parental faith is not based on the promise that any given treatment will lead to a certain recovery, but in the act of testing interventions.

Biomedical treatments for autism, in their pragmatism, tentativeness, and relentless individuality, represent another way of seeing. They present a claim of intimate knowledge against the medical certainties of hereditary disorders and standardized diagnostic questionnaires, or, put differently, a faith that emerges from and is indeed inseparable from the practices that it dictates. The practical and epistemological choices that emerge from this perspective are at once far-reaching and painfully limited. Affective experiences do not circulate along the same channels as other facts: you have to try them out for yourself. As users of an alternative economy, comprised of functional-medicine practitioners and nonconventional forms of testing and laboratories, advocates of biomedical treatments will therefore continue to face skepticism. The "private policy" of healthcare is simultaneously fraught with meaning, based in practice, and often limited to the treatment of one's own children in one's own home or to the clinics of physicians with staggering caseloads, and for that reason, the reasonable experiments of parents and their more far-fetched acts of faith will continue to be indistinguishable to many observers.

Notes

I am grateful to Kaushik Sunder Rajan, Kris Peterson, Susan Lindee, and Martha Herbert, as well as all of the participants in the Lively Capital workshop, for their comments on earlier drafts of this chapter. Any errors of fact, misstatements, or omissions are, of course, my own.

1. One example among many in which researchers have credited their insights to parents is the ongoing research on gut microbiology carried out in the laboratory of Sidney Finegold, who attributed his research program to the observations and research of Ellen Bolte, the mother of a child with autism. Bolte contacted Finegold in the process of seeking a doctor to treat her child with vancomycin, an antibiotic,

based on her hypothesis that some of her son's symptoms were due to neurotoxins produced by gut flora. Both publications resulting from this research list Ellen Bolte as a coauthor. See Sandler et al. 2000, or the earlier study, Finegold et al. 2002. More recently, researchers have provided validation for parent claims that a subset of children eventually diagnosed with autism regress between their first and second birthdays, as in Werner and Dawson 2005, and that children with autism show significant improvement in communication and behavior during fevers, as in Curran et al. 2007.

2. A number of researchers have made this observation, but I am drawing mainly on Martha Herbert's ideas (for example, Herbert 2011).

3. In making the connection between Hays's work and parenting children with autism, I draw on Stevenson (2008, 199).

4. The acronym for Defeat Autism Now!, DAN!, is commonly used in conversations among parents and some practitioners, but is never used in the organization's own materials, where the full name is always spelled out. I have chosen to follow spoken usage here. The organizations that I describe in this chapter, including the Autism Research Institute's Defeat Autism Now! conferences, now called Autism Research Institute Conferences, have in some cases changed significantly since the chapter was written in 2005. For a more current description, see Silverman 2011.

5. See the National Autism Conference website, http://www.nationalautismcon ference.org/; the Autism One website, http://www.autismone.org; the Unlocking Autism website, http://www.unlockingautism.org; the Talk About Curing Autism website, http://www.tacanow.org; and the Generation Rescue website, http://www .generationrescue.org, all of which promote biomedical interventions, although they have slightly different orientations in terms of advocacy. A number of clinical research and treatment centers explicitly research biomedical interventions as well (with varying degrees of emphasis), including Thoughtful House, http://www.thou ghtfulhouse.org; the Center for Autism and Related Disorders, http://www.center forautism.com; and the University of California, Davis, M.I.N.D. Institute, http:// www.ucdmc.ucdavis.edu/mindinstitute. All websites accessed 10 February 2008.

6. Somewhat ironically, Rimland's tentative suggestion has become the basis for a vast number of research programs. Rimland wrote, "If it should turn out that early infantile autism is biologically—not psychologically—determined, then the only logical way of accounting for the parents' unusual personalities and intelligence is biologically. This would mean that basic personality structure may, in at least some cases, be far more closely tied to the biological makeup of the individual than has heretofore been realized. Unless autism can be shown to be largely psychogenic in origin, or the evidence presented by Kanner and others concerning the parents can be disqualified, the conventional view that heredity must invariably act only in general and unspecific ways as a determinant of human behavior must be reconsidered" (1964, 41).

7. See "Appendix: Suggested Diagnostic Check List" in Rimland 1964, 219. The checklist included questions about gut problems; reactions to "bright lights, bright colors, unusual sounds"; "unusual cravings" for certain foods; the possibility of loss

of verbal skills after initial acquisition; and fine motor coordination, along with more conventional diagnostic measures for autism, such as pronominal reversal and insistence on sameness. "Form E-2," a parent assessment of "drugs," biomedical treatments, and "special diets," has been produced and collated by the Autism Research Institute since 1967. They report that they have collected over 27,000 of these forms, designed to contrast the efficacy of a variety of treatments. Also see "Parent Ratings of Behavioral Effects of Biomedical Interventions," available at the Autism Research Institute website, http://www.autism.com/pro_parentratings.asp, accessed 28 July 2011.

8. Rimland maintained that his own son, now a successful illustrator, was atypical from birth and did not appear to be vaccine injured, but he thought that this was not the case for many younger children.

9. At one DAN! conference (fall 2002, San Diego), Rimland told a well-known joke that fell resoundingly flat. After a moment of embarrassed silence, the audience erupted in laughter, realizing that his inability to deliver the joke (a reference to the tendencies toward mild "autistic" traits in parents of children with autism) was in fact the joke itself.

10. During a talk at the Spring 2005 DAN! conference, Dr. Sidney Baker described listening to parental testimony as an "ethos" that involved being conscious of the fact that there are "two people in the room," and he spoke movingly of the need to learn to "translate" the intuitions of parents into "technical language" that the rest of the medical community could learn to respect (April 2005, Quincy, Mass.).

11. The DAN! guide was originally referred to casually as the "DAN! Protocol," but following concerns about appearing to offer a standardized (or clinically tested) protocol as opposed to a series of suggestions to be experimented with on a patient-by-patient basis, the authors have been careful to refer to it as the "DAN! guide."

12. The Specific Carbohydrate Diet was originally developed as a treatment for inflammatory bowel disease by Elaine Gottschall, in the course of researching ways to treat her daughter's intractable case of ulcerative colitis. During the early 2000s, parents of children with autism began experimenting with the diet to treat their children, especially those with severe bowel symptoms. See Gottschall 1994, and for parents using SCD for children with autism, see "Pecanbread.com: Kids and SCD," available online at http://www.pecanbread.com, accessed 20 February 2008. One of the main sites promoting the gf/cf diet is the Autism Network for Dietary Intervention, online at http://www.autismndi.com, accessed 20 February 2008.

13. In a widely reported incident, a child with autism died while receiving an infusion of a chelating agent. Investigators concluded that the doctor in the case had ordered the wrong medication due to confusion about the names of compounds and that the child had not been given one of the conventionally used chelation agents. See Kane 2007.

14. Because their metabolic mechanisms can be described in conventional scientific terms, community members would separate DAN! treatment methods from, for instance, homeopathy or craniosacral therapy, although many families explore such options alongside biomedical treatments. Within conventional medical parlance,

"complementary" treatments are used alongside accepted or mainstream therapies, and "alternative" treatments are used as a substitute for these treatments. Biomedical treatments for autism might be considered complementary, as most parents use them alongside a behavioral treatment program. However, since no other medical options exist (aside from psychopharmaceuticals to control extreme anxiety or antipsychotics such as risperdone [Risperdal] to control self-injurious behavior), and because parents using DAN! approaches see those approaches as central rather than peripheral to their treatment program, the categories do not hold up as well as they do for other medical conditions, such as cancer (see David Hess 2003).

15. Brenda Kerr offers an example of the Microsoft Excel spreadsheet that she uses, on the Autism Research Institute website, http://www.autism.com/ind_track-progress.asp, accessed 1 March 2008.

16. Of the 1,175 autism-devoted groups listed in Yahoo groups as of 25 January 2005, 46 were devoted to the discussion of "biomedical" treatments for autism, meaning that biomedical groups comprise a little under 4 percent of the autism groups on Yahoo—it is important to remember, however, that many of the total number of groups are region-specific support groups or are devoted to specialized topics. Many of my comments on the types of discussions which take place among parents in the community using biomedical treatments for autism are based on "lurking" on the "autism biomedical discussion group," or abmd, a Yahoo-based discussion group, for approximately two years. Parents frequently build close relationships via the listserv, and I have witnessed meetings at conferences between parents who had known each other (and each other's children, by proxy) for years, but had never met in person. New subscribers are given a list of guidelines, including the provision that "this is a DISCUSSION list, not a SUPPORT list," the request that "off-topic" messages on topics such as behavioral therapies should be marked as such, and the instruction that participants should "try to stick to the topic of biomedical factors in autism and avoid ranting about [their] disappointment with the mainstream medical community, the government, etc.," along with the rule against "flaming" or written attacks against another member. Guidelines from "File—abmd List guidelines.txt" from abmd-owner@yahoogroups.com.

17. The most recent (2003) edition includes an additional literature review and discussion by Teresa Binstock, an independent researcher diagnosed with Asperger's Syndrome, who is active in the DAN! community and widely respected by both parents and practitioners.

18. For many years, the Autism Research Institute maintained a list of DAN! doctors, although they eventually abandoned the practice because of the difficulty of reviewing applicants and ensuring that they were reliable and had the requisite training.

19. The campaign was created in collaboration with a public-relations firm. An example of the basic platform can be found in the congressional testimony of Bernard Rimland (Rimland 2003). A form for reporting successful biomedical interventions is available on a website run by the Autism Research Institute, http://legacy.autism.com/treatable/autismistreatable.htm, accessed 3 August 2011.

20. Some laboratories, such as the Great Plains Laboratory and Genova Diagnostics, have actively courted the autism market, while others specialize in functional medicine and include autism among multiple disorders for which they offer work-ups. The head of the Great Plains Laboratory has published a treatment manual that makes use of protocols similar to the DAN! guide and contains an initial chapter citing DAN! doctors and listing many of the hypotheses current in the DAN! community, including concerns about vaccines. See Shaw 2001. In contrast, doctors associated with Metametrix Clinical Laboratory (http://www.metametrix.com), which advertises itself as "Innovative Solutions for Integrative Clinicians," have published a manual on specialized laboratory tests for functional medicine, to encourage and guide their use and interpretation by physicians. See Bralley and Lord 2002.

21. Susan Costen Owens, posting on St. John's List for Autism and Developmental Delay, 6 July 1997, autism@maelstrom.stjohns.edu (shared by and quoted with permission of author).

22. DAN! Bridge Meeting "Save the Date" flyer, "Treatable Features of Autism: Implications of New Findings in Autism Biochemistry and Immunology," for meeting on 15 April 2005, Quincy, Massachusetts.

23. One parent with whom I spoke compared her own research to her experience of the "Jesus Movement" in the late 1960s "Bible Belt" states. Specifically, she related her acquisition of an ability to read and interpret scripture with the faith that revelations would appear by tracking thematic continuities between Old and New Testaments. Her own intensive reading of published texts in biochemistry for their particular relevance to autism (even if their relevance, or the relationships between adjacent fields, was not apparent to the authors themselves) seemed, to her, to have parallels to this practice.

24. Letter from Susan Costen Owens to Jon Shestack, dated 1997 (shared by and quoted with permission of author). The full quote reads: "You and I probably would agree about the need for autism to become a diagnosis of biochemical certainty rather than one built around behavioral things that may change as people respond to a whole host of different sorts of therapy. I think that the only thing you and I may see differently is what may be the fastest and most effective method at getting to a 'physical marker' which could replace current methods of diagnosis. I very much feel that studying the measurable factors in the immune, endocrine and metabolic systems would yield faster results, as opposed to looking at something as incredibly large as the human genome, and looking at only those persons from multi-incidence families who may not represent at all those whose autism appears without precedent in a family."

25. Jon Pangborn and Sidney Baker, founders of DAN!, both use the phrase "biochemically unique" in talks and presentations, although they may be implicitly or explicitly referencing the book by biochemist Roger Williams on the differences in vitamin requirements among individuals (Williams 1956). Cited in Rimland 1976.

26. Martha Herbert, personal communication, 19 September 2010. For further development of these concepts, see Herbert 2011.

13

MICHAEL M. J. FISCHER

LIVELY BIOTECH AND TRANSLATIONAL RESEARCH

In memoriam: Judah Folkman, teacher, healer, scientist

As children, we liked to take walks with our dad. Along the way, he would point out trees, buildings, stop signs, and ask us, what else could those objects be? A lamppost would be a toothpick for a giant, we'd exclaim. Or a tree could be his broccoli. Dad called this game symbolism, showing us the world as a fascinating place of potential.
LAURA AND MARJORIE FOLKMAN, EULOGY FOR JUDAH FOLKMAN,

20 JANUARY 2008, TEMPLE ISRAEL, HARVARD MEDICAL CENTER

For the spring 2008 class, Introduction to Global Medicine, at the Harvard Medical School, I asked Judah Folkman, pioneer and champion of the field of angiogenesis inhibitors, to update his translational-research lecture to focus on the angiogenesis work that his former Chinese postdocs were pursuing in China. (Angiogenesis is the blood flow to tumors that keeps them growing; if one can find the right combination of angiogenesis inhibitors, one can make the tumors regress.) Folkman was eager to add the China dimension. Always charismatic and a highlight of the course, he never said no.

Folkman died in 2008. Just before he died, the first textbook on angiogenesis was published. At his funeral, the long procession of his postdocs and lab alumni testified to a capillary network into the future, fragile, powerful, and lively as life itself: Folkman's lively capital. They included leaders in many fields, from oncology to tissue engineering and drug delivery, as well as researchers attempting to solve the enigma of LAM (lymphangioleiomyomatosis), which had become one of Folkman's concerns at the time of his death. This chapter, which I dedicate to his memory, draws in part on his translational-research lecture, composed in 1999, when his research had just survived a report, which made headlines, that his results had not been replicable at the National Institutes of Health (NIH), for reasons he outlines in the talk. At the time of the lectures he gave in the late 1990s, he was not yet calling the results of his research "translational," but that buzzword disseminated through the NIH system and came, by 2000, to be the way he described these talks in which he understood perfectly the desire in our global-medicine class to explore the social and ethical issues of the biosciences and biotechnologies.[1]

In this chapter, I attempt to draw together how emergent forms of biomedical laboratory life fit into the series of transformations that medicine and the life sciences are undergoing, from discovery to invention of new biologies, from "slash, burn, and poison" therapies to minimally invasive and regenerative ones, from bounded craft science to "good hands" and "green fingers" reaching "through" the computer screen with interactive graphics to networked databanks.[2] I am looking for new epistemic and material-semiotic objects, sites of deep play, gearshifts between macro-, molecular, and nanoworlds (among proteins, enzymes, polymers, and cells), shifts and differences between in vitro and in vivo environments, and among experimental systems provided by nematodes, flies, zebra fish, mice, and humans. I am interested in shifts of scale from laboratory products to clinical-trial prototypes to market-scale volume manufacturing with high-quality control. I am interested in the translations (and failures of communication) that occur between fields of expertise, in the "translation" from bench to clinic, and in the translations across medical systems, national competitive ambitions, and the dances of cooperation and competition that fuel (and slow) the global republic of science. This is a large agenda, toward which this chapter can provide only a few immortalized cell lines to experiment with, to watch as they might culture into robust repair systems and healthy growth (or not).

While the nodes of interest are new epistemic and material-semiotic objects, sites of deep play, and shifts of scale or frame, the substance of interest is the people at these sites and their social and cultural relations. From my perspective, as an anthropologist, a criterion of acceptability excludes the admission, as a recent review of actor-network theory admits, that the theory doesn't handle personhood or culture or people very well. This is not a call for journalistic "human interest" or "genius scientist" writing, but a meditation on the mix of detailing—organizational, historical, intellectual-genealogical, political-economic, material-technological, as well as the singularities of events, passions, and competitive bootstrapping—required to upgrade ethnographic work on the biosciences so that it can be in conversation with the understandings of practitioners.

Lively capital might be understood to include a least four kinds of capital: venture, corporate, or government capital; intellectual capital, in inventive new legal "structured instruments" and safe harbors for risks of exploration; the symbolic capital created through public-relations promissory ploys; and scientific capital, both in institutional depth and in the play of the scientific imagination. All four of these—financial, legal, symbolic, scientific—are "bankable," at least as discountable investments, and play important roles in the flows, speed bumps, and rechannelings of biocapitalism. They are thus central to each of the "culturings" (*passages* is the biological term for separating out a few cells in a new vessel and medium to ensure continued growth of the cell line) or translational arenas addressed here. Judah Folkman's lectures on translation—ten kinds of ethical and practical problems, a decalogue of lively dilemmas never standing still—form the hinge or centerpiece of challenges in getting anything from lab to clinic.

I then proceed to other kinds of culturings, which complement and expand the ten selected by Folkman: translations from lab skills to stable technique; multidisciplinary translations; translations between academe and the market of small biotechs and big pharma; geographical and transnational translations of researchers and research trajectories.

If you listen and watch carefully around the edges, in the corridors and pubs, and between the lines of scientific reports, you can see Alice in Wonderland other worlds beyond the residues (the numbers, models, laws, arguments) by which we claim to understand what science is. These other worlds are lively: full of quirks and rituals, superstitions and fetishizations, all-too-human transferences, competitions, and collaborations, mentoring styles and politicking games, translations and shifts in scale, and financial incentives and other surprises. The other worlds, no less than the laws of nature, the flow of funds, and the organization of laboratories, are what make science work. The interplay (and incompatibilities) between the way scientists speak and the way engineers speak runs through technoscientific exchanges; one is improvisational, like jazz or the neutral third microtones (*koron*) in Persian music, while the other is functional and standardizing. Their interplay is most evident in fields like synthetic biology, where biologists and software engineers must interact, the first being attuned to the lively biological world that incessantly make experiments fail in new, surprising, often illuminating, ways, the other being intent on creating tools and building blocks that will always work. Such counterpoint interactions are necessary throughout the life sciences for knowledge to advance. The music of science talk is enchanting. Much of it is lost in translation in the writing of sociologists (who standardize and anonymize) and journalists (who substitute metaphors for detailed connections). It is ethnographically valuable, at least for a time, to transcribe the songs of science, paying attention to their metalinguistics, their jousting encounters, their impassioned bets and investments, their tactile and rhythmic physical dance, their joy, and their imaginative harmonics, not only their content.

Translation (across science fields and technological scales; from bench to clinic, or from green fingers to stable techniques and scaleable production; and from experimental therapy to standards of care) and *capital* (financial, legal, symbolic, scientific) are the two linchpins of both the Alice in Wonderland worlds and the indirection of scientific opportunistic development. Both are lively, fluid, ever-moving.

The Decalogue and the Network

Translation to the clinic proceeds along a pathway strewn with obstacles and it's actually harder than making the discovery because it involves other people, and involves all their different views.

JUDAH FOLKMAN, 21 SEPTEMBER 1999

In a wonderfully comic but serious account of the changing sites of medical discovery, Judah Folkman lays out ten reasons, a decalogue, for why it is nearly impossible to get discoveries from the lab into the clinic.[3] He begins with a quick survey of the shift from discoveries made by physicians at the bedside in the nineteenth century and early twentieth, to the need for laboratories and the rise of the physician-scientist often fostered by training at NIH in the mid-twentieth century, to the growing difficulties involved in maintaining the dual career of physician-scientist and the disjunctures caused by new discoveries coming increasingly from basic scientists without much clinical experience and unaware of many important clinical clues.

It is no accident, he suggests, that angiogenesis research began in a surgical laboratory (his), not in a molecular laboratory.

[This is] because in a molecular laboratory cancer is studied in a dish, and so, my very good friend and colleague Bob Weinberg thinks all his life of cells that are flat, and I think all my life that cells are crowded like they are in cancer, and then we talk about it occasionally, and now he's studying angiogenic genes. This is why there came the development of the HST program, the M.D./Ph.D. programs. It came from this problem of the two cultures that are getting further and further apart, and so the attempt is, can you, your generation, solve the problem of the disappearing physician-scientist that was a very fruitful career but seems no longer possible. Physicians worry about leaving the whole next century just to the pharmaceutical companies without the clinical clues going on.[4]

Folkman describes a series of different career paths among his colleagues who maintain both labs and clinical practice, and the difficulties they face.[5] But chief among the difficulties is "translation." Indeed, he estimates, some 40 percent of the medical-school faculty would say their obligation is only to do basic work in the laboratory and to publish. "They say you have no obligation to get into the problems of translating to the clinic because it is a political morass. . . . [T]he other half of the faculty feels that if you find something that could improve the care of patients, it would be unethical not to help in some way guiding it into the clinic." As his case study, Folkman uses his own experience with

a new protein that has anti-cancer activity, that was in fact published in *Science* on Friday [17 September 1999], discovered in our lab, a very powerful internal fragment of antithrombin 3, a clotting protein. The fragment itself is a potent angiogenesis inhibitor, very powerful anti-tumor agent, no toxicity, many tests in animals, the whole sequence is known, the genetics is known, and it's actually being made now as recombinant in large quantity. So that was published. So we say, well, that's a really good protein. That should be in the clinic. There are many patients dying of cancer—do you know how many there are? . . . If you take out the suffering, the death rate, which we can count, is 1,500 a day in the U.S. That's one a minute. If it's melanoma, it's one an hour. It's 565,000 deaths per year. The number of new patients with cancer, newly diagnosed, is 1.2 million, and in Europe twice that. The number of current cancer patients being treated is eight million.

The decalogue of obstacles includes (1) patenting, (2) the break up of collaborations, (3) "their culture has to fit your proteins," (4) internal company rejection ("it's always in the cafeteria"), (5) transferring skills (the Stradivarius problem), (6) Fridays are cancellation days, (7) clinical trials, (8) "for your little protein, change a million dollar manufacturing process?," (9) physician resistance, (10) stock manipulations. (There are, of course, more, but ten obstacles were all that could be addressed in an hour-long lecture.)

1. Patenting: "All my postdocs say, 'I don't want a patent on it; I just want to publish it, and I don't want to sit and talk with lawyers about patents.' Why do you have to patent it?"

A student ventured, "To exclude. If a company wants to take it, they want to know that they can exclude everyone else from taking it."

Folkman: "Correct. So we have ten micrograms. We have the gene. We have the CDNA. We have all our assays. And we say to a company, in this case Genzyme in Boston, will you make this for us? We would like ten kilos, enough to treat a hundred patients for one year each. No problem, do you have a patent? No patent, they won't even talk. They hang up. They won't even have a meeting. So you have to have a patent."

But that's hardly the end of the story. Nor is it merely that an ethical shift has to occur institutionally and professionally, but the returns (profits, benefits) and risks are modulated by time.

> So, the medical school and the hospitals therefore each has an office of technology transfer staffed with patent attorneys. . . . Before 1974 the school did not have a patent policy. The policy of the school was publish it, give it to the public domain. MIT has had a patent policy since 1900. So when Harvard finally got a patent policy, they got a big telegram from Jerome Wiesner who said, "Welcome to the twentieth century." [Laughter.] He was the Provost of MIT. The view was that it's not ethical for a physician to patent something because you shouldn't be making money based on the patient's illness. Then they found out that it was actually okay, because most of the patents don't make any money. In fact, nobody makes much
>
> patents,   money on these kinds of patents. And the reason is—it's different
>
> money,   than MIT: if you patent computer software, you can make money—
>
> and time   but if you patent a drug, by the time the FDA has approved it, eight

years, by the time it's gone into the market and has been allowed to be used, it's fifteen years, the patents have run out. So, it's a complicated problem.

2. The Break Up of Collaborations.   The second problem comes with the assignment by the lawyers of one of the coauthors of a scientific paper with the status of inventor. This assignment of legal inventorship is a classic transition from gift economies of affluence (it is immoral to own knowledge, knowledge should be public goods) to capitalized economies of scarcity (patents need to be defended, assigning only one person the status of inventor is a defense against weakening of exclusive claims, and property rights are a moral foundation of society), the accent being on the ethical transitions.

> They won't allow coauthors because they say if we put coauthors on it, the patent is weak and somebody can challenge it and win on infringement, because they can prove these other people didn't have anything to do with the idea. They may have helped with the chemistry, they may have done other things. So after the attorneys have gotten through ruling that one of the coauthors is an inventor and the others are not, the whole group breaks up. . . . If you had collegiality, it goes away in one day. . . . It's like wedding invitations: if you send out wedding invitations and you forget Aunt Minnie, she never talks to you. So you have to be careful.

A student asks if there are not equivalents to prenuptial agreements. And Folkman agrees.

> Yes, exactly. So it turns out, if there is income from a patent in the future, from licensing agreements or other things, there are, there can be. But the philosophy of this school and other medical schools is that the purpose of a patent is to bring something to the social good, not to make income for the school, because it's a poor way to do it for a medical school. It's a very good way for an engineering school. So it turns out that if there is possible money involved, what can you do? What we do is explain to all the authors that we will simply ask the attorneys to write an agreement that if there is any money to the inventor, it will be divided equally among the coauthors. That's been our tradition. That actually works.

Another medical student is still unconvinced, "You said it would ruin the collegiality, but why?"

> Well, it's amazing. You have four authors and you have the first author, and the first author may not be the inventor, because there are different rules [about authorship]. So the first author is really upset that the inventor, who is going to get possibly all the benefits of being the inventor, because nobody knows what they are—you license the patent, they have to pay a licensing fee; you get into the clinic, they have to pay a milestone fee—it goes to him. And what he did was do the sequencing. But the first author thought of it. But for some reason, the attorneys determined that [it was] the sequencing that was the first demonstration of the principle—it's a whole different vocabulary. . . . The last one, we've had a ten-year fabulous collaboration with a professor at UCSF. We have a merit award that's gone ten years. We worked together on the phone; we sent things back and forth; we published together. And then they sent us some of their cells from which we found a totally unexpected new protein. Harvard patented it, and we said coinventors. The attorney said, "Nothing doing": just because they sent you the cells doesn't mean they're inventors. So it made a lot of hard feelings, and we wrote out that we just would equally distribute [whatever proceeds if there were any].

Here the scientists are at the mercy of the attorneys, even if they want to be amicable, and the attorneys in turn are thinking ahead about challenges to the patents: in Folkman's words, a patent "will always be challenged if it is worth anything." It is neither that Harvard is playing hardball, nor that the attorneys are being unmindful of their clients' egalitarian feelings. Rather, if a patent is valuable, "It will be challenged and infringed upon, and they will say, 'Sue us.'" In court tests, patents show their weaknesses and can be invalidated. For a research-group leader, these "facts of life" mean the need always to repair relationships. "I just tell you the end result is you lose colleagues. You have to get them back. So a lot of my colleagues . . . refuse to patent anything."

3. "Their culture has to fit your protein." Academic scientists need companies to manufacture larger quantities than they can produce in university laboratories and to bring it through the regulatory process.

First you have to find a company that—their culture has to fit your protein—that sounds funny. It turns out that big companies prefer to make small molecules, and small companies prefer to make big proteins. So the biotech company in Kendall Square—there are a hundred twenty-nine biotech companies hanging on for their life in Boston. They have no income. They have three more years of funding. The burn rate is going to close them in two years. They have a fabulous new protein, a professor who invented it . . . and they are trying to make it. And then there is Merck which has a rule—only small molecules that you can swallow because they don't want to get into the problems of big molecules. So they'll look at your protein. They'll say "too big" or "too small."

Because "discovery is unscheduled," debts exert pressure to find a company whose culture fits your protein.

I'll give you an example. In 1991 Michael Roddey in our lab discovered angiostatin, an internal fragment of plasminogen, one of the first of the angiogenesis inhibitors that could regress a tumor. All previous ones could hold them or slow them, and we were simply going to continue working with it after we published it, but here's what happens when you have a new discovery in a laboratory and you're funded by National Cancer Institute grants. You go into debt right away because it is unscheduled. Suddenly you have to have fermenters and you have to buy different equipment and you've got to have new mice and you have to hire somebody who does protein chemistry. So what you do is, you take money from all the other grants, and immediately you write a grant and hope that in nine months it will be funded and you'll pay back. So we did that, but the study section said you will never be able to purify this, it's parts per billion, it is so powerful. It'll be hopelessly impossible to purify it. So they turned the grant down. So now, nine months down the road, we had a $250,000 deficit in a big lab, and the hospitals won't cover that. No one will cover that. So what they do is start turning off your heat, light, salaries [laughter], and you can sit up there and look out the window. So what do you do next? We thought, this is a terrific discovery, just because the peer reviewers don't understand it, we'll go ahead anyway. So what would you do?

grants, unscheduled discoveries

and debts

On cue, a student offers the salvationary word *companies*.

Right. So I wrote to a whole series of companies, all of whose CEOS
had been my students. Skolnick, president of Merck, president of
Upjohn, president of Lilly, all Harvard Medical School. And I got
appointments right away. And they said, oh, I remember your lectures and everything [laughter]. However (this is 1991), what would
we do with an angiogenesis inhibitor? Nobody understood it. Only
the scientists understood it; it had not yet permeated to the pharmaceutical companies. So after a year of that, I went around to the
biotech companies. They understood it right away, they have the
most advanced, youngest scientists; they're not yet so conservative,
because they haven't made any money. So they said, terrific idea,
but we're broke. They're all in debt. They fear it will take resources
away from their idea. So a company is a zero-sum game.

networks

risk-
taking
and debt

In the case of angiostatin, there finally was one company, EntreMed, "a tiny
company, about eight people, and they had just become capitalized, nothing
to do yet. So they took on this project . . . and in the bargain we got endostatin and a couple of other things that were discovered because [EntreMed]
gave a grant to Children's Hospital."

In the case of anti-angiogenic antithrombin, the discovery reported in
Science, 17 September 1999, the match with Genzyme came about in a different way.

It's a fragment of antithrombin, and when it was discovered to be
such a powerful inhibitor floating in the blood, and isolated and
purified, and we sequenced it, we looked around the world for companies that make antithrombin the parent, so we could buy that
and make it from the source rather than purify it from the blood.
There are only two in the world and the biggest one is Genzyme on
the Charles River, opposite the [Harvard] Business School. So we
went over to them, and it turns out that they've been making antithrombin 3 for nine years and selling it in Europe, and testing it in
the U.S. It's for people who have a deficiency of this protein. If you
don't have enough antithrombin 3, you have free thrombin floating
around, you get clots in your head and your lungs, you get pulmo-
nary emboli in your legs. These people were always clotting, so they
are given intravenously once every two weeks a protein called antithrombin 3 until their level comes up to normal. It is a sort of bou-
tique thing because there are not that many hundreds of patients.
. . . Unbelievable because, see, they make antithrombin 3 in goats

sourcing
match

transgenic
goat
factory

. . . transgenic goats. They have transgenic goats, the only ones in the world. So they have taken the gene and hybridized it to casein, which is the milk protein. It is a huge sterile farm, blessed by the FDA, no viruses. Then they milk the goats and they get five hundred milligrams per liter. The best fermenter is seventy-five [milligrams per liter] if you are lucky, with *E. coli*. Each goat is a factory. So they can make any protein by this technology, and they hold all the patents. So, they are making antithrombin 3, and they take the milk and ship from Framingham to the Charles River [plant] where they purify it. Now they have to pasteurize it and that gives them: seventy-five percent of it active. But twenty-five percent is inactive and has to be thrown out.

technique
match

That's our protein! So what they said was, our protein is fifty-three kd, our technique the internal fragment, theirs is fifty-eight kd. The way you get there is sixty degrees heat. Match. They heat it a little longer, add citrate, and they've got ninety percent. So they said, why of course we would love to work with you! [Laughter.] Let's have a meeting tomorrow. There was the president: it's his company, he founded it.

biotech
to pharma
threshold

It was a tiny company of four people, now it's a six hundred million dollar company. Six hundred million dollars is the danger point: that is where you can still be bought out and swallowed up by Merck or somebody. But if you get to a billion, it is too big and you can go all the way to be a pharmaceutical firm. Like Genentech. They made it, but most everybody else was eaten up. So he said, hmm. . . . So it is a very good match.

4. Company cultures ("it's always in the cafeteria").   Getting the leadership of a company to adopt the project is not a guarantee that the rest of the company will go along.

zero-sum
games

There are company cultures, and within the company there are certain groups that do not want to work on that idea . . . a company is a zero-sum game, and if the president says, oh, we're going to work on this new protein, it's as if he brought home a second wife. They say, wait a minute! [Laughter.] The president says we're all going to get behind this. Oh yes, in his presence, everybody is gung ho. But then comes the undermining. It's extremely subtle. It's always in the company cafeteria. I hear it's not working. That's the rumor. Then they talk to the press and they say, well, it's from a lab, you can't reproduce the results. It's incredible. You have tremendous, tremendous rumors. And they're afraid that resources will be siphoned off.

The pattern is "so common that some companies have solved it a different way. Johnson and Johnson for example."

risk shelters

If you have a new idea in Johnson and Johnson, and you wanted to do it—[I'm speaking of] three years ago, it's now public. . . . What they do is form a new company. They go around to all their young stars and they pick a young, brand-new vice-president and they say how would you like to be president of this company? It's going to be in Belgium and you're going to have a hundred staff and a total budget for three years. And if you can make it, it'll be sold worldwide and you will be established in the big company. . . . And so Johnson and Johnson has a hundred and seventy-one companies. . . . They never started in-house.

Folkman's experience at Bristol-Myers was a negative case example.

We went to them with angiostatin and there was a whole group that just didn't want to do it. They said we should make cytotoxic chemotherapy. Taxol, Taxol. We're making a billion and a half dollars. We should not go into angiogenic therapy. The president kept saying: have you ever read about IBM? They were making mainframes. How long do you think we will be able to make Taxol if these [angiostatins] come in? He kept saying we have to have this in the company. He forced it through, but it got pretty much undermined. So it is really amazing what you learn from all this.

5. The Stradivarius problem: transferring skills.   Next is the classic problem, in science, of teaching another lab or a company, problems joked about in corridor talk and in such terms as having "good hands," and analyzed more fully in the chemist Michael Polanyi's lectures on *Tacit Knowledge* (1966), the sociologist of science Harry Collins's account of physics labs (1974), and in the historians of science Steven Shapin's and Simon Shaffer's account of the seventeenth-century experiments conducted by Boyle and the Royal Society (1985). The transfer of skills is a standard problem in scaling up from experimental systems to industrial production, from systems that are still return-

ing knowledge through their instability and need for skill to reliable, highly quality-controlled processes.

> So now they say, gee, we'd love to work with you, bring your proteins and your genes and all your technology to our lab. They have never grown endothelial cells. No problem: we'll teach them. It is a big problem. It has to be idiot-proofed, because you write out the methods and . . . they say it doesn't work. How could that be? You have published it. They call you up, you have given them great detail; they say, we cannot reproduce your results. Other people may call. If you are a basic scientist, Ph.D., and you get a call, we can't reproduce your results, it is a chilling, sleep-losing experience. It implies that you have manipulated the data. That is the implication. And many of them are angry and they call collect. [Laughter.] So someone has taken that call and you see them, they're pale white the next morning. They're drained, they can't do any work. They are worried that their theory is over. Now to a surgeon it is a source of pride. It's just the opposite. [Laughter.] Surgeons in the locker room refer to this as the Stradivarius problem. Too bad about surgeon Jones, that his mortality is twice as high as mine. . . . They see it as a skill problem, and most of the time that is what it is.

For example,

> We said, okay, here's how you grow endothelial cells. They had very good gene people, they had molecular biologists, they had done tissue culture. But not endothelial. And so we wrote it out exactly for them. And they did not grow. They have grown all over the world. There are twenty papers a week, and they did not grow at Genzyme. And it turns out the method said, make up the media and the trypsin and the gelatin coating with glass-distilled water. Deionized distilled water kills endothelial cells slowly, so you don't know it for about three weeks. . . . It's been known for twenty years. We went down the check sheet. They said, we don't have glass-distilled water at Genzyme. The whole company is on deionized water, and they could not get anybody to order a glass still. So we said, okay, here's a bottle of glass distilled water from our place, and we made a CARE package and they took it to Genzyme in Framingham, and it worked. Then there was a meeting and they decided to order glass stills.

In a second example, the process of troubleshooting generates new knowledge.

> Then we were trying to teach people how to inject these into mice which have tumors growing. The tumors grow at a certain rate; you measure them every day; then you give the inhibitor and [the growth rate] should come down. This is now the National Cancer Institute, and they said . . . it's not working. Can't reproduce your results. They also talked to the press and everybody else because they don't want to be at fault. . . . So we had to troubleshoot, and what do you suppose was the problem? It didn't work for six months, then they came up to our place in January and it worked fine. They stayed nine days in a hotel, came every day, injected theirs, we injected ours, and it was perfect. Theirs were actually better. Now try to troubleshoot that. You don't know, are they coming in on weekends? They are supposed to: they're not supposed to miss any days. . . . Are they having different people inject? . . . There's a million things. It took us about five months to out what one of the problems was: . . . room temperature.

expensive
mice

> Mice have their tumors grown on their backs so you can see them. We do have mice with the tumors inside, of course, to prove that it works, but that's really expensive because you sacrifice one every day to measure it, and mice are expensive. This way you just measure. They grow on the skin and the blood vessels come from the skin vessels. So now what do you think is the problem?

literature
experiment

> Temperature control in mice: mice cannot sweat, mice cannot shiver. Those are your two main ways of controlling your temperature. . . . A group at MIT published a paper some years ago which is very striking: they built a long cage and took a copper bar that is one meter and at one end it was zero degrees and at the other end it was fifty degrees, and it had a complete gradient, and they put mice in and they all ran to twenty-six degrees. They just huddled around twenty-six degrees. [The investigators] put in a baffle and pushed the mice to twenty-eight degrees, and the mice climbed over, and went to twenty-six. They pushed them to twenty degrees and they ran for twenty-six degrees. In fact, when they were forced to be at twenty, they all huddled; they wouldn't move, trying to save their heat. They love twenty-six degrees. Simple experiment: no gels, no PCR. [Laughter.] It turns out that because of that, our rooms are

always at twenty-six degrees. One of the [NIH] rooms we went to

experience

was at twenty degrees, and we said, we can tell this is not right, because in our rooms we sweat when we are in there for hours inject-

experiment

ing mice. These rooms were air-conditioned. So we did an experiment: we set one room to twenty degrees, and in the twenty-degree room the tumors did not grow, because the animal, in order to keep his core temperature at thirty-seven degrees, has an outer coat, and in that outer coat the only thing he can do to save heat is to close off

theory

the blood vessels, vasoconstrict [as our fingers do when they turn blue from the cold]. . . . These animals are vasoconstricting and the

confirming

tumors can't get any blood, so they don't grow: it is another proof of anti-angiogenic therapy, a beautiful proof.

Now do you know what it is to get the room temperature changed at a government facility? [Laughter.] You have to fill out a form, that has to be taken to your supervisor, who has to get an appointment with building maintenance, that has to go to . . . can't just go in like

network

we do in the laboratory, take off [the lid], unscrew it, and change the temperature. MIT is the most liberal [in this regard]. Bob Langer was a student of ours in 1975, and we were making heparinase, because you couldn't buy it, to try to understand the fragments of heparin, and he had a bacteria that made it, and it was very hard and very expensive. He discovered if you added a little bit of heparin, that induced it to make heparinase. It was a beautiful thing. They needed a big column that was about eighteen feet high to make this, so they drilled a hole in a laboratory at MIT, and they had this column, sticking up through the ceiling into another lab, making heparin. It was a fantastic night and day thing, very exciting. The president came by because he had heard about it, and [Bob] gave him this talk, and he got more and more interested. He was sitting down and asking how many liters. Finally he got up. He had come for disciplinary purposes, [but] he said, Langer, just don't let it happen again. So they are very liberal there. You couldn't drill a hole anywhere at Harvard that I could imagine.

6. Fridays are cancellation days: time-investment gambles.

Every Friday they cancel some projects in all big companies. . . . In a large company with many competing drug candidates in the pipeline, they cannot have more than eight at any one time. . . . If they have some other candidate like an antibiotic or an antihypertensive that is suddenly doing well in the clinic and is closer to market, and if the company puts all their resources behind this other drug now, it will be better for their balance sheet and the analysts' sheet . . . then the decision is to stop everything else. . . . And suddenly your little project, which is so fantastic, is stopped for two or three years, and there is nothing you can do about it.

In dealing with companies there are ways to write contracts and licensing agreements.

You have to negotiate for every foreseeable possibility, which you have to think of, because they don't offer it. You have to say, suppose you don't make a particular milestone by two years from now. Suppose they decide to sit on it to keep competitors out. It's a business decision, not a moral or any other kind of decision. Then we get the right to take it back and there is a penalty.

In the case of negotiations with EntreMed, the company licensed angiostatin in 1991.

time Actually it's a world-record time. They licensed it in 1991. We only had a few micrograms, so at that time there was no gene or anything. By '94 angiostatin was published in *Cell* [October 1994], it took all that time. Nobody knew how to make the recombinant. . . . By '96 we had the recombinant. And then they could make it. The problem then was, what in? Yeast? *E. coli*? You've got to try all these things and look at stability, cost, scale-up, and that took another year. And it turned out that endostatin, which had been discovered later [*Cell* 1997] was easier, and will be out in the spring. So that's about right. People forget. Proscar, Merck's drug for prostate [took] twenty-one years from the time it was discovered in the laboratory to the time they could figure out how to make a structure that was stable in a bottle. It is really complicated. That's why those drugs are expensive in the beginning. There is no four-year drug. The AIDS drugs, which were done at warp speed, were eight years, and they had everybody working on two shifts at Merck. They paid exorbitant fees for overtime. And still there is a shortage that is massive. They can't keep up with it, they can't produce it enough. People are on waiting lists. Nobody in Europe can get it. It's a big
lottery problem. Herceptin, the same way: there's a lottery, a breast-cancer

lottery so you can get this good drug or not. Genentech is trying to build a factory, but they can't make it fast enough. It takes a long time; once they're up and going, it's okay.

7. Clinical Trials.   Clinical trials present a series of complicated ethical and practical problems. "The entire course could be on the ethical problems in clinical trials," Folkman begins. There are a "whole set of medical ethical rules," but the set of ethical problems are not solved to everyone's satisfaction. Placebos, dose escalation, deployment of hope, selecting who gets into trials, patient migration to higher-dose trials (so earlier trials cannot be completed), protocol violations, protecting physicians who decide who gets into the trials, shortages, and difficulties of timely scaling up as soon as good results are assured—these all present complexities and dilemmas.[6]

Folkman is dramatically masterful in the way he poses an obvious solution, only to raise a question about it, and in how he laces answers with detailing so vivid, listeners are unlikely to forget, not the rule, but the trade-offs that produce the rule. Among the enticing loose ends: the ethical rules don't work in entirely the same way in Europe as they do in the United States.

placebos   For example, the best scientific way to do a study of compound A is to have a placebo, because many people, as soon as you give them a pill there is a certain improvement, and you can read that as part of the drug you are testing. So normally half of the patients are not treated. They are given nothing because there was nothing before, if this is a new drug. There is no alternative. What is the problem with that?

Female student: If the drug was a cure, all these people are dead because they did not get it.

Folkman: Right. So if you have diseases like skin diseases and you are trying to test a new drug for psoriasis, placebos are ethically allowed by the FDA. So one patient can get the actual drug and the other patient will get a paste, and then they have crossovers, so that both patients have a chance at it. After the patient with the placebo has had three months, they get to try the drug. And it may be that

cancer   neither is working. But what about cancer?

Male student: I think it depends on how old the patient is and how far along the cancer is and what kind of cancer.

Folkman: Right. It depends on the clock, because the clock is now ticking very fast. Different than our clock. So the point is that when the disease is moving fast and is fatal, they generally, the ethics say: no placebo. The reason is that by the time a cancer patient can even be a candidate for a clinical trial, they have to have become refractory to all conventional drugs. By that time an adult would weigh eighty pounds, they are very sick, they have no hair, they are vomiting, they have metatastic cancer, they have pain all night, they cannot sleep without drugs, they are in the last year of their life, and they cannot be on a placebo, ethically in this country. In Europe, they can, and there are placebo trials in Europe. But in this country the Food and Drug Administration will not allow it. You would have to make a very good case. Okay, but then, how do you get any data?

Europe

Another student asks about angiostatin, which seems to have no side effects versus a drug that has hard-to-tolerate side effects.[7] Folkman responds with a possible conflict between law and ethics.

Okay, so angiostatin has not yet been in clinical trial, but in all the animal studies and all the monkey studies and all the FDA studies, there have been zero side effects with either angiostatin or endostatin. So therefore the FDA has said it is so safe—they have found nothing in years of testing in monkeys—they will let it go ahead into the clinic earlier than usual, but they still will not allow a placebo, because they feel that it may work, and some patients would be untreated. So if a patient is going to get into a clinical trial and be off all other drugs, they do not want a placebo. Anyhow that is just the law, and, of course, it can be argued, but right now that is the law and the ethics sort of disagree.

"But the worst problem," Folkman continues, "is dose escalation," how to tell patients that they are not getting effective treatments, how to manage patients cherry-picking trials, and how to protect the physician committee making the random selections.

The FDA says, new drug, we have never tested it, yes it is safe in mice and rabbits and monkeys, but we have never tested it in humans. So they say, tell us the effective dose in different animals, and they figure out the effective dose per kilogram or meter squared. And they say, lower it fifty times and start treating patients, and every week you may go up five percent. Dose escalation. What is the problem here?

B.G.: They think they are getting treated.

Folkman: Right. So you have to tell the patient that we do not know the dose, and the FDA does not want to start out at what you think is an effective does for animals, because it could be a toxic dose for humans, and you could have some sudden deaths. So you start low. You tell the patients in writing and verbally and holding their hand: look, you are the first in the world. We have been through all this before with other drugs in other trials. This is a low dose, but each week we are going to ramp up, and you tell us if you have any side effects at all, keep a little diary, and when we get to an effective dose, we will let you know, but you have to know that we are on now what we think is not an effective dose.

animal models are not human models

Student: How do you prevent them from taking more?

Folkman: They get it intravenously in the hospital. They cannot just go home and take it. So you have this ethical problem: you have to tell them and they have to know it, but what you are relying on is their hope that it will work for me at that dose, because they are desperate. This is a very difficult ethical problem for a physician. I think it is one of the worst ones. You are giving a subtherapeutic dose and you know better. When you do it in children, it is terrible.

hope

"An even worse problem: who do you select?" Folkman recreates the multiple building pressures.

The Dana Farber announces that on September 27 or before October 1, they will begin endostatin [trials]. They have enough to treat one hundred patients for one year. It costs seven million dollars, the first trial. If it works the price will go right down because you can build a factory. This is all done in a pilot plant.

Student: With something like cancer that is fast and fatal, could you scale it up faster if it seems to be . . .

Yes, Folkman responds, pointing out that there are milestones that can be moved up if things are going well, but, he continues, "here is another interesting [problem]": patient migration to higher-dose trials.

They started in Boston on an earlier drug of ours, at very low dose. The FDA says the tumor name does not matter: all solid tumors [can be included] because it is not like Taxol for breast cancer. It is endothelial cells. So try it on all tumors. The dose [increases] are going slowly: after eight months, they are still not up to the effective dose, but they say it looks safe. They had their meeting, and we will open another study in Houston. [Houston] will start [at the dose level] where Boston left off, or where Boston is, and they will increase the dose at a little faster rate. We will open a third trial in Wisconsin, and a fourth . . . because it is safe. They just do not want a lot of unsafe ones going. But then what happens these days?

It is an extraordinary new thing. All cancer patients are on the Internet. There is a colon cancer club, a breast cancer club, a prostrate cancer club, and they have about a hundred thousand people in them. So now everyone wants to be in the Houston trial, not the Boston one.

The students are compelled by the urgency of cancer patients versus the slowness of trials. A students asks about what would happen if someone "accidentally" increases a dose, especially if it works. People laugh, and although Folkman first jokes that law is interested only in process, while medicine is interested in the results, he proceeds to remind everyone of the seriousness of protocol enforcement.

If you make any violation, they call it a federal citation that one of your doctors has gone off the protocol. . . . They first stop the study at the institution. They prevent all further studies at that institution, and they start taking away grants. Very stiff penalties. So it is very tightly controlled. Even if you give an abstract at a meeting and you have not gotten permission from the FDA, say ten patients were on it and they all had complete regressions, but you didn't check it with the FDA and you go to a public meeting, they stop the study because they say, if you are going to be on an FDA-approved study, you have to follow these rules.

The students probe: what about surgery?

Surgical techniques are done on the spot if you are in a difficult situation. However you cannot just do an experiment on somebody because you think it is a good idea. You would lose your license, your hospital position, and you would probably be sued. There are very strict guidelines to keep physicians from doing that. If you want to try something, you go and get permission, but it is a long process. Your peers have to approve it. [In the question period, a student posed a question about performing fake surgeries as a placebo control, raising an interesting difference between Europe and the United States, with direct implications for angiogenesis research.][8]

What about going to another country?

> So they test in other countries and it gets into the *New York Times*: so and so is testing on poor patients in Third World countries. It does not look good. Many of my colleagues have suffered enormously from that because they wanted to know earlier. It is a good way to end your career.

How do you protect the physicians who select the trial patients?

> When you are selecting patients, if you say, we will just take volunteers, the FDA sets the end points. They say it must be patients with any of these cancers who have become refractory to all known therapies. So, they have had Taxol and [the cancers] have grown through it; they have had radiation therapy, and [the cancer] is still growing. They have had Adriamycin, Cytoxa, platinum, it is still growing. Nothing is working and they have been given up. They are now called terminally ill. Second, they must have an end point, something you can measure. Their prostate-specific antigen is going up in their blood, their CEA for colon cancer is going up, or ovarian marker, or spots on a film that are increasing every month that you can see. There are thousands of patients that have those, so they all sign up.
>
> Then a poor committee has to pick them randomly. But Professor X at Mass General has a thirty-year-old daughter—these are true cases—with ovarian cancer, could you please take her first? A donor who gave a chair to Harvard says could you please take his son first? An NBC announcer, Katie Couric—this is public knowledge—and Sam Donaldson, public knowledge—could you take me first? So now you have all these problems and of course you have to take everybody in a row and you have to take them randomly, because the newspapers are just waiting to say that Harvard treats only VIPs, Harvard treats only its donors. Front page, *Wall Street Journal*. So the physicians on the committee that is selecting are anonymous. They are like the FBI witness-protection program. They're protected from being deluged by the patients.

When time is running out, shortages exacerbate the problems.

Let's say [the clinical trial] works. Herceptin is an example. It worked. It actually turns out to be an angiogenesis inhibitor, but it was not known as that. It is an oncogene inhibitor: it stops breast cancer, has some side effects, but is better than any other breast-cancer drug we have. It was announced in May a year ago at the American Society of Clinical Oncology. They said the clinical trials were opened up and it worked beautifully.

The company is Genentech: their stock has doubled. They had only enough for a thousand patients, because they cannot afford to build a factory until it works. So they start to build it. But the factory is not going to be producing for two more years. So there is a waiting list. So poor breast-cancer patients are on a lottery, a national lottery, and there they are, with their tumors growing, and they are terrified. And each one calls you. We get two hundred calls a day. Once you have a fatal disease and your clock is speeding up, what happens is people narrow their goals. I just want to see my daughter married. I just want to see my son graduate. I just want to see my first grandchild. If I could just make it to that. They do everything they can, and they're on every lottery, they're everywhere, but there are shortages.

time and they are terrified. And each one calls you. We get two hundred

It is really horrendous, because there is no way to speed these up. It is all economics: these are ethical and economic problems. Physicians just hate to get into that, because our view is, we should be able to treat anybody we want with the best there is, with anything we have.

Student [urgently]: Why can't you just bring in some venture capital to help bring up the factories?

Folkman: Oh, you can. It comes in. As soon as it is announced that it works, the venture capital pours in. The problem is building it and getting production and having FDA inspections is an enormous effort and cannot be done like building a high school. You have to have fermenters and you've got to have quality control.

I'll give you an example. Monsanto, one time, was racing to do some veterinary thing and they built fermenters and suddenly all of the drug had polio virus in it. They had to pull out. It turned out that in the pipes they built—they built these fermenters like making beer—they were making a protein, this was twenty years ago. They had stainless-steel piping and every month they would flush it all off and push steam through it to sterilize it. But the pipes had angles in them and in those little dead ends, there were pockets that didn't get the steam. They finally figured it out and they had to make all the pipes curved. It almost broke the company.

8. "For your little protein, change a multimillion dollar manufacturing process?" The eighth problem involves the manufacturing complications, without which you do not have a therapy.

A drug can fail in the clinic for reasons that have nothing to do with the beautiful science. The drug works: it stops endothelium, the blood vessels stop, the tumors regress in all animals and people, and you do everything you can, and you put it in a patient's intravenous and it does not work. The tumors are growing. It takes a while before everybody figures out that this particular protein likes to stick to this polyvinal plastic that happens to be in the tubing and this is a chromatography column. Now what do you do? First it took a long time to figure that out, although with experience that is what you go for first, but good judgment means you made a lot of bad judgments first. So what do you do?

You can either change the protein or change the tubing. Try that. There are only three tubing manufacturers in the country. Abbott is the biggest. Change our tubing? Are you out of your mind? For your little protein, we're going to change a multimillion-dollar manufacturing process which has plasticized this, and we worked it out, and it is FDA approved, and we haven't had any problems? We're not changing our tubing. Make your own tubing.

FDA and Catch-22

So change the protein so that it doesn't stick? Should you change the icing? If you could do it, the FDA will say it is a new protein. You have to start all over! It is like Catch-22. When I retire, I'm going to write this book because you have got all these government agencies. So, it can stick to plastic, it can precipitate into dextrose. And you have to work all that out.

As with the polio in the pipes, details can have consequences, and again changes tested in laboratory animal models are not good human models.

allergies and little things

But the worst problem is that all the [laboratory] mice are identical twins and people are not. If you tested in field mice, you would be fine, except it would take forever. So people get allergies that you could not even have predicted. So there are little things.

The most dramatic event to be discussed came next, a long-standing problem that had affected the NIH's inability the previous year to replicate at first Folkman's experiments, and it got solved by a member of the class: shipping.

> Finally [in all the back and forth with the FDA, NIH, and experimental biological materials], there is the problem of shipping it. The most amazing thing is that all proteins and drugs are shipped on dry ice because they are stored at minus eighty, that's the best place to keep them, but when you ship them on dry ice—and we learned this to our horror—that can in fact damage the protein. How could that be? You can freeze it in the lab and get away with it, take it to minus eighty in freezers. Up to a certain time it works fine. Dry ice is minus seventy-five, minus eighty. You ship it and [the protein] is dead in about four hours. It doesn't work.
>
> *Student*: Ph . . carbon dioxide . . .
>
> *Folkman*: How did you know? You guessed it?
>
> *Student*: I've heard it somewhere that it is a problem. The buffer with the carbon dioxide.
>
> *Folkman*: Right. It has been a problem for years and years. The paper was just published the end of August in *Nature Medicine*. It turns out that no matter what, first of all, people don't put dry ice in glass because the glass will crack when it warms up and they do not want the drug to be contaminated. So they put it in plastic. All plastic, carbon-dioxide gas goes right through it. So this simple experiment published in *Nature Medicine* was: there was a company and they were puzzled that they would send people titers of adeno-viral vectors at ten to the minus eleven, but when it arrived it was only ten to the fourth titer units, and they could not figure out what the losses were. So they said something is happening with Federal Express. They are leaving it on the dock or something. So they put it in a Federal Express box and put it in a car and drove around for two days, and one day, and it turns out that three hours is when it becomes inactive. It took a long time to discover what you just said, so I wondered if you had invented it yourself or read it. [Laughter.] One's genius, the other's just a good memory. So it turns out they put pH markers in, they put buffer, they buffered it at nine, and it dropped to five, and that is as low as things could go in three hours. So any protein that cannot stand that change gets agglutinated or

unfolded or damaged and denatured, and that happens to a lot of them.

Students were intrigued and began discussing solutions. To ship in liquid nitrogen is not so easy. To temporarily ship in glass has not worked because the screw top cannot be tight enough and the carbon dioxide goes up in there. Gas masks work on different problems. Folkman said a number of inventors had appeared since the article.

An inventor in France read the article and developed a one-square-foot, minus-twenty, electrically driven, battery-operated thing that you can carry and ship in, and it will hold minus twenty for about three days. It is beautiful. We said, could you bring it to us, or we blown up will come and see it. He said I will bring it to you. We want to buy it in the and test it. He was arrested in Paris trying to get on the plane with airport it because they thought it was a bomb [laughter], so they blew it up. Honest to God! We get this call at six in the morning about three weeks ago. This is Perletti. I am stuck in Paris and I do not have my icebox. I'll make another one. He'll probably mail it next time. But minus eighty is hard. If anybody can invent a minus-eighty shipping canister that is not carbon dioxide, that would be fabulous.

So where are the MIT students? You guys and ladies ought to be able to do that in the shower.

Female student: Will I get tenure?

Folkman: I would certainly help, I would come before the board for you. If you solve this problem, you have solved a huge problem for the pharmaceutical companies. The reason is, they never had proteins this problem until they had proteins, but a lot of these proteins do do things things the small molecules do not. They're not toxic, they circulate small in your blood, they're stable, they don't hurt you, but give Adria-molecules mycin, the heart is damaged, the hair falls out. So that is another don't problem. And if you solve it, my number is 355-9661. Call me first. [Laughter.] You patent it, we will help you develop it. You can be the inventor, and we'll be the coauthor.

Same student: I just want tenure.

In fact, the problem was solved by a student who came up after class that day. Within months they had the device patented, prototyped, and ready to go.

9. Physician Resistance.   A ninth problem is having to "educate physicians when you have something really new, because they are threatened, because it takes so long to learn how to do anything really well in the clinic." Key examples: Ignaz Semmelweis trying to teach people in Prague to wash their hands before delivering a baby, to prevent the transmission of streptococcus (childbed mortality was 30 percent), and being kicked out of the hospital for his efforts; Howard Florey, in 1941, unsuccessfully trying to persuade military physicians in England to use penicillin for soldiers dying of streptoccal septima, and having to come to the United States before people would listen; resistance to using Tagamet for bleeding ulcers at Massachusetts General Hospital, in the 1960s (when Folkman was chief resident), instead of gastric surgery (a procedure that no one performs or knows how to do any more); and Proscar, the prostate drug, made by Merck, that did the same for prostatectomies, but only because Merck mounted an ad campaign showing older men who insisted they would go to other doctors if their primary physician refused to prescribe it for them.

Another minor problem is introduced by the insurance companies, which can continue to define a treatment as experimental; as long as a drug is considered experimental, insurance companies are not required to pay for it. More insidious is another element of the market.

10. Stock Manipulations.   "Then, finally," Folkman concluded, "there is a worse problem. And that is if it is a small company like EntreMed, and it starts to work, or they have some success, what happens to them in the business world? There are short sellers. These are not things you learn in medical school."

> It is amazing. EntreMed is a small company. It is hanging on. You put on the Internet under Yahoo Finance a rumor—have you heard—you put it as a question, then the SEC cannot hurt you—have you heard that they are having trouble manufacturing, scaling up, or whatever. It is not a statement, it is just a question. Next guy signs in, is that true? Thousands of hits, is that true? Is that true? Suddenly their stock is down and they have to sell the company, or it can be bought, or they can buy the stock and it will rise on another rumor. Have you heard about this new publication that says it is okay? And that goes on all the time and is going on now. It is incredible. There are a lot of greedy people.

Skills Translations: Green Fingers, Good Hands, Tacit Knowledge

When Folkman talks about the Stradivarius problem, he is referring to the translations of skills across labs. Some of the difficulties of this can be solved by material troubleshooting (relating to lab temperature, nature of the growth medium, packaging of samples, etc.). But sometimes the problem of translating skills can be more nebulous, what crystallographers call the "green fingers" problem.

"Green fingers" for chemists and structural biologists, like "green thumbs" for gardeners and "good hands" for experimental physicists, or tacit knowledge and skill in the lingo of philosophers of science, open the doors to the Alice in Wonderland human worlds of science.[9] Green fingers, an Italian postdoctoral crystallographer says, are like cooking skills: "For one person it crystallizes, for another it doesn't" (interview with EP, 7 July 2005). The same batch, the same protocol, and some manage to push the expression to a very high yield, while others cannot manage to get such high yields. Crystallographers have their rituals of procedure which they cannot explain. There are, the postdoc concludes, "whole layers of details pertaining to the way people work." It is, he repeats, like cooking, a rhythmic feeling for what goes together in what proportions with what motions. Labs are full of people who do not have green fingers. The postdoc continued: "I have a friend who is an M.D.: you can tell, you know, the way he sets up his experiments, you can tell they are not going to work." They were talking one day about food, and his friend said, "'Oh, you Italians, you have a tradition of good food,' and he asked me, do you cook? And I said yes, I like to cook. And he said oh no, I don't like to cook, and then it made sense. He said to me in my family we don't have any cultural tradition of food, you know, we don't have, that's what he said to me. At that point I realized where the connection came."

The dance of embodiment is how structural biologists teach their craft; they speak with their bodies, folding their arms and sometimes their whole bodies to mimic their molecules, to capture both their structural folds and their movement over time. The structural biologist Jiahuai Wang demonstrates with his hand (p. 412) how his team solved a key component of the structure of human CD4, the receptor on helper T-cells for the human immunodeficiency virus, HIV (Wang 1990). Structural biologists are trained to use their bodies as well as their hands to demonstrate and viscerally imagine protein folds (Myers 2007). This works in pedagogy, in discussions among colleagues, and most creatively in solving new protein structures by "reach-

1. Jiahuai Wang, a structural biologist, uses body language to demonstrate a structural discovery regarding human CD4. Photo by Michael Fischer.

ing through" the computer screen and feeling how the folds must fall into place, a "feeling for the organism," as biologists at least since August Weismann have said.

Folding the body in mimesis of the protein foldings associated with solving protein structures is a distinct skill from that of crystallizing proteins ("having green fingers"). Not everyone with the one skill has the other. Various imaging models are also used, including atom-by-atom balls, electron-density maps, and cartoons such as the one on the monitor behind Wang. The cartoons are particularly useful for describing folds. Nowadays these are three-dimensional, rotatable models. Wang is showing where a ridge on CD4 provides a hotspot binding site for HIV glycoprotein 120, the initial step in viral entry leading to a fusion of viral and T-cell membranes. It involves the key binding residue Phe43, at the top-right corner of the CD4 molecule (highlighted on the screen), forming a mini-beta ribbon. Glycoprotein 120 competes with MHC (major histocompatibility complex), binding a thousand times more strongly than MHC, covering more surface area of CD4, and thus disabling the helper T-cell's ability to fend off the virus. Hence the name AIDS, acquired immune deficiency syndrome.

The Italian postdoc, with slight embarrassment, admits, "I designed a dance based on the way that my molecule worked. Sometimes I'll make my friends laugh about it, because I dance the way the molecule moves." He points out that capturing movement or "resolving time" is a particular challenge.

> There is a lot of interest in movement and film now in structural biology. Because up to now, until a few years ago crystallography gave a very rigid picture of molecules which we know is false, because

molecules in real life can be moving, and actually it is through this movement they accomplish what they are supposed to do. In some cases it is just vibration around an equilibrium position. . . . But in the case of machineries, like splicers, big machines that form a huge task, like photosynthesis . . . they have movements that are functional. . . . It's the movements that make these things work.

Techniques of capturing movement, of resolving time, and the software to manipulate and simulate constraints and possibilities include fluorescent tagging and laser-beam pulses that allow you to follow the changing conformations of a molecule over time or the unfolding synthesis of a molecule, "what we call 'time-resolved crystallography.' . . . Basically everything that we normally collect by rotating the crystal can be collected in just one go." Imaging often requires collaboration both with wet-lab biologists to verify the structure and with other modes of imaging. Crystallographers work with three-dimensional models on computer screens. When crystallographers are brought together with nuclear magnetic resonance people and their images, "instead of ending up with a single model which is supposed to be accurate, they come up with twenty models."

Instead of a picture of a molecule, they have a picture of twenty molecules that look more or less the same, but there are some areas in which they are highly undetermined, show variations. It means that they are either flexible, or the signal is too low for crystallography. But it gives you a picture of the behavior of a molecule in solution, which means more degrees of freedom for the molecule because there is no backing constraint for the crystal. This means more [you see more] dynamic behavior. . . . There is no single imaging technique that can explain everything, so you have to go with a combination of things, and that is why these kind of movies are very interesting. Microscopy being combined with crystallography. That's *beautiful*.

Aesthetics, tactile skills, and complementarities across expertises and imaging devices are among the fields of translation that are hard to write down as simple protocols or textbook knowledge, but they are part of the seductive joy of science when they work to produce epiphanies of intuition, fit, sense of rightness, and even psychological (if not immediately provable) certitude.

Thematic Centers, Informatics, Human Disease, and Market Translations

People have no idea what it means to have a laboratory information lab. . . . It is a sociological challenge . . . One of the problems is that infrastructure management is not very fungible.

MEMBER OF THE THEMATIC CENTER FOR COMPUTATIONAL BIOLOGY,

MASSACHUSSETTS GENERAL HOSPITAL, 2004

If "translational research" was an NIH-funding catchphrase of the late 1990s, in the early twenty-first century the catchphrase was "multidisciplinary research," to encourage architecturally facilitated synergies across disciplinary silos. Massachusetts General Hospital (MGH) was one of the first large institutions to have a chance to experiment with five new thematic centers in a new research building built in 2004–5. The promissory notes for the creation of these centers are bets (gambles), stakes (assets on the table), and investments (capital) on the future course of the medical and life sciences. Regenerative medicine, computational biology, genetics and genomics, minimally invasive optical technologies, and systems biology were the promissory names of the five thematic centers.

Academically, they grew out of a unique institutionalized leadership for planning, the Executive Committee for Research (ECOR). Financially, they grew from the rare, at best once-a-decade, opportunity provided by a developer-leasing scheme to create a new research building near the hospital. Procedurally, they grew from choosing how best to fill the rental space without huge startup costs. Space, I was repeatedly told, is more valuable than money. After all, MGH is a national leader in raising research funds, but space in close proximity to the hospital is hard to come by. The space could easily have been absorbed by existing departments, or become contested arenas of space wars. ECOR saw an opportunity not merely to mediate competing demands, but to try to build new cross-disciplinary and cross-departmental knowledge. New space can also be a huge financial burden if revenue does not come in fast enough to pay for the large rental or construction costs. MGH was well aware that Beth Israel Deaconness Medical Center had almost ruined itself in this way. MGH itself had also had a previous near-death experience. The now extremely successful, if geographically isolated, Charlestown laboratories had required a five-year, expensive burn period to become fully sustainable. The trick, then, was to fill the research building with a mix of established research groups that would be given new multidisciplinary mandates as part of thematic centers, and the ability to hire new staff or to bring in new labs with established fundraising records. The politics of the older department heads had to be carefully negotiated through the persuasiveness and incentives provided by ECOR.

The collaborations at issue are not just across labs, or with members of a lab with different disciplinary expertises, but across disciplinary formations as understood in traditional medical departments. Multidisciplinarity in this sense has a variety of models, often partially shaped by institutional and financial conditions as much as by pragmatic needs for collaborations or by the shifting ideologies of scientific information flow (ideally free, but full of slowing tactics of intellectual property and competition). There was, for instance, a period in the early history of molecular biology at Berkeley, UCSF, and Stanford when clinical science was discounted and basic science privileged, with programs such as what would become MIT's highly successful Health, Science, and Technology program being dismissed as not leading to basic-science discoveries. Today, under changed government-industry-academic relations, in a post-Bayh-Dole Act, post-Chakrabarty Supreme Court decision, and post-venture capital era, the emphasis is on ensuring that basic science always keeps an eye on medical therapies to relieve patient suffering and create value.

The Computational Biology Thematic Center provides an example. Although computers and molecular biology have been intimately intertwined since the 1950s—with the needs of crystallographers, in particular, affecting the early development of computer technology, which was also more generally affected by the migration of physicists into the early formation of molecular and structural biology—what today is called computational biology, "an algorithmic attempt to identify different kinds of determinants of sequence effectiveness" (interview with Seed, 2004),[10] has a different genealogy, derived less from physics and computer science, and being rather a development by biologists like Brian Seed, the first director of the center, who picked up software skills, initially to interrogate DNA sequences for gene locations and restriction enzyme sites.[11] Only recently have electrical engineers and computer scientists drifted into computational biology. Seed works with computer scientists on this effort to develop postmicroarray technologies. In 2004, Seed had about fifteen software engineers in his lab and was interested in developing postmicroarray technologies. He argued that there was a need at the NIH level to support an infrastructural transition, something that required mobilizing resources beyond the conservative drag of the peer-review system.

Toward this endeavor, the NIH simultaneously launched a series of initiatives, including two five-year grants to MIT, totaling some $19 million, for a new computational and systems biology doctoral program (Sasha Brown 2004, 1, 5). Harvard also initiated a new Systems Biology Department, in which Seed also spent time. The initial four students were admitted to the MIT program in fall 2004 and were expected to help invent the field by making rotations in research groups in the various component and adjacent fields, rather like the rotations medical students make in becoming doctors. The idea was to overcome the fragmentation of specializations. Sabrina Spencer, one of the new graduate students, said, "It's like zooming out as opposed to in. If you only study biological systems using a reductionist approach and look only at very small pieces of the whole puzzle, you might miss the larger dynamics of what is going on" (ibid. 5).

Seed's lab is a model of this sort of zooming out and in, of focusing neither on a single problem, nor on computation for its own sake, but on systems biology in the sense of remaining open to the unpredictable timing and serendipity of discovery, as well as to the search for new tools, be they found through collaborations with companies, such as NeoGenesis, that have proprietary libraries of compounds and high-speed screening technology (size-exclusion chromatography); new "second harmonic imaging," as in a collaboration with Rakesh Jain's lab, designing humanized mice more efficiently in order to demonstrate the potentials of directed protein evolution, to find new platforms for rapid prototyping of antibody-expression systems, and to look at antigen concentrations in real time; or finding the next generation of informatics tools that will displace microarray profiling technologies (the revolutionary advance of the last decade), allowing comprehensive inventorying of the abundance of RNA in different cell types.

Seed's research informatics group's first project was to select seventy base pair oligonucleotides that would uniquely identify an RNA transcript, and do so in a more efficient way than the various competing algorithms in current use, all of which are based on global scores of sequence similarity using BLAST (Basic Local Alignment Sequence Tool). But much better would be if one had an algorithm that gave the longest run of perfect complementarity, rather than a global score that cannot distinguish perfect complementarity from distributed complementarity. The degree of hybridization depends on the former. Using the work he had earlier done on sequence analysis, Seed figured out such an algorithmic analysis, and his software engineer, Seth Shalway, "built a little engine for picking seventy base pair oligos." They published it, and "because we were funded by the National Heart, Lung, and

Blood Institute, we took that and actually made a collection of oligos," which are now used by other institutions (interview with Seed).

That was step one. The next step was to deal with the fact that the DNA is still stuck on a solid surface, and so it cannot hybridize as well as if it were in free solution, able to assume whatever conformation it needs to, and in any case only so much can be put on a spot. In addition, the microarray facility proves to be a lot of work.

Despite this informatics work, Seed's focus was not on the informatics itself, but on human healthcare, "actually making a difference." The collaboration with NeoGenesis was an effort to directly target particular diseases. This is another example of academe-industry collaboration, and this, Seed mused, much like Folkman, "is an extraordinarily difficult thing to contribute to because there are so many reasons for failure" (interview with Seed). Seed's own history in discovering and licensing the "protein fusion" idea behind the injectable arthritis drug Enbrel is a model of the mix of informed strategy, serendipity, and what he calls "unscheduled discovery." As usually told, Seed was working on "decoy" proteins to which the HIV virus might bind and which might in this sense "fool" and disable the virus. In 1997, MGH licensed the idea of "protein fusion" to Immunex (later bought by Amgen), which applied the concept to block a protein that causes destruction of the joints. Enbrel, now the leading anti-rheumatoid arthritis drug (with revenues of $800 million a year) in 2004 was bringing in almost half the total royalties earned by Partners Health Care (the merged MGH and Brigham and Women's hospital system).

But the backstage story of Enbrel is much more diverting than this summary account. It is a classic case of searching for one thing, but discovering tools useful for many other things. The idea of using a decoy, however, was the least of it. "The idea of using a decoy, I thought, was trite, and I wasn't interested in it," says Seed. Viruses, like influenza, invade the immune system by creating their own decoys, using dispensable residues, loops that can be changed at will to miscue the immune system, but don't affect the functioning of the virus. What the virus cannot change without affecting its own functioning is the receptor it binds to. So Seed's idea was to turn the receptor into an antibody. It didn't work, partly because the receptor-antibody did not work sufficiently like an antibody, but also because the virus had ways of defeating the immune system at a different level by blocking deposition of complement, a collection of plasma cells that cause the lysis of targeted mammalian cells. So if you have a viral infection, infected cells express fibrous protein on their surface and an antibody binds to that. If enough antibody binds, patching the surface, then complement comes and creates a hole in the cell, causing lysis of the cell and killing it. That, rather than just using antibodies to gum up the works, is how the body protects against infection. So the idea was to use or enhance this mechanism.

Although it did not work to block HIV, it proved useful to have these structures as tools in the lab to purify cell surface proteins, to interrogate what their receptors were, "all kinds of practical purposes for these things, just because they were good to latch onto" (shades of Levi-Strauss's *brico-leur*). "Leander Loeffler and Patricia Kendo were in the lab and they decided to make these cytokine receptor fusions, protein fusions. Because they were in a collaboration with Immunex, their fusions became useful" (interview with Seed).

The search for HIV treatments continued. Again another discovery, this time about how the T-cell receptor works.

> So then we thought, fine, so then we can't use antibodies. What if we could take the cells' natural cellular machinery for recognizing infections and focus that on HIV. So the next thing we did was make chimeric T-cell receptors, again at the same [site]: CD4 on the outside. It's a transmembrane protein, CD4 on the outside and the internal apparatus T-cell receptor on the inside. We could then create T-cells that would recognize HIV-infected cells and kill them. And this is another kind of backward thing, it turned out to be a big advance, we actually managed to get it published in *Cell*, because it was a big advance in understanding how the T-cell receptor works. But that is not why we were doing it. We were doing it because I wanted to try to cure AIDS. And so far, after taking three or four swings at this one, I haven't been able to do it at all. And actually only three or four years ago I decided, you know what, I better start working on cheap protease inhibitors because, you know, we are not going to get there with a vaccine anytime soon anyway. So anyway this began a tangent, but about two years ago, three years ago, [we] started to create a venture in China to see what we could do to make cheap protease inhibitors that would be distributed to the Third World. I think that is really a missing ingredient. I mean it is pretty easy to make reverse transcriptase inhibitors, but to have a really effective cocktail you would like to have a pi, protease inhibitor. (interview with Seed)

Not only is China an emerging market with national biotechnology competitive goals, but the large numbers of Chinese postdocs who have worked in U.S. biology labs, some of whom have returned to start their own labs in China, provide a potential collaborative platform both for scientific discovery and to carry out development, translational research toward clinical therapies. But in an investment market in China where high returns can be made in real estate and other ventures, it is difficult to raise money, and other perhaps nonprofit organizational mechanisms will need to be designed.

The Enbrel story, in a quite different way from Folkman's angiogenesis story, underscores the difficulties and lively diversions of translation from bench to clinic. Seed put it this way: "We've had a number of things that have had what you might call spin-off potential. We were aiming to do something that was actually pretty clever, and we ended up doing something that was substantially less smart but that was much more successful" (interview with Seed).[12]

Translating Knowledge Workers and Brain Circulation

Both the Folkman and Seed labs, like many other U.S. labs, have had postdocs who have come from China, and some have returned there. There is a growing sense that, particularly in the mathematically dependent postgenomics fields, with their promise of targeted clinical trials and personalized medicine, much new work will be done outside the First World, in China, India, and elsewhere. Already there are both national and transnational initiatives, such as Biopolis in Singapore, the revitalization of an Asian-Pacific network within the Human Genome Organization (HUGO), and the scaling up of interests in translational medicine, drug discovery, and genomics in India. The genetics and genomics quests, and their various applications in agriculture and medicine, have become global, there are cost advantages to doing work in Asia, and while brain drain remains a critical problem in many places, brain circulation is also becoming prominent.

The case of a Chinese-born geneticist working in a Massachusetts General Hospital research lab provides one of many such trajectories across the globe, sometimes from East to West, and increasingly now a return flow from West to East. Unlike most scientists, who usually speak in anticipatory terms of new worlds that genetics and genomics will reveal, this geneticist has a more nuanced view because he is in danger of becoming a casualty of the unforgiving competition. He characterizes genomics as a speculative bubble that has come and gone, leaving behind a few tools, as well as more complicated, and more interesting, problems, and he speaks wryly of himself as a casualty of the speculative bubble of biotechnology companies, as well as of the branching pathways of plant and human genetics, Chinese and U.S. biology. He has a doctorate in genetics from Wisconsin, which he followed with postdoctoral study at Stanford. He was trained to work on individual genes, including some important human disease genes, using mouse models.

Then the Human Genome Project and genomics arrived with the vision of working not gene by gene, but on the whole genome. "In only two or three years, many new companies came up. . . . I think it was just two years for the booming, . . . and then [came] the crash. . . . Only a few are still there."[13] While companies like Millennium, Incyte, and Human Genome Sciences survived the crash, "actually even Millennium started as a genomics company, but now if you go inside and look what they are doing, genomics is phasing out." After Stanford, the geneticist had worked for two different companies, one a Monsanto subsidiary (Cereon Genomics), then headed a research group of eight at Genome Therapeutics, but both companies were shut down. He is one of several refugees taken into the S. lab, with the idea of retooling and pursuing ideas in an academic setting until they are sufficiently developed to have a chance in the private sector.

It is a somewhat different model from traditional incubators set up by universities, where companies are given subsidized space to get them off the ground, but still depend on their own business models. In the aftermath of the venture-capital fueled biotech boom and bust of the 1990s, the S. lab provides a safe haven and experimental space both for those who want to stay in an academic environment and for those who are looking for new technologies that are well-enough developed to launch new companies. It is a potentially fertile environment that brings together different kinds of experienced hands, and accommodates branching goals along the bench-to-clinic pathways.

The geneticist has taken over a project from someone who had recently left the S. lab to join a local biotech company down the street. He suspects that an already existing alliance with yet another company that many members of the lab are working on will slowly wind down. Therefore, he has moved to the new project, on which he works alone, which has to do with using RNAi–RNA interference, which inhibits transcription of genes and thus their expression, in a technology developed at MGH, to find protein products that might help patients with red-blood-cell deficiencies. These erythropoietin (EPO) deficiencies are caused by an autoimmune reaction that causes EPO to bind to its own antibodies instead of to the IL-4 cytokine. To create libraries of possible variant targets, RNAi displays work more powerfully, the geneticist says, than the older phage displays, which required one to work with virons containing an inserted target gene that was to be amplified like a clone, without necessarily knowing exactly what variant one was amplifying. With the RNAi displays, one can generate a much larger library of engineered variants of IL-4, pull out the protein along with its messenger RNA (mRNA), and amplify the protein until one is recognized by the EPO receptors. The idea is that one can create libraries of protein variations, experiment with them until something works (in this case, an IL-4 that can function like EPO, but not be recognized by its auto-antibodies), and be able to identify exactly which ligands and receptors work.

To make headway, the geneticist hopes that the lab will support him for three years, and if the experiments lead to successful results, perhaps he can then acquire grant or corporate support on his own. In the early days of biotech, many academics found relief from the treadmill of grant-writing, university bureaucracy, and departmental politics by going to well-funded new labs in biotech companies founded by scientists (Rabinow 1995). The geneticist, however, understands recent corporate life to be full of meetings and sudden project cancellations or changes of direction, while he understands academia to provide more space for thinking and reflection. The geneticist's perspective is informed not only by his U.S. experience, but also by his Chinese network. One of his roommates in graduate school in China is now the director of the Beijing Genomics Institute, and the vice president of the Chinese Academy of Sciences. The geneticist says, "I know they are doing quite well, well probably on their standard. . . . Before I came to the United States, I was working in the Chinese Academy of Sciences. I did my masters degree in [plant] genetics."

"On their standard" elaborates into an interesting set of observations, rather than being merely a token of deferential politeness to an American interlocutor or a simple evaluation. First, the geneticist notes, the institute is associated with the Chinese human genome project, but neither owns it nor is primarily interested in it. The institute works mainly on the rice-genome sequences "and one percent of the human genome, something like that." In fact, he goes on, "biology in the United States means more about health and medical things, [whereas] biology in China means more agriculture." But this, too, has deeper mappings. I ask him, as a geneticist, if the techniques and technologies in agricultural biotechnologies and medical biotechnologies are not very close: "Yes, close. Um. I would say it this way: You know, the concepts, the general things, are the same. Genetics is the same for all organisms. But for historical reasons, some organisms are used as models, bacteria first, then fungus, *C. elegans*, *Drosophila*, and then mammalian cells. All this information accumulated [and so] now today the genetics of the human is [known as or] more deeply than [that of] *Drosophila*, than *C. elegans*." But now comes the more interesting claim, "For plants, they fall out of this loop. . . . In the U.S., plant genetics does not attract many people. Of course a lot of people work on [agricultural genetics], but not proportionally [relative to the] whole field of biology." I probe again, asking if this impression might be a function of the fact that most agricultural biotechnology is being done privately inside corporations such as Monsanto. In response, the geneticist utters a telltale "Um hum"—an affirmative, but delivered in a way that signals that the situation is more complicated than simply a question of the corporatization of agricultural biotechnology.

Transnational translations are also about research priorities in different contexts and the consequences of those differences.

> Um hum. Actually yesterday one of my friends, an old friend, called me. She is working—actually she was working in Monsanto, but now she switched to another company, Simplex, something like that, a big company that works on tomato. She has the same background [as I do]. We were classmates in college, thirty years ago, and we are still very close. But she has been working with plants continuously. I [took] a lot of plant courses, and was quite good at that. But now I have moved to this [U.S.] environment, follow the mainstream here, and so I work with the human system here. We had a good talk on the phone, about half an hour. She tried to ask me something about what I am doing that could be useful for her. In her mind I am kind of one step further than what they are doing. It's not that she is behind, it is the area. So I told her in the mammalian system we do things this way. . . . For instance, I told her about an experiment with apoptosis, and [suggested that] probably the general principles can be used in agriculture as well.

Such crossings of agricultural and medical biotechnological skills and competencies form part of the contemporary ethical plateau of multiple technologies intersecting in uneven ways. The geneticist's situatedness was resonant: he was about to lose another job, and his skill sets from his various experiences in plant genetics, human genetics, genome companies, and academic research would all have to be in play for his next move.

One of the 26,500 Chinese students who received doctorates in the United States between 1985 and 2000 (twice the number of Western Europeans who did so), the geneticist is part not only of the transnational network, but of the disjunctions in national priorities (between the agricultural work he did in China, and the biomedical work he does in the United States), and of the struggle for secure positions among highly trained scientific workers who do not yet have their own labs.[14] His situatedness is particularly useful in deconstructing and reconstructing the more abstract hopes that reside in the logics of the technologies, business models, and scientific speculative visions.

Asian Translations

The most active or interesting chapter [of the Human Genome Organization], or whatever you want to call it, is the Asia-Pacific.
EDISON LIU, HEAD OF THE GENOME INSTITUTE OF SINGAPORE,
FORMER DIRECTOR, DIVISION OF CLINICAL RESEARCH, U.S. NATIONAL
CANCER INSTITUTE, CURRENT PRESIDENT OF HUGO

Singapore's Biopolis provides a concluding global node and perspective on the emergent threads of interaction and competitive cooperation described in this chapter, as well as a platform for reflections for future ethnographic work and methods. Scientific labs are, of course, nodes in national competitions, but they are also among the most international of institutional spaces, especially at the postdoc and graduate-student level, as can be seen in the indexical backgrounds in earlier parts of this chapter, but especially with a quick perusal of the names listed in the acknowledgments included at the end. While internal lab struggles with language and other intercultural relations has not been a topic of attention here, and collaborative mentoring lineages across national boundaries only slightly more so, the relations of competitive collaboration have been pervasive. While cultural-studies and immigration-policy scholars talk a lot about the positive and negative aspects of emergent global multiculturalism (conflicts as well as expansion of intellectual and moral horizons), scientific laboratories are sites where such work is actually accomplished daily.

Singapore emerged into the headlines not only with its ambitious, ten-year, $3.5 billion initiative in biomedical development (2000–10), but also with its success in luring to its state-of-the-art laboratories and informatics infrastructure (of the sort described by Brian Seed) high-profile scientists including Alan Colman, of Dolly-cloning fame; David and Bridget Lane, of p53-gene fame; Edward Holms and Judith Swain, from the University of California, San Diego; Neal Copeland; Nancy Jenkins; Edison Liu, from NCI; and Bing Lim, from Harvard, who runs two labs simultaneously, one at Harvard and one in Biopolis. Singapore's array of institutes includes direct translational research as well, including building a GMP (good-manufacturing-practices certified, clean room, quality controlled) facility for stem-cell-therapy production.

But it would be a mistake to underestimate the Singapore experiment as merely a buying of talent and facilities. Central to the economic planning of Singapore is an effort to upgrade its skilled work force; outpaced by cheaper labor markets, it can no longer compete in even the higher end of integrated-computer-chip assembly and testing. That Pfizer's Asian clinical-trials division for the region is based in Singapore is another index of Singapore's positioning itself as a biotechnology development node. Singapore's research-granting organization A*STAR (Agency for Science, Technology, and Research) provides student fellowships and scholarships to attract talent from China, Vietnam, and elsewhere as well. And Edison Liu considers it his role, both in Singapore and in international organizations such as HUGO, to create new synergies and coordination. One of the first big conferences Liu convened as president of HUGO was the thirteenth global Human Genome Meeting (in September 2008), in Hyderabad, with co-sponsorship of India's Council of Scientific and Industrial Research (CSIR) and with its new director general and former director of the CSIR Institute of Genomics and Integrative Biology, Samir K. Brahmachari.[15]

Edison Liu has a multitiered vision of developments and how to make competitors into colleagues and allies, at least in fields like genetics and genomics. "Virtually every country [in Asia, and now in Latin America, emerging countries] is doing [genomics] as an economic driver."[16] There are various reasons. "Sociologically . . . [one] reason genomics is an ideal entry point for these countries is because their base is engineering, physics, and math. Who are the best genomicists? Mathematicians. Eric Lander, you know, is a mathematician. And the field is replete with physicists and mathematicians. . . . Furthermore, young aggressive people can go into it and actually make a difference by just computational fooling around."[17] But there is a more subtle reason having to do with the fears of exploitation marking post-colonial history and the troubling debates about marketing populations as relatively homogenous or unique for data mining for genetic markers of disease and drug targets.

> If you look at biological development in many of these countries, you can ask which are the biological fields that seem to do very well on a competitive basis. It is always the genetics-based [ones] like population genetics, *Drosophilia* genetics, whatever, but genetics-based, and the reason is, it is the mathematics. But it is also because, you know . . . recall that during postcolonial days, one of the biggest raps on genetics . . . was fear of exploitation from First World universities who come and take their blood and publish and do anything they want and make claims like, oh, this society is weaker because it has this gene more often.

Haunted by the legacy of the Nazis and this postcolonial history, and by the fears of indigenous populations who have suffered similar exploitation at the hands of their national governments, discussions around disease associations can be troubling, and defensive barriers must be overcome.[18] Liu amusingly describes a meeting where another subtext, of course, is always also the competition for the patents for the identification of the genes, both for the glory and the results: "The Japanese were saying, give me your DNA, and I'll work it out for you. The Chinese were stone-faced, knowing that sooner or later they will be able to do it all themselves and do a lot better than the Japanese. The Koreans would nod their heads, you know, gently, but afterwards sabotage anything the Japanese would want to do. And the Taiwanese and Chinese wouldn't talk to each other."

So Liu suggested they pursue a genetics study that would not involve disease associations, also because phenotyping disease is costly (economics again) and there would be great disparities. "Vietnam, Cambodia, Mongolia, . . . they can't do any of this stuff, even Malaysia at one point and Thailand had difficulty, and Indonesia for sure. . . . So, we said, O.K., let's remove anything that has to do with disease and just talk about how we are related to each other as Asians and how human migrations may have taken place." Beautiful: using migration studies, not just for itself, but as a tool to generate genetic knowledge and collaboration. In this context, rather than prioritizing discrete populations with high rates of disease, where the bigger the number of participants the greater the advantage, researchers focused on cross-sections of diversity, discounting the alleged homogeneity of populations in (richer) Korea and Japan (touted earlier as advantages in seeking distinctive genetic signatures of disease markers). The resulting study, the

Pan-Asian SNP initiative, with eleven countries, seventy-four ethnic groups, and the construction of a 50K SNP chip, found that instead of the traditionally hypothesized "two waves of migration," there was really only one wave.[19]

Culturally and sociologically, Liu suggests, his effort is to foreground Asian traditions of host-guest relationships. Countries enabled with technology and financial resources volunteer as hosts. Other countries choose their hosts and bring their DNA with them, keeping the chain of custody always with their national scientists, but sharing data. This format, together with a policy that designates the use of surpluses generated by HUGO Asia-Pacific meetings to help fund travel, creates a network of friends and colleagues, and "an operating system, you know, a social construct" for cross-cultural, cross-national science and technology development.

All these sociological, cultural, and organizational observations and hypotheses are working tools for science managers such as Liu, John Parish (the leader of MGH's Executive Committee on Research in the founding period of the thematic centers), Seed and other thematic-center directors, and lab heads like Bim Lim, Judah Folkman, and Jiahuai Wang.[20] Their language, insights, and strategies provide a potential space for anthropologists of science and technology, who help translate these worlds into worlds of humanists, philosophers, social scientists, and various public-policy venues for discussions of the social, ethical, and legal implications for publicly supported and nationally prioritized science and technology. As in many professional arenas, broad generalizations and buzzwords serve as signposts or short-hands, but it is access to details rich enough to unpack disputes, interests, passions, and practicalities that is required for such conversations to make a difference.

Conclusions

There are nay-sayers all over the place. It's very interesting to watch. I'm going to publish a book on the physiology of nay-saying some day.
JUDAH FOLKMAN, 1999

In 1992, in his comments on "the aforementioned so-called human genome"—that is, right at the beginnings of the Human Genome Project—the philosopher Jacques Derrida asks in a condensed way many of the frequently raised ethical, legal, and social questions about our integration of software and databanks, the outrunning of our legal theory of patenting, the exceeding of the opposition between private and public by new forms of ownership, the appropriation of knowledge and technical ability through the very compilation and constitution of these databases, and the degree to which we might foreclose the possibilities of a future system of health-care that might privilege even further the rich over the poor.[21] At issue is whether decisions we make are made freely or are constrained under the guise of extending old norms (what the philosopher and chairman of President George W. Bush's President's Council on Bioethics in 1997 infamously proclaimed as the "yuck factor"), or more generally are our decisions freely made or determined by path-dependency (by prior decisions).[22] Do we know what we are gesturing toward when we speak of the human genome: why are we selecting certain bits of information for what is human, and if we select differently, will the definitions change? Or, in more abstract terms, we seem to have no clear criteria for distinguishing invention from discovery, one of the practical problems in today's patent competitions.

Derrida urges a stance of "decidedly keeping watch" against the appropriation of knowledge and power, as well as a firm defense of the ethics of the Enlightenment and of scientific research. No simple antitechnology romanticism here. As he says in an essay on "Nietzsche and the Machine," "Life is a process of self-replacement, the handing down of life is a *mechanike*, a form of technics. Not only, then, is technics not in opposition to life, it also haunts it from the very beginning" (Derrida 2002, 244). If there is an urgency to ban or permit certain lines of research, there is the counter-vailing calming effect of knowing that science is lively and takes time, and in that given time, while "decidedly keeping watch," "a vast political and legal consciousness is indisputably in the process of rising to meet this incredible progress of knowledge" (ibid. 210).

In 1992, as Derrida pointed out, while the ability to map the genome was not yet the ability to reliably manipulate the genome, and while often it is difficult to distinguish inventions from discoveries, most important is the vagueness of our understanding of what being human is, and thus what ends our discoveries and inventions are to achieve. Like most scientists, Derrida affirms the right to know as fundamental to being human: "Purely scientific research and its breathless desire to know must not, in principle, be opposed by any limit. . . . There is something here that is properly human. . . . There is an ethics of the Enlightenment here that must remain unconditional" (ibid. 202–3).

One hears here a harmonic echo of Judah Folkman's refrain that he planned to write a book, after retirement, about all the nay-sayers. The two imperatives are not the same, but they are mutually supportive: Derrida's insistence that the will to know is fundamental to being human, and Judah Folkman's insistence on patiently working through the various obstacles to getting new healing technologies from the bench to the clinic.

What is it to be human? asks Derrida.

> Is man the being that possesses a knowledge about itself? With the possibility of self-fashioning? . . . There are two competing definitions of man . . . one that knows its own norm, that knows itself, that knows its normativity, its normality, and that draws the consequences from this knowledge, as a scientist, techno-scientist, etc.; whereas in another way, this being is one that at least asks itself the question of ethics, of freedom, of responsibility, where not only a norm and a knowledge of this kind are lacking, but further, must be lacking . . . for a responsible decision to be made. (Ibid.)

We are, he agrees, haunted by the archives of past genocides where so-called genetic knowledge was misused, and he contemplates the possibility that genetic manipulation will generate social inequalities, stigma, stratification. And yet our futures are underdetermined, open to the consequences of our own decisions.

Debates over genetically modified foods in Europe became as heated as those over stem-cell research in the United States, exposing religious and other presuppositions (archives of history) that seem to ill-fit a world transformed by new knowledges, new material-semiotic objects, new appreciations of ecological, hormonal, genetic, and other flows, circuits, interconnectivities, interactions, disruptions, and accumulations. Indeed, Derrida teasingly rewrites, as it were, Mark Twain's *A Connecticut Yankee in King Arthur's Court*, asking whether prehistoric human beings, or even sixteenth- or seventeenth-century ones, would recognize us—"people who go to the Moon, freeze their sperm, open a virtual space, etc." (ibid.)—as human beings like them.

At issue is the aporia of whether the human is what we have been or what we will be. Crimes against humanity may be the literal killing of people, but they also act against the freedom of knowledge, which is part of the essence of being human. Derrida is split, he says, between two feelings: on the one hand the anxiety about the accelerating rate of decisions about the processing of the human genome, but on the other hand a "calming effect, relativizing or demystifying in some sense" (ibid.). I take this to mean that as we learn more about our biological connections with the world, we come to feel less estranged. He continues not merely hopefully that "a vast political and legal consciousness is indisputably in the process of rising to meet this incredible progress of knowledge," but even more "one can and one must resist—we must arm ourselves for this—the phantasm and the pathos of ignorance, which ignorance and disinformation may produce or propagate" (2002, 209–10).

The genetics and genomics quests, and their various applications in agriculture and medicine, have since 1992 become global, but not therefore singular: national contexts and priorities differ; the republic of science is a landscape full of competition and intellectual-property minefields, as well as of sites of cooperation and complementarity.

One of the challenges for contemporary ethnographic work in high-tech industries and sciences is finding appropriate vehicles beyond "genius scientist," laymen's metaphors drawn from other fields, or simplistic reassurance against fears of science and technology that fail to deepen understanding. In this chapter, I have tried to focus attention on the different scales and translations that are required, from the level of green fingers, tacit knowledge, or "skills" in handling experiments to more stable tools; from postdoctoral career journeys across countries, labs, and sometimes fields, to the negotiations of lab management internally, to translation to the biotech-development markets and into actually used therapies; and efforts to foster multidisciplinary and cross-national synergies.

Of key or nodal interest are (1) new *epistemic* and *material-semiotic* objects (like the 50K chip of the Pan-Asian SNP project, or the genome itself as databank object), which change the way we think and talk about the objects and the social relationships they reconfigure; (2) sites of *deep play* (investments that are financial but also political, economic, organizational, public relational, and symbolic), and *ethical plateaus* (where different technologies intersect and compete, creating moral concerns about, for instance, social welfare and the long-term restratification processes networked databanks may create, or alternative efforts to create open-source biology databanks); and (3) *shifts of scale* that these epistemic and material-semiotic objects, deep play, and ethical plateaus foster.[23] But along with these sites and objects, strategies and symbols, the substance of interest in this chapter has been (4) the kinds of people being created at these sites: their technological lives, scientific arts, and cultural integrative perspectives. For increasing numbers of people, these are our family members, our children, and our neighbors, whose anxieties, obsessions, joys, and woes we share even when we do not completely understand.

For Judah Folkman and Jacques Derrida this kinship relation was expressed through a special modality of friendship, something that Derrida and his philosophical interlocutors, Emanuel Levinas and Hélène Cixous, wrote extensively about in both their concrete relations with one another and more generally as an attitude more important than any foundational philosophies of usually quite ethnocentric ontology. Judah Folkman lived this modality in his extraordinary commitment to answering phone calls and in taking personal interest in thousands of pleas for help. His was not a distanced professional career, but a life dedicated to a physician-scientist's calling, intimately tied up with his own family romance.[24]

There are many moving parts in the call for upgraded ethnographic accounts of technological lives, scientific arts, and cultural rearticulations. Anthropology over the past century has invoked many frames and metaphors for focusing attention on such connections, both in detail and in setting (see Fischer 2007a). Single metaphors or mapping procedures are unlikely to succeed, as each captures only one or two facets. Some of these are *complex holism*, or the obligation to see things in their historical, cultural, and social contexts; *methodological functionalism*, as the obligation to ask how a change in one element changes other elements in a material sociocultural complex; *ecological relations*, as questions of sustainability, homeostasis, or threshold transformations; *constitutive multicausal conjunctures* of social causes, cultural logics, and psychological motives, as in Max Weber, the Annales School of Historiography, or Talcott Parsons's cybernetic models of social, cultural, and psychological systems; *complex models*, as in the debates over surface and deep linguistic structures generalized to culture, with relatively small numbers of generative rules able to produce infinite varieties of surface enunciations and patterns, and more recent variants couched in terms of algorithmic reproduction generating unexpected fractal, tiling, and other patterns; and *multi-locale or sited ethnographic strategies* to access different scale sites in widely distributed, differentiated, and stratified processes. One of the most recent of these terms is *assemblages* (Marcus and Saka 2005), a term taken from the philosophers Deleuze and Guattari (1987 [1980]) elaborated by de Landa (2006), and adopted by some anthropologists (c.g., Ong and Collier 2005), where it represents a movement away from organismic metaphors toward mechanical and engineering ones, in a pendulum movement reversing that of the late nineteenth century and exemplified in today's synthetic-biology goals (to find standardized building blocks in biochemistry, genetics, and mathematical logic with which to synthesize new life forms), rather than in the holistic goals of systems biology (to work within the messy, wily processes of wet biology, whole organisms, and habitats).

This difference between synthetic and systems biology itself signals the importance of keeping mapping tools multiple. The intent of the term *assemblages* is to loosen the tight coupling of elements in some of the criticisms of particular metaphors or linguistic terms with multiple meanings (such as the overly mathematical metaphor of function, where one function is defined in a deterministic relation to another, as opposed to methodological functionalism; or closed-systems analysis, as in thermodynamics, as opposed to open ecological systems). While Deleuze's and Guattari's usage of assemblages draws from host-parasite coevolution, genetic exchange across species, and rhyzomic proliferation, de Landa seems to draw from mechanical engineering (exteriority of relations, mixtures of roles or functions from material to expressive, variable ratios and degrees of boundary strength).

What I find ethnographically interesting are the sites from which these various metaphors are taken and with which they are traded. These are not just the organism-mechanism debates of the eighteenth and nineteenth centuries, but articulations among different contemporary scientific subfields, imaginaries, problematics, and logics. We find ourselves in a new world of *la pensée sauvage*, with wondrous concrete logics and transformations. Levi-Strauss's metaphors were those of the nebula (as it "gradually spreads, its nucleus condenses and becomes more organized") and germinal molecules ("sequences arranged in transformation groups . . . [that] join up with the initial group and reproduce its structure and determinative tendencies. Thus is brought into being a multi-dimensional body, whose central parts disclose a structure, while uncertainty and confusion continue to prevail along its periphery" [1964 (1970): 3]).

Knowledge and object, discovery and invention, discursive structures and subjectivities—these do not just mirror each other, but diffract, generate, and recompose. This is as true for new biomedical technologies (such as the relations between crystallography and the wet lab) and for medical science translations into the clinic, as it is for anthropological knowledges of them.

As Judah Folkman challenged medical students, "There are four thousand brain diseases we don't know anything about. They are listed in the admissions book in hospitals. We haven't got a clue about them."

> We're all accused of raising expectations . . . never tell a patient
> that he doesn't have any time to live . . . they go into a deep depres-
> sion. What we say is, we're worried . . . but we've got all this work
> going on, and here's our goal: our goal for you is, we've got to get
> it stabilized, doing this, this, and this. You map it all out. You can

make hope, and that's what part of medicine is if you can't treat. There are twenty-six thousand diseases, how many do you think we have cures for? . . . When Lewis Thomas entered Harvard Medical School in 1932, the only thing they had was digitalis, no other drugs. . . . And they could do surgery. They could cure four diseases. [1999]

Instead of nay-saying, the clinician-scientist looks for clues the oncogeneticist misses, scans the experimental literature for effects of other titrations, other interactions, other combinations. "A lamppost would be a toothpick for a giant. A tree could be his broccoli. The world as a fascinating place of potential."

Notes

I gratefully thank Judah Folkman, Jiahuai Wang, Brian Seed, John Parish (the head of ECOR in the crucial years, and of the Center for Integration of Medicine and Innovative Technology [CIMIT]), Edison Liu, and the members of their respective labs and teams for their time, generosity, tutorials, formal talks, and welcome into their labs. I am particularly grateful to the structural biologists Azin Nezami, Rob Mjeirs, Emilio Parisini, and Georgio Skiniotis; to Julian Banarjee, Guo-Hong Feng, Ed Fritsch, Glen Cho, Glen Short, Summer Xuichin Wei, Jha Wolf; Daine Shao, Minh Le, Gugi Guo, Senthil Raja; and of course my fellow anthropologists and historians in this enterprise, particularly Natasha Myers, Aslihan Sanal, Wen-Hua Kuo, Kim and Mike Fortun, Chris Kelty and Hannah Landecker, Byron and Mary Jo Good, and Kaushik Sunder Rajan. Many thanks as well to the participants of the three Lively Capital workshops organized by Kaushik Sunder Rajan at the University of California, Irvine. None of the above, obviously, are responsible for anything that I got wrong.

1. For over a decade now, Byron Good, Mary Jo DelVecchio Good, and I have been teaching a class on social and ethical issues in the biosciences and biotechnologies at the Harvard Medical School in the Joint MIT-Harvard Health Science Technology (HST) track, fulfilling a social medicine requirement. In 2008 we transformed the course into a global-medicine course to fulfill a new HST requirement for the HST track in the Department of Global Health and Social Medicine.

2. See Fischer 2003, chapter 9 (305–69), for how emergent forms of biomedical laboratory life fit into the transformations from "Las Meninas" (the perspective of a neutral observer surveying the external world) to "Ian Hunter's Robotic Surgical Lab" (the perspective of always already being inside our means of knowing), and Rheinberger 1997, for an argument about how the line between discovery and invention has shifted such that we now write, not merely discover, biologies. Computer- and robot-assisted surgical systems (such as the Da Vinci Surgical System, developed by Intuitive Surgical of Mountain View, California) provide a counterpoint to Velás-

quez's early-modern painting *Las Meninas*. Just as Michel Foucault, Norbert Elias, and Svetlana Alpers have used the Velasquez painting to meditate on the epistemic transitions from the Renaissance to the modern period, so, too, we might use images of such surgical systems to meditate on the transition into informatics-immersive environments and networked databanks that are changing our current life worlds.

3. Unless otherwise indicated, all the quotes that follow are from Folkman's talk.

4. For the history of the first twenty-five years of the Harvard-MIT HST program, see Abelmann 2004.

5. See also the marvelous interview with the oncologist Irene Kuter on the challenges of being a physician-scientist (M. Good, I. Kuter, et al. 1995).

6. See also Petryna 2005; Petryna, Lakoff, and Kleinman 2006.

7. For a dramatic physician-scientist's account of the terrors and dilemmas of clinical trials with highly toxic, poorly tolerated, drugs, see Steven Rosenberg's 1992 chronicle of the NIH early trials with interleukin-2 (IL-2).

8. The student asked whether it was ethical to perform a fake surgery in order to compare. Folkman answered that it generally wasn't considered ethical to do so, but that it had been done in Germany and that a classic paper using the results had been published in *Circulation* (February 1996), showing that when the gene for angiogenic factor was injected into the hearts of twenty patients with severe heart disease for which nothing more could be done, they grew new blood vessels and their angina disppeared. The success was picked up in Boston and Atlanta. Then it turned out the German researchers had also had twenty matched patients: they had heat-killed the protein and randomized the patients. That, Folkman noted, would be a criminal case in the United States. It has generated heated discussion, since it is known that some people's angina ceases of its own accord, so without a placebo, the trials could have continued indefinitely. Folkman concluded, "So these are the big dilemmas. Trying to improve medicine really has ethical pitfalls. Trying to practice it has one set of ethical dilemmas, but trying to prove [efficacy] has another."

9. On skill and tacit knowledge in science experiments, as opposed to straightforward protocol- or rule-driven activity or simple rational design and replication, see Polanyi 1966 and Collins 1974, for early reference points in a growing literature; Shapin and Shaffer 1985, for a historical starting point in the modern European discussions on the nature of experimental proof; Rheinberger 1997 and Rheinberger 2006, for more recent approaches in a Massachusetts General Hospital molecular biology lab and in philosophy of science discussions.

10. This and following quotations are from a series of interviews with Brian Seed in 2004.

11. "Crystallographers were the first life scientists to make use of computers, initially [to alleviate] the massive labours they faced with calculation, [and] later [to] reduce the physically labourious process of data collection, and only after that, for facilitating computer graphic representation and manipulation" (Myers 2007, 12). Myers draws on de Chadarevian 2002 and works cited therein by Francoeur and Segal, Siler and Lindberg, as well as research in the MIT archives, including the work

of Cyrus Leventhal in the 1960s Project Mac with interactive graphics ("Molecular Embodiments and The Body-Work of Interactive Molecular Graphics").

12. On the rhetoric of smartness in a different professional field, also structured by neoliberalism (and also powerfully structuring that neoliberalism), see the work of Karen Ho (2009) on the recruitment and use of Harvard, Princeton, and Stanford graduates in investment banking.

13. This and following quotes are from the interview with CS 2004.

14. Brumfiel 2005, 278; the original source of the student statistics was the National Science Foundation. Brumfiel's article, in *Nature*, is about a protest against unfair treatment at Yale. A paired article in the following issue (1 December 2005) reports on worries that post-9/11visa restrictions could threaten the labor pool of graduate students and postdocs working in U.S. biology labs.

15. India's growth and challenges in the global biotech sector since its accession to the WTO's intellectual-property patent regime, TRIPS, has been impressive. At the international conference of HUGO, held in Hyderabad in 2008, there were over a thousand participants, the vast majority Indians, including students, thanks to vigorous advance programming by the co-sponsoring organizations, HUGO and India's CSIR, and their respective leaders, Edison Liu and Samir Brahmachari. Among the challenges for startup firms are investment needs that are below the minimum thresholds set by large venture-capital funds to get sufficient returns. Another challenge is the recent reentry into India of multinational pharmaceutical companies' research-and-development units, which can affect salary levels, threatening to reduce them below the levels required to retain lead investigators (Frew et al. 2007). Shantha Biotechnics (Hyderabad), founded in 1993 at Osmania University, for instance, now supplies almost 40 percent of the global requirements of the United Nations Children's Fund (UNICEF) for hepatitis B vaccine (India's first domestically produced and marketed recombinant-DNA product). Pune-based Serum Institute of India is said to be the world's largest manufacturer of measles and the diphtheria, pertussis, and tetanus group of vaccines for half the children immunized globally (ibid.). As blockbuster drugs come off patent, India's generics industry is well positioned to garner significant market share. There is a growing field of clinical-trial contract organizations. A number of Indian companies have established subsidiaries in the United States to help them access capital and expertise, and, domestically, companies that originated in other fields are beginning to acquire and invest in biotech research units. Entirely new product development is still in its initial stages, but interest, energy, and scale will make India, like China, a global health biotech player. The William J. Clinton Foundation in 2003 announced it had joined with four companies worldwide to supply anti-AIDS medicines in Africa: three of the four—Cipla, Matrix, and Ranbaxy—were Indian companies (Bhandari 2005, 15). Also in 2003, Ranbaxy replaced Roche as the lead partner in the private-public Medicines for Malaria Venture. Ranbaxy has acquired production units in France (RPG Aventis), the United Kingdom, and Germany. It has a collaboration with GlaxoSmithKline in which Ranbaxy is entrusted with discovery and early development through clinical trials of new chemical entities. For a detailed account of the strategies, the training, and the

recruitment of leaders of Ranbaxy, see Bhandari 2005. Ranbaxy has now been partly sold to a Japanese company, Daitchi Sankyo, so there are interesting emergent dynamics of corporate competition in this sector within Asia now.

16. Interview with Edison Liu in 2009. Following quotes unless otherwise noted are from interviews with Liu.

17. See also Kuo (2005) on the Taiwanese cadre of biostatisticians, U.S.-trained, and with experience in the U.S. FDA, who returned to Taiwan to help with the Taiwan national priorities in biotechnology.

18. Not only have India and other countries introduced restrictive laws preventing samples and data from leaving the country, but researchers have been sensitized to react negatively to using particular populations for disease studies for fear of stigmatizing them or having them react defensively for fear of possible stigmatization. Restrictions in Japan including strong confidentiality rules, for instance, it has been suggested, are among the reasons that Japan is weak in epidemiology, and yet insists on its biological uniqueness as a reason to counter European and U.S. pharmaceutical companies' desire for global clinical trials. International organizational structures such as the agencies of the United Nations, Liu observes, have been good venues for working out consensus and templates for handling such ethical questions, so, for instance, countries such as the Philippines, Indonesia, and Malaysia deal with indigenous populations in similar ways. But these agencies tend not to be good at operational working groups, falling into cronyism and bureaucratic involution. For operational purposes, better to have the mix of private and public national initiatives ("from within") such as, he suggests, the new developments stimulated under CSIR in India.

19. The SNP chip is a silicon wafer substrate that has DNA sequences tagged to it, on which hybridizations can be performed to compare two sets, and "states," of genetic samples, to see which genes are selectively regulated in response to certain events, or predispositions to events. These events are usually biochemical interactions that trigger certain genes being turned on or off, or trigger cascades of biochemical interactions constituting a biological pathway. In other words, the chip itself maps clusters of genes to provide broad views of gene expression. An SNP is a single nucleotide polymorphism. These are single base variations in the genetic code that occur about once every 1,000 bases along the 3-billion-base human genome. Knowing the locations of these closely spaced DNA landmarks both eases the sequencing of the human genome and aids in the discovery of genes variably linked to different traits.

20. See also Rabinow's account of Tom White as a scientist-manager (Rabinow 1995).

21. "The Aforementioned So-Called Human Genome" was a set of remarks originally made at a conference in 1992; the proceedings were published in France in 1996 and translated into English in 2002. The original date is interesting, since it reveals that the comments came relatively early in the maneuvering around the Human Genome Project (1990–2003). The first meetings on the feasibility of such a project were held in 1985, by the U.S. Department of Defense, in Santa Fe and at the

University of California, Santa Cruz; and the first initiative was started in the Department of Energy's national laboratories the following year. In 1988, the NIH and the Department of Energy agreed to jointly pursue the project, and HUGO was formed to coordinate international efforts. Derrida, however, had long been interested in the ways in which computer programming and molecular biology were rearranging our conceptual categories, even in his first major work, *Of Grammatology* (1974 [1967]).

22. Kass 1997. The issue was stem-cell research. The President's Council was convened to bolster President George W. Bush's and the right-wing's desire to ban or at least not provide federal funds for stem-cell research. The best that Kass could come up with was a peculiarly reactionary and ethnocentric gut feeling of repugnance to whatever is new that should be treated as a moral compass. In Derrida's more analytic terms, such a response is a binding to what we have been, rather than an openness to what is essentially human: the will to know.

23. There are numerous public-domain repositories for various kinds of biological data, though often restricted in various ways for reasons pertaining to intellectual-property rights. On the creation of a SNP consortium agreement to create an upstream commons because cross-licensing was becoming burdensome to research and innovation, see Sunder Rajan 2006. Better profits in that case could be expected from downstream products if upstream data was freely available. In the synthetic-biology community, there is an effort to make biobricks (biochemical building blocks, or their algorithms) open source from the beginning. On the creation of open-source forms more generally, see Kelty 2008; Coleman 2005.

24. He always managed to end his slideshows on his work with a picture of his granddaughter holding a book or article about angiogenesis, and he never failed to relate how he was the son of a Midwestern rabbi, who learned his calling while paying hospital visits with his father. The stories told at his funeral filled in many other relationships, including his siblings' awe at his incessant drive for exploration and experiments in his basement lab as a child, complementing his daughter's memories of how he taught them to think of the world otherwise, as full of potentials different from the given realities.

KAUSHIK SUNDER RAJAN

EPILOGUE

THREADS AND ARTICULATIONS

The chapters in this volume traverse an array of empirical material (genomics, pharmaceutical marketing, intellectual property, environmental science, clinical trials, and patient advocacy, to name but some); draw on practices that are happening around the world (North and South America, Europe, Africa, South and Southeast Asia are all represented); and adopt a variety of disciplinary approaches. I wish in this brief concluding overview to thread together some of those themes. The attempt here is not to create a unifying framework. Indeed, I think that an attempt to do so would be an impoverishment that would not do justice to the richness and range of empirical material that the authors are contending with. Rather, I wish to highlight some points of convergence, divergence, and distinction that mark this collection.[1] I wish to do this by working through nine specific points that resonate for me in the chapters collected here.

These concern a series of questions that are empirical, methodological, and political (where these three categories cannot be easily teased apart from one another). Empirically—what are these things that we are seeing in the world of technocapital today, and in what ways are they new or do they force a recalibrating of the vocabulary of social theory that we have inherited? Methodologically—*how* do we see these things and write about them in ways that make meaning? And politically—what larger interventions might these empirical and conceptual interventions have? These constitute a question of history, a question of epistemology, and a question of praxis.[2]

1. The chapters in the volume raise a methodological question: *how* do we make sense of the emergent present? If there is a unifying methodological distinction (one that is certainly haunted by an ethnographic sensibility, but is by no means a simple disciplinary mark of anthropology on the volume), then it is the way in which each chapter is grounded in *particularities*, and the attempt in bringing this collection together is to juxtapose these particularities in order to get a thicker picture of the global constitution of the life sciences and capital. There is an absolute importance to each chapter of the cases that constitute it; larger concepts derive up from the cases, rather than frame them in advance. In discussions at the workshop, Donna Haraway made the following point as we concluded our conversations and thought of outcomes and future directions: "Each paper has had a moment of precision that has given me a little whiplash. I want to ask how what I'm doing is very different from what other people are doing. . . . I want conversations that aren't necessarily about the large concepts. I care about the cases. . . . I don't want to reduce it simply to discourses of political economies or science studies, but the stuff of the earth under question."[3]

Therefore, Travis Tanner's chapter is not a general story of population genomics, but a specific one, involving particularly brutal histories of indigenous populations. In Kris Peterson's and Joseph Dumit's cases, the particularities are geographic (Africa as a site of extreme dispossession in the former, the United States as a site of extreme therapeutic marketing and consumption in the latter). For Sheila Jasanoff, the mechanism of modernist legal purification is not at issue as much as the outcomes of *particular* legal purifications (in the United States versus in Canada, for example). The particularities in Elta Smith's case are material and concern the sort of research object that rice is. Tim Choy insists on a "poetics of place" in terms of the particularities of Hong Kong, which itself is constituted by many particular microenvironments. Wen-Hua Kuo's particularity is Japan's refusal to standardize its drug registration according to ICH protocols, insisting instead on drug trials being conducted in Japan in order to be registered to market to Japanese populations (a very different strategic resolution compared to the particular cases of Taiwan and Singapore). The history of Lacanian psychoanalysis grounds a particular response to psychiatric genomics in Argentina in Andrew Lakoff's piece. And the particularities of autism as a disease category are crucial to Chloe Silverman's analysis (part of the particularity being a consequence of the parents also often presenting as quasi-symptomatic, thereby blurring the boundary between patient and parent advocacy).

2. A second methodological emphasis is historicist. The authors are concerned with marking the historical transitions in capitalism that become particular to biocapital. Perhaps most important is the particular salience of speculative capital at this historical moment, which again is emphasized in a number of the chapters.

This is not to suggest that speculation itself is new to capitalism. Indeed, speculative capital is the central subject of analysis of volume 3 of *Capital* (Marx 1974 [1894]), and Giovanni Arrighi (1994) has shown that the speculative nature of mercantilism precedes even the Industrial Revolution. What is crucial here is the particular salience of speculation, when it is possibly more intense and less coupled to a material basis in profit than at any other time in the history of capitalism. It is not that abstraction replaces materiality, but rather that the abstractions that represent value are more and more distantly coupled (ontologically and temporally) from their materialist bases.

This leads to an interesting temporal incongruence in the chapters, which, by virtue of their emphasis on the particular, are historicist, but which, by virtue of tracing actors who are themselves invested in futures of all sorts, in themselves chart a speculative terrain. The temporalities of the emergent present are caught between the promissory futures of speculative capital and the historical, material conditions of production that give rise to these presents in the first place. A whole range of historical transitions in *institutional* forms are traced in these chapters—for instance, the cooptation of scientific research by the market (Smith's chapter), or new relationships between genome companies, hospitals, and states globally (Lakoff). Also marked are historical transitions *within* the life sciences, with genomics a crucial focal point for a number of the contributors. The particular epistemic transition marked by genomics is the way in which the life sciences, in some materially embodied way, become information sciences. Informatics, indeed, occupies a central place in Smith's and Kim Fortun's chapters. In Lakoff's piece, genomics is a key new epistemology, but one whose operation in the context of mental illness involves the prior definition of what bipolar disorder is—a question that, it turns out, does not have an obvious answer. Interdisciplinary epistemologies appear in Fischer's chapter. There is the emergence of new biosocialities, such as the parent-advocates who become "lay" scientific practitioners, as described in Silverman's piece.[4] There is the historicist tracing of particular conjunctures in Jasanoff's chapter, where particular resolutions of public and private occur in response to particular historical conjunctures of biotechnology that are situated in terms not just of the epistemology of the life sciences or the institutional contexts within which they evolve, but also in terms of state and legal rationalities that are not uniform even among states that subscribe to similar legal forms and precedents. In addition to all these historical ruptures are also historical parallels and continuities (between the dispossession of land and of genetic material in Tanner's piece, for instance), all of which contribute to the historicist yet speculative grounding of emergent phenomena that sees expression in the volume.

3. I have alluded to three registers of history that are at play in the volume: the marking of historical ruptures, parallels, and continuities. But a fourth historicist register looks at the historical conditions of possibility that allow certain types of emergence to occur. These historical conditions of possibility are invariably grounded in what David Harvey (2003) refers to as "accumulation by dispossession," in some kind of original violence in space or time. Tanner shows this with respect to the original dispossessions of indigenous populations, and Peterson in the context of the "emptying out" of material space of Africa. These dispossessions are ongoing—the conditions of possibility for the functioning of speculative capital, in this sense, are not just constituted by one-off historical events marked by violence or expropriation.[5] In addition to this original violence—the continuous destructions required for "the new" to come into being—is an ongoing violence of one-sided representational politics, which is at the heart of many of these stories. Much of the violence of capital indeed operates at the level of representation: who controls the means of production, and who gets to speak for whom, in technocapitalist worlds where modes of production and emergent knowledge forms are increasingly entangled.

There is also the constitutive violence of intimacy that underlies the quality of the encounter value that Haraway writes about. Haraway's stories of encounter value are, at their core, not simply warm and fuzzy stories of dogs technologically inserted in new ways into human hearth and home. The encounter value that animates decisions about dogs' life and death are visceral, poignant, and painful, and extend beyond the bounds of domesticity into, for instance, the prison (for example, the globally disseminated images of the role of guard dogs in Abu Ghraib). This violence is noteworthy not simply because affective encounters, and the obligations they entail, are inherently violent, but because, again, there is a one-sided representational politics to these encounters—decisions about dogs' life and death, even when dogs are deeply inserted into human family structures, are always decisions that are made only by humans. Just as, indeed, decisions about therapeutic consumption practices in the United States are invariably dictated by the interests of the pharmaceutical industry; those about the bounds of environmental discourse invariably made by the security state (the democratic potentials of the Internet notwithstanding); those about aid to Africa invariably made by agents such as American governments, leading economists, or the CEOs of large multinational software companies; and those about rice research priorities by agribusiness or by global donor organizations with First World research agendas in mind. Haraway's diagnosis of the violence constitutive to encounter value is not some simple call to consider the dog's "point of view" in making decisions about its life and death—such a liberal political response, indeed, would be an absurdity. Rather, it is to show, in a manner similar to that of many of the other contributions to this volume, the ways in which capital is animated by complex affective processes that are very difficult to reduce to the dynamics of an exchange mechanism, and further, the ways in which these processes are hardly innocent, even when they might appear to be so, if one pushes an analysis of affect in the deep, historically, and biologically contextualized sense that it ought to be. In that sense, Haraway's political lesson is very similar to those that Peterson and Tanner in particular are trying to emphasize, while pushing the boundaries of exchange, kinship, representation, and affect into the domain of transspecies encounters. What is at stake, in Haraway's words, is marking the "relationalities in companion-species worlds," none of which are "innocent, bloodless, or unfit for serious critical investigation." If there is some "fundamental" logic of capital that persists through these stories of the multiplicity of capitalist forms and logics, then that perhaps has to do with the consistency of original and representational violence that structures them.

4. The chapters pose intriguing questions of space, perhaps most interestingly asking what constitutes the "global." For Peterson, the global takes form in Africa, which is simultaneously the material site of dispossession and the target of global trade, aid, and capital flows. This echoes the arguments of scholars such as James Ferguson (1999), who shows how a theorization of modernity must crucially account for Africa (normally diagnosed as precisely lacking such a modernity, but in fact in the midst of a "decline" that is symbolized by the failure of the dreams and hype of modernity that fueled African economies in the 1960s and 1970s), or Achille Mbembe (2001), who argues that to capture the essence of biopolitics, one needs to look not toward advanced liberal societies that serve as the empirical start-

ing point of Foucault's analyses, but toward Africa. But like capitalism, there are multiple globalizations reflected in the volume. For instance, if the dominant form of globalization in Africa manifests through trade with and aid from advanced liberal societies, then the generics market there is also constituted through the globalization of Indian pharmaceutical companies, which strategize Africa as their "global" market site.

The global is constituted still differently in the case of environmentalism (Kim Fortun's and Choy's chapters), where, as a category, it implies porosity and the ways in which the borders of the nation-state get transcended either by information (Fortun) or by pollution (Choy). "The global," here, is a network, but not one that is seamlessly constituted. Choy shows how this network is, to borrow a term from Anna Tsing (2004), *frictioned*, while Fortun shows how the network is simultaneously constituted at multiple scales. Indeed, part of the way in which globalization is frictioned is precisely through the acts of scale-making that constantly animate it (Tsing 2000). Therefore, Choy's "global" story is a story of Hong Kong (itself an odd sort of locality, one that is always already global in a number of ways), but one that has global, regional, national, local, and sublocal scales interacting with one another.

Fischer's global is located at sites such as Harvard and MIT, or at Singapore's "Biopolis," which, like many other technoscientific institutions, are nodes at which "transnational ligaments of technoscience" get articulated. Once again, questions of scaling become central—in Fischer's stories, there are at least at three interfaces: between lab and clinic, biology and informatics, and university and market. These interfaces are also sites at which friction is produced—the scaling up of laboratory research so that it works in clinical settings is hardly a seamless process, as things that work in one setting often work very differently, or not at all, in another. Similarly, the interdisciplinary coming together of biology and informatics sees the coupling of different knowledge systems and labor practices in ways that have to be constantly worked through by those engaged in the activity of bioinformatics, just as the transition of the life sciences from university settings to market settings is not simply a cooptation of the former by the latter, but constantly sees competing value systems at play, often manifesting in different incentive and normative structures.[6]

These scalar transitions also have to be located against global imperatives of standardization. If the work of scale-making is always frictioned, then constitutive to that work are impulses of homogenization. Hence Kuo's example of attempts at standardizing clinical-trials regulations in Asia in accordance with "global" standards, which itself sees incongruent articulations when one compares Japan, Taiwan, and Singapore, each drawing on different registers of nationalist sentiment, regional power plays, and global ambitions and imaginaries. These standardizing imperatives and epistemic attempts at establishing universality are also seen in Lakoff's piece. If in Kuo's chapter standardization is difficult because of different strategic interests of different state actors, in Lakoff's it is difficult because while technical practices are easy to standardize across borders, epistemology is much harder to do so, not least because of different procedures of medical-record collection in different places. Part of the historical transition that this speaks to is one within the life sciences, from psychoanalysis to molecular medicine, but it is an extremely frictioned transition. More generally, standardization imperatives are seen to exist in contradictory tension with the individualizing imperatives of neoliberal capital, which is particularly marked in the U.S. case that Dumit describes.

Many of the above conditions of the global engage conceptions of the nation-state — most explicitly in Kuo's case, but also in Lakoff's (the flow of biology and capital between Argentina and France) — or other forms of ter ritory (the "city-state" of Hong Kong, the continent of Africa). But there are flows of various sorts that are described in this volume which are not necessarily mediated through the form of the nation-state or similar territorial entity. In other words, the global is never simply a concept that is one step up from the state — global capital is fluid at multiple scales, only some of which are congruent with the scales at which state power is able to regulate it.

5. A number of the chapters are concerned with questions of property, which also implies spatial demarcations (in this case, between public and private). These are not binary categories, but hybrid, as shown in Smith's chapter. There is a relationship of the notions of public and private to the materiality of information — a crucial question is often whether information can or should be made public. This politics of the public domain does not just have to do with property, but also, in Kim Fortun's case, with security.

Property rights mediate the relationship of capital to the state in very particular ways, by bringing it into the purview of state law and rationality (as seen most markedly in Jasanoff's chapter). In that sense, they reflect the territorializing impulses of capital. As Jasanoff shows, particular state rationalities in relation to intellectual property can vary greatly; what is common is the formal relation between the state and capital that gets mediated through the form of property. If property territorializes capital by locating it in land, then the abstraction of "economic development" is that which deterritorializes property, since it, in Jasanoff's words, "overflows the territorial limits of land; it converts place, which sits still, into money, which moves."

6. A common thread of praxis that weaves through the volume is the attempt to denaturalize capitalism. In Sheila Jasanoff's chapter in particular, naturalization and denaturalization are quite literally what are at stake. Determining the bounds of intellectual-property protection involves deciding what is "natural" and what is "cultural," a process that, as she shows, hardly plays out in ways one would naturally expect.

If one component of the exercise in denaturalizing capitalism involves denaturalizing the law, then another involves a denaturalization of liberal notions of ethics that underlie the bioethical enterprise. Tanner, for instance, shows the insufficiency of such an ethics in situations that are overdetermined by violent prehistories of expropriation. "Ethics," in the case of the Human Genome Diversity Project (HGDP), which Tanner studies, was framed entirely in the context of informed consent, which takes individual autonomy to be the ground on which the ethical can be conceived. Such a particular grounding of ethics, as Tanner contends, often does not do justice, which is altogether a more complicated and indeterminate notion.[7]

It is through an insistence on the accumulative tendencies of capital that Tanner critiques informed consent. For him, the issue is not the content of consent, but its form. And so, the HGDP is not rendered problematic because its informed consent procedures were somehow insufficient or culturally insensitive; it is because informed consent, as a liberal contractual procedure, is in itself insufficient to account for the deep historical violence that has consistently been inflicted on indigenous populations, who were the major opponents of the HGDP. The problem is not with the terms of the contract, but the contract itself—a contract that, by its very existence, naturalizes relationships between the capitalist state and native populations as one between "free" contracting agents, and in the process effaces a history of dispossession that, as Tanner insists, is ongoing. Tanner therefore contrasts ethics as the calculable, contractual, rule-governed, liberal form of interaction, to justice, which, as Jacques Derrida (1994) articulated it, is a notion that is both deeply informed by the realities and traumas of history, and inspired by the hopes and fears of a promissory future. It is the temporal incongruence of the notion of justice, always already embedded in the past and the future, that Tanner tries to capture, contrasting the hyperfuturity that genomics is associated with to a "primitive" accumulation which itself continues into the future.

Of course, theories of justice, too, can be grounded in liberalism, as they most famously are by John Rawls (2005 [1971]). Indeed, the fundamental guiding principles on which informed consent for human subjects experimentation in the United States is based are provided by the 1979 Belmont Report, which stresses respect for persons (basically, a concern with individual autonomy and the protection of persons with reduced autonomy), beneficence (a calculus that maximizes the ratio between benefit and harm), and justice (defined in terms of equitable costs and benefits).[8] In the calculus of liberal ethics, justice is precisely that—a calculus, something that can be adequately represented. The Derridean ethics that pervades this volume, and is explicated in Tanner's and also Mike Fortun's chapters, is in contrast to one where justice is that which *cannot* be adequated. Ethics is not being abandoned as a useful category, or even as a category of praxis (there is a belief that the practice of the life sciences needs to be ethically grounded). The attempt here, most explicit in Mike Fortun's paper, is to recover an ethics that is contrasted to the moralizing impulses of bioethics.

The chapters here are therefore often concerned with the insufficiency of the very form and epistemology of liberal ethics. But they are also concerned with the institutional manifestations of bioethics, particularly the instrumentality that pervades an ethical enterprise when it becomes constitutive to corporate agendas also (as Smith shows in her chapter). Ethics is about decisions—which research to prioritize, why, for whom? Who gets to make those choices? There is a crucial relationship between ethical questions and questions of representation, questions that can only be asked if the ethical is not reduced to the moral, on the one hand, and if the ethical does not become an instrument that reproduces larger social and institutional relations of production and power hierarchies, on the other.[9]

7. *Investment*, a central notion in the volume, points to the monetary and affective registers of value that are simultaneously at stake in a number of the chapters. Indeed, affective and subjective dimensions of biocapital are a central part of the analysis that this volume undertakes. Some of the affect is because of the ways in which circuits of exchange are always embedded within circuits *and* epistemologies of kinship. After all, one of the things that genetics speaks to from the beginning is kinship, which is one of the categories whose reconfiguration therefore comes to be most at stake through contemporary advances in the life sciences. As seen in Donna Haraway's analysis, kinship is marked both at such epistemic levels and at familial

levels, with the locus of technocapitalist interactions often being at the level of family ties (the human-canine companionate family in Haraway's account being one such example).

There are other elements of affect that operate independent of an immediate kinship equation. For example, there is a political economy of fear that pharmaceutical marketing speaks to (Dumit), or that is part of the security discourse surrounding ecological informatics (Kim Fortun).[10] There are political economies of desire, as seen in the ways in which the affective economy of the promise works through the discursive apparatus of hype (Mike Fortun) or the sense of desire among professional scientists in Fischer (which gets manifested not just toward the material consequences of their research, but also toward an aesthetics of research). And there is the love that structures stories of autism research in Silverman's chapter, where that love is tied into the irrationality of deep monetary and emotional investment into disease cures in ways that seriously threaten the disinterested norms of science, and into the necessary violence of experimentation on children that is a necessary part of the search for a cure. These registers of affect point to the complicated layers of obligation, commitment, and desire that lie at the heart both of contemporary life-science research and of the mechanisms and logics by which capital operates. A simply structural analysis of either technoscience or capital, the chapters argue, is not enough to understand what is emerging here. These forms of affect are hardly innocent, and contain within them a necessary violence that is as integral to these forms of biocapital as the structural violence that Tanner or Peterson describe is.

8. If lively capital is marked by structural and affective violence, it is also marked by a series of absurdities that a number of the papers point to. The cataloging of absurdity is central to the project of the denaturalization of capital. Consider Dumit's chapter, for instance. What might seem like a dystopian portrayal of the power of pharmaceutical marketing in shaping neoliberal consumption and selfhood is in fact an attempt to mark a historical transition, with "medicine as an arm of capital" now becoming "an industry in itself." Central to this industry of biomedicine is the clinical trial, which serves a function analogous to that of machinery in industrial capital. It is through the mechanism of the clinical trial that risk thresholds become scientific, factual, and authoritative. But in reality, Dumit shows, they are arbitrary and do not arise in any way consequent to the experimental procedures that constitute the trial itself. Therefore, far from pointing to seamlessness, what Dumit does is show the sheer absurdity of pharmaceutical logic. It is logic, however, that is already real. And yet its "reality" operates in the realm

of potential, its potential to grow markets and treat diseases (both these imperatives collapsing seamlessly on one another). Public health becomes market opportunity, in itself an absurd proposition, and yet one that is immediately naturalized and disseminated as inevitable, by physicians, by the pharmaceutical industry, by the press, by government regulatory agencies. It is this double effect of absolute presence and absolute absurdity, chillingly Kafkaesque, that Dumit is trying to present. And that "chill" is central to the affective politics of Dumit's chapter, as he shows how the operation of pharmaceutical logic can never occur through purely discursive mechanisms without calling into being a series of fetishes and fears that are at the heart of biopolitics.[11]

9. Finally, there is the question of praxis. If there is a certain consonance in the analytic and methodological moves of the authors in the volume, some of which I have tried to thread here, then there is also a range of political positions that they adopt. One is affective and subjective—part of Dumit's attempt in tracing pharmaceutical grammar in the manner that he does is not just to diagnose the structure of fear-mongering that is constitutive to that grammar, but to evoke an affect of fear in the reader. Such a provocation of affect is precisely a part of the act of denaturalization of capital that he is engaged in; the key here is to insist that we cannot take pharmaceutical discourses for granted. These affective encounters often demand new epistemologies in themselves, new ways of looking at ourselves in relation to life, the life sciences, and capitalisms (for instance, the re-cognition of trans-species encounters that Haraway calls for in her chapter).

A second is a political position against liberalism, which is articulated through a sentiment that depends deeply on Derrida's deconstructive reading of liberal notions of ethics, and calls into account instead a sense of justice that is both promissory and incalculable. As with the affect of fear, which Dumit simultaneously diagnoses as constitutive to the grammar of capital and invokes as a political response to that grammar, a promissory ethics is both diagnosed as a constitutive feature of speculative technocapital and invoked as a necessary political response to it. One sees this dual relationship to promise, both critical and embracing of it, in Kim Fortun's diagnosis of the possible promissory futures for environmental politics that are provided by ecological informatics. A promissory ethics is also what Mike Fortun is trying to recover in the context of genomics, when one is deeply entangled in the hype surrounding the technoscience in ways that make disentangling fiction from reality in a rational manner basically impossible. And yet one cannot dismiss the discursive apparatus of corporate genomics—one has to adequately respond to the hype and make it accountable. Cynicism, or what Fortun calls an "ethic of suspicion," is just not an effective or engaged enough response in this context—one needs to suspend disbelief without so completely buying into the hype that one suspends the ability to be critical of particular actions taken by particular people or organizations at particular places and times.

In parallel ways, the promissory investments of parent-advocates in Silverman's piece are shown to potentially reconstitute forms and modes of knowledge production—a political economy that parallels the corporate political economy of hype that Mike Fortun speaks of, but where the investment is in terms of the hope for a cure for autism. Fischer writes about the promissory potential of new interdisciplinary and translational endeavors that are entangled in hype and hope (the hype attendant to all high-profile institutional arrangements that see this level of monetary investment; the hope that new research agendas that transcend traditional disciplinary and epistemic confines can emerge).

The ethical position in these chapters then becomes one of thinking about how to keep open the promise of the life sciences and new biotechnologies (which, in a Derridean sense, is a "promise of justice"), while holding its promises accountable (Derrida 1994). The politics here is one that avoids a knee-jerk technological positivism, but that deeply wishes to keep open the promissory possibilities that technoscience provides. It is a politics that is animated by an ethics that is at once experimental and that demands accountability. It is deconstructive at its core, not passing judgment on new technologies, but rather working through and opening up their experimental potential, and yet echoing Derrida's own insistence that the ethic of the Enlightenment is unconditional.[12] There prevails in this volume a promissory register, a belief that biotechnologies, if democratic, open-ended, and made capable of realizing their experimental potential, rather than closed, proprietary, and subjected to the dictates of hegemonic state and corporate power structures, can in fact make the world a better place. There is also a related sensibility that intellectual projects such as those reflected in this volume have the potential to contribute to this betterment, by serving as interlocutors and auditors of these new technologies and their practitioners in more promising ways than economics or bioethics, for instance.[13]

The real question for most of the authors is not one of being in favor of new biotechnologies or against them, or being an enthusiast of capitalism or wishing for its dissolution. Rather, it is one of how to seriously embrace the promise that the life sciences and biotechnologies present to us in ways that are attentive to the histories within which the development of such science and technology has taken shape. These are histories that have, first, been heavily overdetermined by systems and logics of capital in recent times. These systems and logics of capital have been multiple and particular to different places and times, making it impossible to suggest a unitary logic of capitalism or globalization that one can grasp or hold onto. At the same time, there have been certain fundamental aspects to capital that recur—most notably, that capital constantly engages affect and therefore is deeply subjective, and that capital, like its affective manifestations (obligation, commitment, fear, desire, or love) is necessarily violent. What becomes crucial, then, is not the diagnosis or attribution of an essential nature to the life sciences or to capitalism, but an exercise of cataloging the representational politics that are at play in the merger of the two. This is a politics that is often painfully familiar and reproductive of old power hierarchies (of the state or of corporations "taking over" the domain of biocapital), but which also, because of the contradictions inherent to biocapital, offer possibilities for alternative representational possibilities.

Particularity; materialism; promise; history; violence; the global; property; ethics; investment; affect; subjectivity; absurdity; politics; praxis. These then are some of the keywords that thread this volume together in spite of its purposeful absence of a single topical, conceptual, or disciplinary frame. In this brief conclusion, I have tried both to thread together and to highlight the incongruence across the papers in this volume, as they attempt simultaneously to describe the rapidly changing worlds of technocapital and to intervene at multiple registers—conceptually, in debates within philosophy and the human sciences; reflexively, in methodological debates within the disciplines and interdisciplines that the authors inhabit; and perhaps in subtle or pronounced ways in the social worlds that constitute the objects of study for the authors here.

Notes

1. This synthesis is deeply influenced by Jean Comaroff's concluding remarks at the Lively Capital workshop, for which I offer many thanks.

2. These three questions were foundational to the method of Marx's philosophy, which is why I would argue that even those chapters in this volume that do not explicitly engage Marx cannot escape a Marxian inheritance.

3. Donna Haraway, transcripts from Lively Capital workshop, 7 November 2004, transcribed by Elta Smith.

4. See also Gibbons and Novas 2007 for an edited collection that specifically studies emergent biosocialities.

5. See also Sunder Rajan 2005 for an elaboration of this logic of expropriation in the establishment of biocapital in the mill districts of Bombay.

6. The engagement with scale and friction as constitutive to global networks is implicitly an argument with actor-network theory, which has been a powerful frame for thinking about scientific practice within STS (for instance, Callon 1986; Latour 1987), particularly with the seamlessness that is constitutive in these accounts to processes of knowledge production, with technoscience often getting reduced to an instrumental act of resource generation.

7. Gail Davies (2006) provides a useful conceptualization of justice in the context of inequities in organ transplantation in London.

8. See http://www.hhs.gov/ohrp/humansubjects/guidance/belmont.html.

9. For the relationship between the ethical and the moral, see Michael Fortun 2000; Kleinman 1999.

10. See also Comaroff and Comaroff 2006, which shows the ways in which fear is central to the fetish that is at the heart of the apparatus of biopolitics, using as a case study the development and use of the police state in post-Apartheid South Africa.

11. I think here of the comedians Jon Stewart and Stephen Colbert, and their extremely effective response to media and politics in the United States by pointing to their utter absurdity. Their mode of critique is not denunciatory, but adopts the form of parody. Importantly, it is parody that takes the absurdities that it makes fun of at face value and very seriously. This lets the absurdities, in a sense, speak for themselves. Douglas Adams, another favorite of mine, adopted precisely such a critical posture in lampooning late-modern capitalist (in his case, British) culture and society in the *Hitchhiker's Guide to the Galaxy* series. Such critiques do not pretend to be outside the material and ideological apparatus of capital that they critique, as denunciatory critiques often do. After all, Stewart and Colbert inhabit the very world of big media that they make fun of. It is because these are critiques that point to absurdity from the inside, and therefore have an element of reflexivity as constitutive to them, that they are so effective.

12. Derrida's chapter that Fischer cites in this regard is "The Aforementioned So-Called Human Genome" (Derrida 2002).

13. This led to an interesting, generative, and as yet unresolved tension (perhaps one that does not have a resolution?) in discussions at the workshop, for it is in this sensibility that the anthropology of science such as it is represented here differs from recent work in the anthropology of finance that has also studied the liveliness of capital in interesting ways, albeit grounded in different empirical material that does not engage emergent epistemologies of the life sciences in ways this collection does. An important representation of that work is Bill Maurer's (for instance, Maurer 2005), who made the following critique at the workshop: "The question of empirical has been troubling me all along. . . . We need to un-sunder evidence and ethics. . . . [This is a] plea for thinking more seriously about [the] ethical form of evidence being deployed. Makes me want to ask why return to Marx of *Capital*. Why is it so comfort-

ing? Why that now? What if we disappear the empirical—what pops up? If we literally dissolve those, what do we see? Worlds where the very question of evidence and [the] real is already obviated? I think one of the things from anthropology of finance and law is that . . . listening in on sts people . . . [it seems that] projects are going to get in there and change the world with tools. If only I had the right hammer, then I could change the world. The world follows the word. . . . For finance people, nothing really matters because if you fail you win. You never really lose."

Legal Cases

Diamond v. Chakrabarty, 447 U.S. 303 (1980).

Dolan v. City of Tigard, 512 U.S. 374 (1994).

Dred Scott v. Sandford, 60 U.S. (19 How.) 393, 15 L. Ed. 691 (1857).

Johnson v. Calvert, 5 Cal.4th 84 (1993).

Kelo v. City of New London, 545 U.S. 469 (2005).

Laboratory Corp. of America v. Metabolite Laboratories. 548 U.S. 124 (2006).

Linda Rubin v. deCODE Genetics, Inc, Lehman Brothers, Inc., Morgan Stanley and Co. Incorporated, Kári Stefánsson, and Axel Nielsen, No. 01 Civ. 11219, U.S. District Court, Southern District of New York (2001).

Lucas v. South Carolina Coastal Council, 505 U.S. 1003 (1992).

Marx v. Computer Sciences Corporation, 507 F.2d 485 (9th Cir. 1974).

Matter of Quinlan, 70 N.J. 10 (1976).

Moore v. Regents of the University of California, 793 P.2d 479 (Cal. 1990).

Nollan v. California Coastal Commission, 483 U.S. 825 (1987).

Palazzolo v. Rhode Island et al., 533 U.S. 606 (2001).

President and Fellows of Harvard College v. Canada (Commissioner of Patents) SCC 76 (2002).

SEC v. Morgan Stanley & Co. Incorporated, No. 03 Civ. 2948, U.S. District Court, Southern District of New York (2005).

Washington University v. Catalona, 490 F.3d 667 (8th Cir. 2007).

Abate, Tom. 2000. "Call for Access to Genome Data." *San Francisco Chronicle*, 15 March.

Abelmann, Walter, ed. 2004. *The Harvard-MIT Division of Health Sciences and Technology: The First Twenty-Five Years 1970–1995.* Cambridge: Harvard-MIT Division of Health Sciences and Technology.

Abraham, John. 2002. "The Pharmaceutical Industry as Political Player." *Lancet* 360: 1498–1502.

Abraham, John, and Tim Reed. 2002. "Progress, Innovation and Regulatory Science in Drug Development: The Politics of International Standard-Setting." *Social Studies of Science* 32, no.2:337–69.

Abu-Lughod, Lila. 1986. *Veiled Sentiments: Honor and Poetry in a Bedouin Society.* Berkeley: University of California Press.

"Access to the Literature: The Debate Continues." 2004. *Nature Web Focus,* http://www.nature.com, accessed 12 April 2008.

ACT UP Paris. 2002. "'Accelerating Access' Serves Pharmaceutical Companies while Corrupting Health Organizations." Press release. http://www.actupny.org/reports/Durban-accessPR.html.

Adams, J. B., and C. Holloway. 2004. "Pilot Study of a Moderate Dose Multivitamin/ Mineral Supplement for Children with Autistic Spectrum Disorder." *Journal of Alternative and Complementary Medicine* 10, no. 6:1033–39.

Adorno, Theodor. 1991. "The Essay as Form." *Notes to Literature,* vol. 1, 3–23. New York: Columbia University Press.

Afsah, Shakeb, Benoit Laplante, and David Wheeler. 1996. "Controlling Industrial Pollution: A New Paradigm." Working Paper No. 1672. World Bank, Policy Research Department, May.

Agamben, Giorgio. 1998. *Homo Sacer: Sovereign Power and Bare Life.* Stanford: Stanford University Press.

Ake, Claude. 1996. *Democracy and Development in Africa.* Washington: Brookings Institution.

Akiskal, Hagop S. 1996. "The Prevalent Clinical Spectrum of Bipolar Disorders: Beyond DSM-IV." *Journal of Clinical Psychopharmacology* 16, no. 2 (supplement 1): 4–14S.

Alder, Ken. 1998. "Making Things the Same: Representation, Tolerance and the End of the *Ancien Régime* in France." *Social Studies of Science* 28, no. 4:499–545.

Plotkin, Hal. "All Hail Creative Commons: Professor and Author Lawrence Lessig Plans a Legal Insurrection." San Francisco Chronicle. 11 February 2002. http://www.SFGate.com, accessed 24 October 2004.

Althusser, Louis. 1969 [1965]. *For Marx,* trans. Ben Brewster. New York: Pantheon.

———. 2006. *Philosophy of the Encounter: Later Writings, 1978–87,* ed. François Matheron and Oliver Corpet. London: Verso.

Amani, Bita, and Rosemary J. Coombe. 2005. "The Human Genome Diversity Project: The Politics of Patents at the Intersection of Race, Religion, and Research Ethics." *Law and Policy* 27:152–88.

American Psychiatric Association. 1994. *Diagnostic and Statistical Manual of Mental Disorders: DSM-IV.* Washington: American Psychiatric Association.

———. 2000. *Diagnostic and Statistical Manual of Mental Disorders: DSM-IV-TR.* Washington: American Psychiatric Association.

Amos, Jonathan. 2002. "Genome Dispute Touches Rice." *BBC News,* 4 April. http://news.bbc.co.uk/2/hi/science/nature/1910889.stm.

Anderson, Benedict. 1983. *Imagined Communities: Reflections on the Origin and Spread of Nationalism*. London: Verso.

Anderson, H. Ross, Antonio Ponce de Leon, J. Martin Bland, Jonathan S. Bower, and David P. Strachan. 1996. "Air Pollution and Daily Mortality in London: 1987–92." *British Medical Journal* 312, no. 7032:665–69.

Andrews, Lori. 2006. "Who Owns Your Body? A Patient's Perspective on *Washington University v. Catalona*." *Journal of Law, Medicine and Ethics* 34, no. 2:398–407.

Andrews, Richard. 2003. "Risk-Based Decision Making." *Environmental Policy: New Directions for the Twenty-First Century*, 5th edn, ed. Norman Vig and Michael Kraft, 215–38. Washington: CQ Press.

Angst, J. 1998. "The Emerging Epidemiology of Hypomania and Bipolar II Disorder." *Journal of Affective Disorders* 50, nos. 2–3:143–51.

Aoki, Keith. 1998. "Neocolonialism, Anticommons Property, and Biopiracy in the (Not-So-Brave) New World Order of International Intellectual Property Protection." *Indiana Journal of Global Legal Studies* 6:11.

Appadurai, Arjun. 1991. "Disjuncture and Difference in the Global Economy." *Public Culture* 2, no. 2:1–24.

———. 2001. *Globalization*. Durham: Duke University Press.

Applbaum, Kalman. 2006. "Educating for Global Mental Health: American Pharmaceutical Companies and the Adoption of ssris in Japan." *Global Pharmaceuticals: Ethics, Markets, Practices*, ed. Adriana Petryna, Andrew Lakoff, and Arthur Kleinman, 85–110. Durham: Duke University Press.

Apter, Andrew. 2005. *The Pan-African Nation: Oil and the Spectacle of Culture in Nigeria*. Chicago: University of Chicago Press.

Arendt, Hannah. 1968. *The Origins of Totalitarianism*. New York: Harcourt.

Arnold, G., S. L. Hyman, R. A. Mooney, and R. S. Kirby. 2003. "Plasma Amino Acids Profiles in Children with Autism: Potential Risk of Nutritional Deficiencies." *Journal of Autism and Developmental Disorders* 33, no. 4:449–54.

Arrighi, Giovanni. 1994. *The Long Twentieth Century: Money, Power, and the Origins of Our Times*. New York: Verso.

Austin, J. L. 1962. *How to Do Things with Words*. Cambridge: Harvard University Press.

Autism Research Institute. 2005. "Treatment Options for Mercury/Metal Toxicity and Autism and Related Developmental Disabilities: Consensus Position Paper." San Diego: Autism Research Institute, http://www.autism.com, accessed 10 January 2008.

Badiou, Alain. 2003. *Saint Paul: The Foundation of Universalism*. Stanford: Stanford University Press.

Bakhtin, Mikhail. 1984. *Problems of Dostoevsky's Poetics*. Ed. and trans. Caryl Emerson. Minneapolis: University of Minnesota Press.

Balibar, Étienne. 1994. *Masses, Classes, Ideas: Studies on Politics and Philosophy before and after Marx*. New York: Routledge.

———. 1995a. "Ambiguous Universality." *Differences* 7, no. 1:48–74.

———. 1995b. *The Philosophy of Marx*, trans. Chris Turner. London: Verso.

Barry, Andrew. 2001. *Political Machines: Governing a Technological Society*. London: Athlone.

Bartfai, Tamas, and Graham V. Lees. 2006. *Drug Discovery: From Bedside to Wall Street*. Amsterdam: Elsevier/Academic.

Barton, John, and Peter Berger. 2001. "Patenting Agriculture." *Issues in Science and Technology* 17, no. 4:43–50.

Barzyk, T., K. Conlon, T. Chahine, D. Hammond, V. Zartarian, and B. Schultz. 2009. "Tools Available to Communities for Conducting Cumulative Exposure and Risk Assessments." *Journal of Exposure Science and Environmental Epidemiology* 20:371–84.

Bateson, Gregory. 1972. *Steps to an Ecology of Mind: Collected Essays in Anthropology, Psychiatry, Evolution, and Epistemology*. San Francisco: Chandler.

Battiata, Mary. 2004. "Whose Life Is It, Anyway?" *Washington Post*, 2 August.

Bayart, Jean-François. 1997. *The Criminalization of the State in Africa*. Bloomington: Indiana University Press.

Beijing Genomics Institute. 2001. "Rice Genome Sequencing: The Science, the International Effort, and the Chinese Contribution." *Rice GD: Genome Database of Chinese Super Hybrid Rice*. http://rice.genomics.org.cn/rice/index2.jsp, accessed 25 April 2008.

Bennetzen, Jeffrey. 2002. "Opening the Door to Comparative Plant Biology." *Science* 296, no. 5565:60–63.

Bergel, Salvador. 1998. "Patentamiento de genes y secuencias de genes." *Revista de Derecho y Genoma Humano* 8:31–59.

Berman, Marshall. 1988. *All That Is Solid Melts into Air: The Experience of Modernity*. New York: Viking Penguin.

Bernstein, Charles. 1992. *A Poetics*. Cambridge: Harvard University Press.

Berton, Justin. 2006. "Hotels for the Canine Carriage Trade." *San Francisco Chronicle*, 13 November.

Bettelheim, Bruno. 1967. *The Empty Fortress: Infantile Autism and the Birth of the Self*. New York: Free Press.

Bhabha, Homi K. 1994. *The Location of Culture*. New York: Routledge.

Bhandari, Bhupesh. 2005. *The Ranbaxy Story: The Rise of an Indian Multinational*. New Delhi: Penguin Viking.

Biehl, João. 2001. "Vita: Life in the Zone of Social Abandonment." *Social Text* 19, no. 3 (September): 131–49.

Bloom, Floyd. 2000. "Rices, Races, and Riches." *Science* 288, no. 5468:973.

Bloor, David. 1991 [1976]. *Knowledge and Social Imagery*. Chicago: University of Chicago Press.

Blumenfeld, Marta, et al. 2002. "Genes, Proteins and Biallelic Markers Related to Central Nervous System Disease." U.S. Patent and Trade Office, United States Patent Application 20020081584.

Bollier, David. 2002. *Why the Public Domain Matters: The Endangered Wellspring of Creativity, Commerce, and Democracy*. Washington: New America Foundation.

Bourdieu, Pierre. 1980. *The Logic of Practice*, trans. Richard Nice. Stanford: Stanford University Press.

Bowker, Geoffrey. 2001. "Biodiversity Datadiversity." *Social Studies of Science* 309, no. 5:643–84.

Bowker, Geoffrey, and Susan Leigh Star. 1999 *Sorting Things Out: Classification and Its Consequences*. Cambridge: MIT Press.

Boyle, James. 1996. *Shamans, Software, and Spleens: Law and the Constitution of the Information Society*. Cambridge: Harvard University Press.

———. 2003. "Foreword: The Opposite of Property." *Law and Contemporary Problems* 66, no. 1.

Bradbury, Jane. 2001. "Teasing Out the Genetics of Bipolar Disorder." *Lancet* 357 (19 May): 1596.

Bradley, Janis. 2006. *Dog Bites: Problems and Solutions*. Animals and Society Institute Policy Paper. Baltimore: Animals and Society Institute.

Bralley, Alexander, and Richard S. Lord. 2002. *Laboratory Evaluations in Molecular Medicine: Nutrients, Toxicants, and Cell Regulators*. Norcross, Ga.: Institute for Advances in Molecular Medicine.

Branson, Douglas. 1998. "Securities Litigation in State Courts: Something Old, Something New, Something Borrowed." *Washington University Law Quarterly* 76:509–36.

Brayne, C., and P. Calloway. 1988. "Normal Ageing, Impaired Cognitive Function, and Senile Dementia of the Alzheimer's Type: A Continuum?" *Lancet* 1:1265–67.

Bremner, G. Alex, and David P. Y. Lung. 2003. "Spaces of Exclusion: The Significance of Cultural Identity in the Formation of European Residential Districts in British Hong Kong, 1877–1904." *Environment and Planning D: Society and Space* 21, no. 2:223–52.

Brenner, Neil. 1999. "Globalization as Reterritorialisation: The Re-scaling of Urban Governance in the European Union." *Urban Studies* 36, no. 3:431–51.

Bright, A. O., and O. Taylor. 1999. "A Socio-Economic and Demographic Assessment of the Extent of Self Medication in Nigeria." *West African Journal of Pharmacy* 13, no. 2:74–77.

Brotherton, Sean. 2008. "We Have to Think like Capitalists but Continue Being Socialists: Emergent Capital, Shifting Ideologies, and Cuba's Changing Health Sector." *American Ethnologist* 35, no. 2:259–74.

Brown, P. 1992. "Popular Epidemiology and Toxic Waste Contamination: Lay and Professional Ways of Knowing." *Journal of Health and Social Behavior* 33, no. 3:267–81.

Brown, P., et al. 2001. "A Gulf of Difference: Disputes over Gulf War-Related Illnesses." *Health and Illness* 42, no. 3:235–57.

Brown, Sasha. 2004. "Biology Program Launched: Computational and Systems Biology Program Is First of Its Kind in U.S." *Tech Talk* 49, no. 7 (27 October): 1, 5.

Brumfiel, Geoff. 2005. "Chinese Students in the U.S.: Taking a Stand." *Nature* 438, no. 7066 (17 November): 278–79.

Burchell, Graham, Colin Gordon, and Peter Miller, eds. 1991. *The Foucault Effect: Studies in Governmentality: With Two Lectures by and an Interview with Michel Foucault*. Chicago: University of Chicago Press.

Burr, Benjamin. 1997. "An International Collaboration to Sequence the Rice Genome." http://web.archive.org/web/20040128094512/http://rgp.dna.affrc.go.jp/rgp/News/Newsletter.html, accessed 25 April 2008.

———. 1999. *Oryza: Newsletter for International Rice Genome Sequencing Project* (January), http://web.archive.org/web/20000817044232; http://rgp.dna.affrc.go.jp/rgp/News/Newsletter.html, accessed 25 April 2008.

Burr, Benjamin, and Takuji Sasaki. 2002. "International Rice Genome Sequencing Project." http://web.archive.org/web/20061002013440/rgp.dna.affrc.go.jp/E/IRGSP/bnl/Guidelines.html, accessed 25 April 2008.

Butler, Declan. 2002. "Geneticists Get Steamed Up over Public Access to Rice Genome. *Nature* 416:111–12.

Butler, Declan, and Peter Pockley. 2000. ". . . as Monsanto Makes Rice Genome Public." *Nature* 404, no. 6778:534–36.

Butler, Judith. 2000a. "Competing Universalities." *Contingency, Hegemony, Universality: Contemporary Dialogues on the Left*, ed. Judith Butler, Ernesto Laclau, and Slavoj Žižek, 136–81. London: Verso, 2000.

———. 2000b. "Dynamic Conclusions." *Contingency, Hegemony, Universality: Contemporary Dialogues on the Left*, ed. Judith Butler, Ernesto Laclau, and Slavoj Žižek, 263–80. London: Verso.

Butler, Judith, Ernesto Laclau, and Slavoj Žižek, eds. 2000. *Contingency, Hegemony, Universality: Contemporary Dialogues on the Left*. London: Verso.

Buttel, Frederick, and Jill Belsky. 1987. "Biotechnology, Plant Breeding, and Intellectual Property: Social and Ethical Dimensions." *Science, Technology, and Human Values* 12, no. 1:31–49.

Caffentzis, Constantine George. 1995. "Fundamental Implications of the Debt Crisis for Social Reproduction in Africa." In *Paying the Price: Women and the Politics of International Economic Strategy*, ed. Maria Rosa Dalla Costa and Giovanna Dalla Costa, 15–41. London: Zed.

Callon. Michel. 1986. "Some Elements of a Sociology of Translation: Domestication of the Scallops and the Fishermen of St. Brieuc Bay." *Power, Action, and Belief: A New Sociology of Knowledge?*, ed. John Law, 196–233. London: Routledge and Kegan Paul.

———. 1998. "The Embeddedness of Economic Markets in Economics." *The Laws of the Markets*, ed. Michel Callon, 1–57. Oxford: Blackwell.

Callon, Michel, Yuval Millo, and Fabian Muniesa, eds. 2007. *Market Devices*. London: Wiley-Blackwell.

Cantor, David. 1992. "Cortisone and the Politics of Drama, 1949–55." *Medical Innovations in Historical Perspective*, ed. John V. Pickstone, 165–84. New York: St. Martin's.

Cantrell, Ronald, and Timothy Reeves. 2002. "The Cereal of the World's Poor Takes Centre Stage." *Science* 296:53.

Carruthers, Bruce, and Arthur Stinchcombe. 1999. "The Social Structure of Liquidity: Flexibility, Markets, and States." *Theory and Society* 28, no. 3:353–82.

Castells, Manuel. 2000. *The Rise of Network Society*. Malden, Mass.: Blackwell.

Centers for Disease Control. 2005. "Are You the Picture of Health?" Screen for Life: National Colorectal Cancer Action Campaign. Centers for Disease Control.

Chadarevian, Soraya de. 2002. *Designs for Life: Molecular Biology after World War II*. Cambridge: Cambridge University Press.

Charan, Ram, and Noel M. Tichy. 1998. *Every Business Is a Growth Business: How Your Company Can Prosper Year after Year*. New York: Three Rivers.

Cheah, Pheng, and Bruce Robbins, eds. 1998. *Cosmopolitics: Thinking and Feeling Beyond the Nation*. Minneapolis: University of Minnesota Press.

Chen, K.C. 1998. "ICH in Taiwan." *Drug Information Journal*, 32:1301–7S.

Choy, Timothy K. 2005. "Articulated Knowledges: Environmental Forms after Universality's Demise." *American Anthropologist* 107, no. 1:5–18.

Clarke, Adele. 2007. "Reflections on the Reproductive Sciences in Agriculture in the U.K. and U.S., ca. 1900–2000." *Studies in History and Philosophy of Biological and Medical Sciences* 38, no. 2:316–39.

Clifford, James, and George Marcus. 1986. *Writing Culture: The Poetics and Politics of Ethnography*. Berkeley: University of California Press.

Coca-Cola Company. 1996. *The Coca-Cola Company 1996 Annual Report*. Atlanta: Coca-Cola Company.

Coghlan, Andy. 1996. "Chinese Deal Sparks Eugenics Protests." *New Scientist* 2056 (16 November): 4.

Cohen, Lawrence. 1998. *No Aging in India: Alzheimer's, the Bad Family, and Other Modern Things*. Berkeley: University of California Press.

———. 1999. "Where It Hurts: Indian Material for an Ethics of Organ Transplantation." *Daedelus* 128:135–65.

———. 2005. "Operability, Bioavailability, and Exception." *Global Assemblages: Technology, Politics, and Ethics*, ed. Aihwa Ong and Stephen J. Collier, 79–90. Malden, Mass: Blackwell.

Cohn, Carol. 1987. "Sex and Death in the Rational World of Defense Intellectuals." *Signs* 12, no. 4:687–728.

Coleman, Gabriella. 2005. "The Social Construction of Freedom in Free and Open Source Software: Hackers, Ethics, and the Liberal Tradition." Ph.D. diss., University of Chicago, Department of Anthropology.

Collier, Stephen, and Aihwa Ong. 2005. "Global Assemblages, Anthropological Problems." *Global Assemblages: Technology, Politics, and Ethics*, ed. Aihwa Ong and Stephen J. Collier, 3–21. Malden, Mass: Blackwell.

Collins, Harry. 1974. "The TEA Set: Tacit Knowledge and Scientific Networks." *Science Studies* 4:165–86. Reprinted in *Science in Context: Readings in the Sociology of Science*, ed. Barry Barnes and David Edge, 44–64. Cambridge: MIT Press, 1982.

Comaroff, Jean, and John Comaroff. 2000. *Millennial Capitalism and the Culture of Neoliberalism*. Durham: Duke University Press.

Comaroff, John, and Jean Comaroff. 2006. "Figuring Crime: Quantifacts and the Production of the Un/Real." *Public Culture* 18, no. 1:209–46.

Cook, Arthur G. 2006. *Forecasting for the Pharmaceutical Industry: Models for New Product and In-Market Forecasting and How to Use Them*. Aldershot: Gower.

Coombe, Rosemary. 1996. "Left Out on the Information Highway." *Oregon Law Review* 75, no. 237:237–48.

———. 1998. *The Cultural Life of Intellectual Properties: Authorship, Appropriation, and the Law*. Durham: Duke University Press.

Cooper, Frederick. 2002. *Africa since 1940: The Past of the Present*. Cambridge: Cambridge University Press.

Cooper, Melinda. 2008. *Life as Surplus: Biopolitics in the Neo-liberal Era*. Seattle: University of Washington Press.

Corelli, Rae, with Brian Bergman. 1997. "Getting the Call." *MacLean's* 110, no. 41 (13 October): 14.

Covello, Vincent T., and Frederick H. Allen. 1988. *Seven Cardinal Rules of Risk Communication*. Washington: Environmental Protection Agency.

Craddock, Nick, and Ian Jones. 1999. "Genetics of Bipolar Disorder," *Journal of Medical Genetics* 36, no. 8:585–94.

Creative Commons. 2007. "History." Creative Commons Wiki, http://wiki.creativecommons.org, accessed 12 April 2008.

Crenshaw, Kimberlé. 1990. "Demarginalizing the Intersection of Race and Sex: A Black Feminist Critique of Antidiscrimination Doctrine, Feminist Theory and Antiracist Politics." *The Politics of Law: A Progressive Critique*, 2d edn, ed. David Kairys, 195–217. New York: Pantheon.

———. 1991. "Mapping the Margins: Intersectionality, Identity Politics, and Violence against Women of Color." *Stanford Law Review* 34:1241–99.

Crichton, Michael. 2006. "This Essay Breaks the Law." *New York Times*, 19 March, WK-13.

Cronon, William. 1991. *Nature's Metropolis: Chicago and the Great West*. New York: W. W. Norton.

Crow, T. J. 1986. "The Continuum of Psychosis and Its Implication for the Structure of the Gene." *British Journal of Psychiatry* 149:419–29.

Curran, Laura C., et al. 2007. "Behaviors Associated with Fever in Children with Autism Spectrum Disorders." *Pediatrics* 120:e1386–92.

Das, Veena. 2002. "The Practise of Organ Transplants: Networks, Documents, Translations." *Living and Working with the New Medical Technologies*, ed. Margaret Lock, Allan Young, and Alberto Cambrosio, 263–87. Cambridge: Cambridge University Press.

Daston, Lorraine, and Peter Galison. 1992. "The Image of Objectivity." *Representations* 40:81–128.

Data Protection Act. 1998. Council Directive No. 96/9/EC of 11 March 1996. *Official Journal of the European Communities*, L77, 27.3.96.

Davies, Gail. 2006. Patterning the geographies of organ transplantation: corporeality, generosity and justice. *Transactions of the Institute of British Geographers* 31, no. 3:257–71.

de Landa, Manuel. 2006. *A New Philosophy of Society: Assemblage Theory and Social Complexity.* New York: Continuum.

Deleuze, Gilles, and Felix Guattari. 1987 [1980]. *A Thousand Plateaus: Capitalism and Schizophrenia.* Translated, with a foreword by Brian Massumi. Minneapolis: University of Minnesota Press.

DeNicola, Lane. 2007. "Techniques of the Environmental Observer: India's Earth Remote Sensing Program in the Age of Global Information." Ph.D. diss., Rensselaer Polytechnic Institute, Department of Science and Technology Studies.

Derrida, Jacques. 1974 [1967]. *Of Grammatology.* Baltimore: John Hopkins University Press.

———. 1989–90. "Force of Law: The Mystical Foundation of Authority." *Cardozo Law Review* 11, nos. 5–6:920–1046.

———. 1994. *Specters of Marx: The State of the Debt, the Work of Mourning, and the New International.* New York, Routedge.

———. 1998. *Monolingualism of the Other: Or, The Prosthesis of Origin.* Palo Alto: Stanford University Press.

———. 2001. *On Cosmopolitanism and Forgiveness.* New York: Routledge.

———. 2002. *Negotiations.* Stanford: Stanford University Press.

Derry, Margaret. 2003. *Bred for Perfection: Shorthorn Cattle, Collies, and Arabian Horses since 1800.* Baltimore: Johns Hopkins University Press.

Descola, Philippe. 1996. *The Spears of Twilight: Life and Death in the Amazon Jungle.* New York: New Press.

Diawara, Manthia. 1998. "Toward a Regional Imaginary in Africa." *The Cultures of Globalization,* ed. Fredric Jameson and Masao Miyoshi, 103–24. Durham: Duke University Press.

Dietz, Thomas, Elinor Ostrom, and Paul Stern. 2007. "The Struggle to Govern the Commons." *Science* 302:1907–12.

DiMasi, J. A., R. W. Hansen, H. G. Grabowski, and L. Lasagna. 1991. Cost of innovation in the pharmaceutical industry. *Journal of Health Economics* (July): 107–42.

Dixon, Bernard. 1988. "Genetic Engineers Call for Regulation." *Scientist* 2, no. 8 (2 May): 22.

Dodier, Nicolas. 1998. "Clinical Practice and Procedures in Occupational Medicine." *Differences in Medicine: Unraveling Practices, Techniques, and Bodies,* ed. Marc Berg and Annemarie Mol, 53–85. Durham: Duke University Press.

Doll, John J. 1998. "The Patenting of DNA." *Science* 280:689–90.

———. 2001. "Staking Claims: Talking Gene Patents." *Scientific American* 285, no. 2:28.

Doull, John. 2001. "Toxicology Comes of Age." *Annual Review of Pharmacology and Toxicology* 41:1–21.

Downey, Gary Lee, and Joseph Dumit, eds. 1998. *Cyborgs and Citadels: Anthropological Interventions in Emerging Sciences and Technologies.* Santa Fe: School of American Research.

Doyle, Richard. 1997. *On Beyond Living: Rhetorical Transformations of the Life Sciences.* Stanford: Stanford University Press.

———. 2003. *Wetwares: Experiments in Postvital Living.* Minneapolis: University of Minnesota Press.

Dumit, Joseph. 2002. "Drugs for life." *Molecular Interventions* 2 (2002): 124–27.

Dumit, Joseph. 2009a. "Normal Insecurities, Healthy Insecurities." *The Insecure American: How We Got Here and What We Should Do about It*, ed. Hugh Gusterson and Catherine Besteman, 163–81. Berkeley: University of California Press.

———. 2009b. "Pharmaceutical Witnessing: Drugs for Life in an Era of Direct-to-Consumer Advertising." *Technologized Images, Technologized Bodies*, ed. Jeannette Edwards, Penny Harvey, and Peter Wade, 37–84. New York: Berghahn, 2010.

Eccles, Robert G., and Sarah C. Mavrinac. 1995. "Improving the Corporate Disclosure Process." MIT *Sloan Management Review*, 15 July.

Edelson, Ed. 2006. "U.S. Experts Issue New Heart Disease Treatment Guidelines." *ExpressScripts Drug Digest.* http://www.drugdigest.org/wps/portal/ddigest.

Edwards, Paul. 1996. *The Closed World: Computers and the Politics of Discourse in Cold War America.* Cambridge: MIT Press.

———. 1999. "Global Climate Science, Uncertainty and Politics: Data-Laden Models, Model-Filtered Data." *Science as Culture* 8, no. 4:437–72.

———. 2010. *A Vast Machine: Computer Models, Climate Data, and the Politics of Global Warming.* Cambridge: MIT Press.

Ehrlich, Jennifer. 2000. "A Breath of Fresh Poison." *South China Morning Post*, 14 August, 13.

Elbe, Stefan. 2005. "AIDS, Security, Biopolitics." *International Relations* 19, no. 4:403–19.

Elyachar, Julia. 2005. *Markets of Dispossession: NGOs, Economic Development, and the State in Egypt.* Durham: Duke University Press.

Ember, Louis. 2007a. "Chemical Security." *Chemical and Engineering News* 85, no. 21 (6 May): 8.

———. 2007b. "Securing Chemicals." *Chemical and Engineering News* 85, no. 46 (12 November): 11.

Engels, Frederick. 1972 [1895]. "Engels to J. Bloch in Königsberg," letter of 21–22 September 1890. *Historical Materialism (Marx, Engels, Lenin)*, trans. Brian Baggins, 294–96. Moscow: Progress.

Enserink, Martin. 1998. "Physicians Wary of Scheme to Pool Icelanders' Genetic Data." *Science* 281:890–91.

Environmental Defense Fund. 2003. "About Us" Environmental Defense Fund, http://www.edf.org, accessed 13 January 2009.

Epstein, Steven. 1996. *Impure Science: AIDS, Activism, and the Politics of Knowledge.* Berkeley: University of California Press.

———. 2007. *Inclusion: The Politics of Difference in Medical Research.* Chicago: University of Chicago Press.

Erickson, C. A., et al. 2005. "Gastrointestinal Factors in Autistic Disorder: A Critical Review." *Journal of Autism and Developmental Disorders* 35, no. 6:713–27.

Escamilla, Michael A., et al. 1999. "Assessing the Feasibility of Linkage Disequilibrium Methods for Mapping Complex Traits: An Initial Screen for Bipolar Disorder Loci on Chromosome 18." *American Journal of Human Genetics* 64:1670–78.

Espeland, Wendy, and Mitchell Stevens. 1998. "Commensuration as a Social Process." *Annual Review of Sociology* 24:314–43.

Ernst and Young. 2006. "Contract Research: Contracted for Trouble?" *R&D Directions* 12, no. 5 (May). http://www.pharmalive.com/magazines/randd.

Evans, Peter, Dietrich Rueschemeyer, and Theda Skocpol, eds. 1985. *Bring the State Back In.* Cambridge: Cambridge University Press.

Express Scripts. 2007. *2006 Drug Trend Report.* St. Louis: Express Scripts.

Fanon, Frantz. 1966. *The Wretched of the Earth.* New York: Grove.

Farquhar, Judith. 2002. *Appetites: Food and Sex in Postsocialist China.* Durham: Duke University Press.

Fassin, Diddier. 2004. "Citizens and Humans: Competing Paradigms in the Production of the Political Subject in South African AIDS." Paper presented at the "Reframing Infectious Diseases" conference, Institute for the Humanities, University of Michigan, 4 December.

Feld, Steven. 1996. "Waterfalls of Song: An Acoustemology of Place Resounding in Bosavi, Papua New Guinea." *Senses of Place,* ed. Steven Feld and Keith H. Basso, 91–136. Santa Fe: School of American Research Press.

Feld, Steven, and Keith H. Basso, eds. 1996. *Senses of Place.* Santa Fe: School of American Research Press.

Feldman Maryann, Alessandra Colaianni, and Connie Kang Liu. 2007. "Lessons from the Commercialization of the Cohen-Boyer Patents: The Stanford University Licensing Program." *Health and Agricultural Innovation: A Handbook of Best Practice,* ed. Anatole Krattiger, Richard T. Mahoney, Lita Nelsen, Jennifer A. Thomson, Alan B. Bennett, Kanikaram Satyanarayana, Gregory D. Graff, Carlos Fernandez, and Stanley P. Kowalski, 1797–1808. Oxford: MIHR/PIPRA.

Felman, Shoshana. 2002. *The Scandal of the Speaking Body: Don Juan with J. L. Austin, or Seduction in Two Languages.* Stanford: Stanford University Press.

Ferguson, James. 1999. *Expectations of Modernity: Myths and Meanings of Urban Life on the Zambian Copperbelt.* Berkeley: University of California Press.

———. 2005. "Seeing like an Oil Company: Space, Security, and Global Capital in Neoliberal Africa." *American Anthropologist* 107, no. 3:377–82.

———. 2006. *Global Shadows: Africa in the Neoliberal World Order.* Durham: Duke University Press.

Finegold, S. M., et al. 2002. "Gastrointestinal Microflora Studies in Late-Onset Autism." *Clinical Infectious Diseases* 35 (1 September): S6–16.

Firn, David. 2000. "Takeovers Cure for German Biotechs." *Financial Times,* 14 June, http://www.ft.com, accessed 14 June 2000.

Fischer, Michael M. J. 1980. *Iran: From Religious Dispute to Revolution.* Cambridge: Harvard University Press.

———. 2003. *Emergent Forms of Life and the Anthropological Voice.* Durham: Duke University Press.

———. 2007a. "Culture and Cultural Analysis as Experimental Systems." *Cultural Anthropology* 22, no. 1:1–64.

———. 2007b. "Four Genealogies for a Recombinant Anthropology of Science and Technology." *Cultural Anthropology* 22, no. 4:539–615.

Fisher, Jill A. 2009. *Medical Research for Hire: The Political Economy of Pharmaceutical Clinical Trials.* New Brunswick, N.J.: Rutgers University Press.

Fong, Tan Shook. 1998. "Development of Clinical Trials in Singapore," *Drug Information Journal*, 32:1199–1201S.

Fortun, Kim. 2001. *Advocacy after Bhopal: Environmentalism, Disaster, New Global Orders.* Chicago: University of Chicago Press.

———. 2004. "From Bhopal to the Informating of Environmental Health: Risk Communication in Historical Perspective." *Landscapes of Exposure: Knowledge and Illness in Modern Environments*, ed. Gregg Mitman, Michelle Murphy, and Christopher Sellers. *Osiris* 19:283–96 [special issue].

Fortun, Kim, and Michael Fortun. 2005. "Scientific Imaginaries and Ethical Plateaus in Contemporary U.S. Toxicology." *American Anthropologist* 107, no. 1 (March): 43–54.

———. 2007. "Experimenting with the Asthma Files." Paper presented at the Experimental Systems Workshop, University of California, Irvine, 13–14 April.

Fortun, Michael. 2000. "Experiments in Ethnography and Its Performance." Available online at www.mannvernd.is.

———. 2001. "Mediated Speculations in the Genomics Futures Markets." *New Genetics and Society* 20:139–56.

———. 2008. *Promising Genomics: Iceland and* deCODE *Genetics in a World of Speculation.* Berkeley: University of California Press.

Fortun, Michael, and Kim Fortun. 1999. "Due Diligence and the Pursuit of Transparency: The Securities and Exchange Commission, 1996." *Paranoia within Reason: A Casebook on Conspiracy as Explanation (Late Editions 6)*, ed. George Marcus, 157–93. Chicago: University of Chicago Press.

Fosket, Jennifer R. 2002. "Breast Cancer Risk and the Politics of Prevention: Analysis of a Clinical Trial." Ph.D. diss., University of California, San Francisco.

Foucault, Michel. 1973. *The Order of Things: An Archaeology of the Human Sciences.* New York: Vintage.

———. 1977. *Discipline and Punish: The Birth of the Prison*, trans. Alan Sheridan. New York: Vintage.

———. 1990 [1978]. *The History of Sexuality*, vol. 1. New York: Vintage.

———. 2001. *Fearless Speech.* Los Angeles: Semiotext(e).

Franklin, Sarah. 1997. *Embodied Progress: A Cultural Account of Assisted Conception.* New York: Routledge.

———. 2007. *Dolly Mixtures: The Remaking of Genealogy.* Durham: Duke University Press.

Franklin, Sarah, and Margaret M. Lock. 2003. *Remaking Life and Death: Toward an Anthropology of the Biosciences.* Santa Fe: School of American Research Press.

Franklin, Sarah, and Susan MacKinnon, eds. 2002. *Relative Values: Reconfiguring Kinship Studies*. Durham: Duke University Press.

Franklin, Sarah, and Helena Ragone, eds. 1997. *Reproducing Reproduction: Kinship, Power, and Technological Innovation*. Philadelphia: University of Pennsylvania Press.

Frew, Sarah E., et al. 2007. "India's Health Biotech Sector at a Crossroads." *Nature Biotechnology* 25, no. 4:403–17.

Fujimura, Joan H. 2000. "Transnational Genomics: Transgressing the Boundary between the 'Modern/West' and the 'Premodern/East.'" *Doing Science + Culture*, ed. Roddey Reid and Sharon Traweek. 71–92. New York: Routledge.

Gaia, Vince. 2002. "Fears over Rice Genome Access." *New Scientist* 18, no. 38 (19 March). http://www.newscientist.com/article/dn2061-fears-over-rice-genome-access.html.

Galison, Peter. 1997. *Image and Logic: A Material Culture of Microphysics*. Chicago: University of Chicago Press.

Gaonkar, Dilip Parameshwar, ed. 2001. *Alternative Modernities*. Durham: Duke University Press.

Geertz, Clifford. 1973. *The Interpretation of Cultures*. New York: Basic Books.

Gelernter, J. 1995. "Editorial: Genetics of Bipolar Affective Disorder: Time for Another Reinvention?" *American Journal of Human Genetics* 56:1762, 1766.

Gellner, Ernest. 1983. *Nations and Nationalism*. Ithaca: Cornell University Press.

Genetic Engineering News. 2000. July issue.

Gershon, Michael D. 1998. *The Second Brain: The Scientific Basis of Gut Instinct and a Groundbreaking New Understanding of Nervous Disorders of the Stomach and Intestine*. New York: Harper Collins.

Gibbons, Sahra, and Carlos Novas, eds. 2007. *Making Biosociality: Biologies and Identities in Formation*. New York: Routledge.

Gilder, George. 1981. *Wealth and Poverty*. New York: Basic Books.

Gilliland, F., et al. 2005. "Air Pollution Exposure Assessment for Epidemiologic Studies of Pregnant Women and Children: Lessons Learned from the Centers for Children's Environmental Health and Disease Prevention Research." *Environmental Health Perspectives* 113 (October): 1447–54.

Ginsburg, Faye, and Rayna Rapp, eds. 1995. *Conceiving the New World Order: The Global Politics of Reproduction*. Berkeley: University of California Press.

Golub, D. R., et al. "Molecular Classification of Cancer: Class Discovery and Class Prediction by Gene Expression Monitoring." *Science* 286:531–37.

Goobar, Walter. 1998. "De quien es esa naricita?" *Siglo XXI* (27 August): 66.

Good, Mary-Jo Del Vecchio, et al. 2005. "Medicine on the Edge: Conversations with Oncologists." *Technoscientific Imaginaries: Conversations, Profiles, and Memoirs*, ed. George E. Marcus, 129–52. Chicago: University of Chicago Press.

Gordon, Avery. 2004. *Keeping Good Time: Reflections on Knowledge, Power, and People*. Boulder: Paradigm.

Gottschall, Elaine. 1994. *Breaking the Vicious Cycle: Intestinal Health through Diet*. Baltimore: Kirkton.

Grace, Patricia. 1998. *Baby No-Eyes*. Honolulu: University of Hawaii Press.

Graeber, David. 2001. *An Anthropological Theory of Value*. London: Palgrave Macmillan.

Gramsci, Antonio. 1992. *Prison Notebooks*, ed. Joseph Buttigeig. New York: Columbia University Press.

Grefe, Edward, and Marty Linsky. 1995. *The New Corporate Activism: Harnessing the Power of Grassroots Tactics for Your Organization*. New York: McGraw-Hill.

Greene, Jeremy A. 2007. *Prescribing by Numbers: Drugs and the Definition of Disease*. Baltimore: Johns Hopkins University Press.

Greener, Mark. 2001. *A Healthy Business: A Guide to the Global Pharmaceutical Industry*. London: Informa Pharmaceuticals.

Grewal, Inderpal, and Caren Kaplan. 1994. *Scattered Hegemonies: Postmodernity and Transnational Feminist Practices*. Minneapolis: University of Minnesota Press.

Grove, Richard. 1995. *Green Imperialism: Colonial Expansion, Tropical Island Edens and the Origins of Environmentalism, 1600–1860*. Studies in Environment and History. Cambridge: Cambridge University Press.

Gupta, Akhil. 1998. *Postcolonial Developments: Agriculture in the Making of Modern India*. Durham: Duke University Press.

Haas, Ernst B. 1990. *When Knowledge Is Power: Three Models of Change in International Organizations*. Berkeley: University of California Press.

Haas, Peter M. 1990. *Saving the Mediterranean: The Politics of Environmental Cooperation*. New York: Columbia University Press.

Hacking, Ian. 1998. *Mad Travelers: Reflections on the Reality of Transient Mental Illnesses*. Charlottesville: University Press of Virginia.

———. 1999. *The Social Construction of What?* Cambridge: Harvard University Press.

Hadden, Susan. 1994. "Citizen Participation in Environmental Policy Making." *Learning from Disaster: Risk Management after Bhopal*, ed. Sheila Jasanoff, 91–112. Philadelphia: University of Pennsylvania Press.

Hamilton, Cindy. 2001. "The Human Genome Diversity Project and the New Biological Imperialism." *Santa Clara Law Review* 41:619–42.

Hamilton, Lynn. 2000. *Facing Autism: Giving Parents Reasons for Hope and Guidance for Help*. Colorado Springs: Waterbrook.

Happe F., A. Ronald, and R. Plomin. 2006. "Time to Give Up on a Single Explanation for Autism." *Nature Neuroscience* 9, no. 10:1218–20.

Haraway, Donna. 1991. *Simians, Cyborgs, and Women: The Reinvention of Nature*. New York: Routledge.

———. 1997. *Modest_Witness@Second_Millennium.FemaleMan©_Meets_Onco-Mouse™: Feminism and Technoscience*. New York: Routledge.

———. 2003a. "Cloning Mutts, Saving Tigers: Ethical Emergents in Technocultural Dog Worlds." *Remaking Life and Death: Toward an Anthropology of the Biosciences*, ed. Sarah Franklin and Margaret Lock, 293–327. Santa Fe: School of American Research Press.

———. 2003b. *The Companion Species Manifesto: Dogs, People, and Significant Otherness*. Chicago: Prickly Paradigm.

———. 2003c. "For the Love of a Good Dog: Webs of Action in the World of Dog Genetics." *Genetic Nature/Culture: Anthropology and Science beyond the Two-Culture Divide*, ed. Alan H. Goodman, Deborah Heath, and M. Susan Lindee, 111–31. Berkeley: University of California Press.

———. 2008. *When Species Meet*. Minneapolis: University of Minnesota Press.

Harbolt, Tami, and Tamara H. Ward. 2001. "Teaming Incarcerated Youth with Shelter Dogs for a Second Chance." *Society and Animals* 9, no. 2:177–82.

Hardin, Garrett. 1968. "The Tragedy of the Commons." *Science* 162:1243–48.

Hardt, Michael, and Antonio Negri. 2000. *Empire*. Cambridge: Harvard University Press.

Harris, Gardiner, and Anahad O'Connor. 2005. "On Autism's Cause, It's Parents vs. Research." *New York Times*, 25 June.

Hartouni, Valerie. 1997. *Cultural Conceptions: On Reproductive Technologies and the Remaking of Life*. Minneapolis: University of Minnesota Press.

Harvey, David. 1996. *Justice, Nature and the Geography of Difference*. Cambridge, Mass.: Blackwell.

———. 2003. *The New Imperialism*. Oxford: Oxford University Press.

Hayden, Cori. 2003. *When Nature Goes Public: The Making and Unmaking of Bioprospecting in Mexico*. Princeton: Princeton University Press.

Hays, Sharon. 1998. *The Cultural Contradictions of Motherhood*. New Haven: Yale University Press.

Health Gap Coalition. 2000. *Questioning the UNAIDS/Pharmaceutical Industry Initiative: Seven Months and Counting*. http://www.healthgap.org/press_releases/00/121300_HGAP_PS_UNAIDS.pdf.

Healy, David. 2002. *The Creation of Psychopharmacology*. Cambridge: Harvard University Press.

Helmreich, Stefan. 2003. "Trees and Seas of Information: Alien Kinship and the Biopolitics of Gene Transfer in Marine Biology and Biotechnology." *American Ethnologist* 30, no. 3:341–59.

———. 2008. "Species of Biocapital." *Science as Culture* 17, no. 4:463–78.

Herbert, Martha R. 2004. "Complexity in Neurobiology: Autism as a Case Study of Gene-Brain-Immune-Environment Interactions in a Changing World." Paper presented at the Second Biennial International Seminar on the Philosophical, Epistemological and Methodological Implications of Complexity Theory, Havana, January.

———. 2005a. "Autism: A Brain Disorder, or a Disorder That Affects the Brain?" *Clinical Neuropsychiatry* 2, no. 6:354–79.

———. 2005b. "Is the Brain 'Downstream'? Implications of Biomedical Findings in Autism." Presentation at the spring 2005 DAN! Conference, Boston, 14–17 April.

———. 2011. "A Whole Body Systems Approach to ASD." *The Neuropsychology of Autism*, ed. Deborah A. Fein, 499–510. Oxford: Oxford University Press.

Herrera, Stephan. 2001. "The Biotech Boom: Revenge of the Neurons." *Red Herring*, 1 October.

Hess, Charlotte, and Elinor Ostrom. 2003. "Ideas, Artifacts, and Facilities: Information as a Common-Pool Resource." *Law and Contemporary Problems* 66, nos. 1–2:111–45.

Hess, David. 2003. "CAM Cancer Therapies in Twentieth-Century North America: Examining Continuities and Change." *The Politics of Healing*, ed. Robert Johnston, 218–30. New York: Routledge.

Hilgartner, Stephen. 2004. "Mapping Systems and Moral Order: Constituting Property in Genome Laboratories." *States of Knowledge: The Co-production of Science and Social Order*, ed. Sheila Jasanoff, 131–41. New York: Routledge.

Hind, Rick, and David Halperin. 2004. "Lots of Chemicals, Little Reaction." *New York Times*, 22 September, §A, 31.

Hirakawa, Mika, T. Tanaka, Y. Hashimoto, M. Kuroda, T. Takagi, and Y. Nakamura. 2002. "JSNP: A Database of Common Gene Variations in the Japanese Population." *Nucleic Acids Research* 30, no. 1:158–62.

Hirsch, Eric, and Marilyn Strathern, eds. 2005. *Transactions and Creations: Property Debates and the Stimulus of Melanesia*. New York: Berghahn.

Ho, Karen. 2009. *Liquidated: An Ethnography of Wall Street*. Durham: Duke University Press.

Ho, Louise. 2000 [1997]. "Storm." *New Ends, Old Beginnings*, 54. Hong Kong: Asia 2000.

Ho, Mae-Wan. 2001. "Breaching the Knowledge Monopoly: China Trumps the West in Sequencing Rice Genome." Rice GD: Genome Database of Chinese Super Hybrid Rice, http://btn.genomics.org.cn: 8080/rice/new3.php, accessed 25 April 2008.

Hogle, Linda. 1995. "Standardization across Non-Standard Domains: The Case of Organ Procurement." *Science, Technology and Human Values* 20, no. 4:482–500.

———. 2001. "Chemoprevention for Healthy Women: Harbinger of Things to Come?" *Health* 5:311–33.

Holden, Kerry, and David Demeritt. 2008. "Democratising Science? The Politics of Promoting Biomedicine in Singapore's Developmental State." *Environment and Planning D: Society and Space* 26, no. 1:68–86.

Holmes, Douglas, and George Marcus. 2005. "Cultures of Expertise and the Management of Globalization: Toward the Refunctioning of Ethnography." *Global Assemblages: Technology, Politics, and Ethics as Anthropological Problems*, ed. Aihwa Ong and Stephen Collier, 235–52. Oxford: Blackwell.

Hong Yun-Chul, Jong-Han Leem, Eun-Hee Ha, and David C. Christiani. 1999. "PM_{10} Exposure, Gaseous Pollutants, and Daily Mortality in Inchon, South Korea." *Environmental Health Perspectives* 54:108–16.

Huber, Peter. 1993. *Galileo's Revenge: Junk Science in the Courtroom*. New York: Basic Books.

———. 1998. "Saving the Environment from the Environmentalists." *Commentary* 105, no. 4 (April): 25–30.

———. 2000. *Hard Green: Saving the Environment from the Environmentalists: A Conservative Manifesto*. New York: Basic Books.

Hughes, Sally. 2001. "Making Dollars out of DNA: The First Major Biotechnology Patent and the Commercialization of Molecular Biology, 1974–1980." *Isis* 92:541–75.

Hwang, Woo-suk, et al. 2005. "Dogs Cloned from Adult Somatic Cells." *Nature* 436, no. 7051 (4 August): 641.

Ingram, Alan. 2007. "HIV/AIDS, Security and the Geopolitics of U.S.-Nigerian Relations." *Review of International Political Economy* 14, no. 3:510–34.

International Conference on Harmonisation of Technical Requirements for Registration of Pharmaceuticals for Human Use. 1997. "The Future of the ICH: Conference Statement by the ICH Steering Committee on the Occasion of the Fourth International Conference on Harmonisation."

International Rice Genome Sequencing Project. 2008. http://rgp.dna.affrc.go.jp, accessed 25 June 2011.

Jacob, François. 1982. *The Possible and the Actual*. Seattle: University of Washington Press.

———. 1988. *The Statute Within*. New York: Basic Books.

———. 1993 [1973]. *The Logic of Life: A History of Heredity*. Princeton: Princeton University Press.

Jain, Sarah Lochlann. 2007. "Living in Prognosis: Toward an Elegiac Politics." *Representations* 98, no. 1:77–92.

James, S. J., et al. 2004. "Metabolic Biomarkers of Increased Oxidative Stress and Impaired Methylation Capacity in Children with Autism." *American Journal of Clinical Nutrition* 80:1161–67.

Jameson, Fredric, and Masao Miyoshi. 1998. *The Cultures of Globalization*. Durham: Duke University Press.

Jamison, Kay Redfield. 1997. *An Unquiet Mind*. New York: Random House.

Jasanoff, Sheila. 1992. "Science, Politics and the Renegotiation of Expertise at EPA." *Osiris* 7:195–217.

———. 1995. *Science at the Bar: Law, Science, and Technology in America*. Cambridge: Harvard University Press.

———. 2004. "Ordering Knowledge, Ordering Society." *States of Knowledge: The Co-production of Science and Social Order*, ed. Sheila Jasanoff, 13–45. New York: Routledge.

———. 2005a. *Designs on Nature: Science and Democracy in Europe and the United States*. Princeton: Princeton University Press.

———. 2005b. "Let Them Eat Cake: GM Foods and the Democratic Imagination." *Science and Citizens*, ed. Melissa Leach, Ian Scoones, and Brian Wynne, 183–99. London: Zed.

Jones, Arthur F. 1934. "Erin's Famous Hounds Finding Greater Glory at Rathmullan." *American Kennel Gazette* 5, no. 5 (31 May).

Juma, Calestous. 1989. *The Gene Hunters: Biotechnology and the Scramble for Seeds*. Princeton: Princeton University Press.

Jyonouchi, H., S. Sun, and H. Le. 2001. "Proinflammatory and Regulatory Cytokine Production Associated with Innate and Adaptive Immune Responses in Children with Autism Spectrum Disorders and Developmental Regression." *Journal of Neuroimmunology* 120:170–79.

Kahn, Joan Y. 1978. "A Diagnostic Semiotic." *Semiotica* 22:75–106.

Kahn, Jonathan. 2006. "Harmonizing Race: Competing Regulatory Paradigms of Racial Categorization International Drug Development." *Santa Clara Journal of International Law* 1:34–56.

Kane, Karen. "Drug Error, Not Chelation Therapy, Killed Boy, Expert Says." *Pittsburgh Post-Gazette*, 18 January 2006, http://www.post-gazette.com, accessed 4 August 2007.

Kanner, Leo. 1943. "Autistic Disturbances of Affective Contact." *Nervous Child* 10: 217–50.

Kass, Leon. 1997. "The Wisdom of Repugnance." *New Republic* 216, no. 22:17–26.

Kassirer, Jerome P. 2005. *On the Take: How America's Complicity with Big Business Can Endanger Your Health*. New York: Oxford University Press.

Kay, Lily. 1996. *The Molecular Vision of Life: Caltech, the Rockefeller Foundation, and the Rise of the New Biology*. Oxford: Oxford University Press.

———. 2000. *Who Wrote the Book of Life? A History of the Genetic Code*. Stanford: Stanford University Press.

Keating, Peter, and Alberto Cambrosio. 2000. "Biomedical Platforms." *Configurations* 8:337–87.

Keller, Evelyn Fox. 1984. *A Feeling for the Organism: The Life and Work of Barbara McClintock*. New York: Times Books.

———. 1992. *Secrets of Life, Secrets of Death*. New York: Routledge.

———. 1995. *Refiguring Life: Metaphors of Twentieth-Century Biology*. New York: Columbia University Press.

———. 2002a. *The Century of the Gene*. Cambridge: Harvard University Press.

———. 2002b. *Making Sense of Life: Explaining Biological Development with Models, Metaphors, and Machines*. Cambridge: Harvard University Press.

Kelty, Christopher M. 2004. "Punt to Culture." *Anthropological Quarterly* 77, no. 3: 547–58.

———. 2008. *Two Bits: The Cultural Significance of Free Software*. Durham: Duke University Press.

Kennedy, Donald. 2002. "The Importance of Rice." *Science* 296:13.

Keown, Michelle. 2005. *Postcolonial Pacific Writing*. New York: Routledge.

Kessler, R. C., et al. 1997. "The Epidemiology of DSM-III-R Bipolar I Disorder in a General Population Survey." *Psychological Medicine* 27, no. 5:1079–90.

Kim, Hyung-Jin. 2008. "South Korean Firm Delivers Commercial Dog Clones." USA *Today*, 5 August.

Kinney, P., S. N. Chillrud, S. Ramstrom, J. Ross, and J. D. Spengler. 2002. "Exposures to Multiple Air Toxics in New York City." *Environmental Health Perspectives* 110, no. 4 (August): 539–46.

Kitcher, Philip. 2001. *Science, Truth, and Democracy*. Oxford: Oxford University Press.

Kittay, Eva Feder. 1999. *Love's Labor: Essays on Women, Equality, and Dependency*. New York: Routledge.

———. 2001. "When Caring Is Just and Justice Is Caring: Justice and Mental Retardation." *Public Culture* 13, no. 3:557–79.

Klein, D. F., et al. 2002. "Improving Clinical Trials: American Society of Clinical Psychopharmacology Recommendations." *Archives of General Psychiatry* 59:272–78.

Kleinman, Arthur. 1999. "Moral Experience and Ethical Reflection: Can Ethnography Reconcile Them? 'A Quandary for the New Bioethics.'" *Bioethics and Beyond*, ed. Arthur Kleinman, Renée C. Fox, and Allan M. Brandt. *Daedalus* 128, no. 4:68–97 [special issue].

Knorr-Cetina, Karin, and Aaron Victor Cicourel, eds. 1981. *Advances in Social Theory and Methodology: Toward an Integration of Micro and Macro Level Sociology*. Boston: Routledge and Kegan Paul.

Koerner, Brendan. 2003. "That Pudgy Pooch is an Industry's Best Friend." *New York Times*. 30 November 2003.

Kohn, Eduardo. 2007. "How Dogs Dream: Amazonian Natures and the Politics of Transspecies Engagement." *American Ethnologist* 34, no. 1:3–24.

Kolata, Gina. 2001. "U.S. Panel Backs Broader Steps to Reduce Risk of Heart Attacks." *New York Times*, 16 May, §A, 1.

———. 2003. "The Cholesterol Challenge: Just How Low Can You Go?" *New York Times*, 2 December 2003.

———. 2005. "Beating Hurdles, Scientists Clone a Dog for a First." *New York Times*, 4 August, §A, 1, 10.

Kopnisky, Kathy L., and Steven Hyman. 2002. "Psychiatry in the Postgenomic Era." *TEN* 4, no. 1:27–31.

Kraepelin, Emil. 1904. "Mixed Conditions of Maniacal-Depressive Insanity." *Lectures on Clinical Psychiatry*, ed. Thomas Johnstone, 69–78. London: Bailliere, Tindall and Cox.

Kramer, Peter. 1993. *Listening to Prozac*. New York: Penguin.

Kremer, Michael, and Christopher M. Snyder. 2003. "Why Are Drugs More Profitable Than Vaccines?" National Bureau of Economic Research Working Paper No. 9833. New York: William Wood.

Krigsman, A. 2007. "Gastrointestinal Pathology in Autism: Description and Treatment." *Medical Veritas* 4:1528–36.

Kroll-Smith, Stephen, and H. Hugh Floyd. 1997. *Bodies in Protest: Environmental Illness and the Struggle over Medical Knowledge*. New York: New York University Press.

Krupp, F. 1999. "A Letter from EDF's Executive Director." Scorecard, http://scorecard.goodguide.com, accessed July 2002.

Kuo, Wen-Hua. 2005. "Japan and Taiwan in the Wake of Bioglobalization: Drugs, Race and Standards." Ph.D. diss., Massachusetts Institute of Technology.

Kuo, Wen-Hua. 2008. "Understanding Race at the Frontier of Pharmaceutical Regulation: An Analysis of the Racial Difference Debate at the ICH." Journal of Law, Medicine, and Ethics, 36, no. 3:498–505.

Kuriyama, Shigehisa. 1999. *The Expressiveness of the Body and the Divergence of Greek and Chinese Medicine*. New York: Zone.

Kushak, Rafail I., Harland S. Winter, Nathan S. Farber, and Timothy M. Buie. 2005. "Gastrointestinal Symptoms and Intestinal Disaccharidase Activities in Children with Autism: 49." Abstracts: North American Society of Pediatric Gastroenterology, Hepatology, and Nutrition Annual Meeting, 20–22 October, Salt Lake City. *Journal of Pediatric Gastroenterology and Nutrition* 41, no. 4:508.

La Feria, Ruth. 2006. "Woman's Best Friend, or Accessory?" *New York Times*, 7 December, §E, 4, 7.

Laclau, Ernesto. 2000. "Identity and Hegemony: The Role of Universality in the Constitution of Political Logics." *Contingency, Hegemony, Universality: Contemporary Dialogues on the Left*, ed. Judith Butler, Ernesto Laclau, and Slavoj Žižek, 44–89. London: Verso.

Laidler, James R. 2004. "Through the Looking Glass: My Involvement with Autism Quackery." http://www.autism-watch.org/about/bio2.shtml, accessed 28 July 2011.

Lakoff, Andrew. 2000. "Adaptive Will: The Evolution of Attention Deficit Disorder." *Journal of the History of the Behavioral Sciences* 36, no. 2 (spring): 149–69.

———. 2006. *Pharmaceutical Reason: Knowledge and Value in Global Psychiatry*. Cambridge: Cambridge University Press.

Lalli, Frank. 1995. "Now Only Clinton Can Stop Congress from Hurting Small Investors Like You." *Money*, 1 December. http://money.cnn.com/magazines/money mag/moneymag_archive/1995/12/01/207589/index.htm.

Landecker, Hannah. 2007. *Culturing Life: How Cells Became Technologies*. Cambridge: Harvard University Press.

Landes, William, and Richard Posner. 2003. *The Economic Structure of Intellectual Property Law*. Cambridge: Harvard University Press.

Landow, George P. 1992. *Hypertext: The Convergence of Contemporary Critical Theory and Technology*. Baltimore: Johns Hopkins University Press.

Lang, Martha Elizabeth. 1998. "Welcome to the Plastic City: Community Responses to the Leominster Autism Cluster." Ph.D. diss., Brown University, Department of Sociology.

Latham, A. J. H. 1998. *Rice: The Primary Commodity*. New York: Routledge.

Latour, Bruno. 1987. *Science in Action*. Cambridge: Harvard University Press.

———. 1988. *The Pasteurization of France*, trans. Alan Sheridan and John Law. Cambridge: Harvard University Press.

———. 1992. "Where Are the Missing Masses? The Sociology of a Few Mundane Artifacts." *Shaping Technology / Building Society: Studies in Sociotechnical Change*, ed. Wiebe E. Bijker and John Law, 225–58. Cambridge: MIT Press.

———. 1993. *We Have Never Been Modern*. Cambridge: Harvard University Press.

———. 1999. *Pandora's Hope: Essays on the Reality of Science Studies*. Cambridge: Harvard University Press.

Latour, Bruno, and Steve Woolgar. 1986 [1979]. *Laboratory Life: The Construction of Scientific Facts*. Princeton: Princeton University Press.

Lavery, Brian. 2004. "For Dogs in New York, a Glossy Look at Life." *New York Times*, 16 August, §c, 8.

Leboyer, Marion, et al. 1998. "Psychiatric Genetics: Search for Phenotypes." *Trends in Neurosciences* 21, no. 3:102–5.

Lee, Benjamin, and Edward LiPuma. 2002. "Cultures of Circulation: The Imaginations of Modernity." *Public Culture* 14, no. 1:191–213.

Lepkowski, Wil. 1984. "Bhopal Disaster Spotlights Chemical Hazard Issues." *Chemical and Engineering News* 62, no. 52 (24 December): 19–20.

———. 1994. "The Restructuring of Union Carbide." *Learning from Disaster: Risk Management after Bhopal*, ed. Sheila Jasanoff, 22–43. Philadelphia: University of Pennsylvania Press.

Lerach, William S. 1998. "The Private Securities Litigation Reform Act of 1995 — 27 Months Later: Securities Class Action Litigation under the Private Securities Litigation Reform Act's Brave New World." *Washington University Law Quarterly* 76:597–644.

Lessig, Lawrence. 2004. *Free Culture: How Big Media Uses Technology and the Law to Lock Down Culture and Control Creativity*. New York: Penguin.

Lévi-Strauss, Claude. 1962. *The Savage Mind*. Chicago: University of Chicago Press.

———. 1970 [1964]. *The Raw and the Cooked*, trans. John Weightman and Doreen Weightman. London: Jonathan Cape.

Levitz, Jennifer. 2005. "Desperate Parents Seek Autism's Cure." *Providence Journal*, 27 August, 1.

Levy, S. E., et al. 2003. "Use of Complementary and Alternative Medicine among Children Recently Diagnosed with Autistic Spectrum Disorder." *Journal of Developmental and Behavioral Pediatrics* 24, no. 6:418–23.

Levy, S. E., and S. L. Hyman. 2003. "Use of Complementary and Alternative Treatments for Children with Autistic Spectrum Disorders Is Increasing." *Pediatric Annals* 32, no. 10:685–91.

Lewis, Lisa S. 1998. *Special Diets for Special Kids: Understanding and Implementing Special Diets to Aid in the Treatment of Autism and Related Developmental Disorders*. Arlington, Texas: Future Horizons.

Lewis, Michael. 1999. *The New New Thing: A Silicon Valley Story*. New York: W. W. Norton.

Licking, Ellen, et al. 2000. "Move Over, Dot-Coms: Biotech Is Back." *Businessweek*, 26 March. http://www.businessweek.com/2000/00_10/b3671142.htm, accessed 26 March 2000.

Lin, Marie, C. C. Chu, S. L. Chang, H. L. Lee, J.-H. Loo, T. Akaza, T. Juji, J. Ohashi, and K. Tokunaga. 2001. "The Origin of Minnan and Hakka, the So-called 'Taiwanese,' Inferred by HLA Study." *Tissue Antigen* 57:192–99.

Lin, Yeong-Liang, Herng-Der Chern, and Mong-Ling Chu. 2003. "Taiwan's Experience with the Assessment of the Acceptability of the Extrapolation of Foreign Clinical Data for Registration Purposes." *Drug Information Journal*, 37:143–45.

Lindblad-Toh, Kerstin, et al. 2005. "Genome Sequence, Comparative Analysis, and Haplotype Structure of the Domestic Dog." *Nature* 438:803–19.

Litman, Jessica. 1990. "The Public Domain." *Emory Law Journal* 39:965–1024.

Lock, Margaret, Allan Young, and Alberto Cambrosio, eds. 2000. *Living and Working with the New Medical Technologies: Intersections of Inquiry.* Cambridge: Cambridge University Press.

London, Eric, and Rush Etzel. 2000. "The Environment as an Etiologic Factor in Autism: A New Direction in Research." *Environmental Health Perspectives Supplements* 108, no. S3 (June): 401–4.

Long, Janice, and David Hanson. 1985. "Bhopal Triggers Massive Response from Congress, the Administration." *Chemical and Engineering News* 63, no. 6 (11 February): 53–60.

Lopate, Phillip. 1995. *The Art of the Personal Essay: An Anthology from the Classical Era to the Present.* New York: Anchor.

Lord, C., et al. 2000. "The Autism Diagnostic Observation Schedule—Generic: A Standard Measure of Social and Communication Deficits Associated with the Spectrum of Autism. *Journal of Autism and Developmental Disorders* 30, no. 3:205–23.

Lord, C., M. Rutter, and A. LeCouteur. 1994. "Autism Diagnostic Interview—Revised: A Revised Version of a Diagnostic Interview for Caregivers of Individuals with Possible Pervasive Developmental Disorders." *Journal of Autism and Development Disorders* 24, no. 5:659–85.

Luhrmann, Tanya. 2000. *Of Two Minds: The Growing Disorder in American Psychiatry.* New York: Alfred A. Knopf.

Luxemburg, Rosa. 1968. *The Accumulation of Capital,* trans. Agnes Schwarzschild. New York: Monthly Review.

MacKenzie, Donald, Fabian Muniesa, and Lucia Siu, eds. 2008. *Do Economists Make Markets? On the Performativity of Economics.* Princeton: Princeton University Press.

Mackenzie, Michael, Peter Keating, and Alberto Cambrosio. 1990. "Patents and Free Scientific Information in Biotechnology: Making Monoclonal Antibodies Proprietary." *Science, Technology, and Human Values* 15, no. 1:65–83.

Mackey, Nathaniel. 1992. "Other: From Noun to Verb." *Representations* 39 (summer): 51–70.

MacKinnon, Dean F., Kay Redfield Jamison, and J. Raymond DePaulo. 1997. "Genetics of Manic Depressive Illness." *Annual Review of Neuroscience* 20, no. 1:355–74.

Maiwada, Jude. 2004. "70% of Nigerians Patronize Traditional Medicine Practitioners." *Nigerian Newsday,* 11 August. http://www.nasarawastate.org/newsday/news/culture/10811174558.html.

Mander, Jerry, and Victoria Tauli-Corpuz, eds. 2006. *Paradigm Wars: Indigenous People's Resistance to Globalization.* Berkeley: University of California Press.

Manuel, D. G., et al. 2006. "Effectiveness and Efficiency of Different Guidelines on Statin Treatment for Preventing Deaths from Coronary Heart Disease: Modeling Study." *British Medical Journal* 332:1419.

Marcus, George E. 1995a. "Ethnography in/of the World System: The Emergence of Multi-Sited Ethnography." *Annual Review of Anthropology* 24:95–117.

———, ed. 1995b. *Technoscientific Imaginaries: Conversations, Profiles, and Memoirs.* Chicago: University of Chicago Press.

———. 2005. "Multi-Sited Ethnography: Five or Six Things I Know about It Now." Paper presented at workshop on "Problems and Possibilities of Multi-Sited Ethnography," University of Sussex, 27 June.

———. 2007. "Ethnography Two Decades after Writing Culture: From the Experimental to the Baroque." *Anthropological Quarterly* 80, no. 4:1127–45.

Marcus, George E., and Michael M. J. Fischer. 1986. *Anthropology as Cultural Critique: An Experimental Moment in the Human Sciences.* Chicago: University of Chicago Press.

Marcus, George E., and Erkan Saka. 2005. "Assemblages." *Theory Culture and Society* 23, nos. 2–3:107–8.

Marks, Harry M. 1992. "Cortisone, 1949: A Year in the Political Life of a Drug." *Bulletin of the History of Medicine* 66:419–39.

———. 1997. *The Progress of Experiment: Science and Therapeutic Reform in the United States, 1900–1990.* Cambridge: Cambridge University Press.

Marshall, Eliot. 2002. "A Deal for the Rice Genome." *Science* 296, no. 5565:34–36.

Martin, Emily. 1991. "The Egg and the Sperm: How Science has Constructed a Romance Based on Stereotypical Male-Female Roles." *Signs* 16, no. 3:485–501.

———. 1995. *Flexible Bodies: The Role of Immunity in American Culture from the Days of Polio to the Age of AIDS.* Boston: Beacon.

———. 1998. "Anthropology and the Cultural Study of Science." *Science, Technology and Human Values* 23, no. 1:24–44.

———. 1999. "Toward an Anthropology of Immunity: The Body as Nation State." *The Science Studies Reader,* ed. Mario Biagioli, 358–71. New York: Routledge.

———. 2007. *Bipolar Expeditions: Mania and Depression in American Culture.* Princeton: Princeton University Press.

Marx, Karl. 1959. *Capital: A Critical Analysis of Capitalist Production.* Trans. from the 3d German edn by Samuel Moore and Edward Aveling, ed. Friedrich Engels. Moscow: Foreign Languages.

———. 1974 [1894]. *Capital: A Critique of Political Economy,* vol. 3, ed. Frederick Engels. Moscow: Progress.

———. 1976 [1867]. *Capital: A Critique of Political Economy,* vol. 1, trans. Ben Fowkes. London: Penguin.

———. 1977 [1852]. *The Eighteenth Brumaire of Louis Bonaparte.* Moscow: Progress.

———. 1978a [1867]. "The Communist Manifesto." *The Marx-Engels Reader,* vol. 1, ed. Robert Tucker, 496–500. New York: W. W. Norton.

———. 1978b [1867]. "Speech at the Anniversary of the People's Paper." *The Marx-Engels Reader,* vol. 1, ed. Robert Tucker, 577–78. New York: W. W. Norton.

———. 1993 [1857]. *Grundrisse: Foundations of a Critique of Political Economy,* trans. Martin Nicolaus. London: Penguin.

———. 2009 [1858]. *A Contribution to a Critique of Political Economy.* General Books Club, http://generalbooksclub.com.

Marx, Karl, and Frederick Engels. 1963 [1845]. *The German Ideology*. New York: International.

Maurer, Bill. 2005. *Mutual Life, Limited: Islamic Banking, Alternative Currencies, Lateral Reason*. Princeton: Princeton University Press.

Mbembe, Achille. 2001. *On the Postcolony*. Berkeley: University of California Press.

McCaffery, Richard. 2001. "The Reason to Avoid Biotech Stocks." The Motley Fool, 5 April, accessed 5 April 2001.

McCaig, Donald. 1992 [1984]. *Nop's Trials*. Guilford, Del.: Lyons.

———. 1998a. *Eminent Dogs, Dangerous Me*. Guilford, Del.: Lyons.

———. 1998b. *Nop's Hope*. Guilford, Del.: Lyons.

McCandless, Jacquelyn. 2003. "Children with Starving Brains: A Medical Treatment Guide for Autism Spectrum Disorder." 2d edn. Putney, Vt.: Bramble.

McGinn, Anne Platt. 2002. "From Rio to Johannesburg: Reducing the Use of Toxic Chemicals Advances Health and Sustainable Development." *World Summit Policy Briefs* (WorldWatch Institute, 25 June): 3.

McGovern, Cammie. "Autism's Parent Trap," *New York Times*, 5 June 2006, 3.

McInnes, L. Alison, et al. 1999. "Mapping Genes for Psychiatric Disorders and Behavioral Traits." *Current Opinion in Genetics and Development* 8:287–92.

McMichael, A. J., H. R. Anderson, B. Brunekreef, and A. J. Cohen. 1998. "Inappropriate Use of Daily Mortality Analyses to Estimate Longer-Term Mortality Effects of Air Pollution." *International Journal of Epidemiology* 27, no. 3:450–53.

McPherson, James M. 1990. *Battle Cry of Freedom: The American Civil War*. New York: Penguin.

Meek, James. 2000. "Why You Are First in the Great Gene Race." *Guardian*, 15 November, 4.

———. 2002. "Decode Was Meant to Save Lives . . . Now It's Destroying Them." *Guardian*, 13 October, 2.2.

Melucci, Albert. 1996. *Challenging Codes: Collective Action in the Information Age*. Cambridge: Cambridge University Press.

Merton, Robert. 1942. "The Normative Structure of Science." *The Sociology of Science*. 267–78. Chicago: University of Chicago Press.

Mol, Annemarie, and John Law. 1994. "Regions, Networks and Fluids: Anaemia and Social Topology." *Social Studies of Science* 24:641–71.

Monod, Jacques. 1971. *Chance and Necessity*. New York: Alfred A. Knopf.

Monsanto. 2010. http://www.monsanto.com, accessed 25 June 2011.

Moolgavkar, Suresh H., George E. Luebeck, Thomas A. Hall, Elizabeth L. Anderson. 1995. "Air Pollution and Daily Mortality in Philadelphia." *Epidemiology* 6, no. 5:476–84.

Moore, Donald S. 2005. *Suffering for Territory: Race, Place, and Power in Zimbabwe*. Durham: Duke University Press.

Morey, Darcy. 2006. "Burying Key Evidence: The Social Bond between Dogs and People." *Journal of Archeological Science* 33:158–75.

Morgan, Steve, Morris Barer, and Robert Evans. 2000. "Health Economists Meet the

Fourth Tempter: Drug Dependency and Scientific Discourse." *Health Economics* 9:659–67.

Moss, Giles D. 2007. *Pharmaceuticals — Where's the Brand Logic? Branding Lessons and Strategy.* New York: Pharmaceutical Products Press.

Mulkay, Michael. 1976. "The Mediating Role of the Scientific Elite." *Social Studies of Science* 6, nos. 3–4:445–71.

Mullin, Molly H. 2001. *Culture in the Marketplace: Gender, Art, and Value in the American Southwest.* Durham: Duke University Press.

Murakami, Yoichiro. 1998. *Anzengaku* [On security]. Tokyo: Seitosha.

Murch, S. H., et al. 2004. "Retraction of an Interpretation." *Lancet* 363, no. 9411:750.

Murphy, Michelle. 2000. "The 'Elsewhere within Here,' and Environmental Illness: Or, How to Build Yourself a Body in a Safe Space." *Configurations* 8:87–120.

Myers, Natasha. 2006. "Animating Mechanism: Animation and the Propagation of Affect in the Lively Arts of Protein Modeling." *Science Studies* 19, no. 2:6–30.

————. 2007. "Modeling Proteins, Making Scientists: An Ethnography of Pedagogy and Visual Culture in Contemporary Structural Biology." Ph.D. diss., Massachusetts Institute of Technology, Science and Technology Studies.

Naanen, Ben. 2004. "The Political Economy of Oil and Violence in the Niger Delta." ACAS *Bulletin: The Warri Crisis, the Niger Delta, and the Nigerian State,* no. 68 (fall): 4–9.

Nash, Roderick. 1982. *Wilderness and the American Mind.* New Haven: Yale University Press.

National Center for Biotechnology Information. 2008. "GenBank Overview." http://www.ncbi.nlm.nih.gov, accessed 3 May 2011.

National Health Research Institutes. 2003. *Proceedings of 2003 Symposium on Statistical Methodology for Evaluation of Bridging Evidence.* Nankang, Taipei.

Navarra, Gabriela. 1998. "De la euforia a la depression." *La Nación,* 22 July, §6, 4.

Neal, Andrea. 2005. "Trained Dogs Transforming Lives: A Service Program to Benefit People with Disabilities Is Also Helping U.S. Prison Inmates Develop a Purpose for Their Lives." *Saturday Evening Post* 277, no. 5 (1 September): 64–65.

Nelkin, Dorothy, and Laurence Tancredi. 1989. *Dangerous Diagnostics: The Social Power of Biological Information.* New York: Basic Books.

Nestle, Marion. 2008. *Pet Food Politics: The Chihuahua in the Coal Mine.* Berkeley: University of California Press.

Nguyen, Vinh-Kim. 2004. "Antiretroviral Globalism, Biopolitics, and Therapeutic Citizenship." *Global Assemblages: Technology, Politics, and Ethics as Anthropological Problems,* ed. Aihwa Ong and Stephen Collier, 124–44. Oxford: Blackwell.

————. 2007. "Experimentality: Massive AIDS Intervention in Africa as Military Therapeutic Complex." Paper presented at the Experimental Systems Conference, University of California, Irvine, 13–14 April.

Noble, David. 1979. *America by Design: Science, Technology, and the Rise of Corporate Capitalism.* Oxford: Oxford University Press.

Normile, Dennis. 2002. "Syngenta Agrees to Wider Release." *Science* 296, no. 5574 (7 June): 1785–87.

Normile, Dennis, and Elizabeth Pennisi. 2002. "Rice Boiled Down to Bare Essentials." *Science* 296, no. 5565 (7 June): 32–36.

Novas, Carlos, and Nikolas Rose. 2004. "Biological Citizenship." *Global Assemblages: Technology, Politics, and Ethics as Anthropological Problems*, ed. Aihwa Ong and Stephen Collier, 439–63. Oxford: Blackwell.

Nutley, Caroline ed. 2000. *The Value and Benefits of ICH to Industry*. http://www.ich.org/ichnews/publications/browse/article/the-values-and-benefits-of-ich-to-industry.html, last accessed on 18 July 2011.

Okonta, Ike. 2004. "Death-Agony of a Malformed Political Order." ACAS *Bulletin: The Warri Crisis, the Niger Delta, and the Nigerian State*, no. 68 (fall): 23–29.

Okonta, Ike, and Oronto Douglass. 2003. *Where Vultures Feast*. London: Verso.

Ong, Aihwa. 2000. "Graduated Sovereignty in Southeast Asia." *Theory, Culture, and Society* 17, no. 4:55–75.

Ong, Aihwa. 2006. *Neoliberalism as Exception: Mutations in Citizenship and Sovereignty*. Durham: Duke University Press.

Ong, Aihwa, and Stephen Collier, eds. 2005. *Global Assemblages: Technology, Politics, and Ethics as Anthropological Problems*. Oxford: Blackwell.

Orum, Paul. 2008. *Chemical Security 101: What You Don't Have Can't Leak or Be Blown Up by Terrorists*. Washington: Center for American Progress.

Ostrom, Elinor, Joanna Burger, Christopher Field, Richard Norgaard, and David Policansky. 1999. "Revisiting the Commons: Local Lessons, Global Challenges." *Science* 284, no. 5412:278–82.

Ovbiagele, Godwin. 2000. "Decorous Drug Distribution Channels: Challenges of the Democratic Dispensation." EKO *Pharmacist* (November): 19–38.

"Override the Securities Bill Veto." 1995. *Washington Post*, 22 December, §A, 18.

Palsson, Gisli, and Paul Rabinow. 1999. "Iceland: The Case of a National Human Genome Project." *Anthropology Today* 15, no. 5:14–18.

Page, Theodore. 2000. "Metabolic Approaches to the Treatment of Autism Spectrum Disorders." *Journal of Autism and Developmental Disorders* 30, no. 5:463–69.

Pangborn, Jon, and Sidney Baker. 2005. *Autism: Effective Biomedical Treatments: Have We Done Everything We Can for This Child? Individuality in an Epidemic*. San Diego: Autism Research Institute.

Paoloni, M. C., C. Mazcko, E. Fox, T. Fan, S. Lana, et al. 2010. "Rapamycin Pharmacokinetic and Pharmacodynamic Relationships in Osteosarcoma: A Comparative Oncology Study in Dogs," PLoS ONE 5, no. 6:e11013.

Patel, Molini, and Rachel Miller. 2009. "Air Pollution and Childhood Asthma: Recent Advances and Future Directions." *Current Opinion in Pediatrics* 21, no. 2 (April): 235–42.

Paul, Diana. 1998. *The Politics of Heredity: Essays on Eugenics, Biomedicine, and the Nature-Nurture Debate*. Albany: State University of New York Press.

Pemberton, Stephen. 2004. "Canine Technologies, Model Patients: The Historical Production of Hemophiliac Dogs in American Biomedicine." *Industrializing Organism: Introducing Evolutionary History*, ed. Susan R. Schrepfer and Philip Scranton, 191–213. New York: Routledge.

Pennisi, Elizabeth. 2000. "Stealth Genome Rocks Rice Researchers." *Science* 288: 239–41.

Perino, Michael A. 1997. "What We Know and Don't Know about the Private Securities Litigation Reform Act of 1995." Testimony before the Subcommittee on Finance and Hazardous Materials of the Committee on Commerce, U.S. House of Representatives, 21 October. Securities Class Action Clearinghouse, http://securities.stanford.edu.

Perkins, Greg. 2002. "Principles of Product Research and Development." *Pharmaceutical Marketing: Principles, Environment, and Practice*, ed. Mickey C. Smith, E. M. "Mick" Kolassa, Greg Perkins, and Bruce Siecker, 103–42. New York: Pharmaceutical Products Press.

Peterson, Kristin. 2005. "The Tenofovir Trials in Nigeria: Agency, Knowledge and Science." Paper presented at the Michigan State University Center for Ethics and Humanities in the Life Sciences, 23 February.

Petkova, Elena, with Peter Veit. 2000. "Environmental Accountability beyond the Nation-State: The Implications of the Aarhus Convention." Washington: World Resources Institute.

Petryna, Adriana. 2002. *Life Exposed: Biological Citizens after Chernobyl*. Princeton: Princeton University Press.

———. 2005. "Ethical Variability: Drug Development and Globalizing Clinical Trials." *American Ethnologist* 32, no. 2:183–97.

Petryna, Adriana. 2006. "Globalizing Human Subjects Research." *Global Pharmaceuticals: Ethics, Markets, Practices*, ed. Adriana Petryna, Andrew Lakoff, and Arthur Kleinman, 33–60. Durham: Duke University Press.

Petryna, Adriana, Andrew Lakoff, and Arthur Kleinman, eds. 2006. *Global Pharmaceuticals: Ethics, Markets, Practices*. Durham: Duke University Press.

"Pets Contribute to China's Economy." 2005. *China Daily*, 14 February, http://www.chinadaily.com.cn/english. Accessed 17 August 2008.

Pharmaceuticals Manufacturing Group, Manufacturing Association of Nigeria (PMG-MAN), 2001. "Dangers Posed to Life and the Economy by Drugs and Medicines from Illegal Sources." Presentation at the Public Hearing of the Health and Social Services Committee, Federal House of Representatives, National Assembly, Nigeria, March.

Pharmaceutical Research and Manufacturers of America (PhRMA). 2000. "Why Do Prescription Drugs Cost So Much . . . and Other Questions about Your Medicines." http://www.phrma.org/publications, accessed 11 August 2006.

Pharmaceutical Research and Manufacturers of America (PhRMA). 2006. *Pharmaceutical Industry Profile 2006*. Washington: PhRMA.

Pharmacists Council of Nigeria (PCN). 2007. *PCN List of Registered Pharmacies*. Lagos: Pharmacists Council of Nigeria.

Piasecki, Bruce. 1995. *Corporate Environmental Strategy: The Avalanche of Change since Bhopal*. New York: John Wiley and Sons.

Pickering, G. W. 1968. *High Blood Pressure*, 2d edn. London: Churchill.

Plotkin, Mariano. 2000. *Freud in the Pampas: The Emergence and Development of a Psychoanalytic Culture in Argentina*. Stanford: Stanford University Press.

"PMDA Challenges for Global Drug Development including Japan." Session 281 at the Forty-third Annual Meeting of the Drug Information Association, Atlanta, 17–21 June.

Polanyi, Michael. 1962. "The Republic of Science, Its Political and Economic Theory." Lecture delivered at Roosevelt University, 11 January.

————. 1966. *The Tacit Dimension*. New York: Doubleday.

Pollack, Andrew. 2006. "In Trials for New Cancer Drugs, Family Pets Are Benefiting, Too." *New York Times*, 24 November, §A, 1.

Pomfret, Johan, and Deborah Nelson. 2000. "In Rural China, a Genetic Mother Lode." *Washington Post*, 20 December, §A, 1.

Postone, Moishe. 1993. *Time, Labor, and Social Domination: A Reinterpretation of Marx's Critical Theory*. Cambridge: Cambridge University Press.

Potrykis, Ingo. 2001. "Golden Rice and Beyond." *Plant Physiology* 125 (March): 1157–61.

Public Citizen. 2001. "R$_x$ R&D Myths: The Case against the Drug Industry's R&D 'Scare Card.'" Available at http://www.citizen.org/documents/ACFDC.pdf, last accessed on 18 July 2011.

Rabinow, Paul. 1992. "Artificiality and Enlightenment: From Sociobiology to Biosociality." *Zone 6: Incorporations*, ed. Jonathan Crary, 234–52. New York: Zone.

————. 1995. *Making PCR: A Story of Biotechnology*. Chicago: University of Chicago Press.

————. 1996. *Essays on the Anthropology of Reason*. Princeton: Princeton University Press.

————. 1999. "Artificiality and Enlightenment: From Sociobiology to Biosociality." *The Science Studies Reader*, ed. Mario Biagioli, 407–16. New York: Routledge.

————. 2003. *Anthropos Today: Reflections on Modern Equipment*. Princeton: Princeton University Press.

————. 2004. "Midst Anthropology's Problems." *Global Assemblages: Rationality, Technology, Ethics*, ed. Aihwa Ong and Stephen J. Collier, 4–54. Malden, Mass.: Blackwell.

Raffles, Hugh. 2002. *In Amazonia: A Natural History*. Princeton: Princeton University Press.

Rai, Arti, and Rebecca Eisenberg. 2003. "Bayh-Dole Reform and the Progress of Biomedicine." *Law and Contemporary Problems* 66, nos. 1–2 (winter–spring): 289–314.

Rapp, Rayna. 2000a. "Extra Chromosomes and Blue Tulips: Medico-Familial Interpretations." *Living and Working with the New Medical Technologies: Intersections of Inquiry*, ed. Margaret Lock, Allan Young, and Alberto Cambrosio, 184–208. Cambridge: Cambridge University Press.

————. 2000b. *Testing Women, Testing the Fetus: The Social Impact of Amniocentesis in America*. New York: Routledge.

Rapp, Rayna, and Faye Ginsburg. 2001. "Enabling Disability: Rewriting Kinship, Re-imagining Citizenship." *Public Culture* 13, no. 3:533–56.

Rawls, John. 2005 [1971]. *A Theory of Justice.* Cambridge: Belknap.

Read, Jason. 2003. *The Micro-Politics of Capital: Marx and the Pre-History of the Present.* Albany: State University of New York Press.

Reardon, Jenny. 2001. "The Human Genome Diversity Project: A Case Study in Co-production." *Social Studies of Science* 31:357–88.

———. 2004. *Race to the Finish: Identity and Governance in an Age of Genomics.* Princeton: Princeton University Press.

Regalado, Antonio. 1999. "Inventing the Pharmacogenomics Business." *American Journal of Health System Pharmacy* 56, no. 1:40–50.

Rheinberger, Hans-Jörg. 1997. *Toward a History of Epistemic Things: Synthesizing Proteins in the Test Tube.* Stanford: Stanford University Press.

———. 1998. "Experimental Systems, Graphematic Spaces." *Inscribing Science: Scientific Texts and the Materiality of Communication,* ed. Timothy Lenoir, 285–303. Stanford: Stanford University Press.

———. 2006. *Epistemologie des Konkreten: Studien zur Geschichte der moderne Biologie.* Frankfurt am Main: Suhrkamp.

Rice Information System. 2003. "About RISe." Beijing Genomics Institute, http://rise.genomics.org.cn, accessed 2008.

Richler, J., et al. 2006. "Is There a 'Regressive Phenotype' of Autism Spectrum Disorder Associated with the Measles-Mumps-Rubella Vaccine? A CPEA Study." *Journal of Autism and Developmental Disorders* 36, no. 3:299–316.

Rimland, Bernard. 1964. *Infantile Autism: The Syndrome and Its Implications for a Neural Theory of Behavior.* Des Moines: Meredith.

———. 1973. "The Effect of High Dosage Levels of Certain Vitamins on the Behavior of Children with Severe Mental Disorders." *Orthomolecular Psychiatry: Treatment of Schizophrenia,* ed. David R. Hawkins and Linus Pauling. San Francisco: W. H. Freeman.

———. 1976. "Psychological Treatment versus Megavitamin Therapy." *Modern Therapies: Noted Practitioners Describe Twelve Different Types of Therapy and the Problems Each Can Help You Solve,* ed. Virginia Binder, Arnold Binder, and Bernard Rimland, 194–211. Englewood Cliffs, N.J.: Prentice-Hall.

_____. 2003. "Autism *Is* Treatable: Congressional Testimony of Bernard Rimland, Ph.D., November 19, 2003." San Diego: Autism Research Institute.

Rimland, Bernard, and Stephen M. Edelson. 2003. *Treating Autism: Parent Stories of Hope and Success.* San Diego: Autism Research Institute.

Risch, Neil J. 2000. "Searching for Genetic Determinants in the New Millennium." *Nature* 405:847–56.

Risch, Neil, and David Botstein. 1996. "A Manic Depressive History." *Nature Genetics* 12, no. 4:351–53.

Robbins-Roth, Cynthia. 2000. *From Alchemy to IPO: The Business of Biotechnology.* Cambridge, Mass.: Perseus.

Robertson, A. P. N. D. "Corruption, 'Shadow Revenues' and Capital Flight." Unpublished manuscript.

Robinson, Cedric. 2001. *The Anthropology of Marxism*. London: Ashgate.

Rodman, Margaret. 1992. "Empowering Place: Multilocality and Multivocality." *American Anthropologist* 94:640–56.

Roitman, Janet. 2005. *Fiscal Disobedience: An Anthropology of Economic Regulation in Central Africa*. Princeton: Princeton University Press.

Ronell, Avital. 2005. *The Test Drive*. Urbana: University of Illinois Press.

Rose, Carol. 1986. "The Comedy of the Commons: Custom, Commerce, and Inherently Public Property." *University of Chicago Law Review* 53, no. 3:711–81.

Rose, Deborah Bird. 2006. "What If the Angel of History Were a Dog?" *Cultural Studies Review* 12, no. 1:67–78.

Rose, G. A., Kay-Tee Khaw, and Michael G. Marmot. 2008. *Rose's Strategy of Preventive Medicine: The Complete Original Text*, new edn. Oxford: Oxford University Press.

Rosen, Richard A. 1998. "The Statutory Safe Harbor for Forward-Looking Statements after Two and a Half Years: Has It Changed the Law? Has It Achieved What Congress Intended?" *Washington University Law Quarterly* 76:645–80.

Rosenberg, Charles. 2002. "The Tyranny of Diagnosis: Specific Entities and Individual Experience." *Milbank Quarterly* 80:237–60.

Rosenwald, Andreas, George Wright, Wing C. Chan, Joseph M. Connors, Elias Campo, Richard I. Fisher, Randy D. Gascoyne, H. Konrad Muller-Hermelink, Erlend B. Smeland, Jena M. Giltnane, Elaine M. Hurt, Hong Zhao, Lauren Averett, Liming Yang, Wyndham H. Wilson, Elaine S. Jaffe, Richard Simon, Richard D. Klausner, John Powell, Patricia L. Duffey, Dan L. Longo, Timothy C. Greiner, Dennis D. Weisenburger, Warren G. Sanger, Bhavana J. Dave, James C. Lynch, Julie Vose, James O. Armitage, Emilio Montserrat, Armando López-Guillermo, Thomas M. Grogan, Thomas P. Miller, Michel LeBlanc, German Ott, Stein Kvaloy, Jan Delabie, Harald Holte, Peter Krajci, Trond Stokke, and Louis M. Staudt. 2002. "The Use of Molecular Profiling to Predict Survival after Chemotherapy for Diffuse Large-B-Cell Lymphoma." *New England Journal of Medicine* 346, no. 25:1937–47.

Roy, Arundhati. 2005. "People vs. Empire." *In These Times*, 3 January, 16–19.

Russell, Edmund. 2004. "The Garden in the Machine: Toward an Evolutionary History of Technology." *Industrializing Organisms: Introducing Evolutionary History*, ed. Susan R. Schrepfer and Philip Scranton, 1–16. New York: Routledge.

Sacks, Harvey, and Gail Jefferson. 1992. *Lectures on Conversation*. Oxford: Blackwell.

Salako, Lateef. 1997. "Health for All Nigerians—So Far, So What?" *Nigerian Quarterly Journal of Hospital Medicine* 7, no. 3:199–206.

Samba, Ebrahim Malick. 2004. "African Health Care Systems: What Went Wrong?" News-Medical.Net, 8 December, http://www.news-medical.net.

Sanal, Aslihan. 2005. "Flesh Yours, Bones Mine: The Making of the Biomedical Body in Turkey." Ph.D. diss., Massachusetts Institute of Technology.

Sandler, R. H., et al. 2000. "Short-Term Benefit from Oral Vancomycin Treatment of Regressive-Onset Autism." *Journal of Child Neurology* 15, no. 7:429–35.

Santora, Marc. 2005. "U.S. Praises Program in City for Children with Asthma." *New York Times*, 14 January, New York Region, 1.

Sarewitz, Daniel, Roger Pielke Jr., and Radford Byerly Jr., eds. 2000. *Prediction: Science, Decision-making and the Future of Nature*. Washington: Island.

Sasich, Larry. 2000. "Public Citizen's Health Research Group: Comments." Presentation at the FDA's Meeting on the Prescription Drug User Fee Act, 15 September.

Sassen, Saskia. 2000. "Spatialities and Temporalities of the Global: Elements for a Theorization." *Public Culture* 12, no. 1:215–32.

———. 2001. *The Global City: New York, London, Tokyo*. Princeton: Princeton University Press.

Schacter, Bernice Z. 2006. *The New Medicines: How Drugs Are Created, Approved, Marketed, and Sold*. Westport, Conn.: Praeger.

Scheper-Hughes, Nancy. 1992. *Death without Weeping: The Violence of Everyday Life in Brazil*. Berkeley: University of California Press.

Schiebinger, Londa. 1991. "The Mind Has No Sex? Women in the Origins of Modern Science." Cambridge: Harvard University Press.

Schienke, Erich. 2006. "Greening the Dragon: Environmental Imaginaries in the Science, Technology and Governance of Contemporary China." Ph.D. diss., Rensselaer Polytechnic Institute, Department of Science and Technology Studies.

Schofeld, Ian. 2006. "Developing Countries and Growing Threats." *The Scrip 100*, 26–27.

Schwartz, Joel. 1991. "Particulate Air Pollution and Daily Mortality in Detroit." *Environmental Research* 56, no. 2:204–13.

Schwartz, Joel, and Douglas W. Dockery. 1992. "Particulate Air Pollution and Daily Mortality in Steubenville, Ohio." *American Journal of Epidemiology* 135, no. 1:12–19.

ScienceDaily. 2007. "As Autism Diagnoses Grow, So Do Number of Fad Treatments, Researchers Say." ScienceDaily, 21 August, http://www.sciencedaily.com, accessed 21 August 2007.

Scott, John Paul, and J. H. Fuller. 1965. *Genetics and the Social Behavior of the Dog*. Chicago: University of Chicago Press.

Scrip 100: The Analysis of the Pharmaceutical Industry's Performance and Prospects. London: Infoma UK, 2006.

Sedgwick, Eve Kosofsky. 2003. *Touching Feeling: Affect, Pedagogy, Performativity*. Durham: Duke University Press.

Seroussi, Karen. 2003 [2000]. *Unraveling the Mystery of Autism and Pervasive Developmental Disorder: A Mother's Story of Research and Recovery*. New York: Broadway.

Shalo, Sibyl. 2004. "Built for Speed." *Pharmaceutical Executive* 24, no. 2:40–53.

Shapin, Steven. 1999. "The House of Experiment in Seventeenth-Century England." *The Science Studies Reader*, ed. Mario Biagioli, 479–504. New York: Routledge.

Shapin, Steven, and Simon Shaffer. 1985. *The Leviathan and the Air Pump: Hobbes, Boyle, and the Experimental Life*. Princeton: Princeton University Press.

Shartin, Emily. 2004. "Difficult Choices Variety of Treatments Face Parents of Autistic Children." *Boston Globe*, 25 January, §A, 1.

Shaw, William. 2001. *Biological Treatments for Autism and* PDD, 2d rev. edn. Lenexa, Kan.: Great Plains Laboratory.

Shieve, L., et al. 2006. "Mental Health in the United States: Parental Report of Diagnosed Autism in Children Aged 4–17 Years, United States, 2003–2004." *Morbidity and Mortality Weekly Report* 55, no. 17:481–86.

Shotwell, Alexis. 2006. "Implicit Understanding and Political Transformation." Ph.D. diss., University of California, Santa Cruz.

Sieber, Renee E. 1997. "Computers in the Grassroots: Environmentalists, GIS and Public Policy." Ph.D. diss., Rutgers University.

Sigurdsson, Skuli. 2001. "Yin-Yang Genetics, or the HSD deCODE Controversy." *New Genetics and Society* 20, no. 2:103–17.

Silverman, Chloe. 2011. *Understanding Autism: Parents, Doctors, and the History of a Disorder*. Princeton: Princeton University Press.

Silverstein, Ken. 1999. "Millions for Viagra, Pennies for Diseases of the Poor." *Nation*, 19 July, 13–19.

Sinclair, Jim. 1993. "Don't Mourn for Us," originally published in the Autism Network International newsletter, *Our Voice* 1, no. 3 (1993), based on a presentation given at the International Conference on Autism, Toronto (1993). http://www.au treat.com/dont_mourn.html, accessed 28 July 2011.

Singer, Joseph William. 2000. *Entitlement: The Paradoxes of Property*. New Haven: Yale University Press.

Smith, Mickey C. 1991. *Pharmaceutical Marketing: Strategy and Cases*. New York: Pharmaceutical Products Press.

Smith, Mickey C., E. M. "Mick" Kolassa, Greg Perkins, and Bruce Siecker. 2002. *Pharmaceutical Marketing: Principles, Environment, and Practice*. New York: Pharmaceutical Products Press.

Smith, Wayne, and Robert Dilday, eds. 2003. *Rice: Origin, History, Technology, and Production*. Hoboken, N.J.: John Wiley and Sons.

Sokoloff, Alex. 2006. "Informating Greenpeace: Material Practice, Work Culture and Global Organization." Ph.D. diss., Rensselaer Polytechnic Institute, Department of Science and Technology Studies.

Southern Poverty Law Center. 2001. "Woman's Death Exposes Seamy Prison Scam: Aryan Brotherhood." *Intelligence Report* 102 (summer), http://www.splcenter.org.

Sozialistisches Patientenkollektiv. 1993. SPK — *Turn Illness into a Weapon: For Agitation by the Socialist Patients' Collective at the University of Heidelberg*. Heidelberg: Krrim.

Sperling, Stefan. 2006. "Science and Conscience: Bioethics, Stem Cells and Citizenship in Germany." Ph.D. diss., Princeton University.

Spilker, Bert. 1989. *Multinational Drug Companies: Issues in Drug Discovery and Development*. New York: Raven.

Spitzer, R. L., J. Endicott, and E. Robins. 1978. "Research Diagnostic Criteria: Rationale and Reliability." *Archives of General Psychiatry* 35, no. 6:773–82.

Spivak, Gayatri Chakravorty. 1987. *In Other Worlds: Essays in Cultural Politics*. New York: Methuen.

Stark, Gregor, and E. Catherine Rayne. 1998. *El Delirio: The Santa Fe World of Elizabeth White*. Santa Fe: School of American Research.

Starr, Paul. 1982. *The Social Transformation of American Medicine: The Rise of a Sovereign Profession and the Making of a Vast Industry*. New York: Basic Books.

Stein, Mike. 2001. "Environmental Defense: From Brochureware to Actionware." Interview with Bill Pease. Benton Foundation, 11 April, http://www.benton.org, accessed 1 July 2001.

Stern, Rachel. 2003. "Hong Kong Haze: Air Pollution as a Social Class Issue." *Asian Survey* 43, no. 5:780–800.

Stewart, James B. 1984. *The Partners: Inside America's Most Powerful Law Firms*. New York: Simon and Schuster.

Stolberg, Sheryl Gay. 2005. "In Rare Threat, Bush Vows Veto of Stem Cell Bill." *New York Times*, 21 May, §A, 1.

Strange, Susan. 1996. *The Retreat of the State: The Diffusion of Power in the World Economy*. Cambridge: Cambridge University Press.

Strathern, Marilyn. 1992a. *After Nature: English Kinship in the Late Twentieth Century*. Cambridge: Cambridge University Press.

———. 1992b. *Reproducing the Future: Essays on Anthropology, Kinship, and the New Reproductive Technologies*. New York: Routledge.

———. 2005. *Kinship, Law and the Unexpected: Relatives Are Always a Surprise*. Cambridge: Cambridge University Press.

Strathern, Marilyn, and Eric Hirsch, eds. 2004. *Transactions and Creations*. Oxford: Berghahn.

Sturdy, Steve. 1998. "Reflections: Molecularization, Standardization and the History of Science." *Molecularizing Biology and Medicine: New Practices and Alliances, 1910s–1970s*, ed. Soraya de Chadarevian and Harmke Kamminga, 273–89. Amsterdam: Harwood.

Sunder Rajan, Kaushik. 2002. "Biocapital: The Constitution of Postgenomic Life." Ph.D. diss., Massachusetts Institute of Technology.

———. 2003. "Genomic Capital: Public Cultures and Market Logics of Corporate Biotechnology." *Science as Culture* 12, no. 1:87–121.

———. 2005. "Subjects of Speculation: Emergent Life Sciences and Market Logics in the U.S. and India." *American Anthropologist* 107, no. 1:19–30.

———. 2006. *Biocapital: The Constitution of Postgenomic Life*. Durham: Duke University Press.

———. 2007. "Experimental Values: Indian Clinical Trials and Surplus Health." *New Left Review* 45:67–88.

Sunyer, J., J. Castellsague, and M. Saez. 1996. "Air Pollution and Mortality in Barcelona." *Journal of Epidemiology and Community Health* 50 (supplement 1): S76–80.

Supreme Court of the United States. 1979a. Brief on Behalf of Genentech, Inc., Amicus Curiae. October Term, No. 79-136.

———. 1979b. Brief on Behalf of the Peoples Business Commission, Amicus Curiae. October Term, No. 79-136.

————. 2005. Brief for the United States, Amicus Curiae, *Homan McFarling v. Monsanto Company*. May Term, No. 04-31.

Syngenta. 2011. http://www.syngenta.com, accessed 25 June 2011.

Takeuchi, Masahiro. 2003. "Variance Components Testing in Mixed Effects Models in Bridging Studies." Presentation at the 2003 APEC symposium on statistical methodology, Nankang, Taipei, 15 November.

TallBear, Kimberly. 2005. "Native American DNA." Ph.D. diss., University of California, Santa Cruz.

Taylor, O. 1998. "Dispensing Practices in Private and Government Health Facilities in Lagos State." *Nigerian Journal of Pharmacy* 29, no. 1:63–67.

Taylor, R. B., et al. 2001. "Pharmacopoeial Quality of Drugs Supplied by Nigerian Pharmacies." *Lancet* 357, no. 9272 (16 June): 1933–36.

Tennant, Raymond W. 2002. "The National Center for Toxicogenomics: Using New Technologies to Inform Mechanistic Toxicology." *Environmental Health Perspectives* 110, no. 1: A8–10.

"Texas City Blast of 1947." 2010. *Houston Chronicle*, 14 April, http://www.chron.com.

Thompson, Charis. 2005. *Making Parents: The Ontological Choreography of Reproductive Technologies*. Cambridge: MIT Press.

Tietenber, Tom, and David Wheeler. 1998. "Empowering the Community: Information Strategies for Pollution Control." Paper presented at Frontiers of Environmental Economics Conference, Airlie House, Va., 23–25 October.

Timmermans, Stefan, and Marc Berg. 1997. "Standardization in Action: Achieving Local Universality through Medical Protocols." *Social Studies of Science* 27, no. 2:273–305.

————. 2003. *The Gold Standard: The Challenge of Evidence-Based Medicine and Standardization in Health Care*. Philadelphia: Temple University Press.

Tinnell, Robert, and Craig Taillefer. 2007a. *The Chelation Kid: Episode 129*. Webcomics Nation, 13 September, http://www.webcomicsnation.com, accessed 12 January 2008.

————. 2007b. *The Chelation Kid: Episode 135*. Webcomics Nation, 21 September, http://www.webcomicsnation.com, accessed 12 January 2008.

Toobin, Jeffrey. 2002. "The Man Chasing Enron." *New Yorker*, 9 September, 86–96.

Torrey Mesa Research Institute. 2004. "Accessing Torrey Mesa Research Institute (TMRI) Rice Genome Data." http://www.sciencemag.org/content/suppl/2002/04/04/296.5565.92.DC1/Goffweb2.pdf, accessed 1 January 2004.

Touloumi, G., E. Samoli, and K. Katsouyanni. 1996. "Daily Mortality and 'Winter Type' Air Pollution in Athens, Greece: A Time Series Analysis within the APHEA Project." *Journal of Epidemiology and Community Health* 50 (supplement 1): S47–51.

Traweek, Sharon. 1988. *Beamtimes and Lifetimes: The World of High Energy Physicists*. Cambridge: Harvard University Press.

Tsing, Anna Lowenhaupt. 2000. "Inside the Economy of Appearances." *Public Culture* 12, no. 1:115–44.

————. 2004. *Friction: An Ethnography of Global Connection.* Princeton: Princeton University Press.

Tucker, Robert, ed. 1978. *The Marx-Engels Reader,* 2d edn. New York: W. W. Norton.

University of California, Office of the President. 2003. "Patent Survey Reveals AG Biotech Intellectual Property Ownership." Press release. University of California, Office of the President, 9 September, http://www.universityofcalifornia.edu/news/article/5709.

U.S. Department of Energy. 2008. "Facts about Genome Sequencing." Human Genome Project Information, http://www.ornl.gov, accessed 25 June 2011.

U.S. Environmental Protection Agency. 1998. *Chemical Hazard Data Availability Study: What Do We Really Know about the Safety of High Production Volume Chemicals?* Washington: U.S. Environmental Protection Agency.

————. 2003. "America's Children and the Environment: Measures of Contaminants, Body Burdens and Illnesses," 2d edn. Washington: Environmental Protection Agency, EPA 240-R-03-001, http://www.epa.gov, accessed 30 September 2004.

U.S. General Accounting Office. 2001. *Major Management Challenges and Program Risks: Environmental Protection Agency.* Report No. GAO-01-257. Washington: General Accounting Office.

U.S. Patent and Trademark Office. 2001. "Utility Examination Guidelines." *Federal Register* 66, no. 4 (5 January): 1092–99.

U.S. Securities and Exchange Commission. 1994a. "Concept Release and Notice of Hearing: Safe Harbor for Forward-Looking Statements." Release Nos. 33-7101; 34-34831; 35-26141; 39-2324; IC-20613; File No. S7-29-94. Washington: Securities and Exchange Commission.

————. 1994b. "Concept Release and Notice of Hearing: Safe Harbor for Forward-Looking Statements." Release Nos. 33-7101; 34-34831; 35-26141; 39-2324; IC-20613; File No. S7-29-94. Washington: Securities and Exchange Commission.

————. 2005. "SEC Sues Morgan Stanley and Goldman Sachs for Unlawful IPO Allocation Practices." Press Release. U.S. Securities and Exchange Commission, 25 January, http://www.sec.gov/news/press/2005-10.htm, accessed July 2006.

Uyama, Y., T. Shibata, N. Nagai, H. Hanaoka, S. Toyoshima, and K. Mori. 2005. "Successful Bridging Strategy Based on ICH E5 Guideline for Drugs Approved in Japan." *Clinical Pharmacology and Therapeutics* 78, no. 2:102–13.

Vaidhyanathan, Siva. 2003 [2001]. *Copyrights and Copywrongs: The Rise of Intellectual Property and How It Threatens Creativity.* New York: New York University Press.

Vanelli, Mark, et al. 2002. "Moving beyond Market Share." *In Vivo: The Business and Medicine Report* 20, no. 3:1–6.

Vargas, D. L., et al. 2005. "Neuroglial Activation and Neuroinflammation in the Brain of Patients with Autism." *Annals of Neurology* 57, no. 1 (January): 67–81.

Vezzetti, Hugo. 1996. *Aventuras de Freud en el país de los argentinos: De José Ingenieros a Enrique Pichon-Rivière.* Buenos Aires: Paidós.

Visiongain. 2006. *Japanese Pharmaceutical Market, 2006–2011.* London: Visiongain.

Wakefield, A. J., et al. 1998. "Ileal-Lymphoid-Nodular Hyperplasia, Non-specific Colitis, and Pervasive Developmental Disorder in Children." *Lancet* 351, no. 9103:637–41.

Wald, N., and M. Law. 2003. "A Strategy to Reduce Cardiovascular Disease by More Than 80%." *British Medical Journal* 326:1419–24.

Waldby, Catherine, and Rob Mitchell. 2006. *Tissue Economies: Blood, Organs, and Cell Lines in Late Capitalism*. Durham: Duke University Press.

Walgate, Robert. 2001. "Syngenta Claims Ownership of Rice but Will Give Data Away." *Scientist*, 1 February.

Wang, Horng-Luen. 2000. "Rethinking the Global and the National: Reflections on National Imaginations in Taiwan." *Theory, Culture and Society* 17, no. 4:93–117.

Wang, Jia-Huai, Y. Yan, T. Garrett, J. Liu, D. Rodgers, R. L. Garlick, G. E. Tarr, Y. Husain, E. L. Reinherz, and S. C. Harrison. 1990. "Atomic Structure of a Fragment of Human CD4 Containing Two Immunoglobulin-like Domains." *Nature* 348 (29 November): 411–18.

Waters, M., et al. 2003. "Systems Toxicology and the CEBS Knowledge Base." *EHP Toxicogenomics* 111 (1T): 15–20.

Watts, Michael. 2001. "Petro-Violence: Community, Extraction, and Political Ecology of a Mythic Commodity." *Violent Environments*, ed. Nancy Lee Peluso and Michael Watts, 189–212. Ithaca: Cornell University Press.

———. 2004. "Resource Curse? Governmentality, Oil and Power in the Niger Delta, Nigeria." *Geopolitics* 9, no. 1:50–80 [special issue].

Weber, Leonard J. 2006. *Profits before People? Ethical Standards and the Marketing of Prescription Drugs (Bioethics and the Humanities)*. Bloomington: Indiana University Press.

Weber, Max. 1978. *Weber: Selections in Translation*, ed. Walter Garrison Runciman. Cambridge: Cambridge University Press.

———. 2008. *The Protestant Ethic and the Spirit of Capitalism*. Miami: BN Publishing.

Weber, W., and S. Newmark. 2007. "Complementary and Alternative Medical Therapies for Attention-Deficit/Hyperactivity Disorder and Autism." *Pediatric Clinics of North America* 54:983–1006.

Weiner, Charles. 1987. "Patenting and Academic Research: Historical Case Studies." *Science, Technology and Human Values* 12, no. 1:50–62.

Weiss, Elliot J., and Janet E. Moser. 1998. "Enter Yossarian: How to Resolve the Procedural Catch-22 that the Private Securities Litigation Reform Act Creates." *Washington University Law Quarterly* 76:457–507.

Wellman, Carl. 1998. *The Proliferation of Rights: Moral Progress or Empty Rhetoric?* Boulder: Westview.

Werner, Emily, and Geraldine Dawson. 2005. "Validation of the Phenomenon of Autistic Regression Using Home Videotapes." *Archives of General Psychiatry* 62: 889–95.

White, J. F. 2003. "Minireview: Intestinal Pathophysiology in Autism." *Experimental Biology and Medicine* 228, no. 6:639–49.

Wilde, Oscar. 1990. *Complete Fairy Tales of Oscar Wilde*. New York: Signet Classics.

Wildenauer, Dieter B., et al. 1999. "Do Schizophrenia and Affective Disorder Share Susceptibility Genes?" *Schizophrenia Research* 39:107–11.

Williams, Roger J. 1956. *Biochemical Individuality.* New York: John Wiley and Sons.

Winner, Langdon. 1978. *Autonomous Technology: Technics-out-of-Control as a Theme in Political Thought.* Cambridge: MIT Press.

———. 1988. *The Whale and the Reactor: A Search for Limits in an Age of High Technology.* Chicago: University of Chicago Press.

Winslow, Ron. 2004. "For Bristol-Myers, Challenging Pfizer Was a Big Mistake: In Rare Head-to-Head Study, Lipitor Beats Pravachol at Reducing Heart Risk: A New Push on Cholesterol." *Wall Street Journal,* 9 March, §A, 1.

Wolpe, Paul R. 1990. "The Holistic Heresy: Strategies of Ideological Challenge in the Medical Profession." *Social Science and Medicine* 31, no. 8:913–23.

———. 1994. "The Dynamics of Heresy in a Profession." *Social Science and Medicine* 39, no. 9:1133–48.

Wong, Chit-Ming, Stefan Ma, A. J. Hedley, and T. H. Lam. 2001. "Effect of Air Pollution on Daily Mortality in Hong Kong." *Environmental Health Perspectives* 109, no. 4:335–40.

Wong, Ellick. 2003. "The Regulatory Environment and Clinical Trials in Southeast Asia." *Drug Information Journal* 37:155–63S.

Wong, Gerard B. N., and John C. W. Lim. 2003. "Drug Regulation by Singapore's Health Sciences Authority: The Role of the Centre for Drug Evaluation." *Drug Information Journal* 37, no. 4:15–25S.

Wong, Helen H. L., and Ronald G. Smith. 2006. "Patterns of Complementary and Alternative Medical Therapy Use in Children Diagnosed with Autism Spectrum Disorders." *Journal of Autism and Developmental Disorders* 36, no. 7 (October): 901–9.

Wong, Nadine. 2000. "Judgment Day Is Here for Biotechs." The Worldly Investor, http://www.worldlyinvestor.com, accessed 20 July 2000.

Wong, T. W., W. S. Tam, T. S. Yu, and A. H. S. Wong. 2002. "Associations between Daily Mortalities from Respiratory and Cardiovascular Diseases and Air Pollution in Hong Kong, China." *Occupational and Environmental Medicine* 59, no. 1:30–35.

World Bank. 2004. *World Bank Annual Report.* Washington: World Bank.

World Health Organization. *World Health Report.* Geneva: World Health Organization.

World Health Organization Statistical Information System. 2001. *Statistics by Country: Nigeria.* Geneva: World Health Organization.

Xi, Xi. 1997. *Marvels of a Floating City and Other Stories: An Authorized Collection,* trans. Eva Hung. Hong Kong: Renditions Paperbacks.

Xu, X. P., D. W. Dockery, and J. Gao. 1994. "Air Pollution and Daily Mortality in Residential Areas of Beijing, China." *Archives of Environmental Health* 49:216–22.

Yakujinippo [pharmaceutical news], 1 January 2001.

Young, Allan. 1996. *The Harmony of Illusions: Inventing Post-traumatic Stress Disorder.* Princeton: Princeton University Press.

Young, John. 1994. "Using Computers for the Environment." *State of the World 1994:*

A *WorldWatch Institute Report on Progress toward a Sustainable Society*, ed. Lester R. Brown et al., 999–116. New York: W. W. Norton.

"Your Company's Most Important Asset: Intellectual Capital." 1994. *Fortune*, 3 October, 68–74.

Zalik, Anna. 2004. "The Niger Delta: 'Petro Violence' and 'Partnership Development.'" *Review of African Political Economy*, no. 101:401–24.

Zimmerman, Jack. 2003. Preface, *Children with Starving Brains: A Medical Treatment Guide for Autism Spectrum Disorder*, by Jacquelyn McCandless. Putney, Vt.: Bramble.

Žižek, Slavoj. 1994. "How Did Marx Invent the Symptom?" *Mapping Ideology*, 296–332. London: Verso.

———. 2004. "The Ongoing 'Soft Revolution.'" *Critical Inquiry* 30, no. 2:292–323.

ABOUT THE CONTRIBUTORS

TIMOTHY CHOY is an associate professor of anthropology and of science and technology studies at the University of California, Davis. He is the author of *Ecologies of Comparison: An Ethnography of Endangerment in Hong Kong* (2011). His next project is on the politics of atmosphere.

JOSEPH DUMIT is the director of the program in science and technology studies and a professor in the department of anthropology at the University of California, Davis. His research focuses on the anthropology of science, technology, medicine, and media. He is the author of *Picturing Personhood: Brain Scans and Biomedical Identity* (2004) and the coeditor of *Cyborgs and Citadels: Anthropological Interventions in Emerging Sciences and Technologies* (with Gary L. Downey, 1997), *Cyborg Babies: From Techno-Sex to Techno-Tots* (with Robbie Davis-Floyd, 1998), and *Biomedicine as Culture: Instrumental Practices, Technoscientific Knowledge, and New Modes of Life* (with Regula Burri, 2007). He was associate editor of *Culture, Medicine and Psychiatry* for ten years. He is currently finishing a book on pharmaceutical marketing and clinical trials called *Drugs for Life*. He has begun work on new projects on the history of flowcharts and psychotic computers, geologists and 3D immersive visualization environments, and comparative anatomies across therapeutic massage modalities.

MICHAEL M. J. FISCHER is Andrew W. Mellon Professor in the Humanities and a professor of anthropology and of science and technology studies at MIT; he is also a lecturer in the department of global health and social medicine at the Harvard Medical School. He recently coedited *A Medical Anthropology Reader: Theoretical Trajectories and Emergent Realities* (with Byron Good, Sarah Willen, and Mary Jo DelVecchio Good, 2010). He is the author of *Anthropological Futures* (2009), *Mute Dreams, Blind Owls and Dispersed Knowledges: Persian Poesis in the Transnational Circuitry* (2004), *Emergent Forms of Life and the Anthropological Voice* (2003), *Debating Muslims: Cultural Dialogues in Postmodernity and Tradition* (with Mehdi Abedi, 1990), *Anthropology as Cultural Critique: An Experimental Moment in the Human Sciences* (with George

Marcus, 1986), and *Iran: From Religious Dispute to Revolution* (1980). He is currently working on multiple projects relating to the growth of the biosciences, biotechnologies, and translational medicine in Asia.

KIM FORTUN, a professor in the department of science and technology studies at Rensselaer Polytechnic Institute, is an anthropologist of science and environmentalism. She is the author of *Advocacy After Bhopal: Environmentalism, Disaster, New Global Orders* (2001). Her more recent work analyzes developments in toxicology, exposure science, and other environmental-health sciences in the United States. Fortun also contributes to *The Asthma Files*, a collaborative and experimental digital ethnographic project.

MIKE FORTUN is an associate professor in the department of science and technology studies at Rensselaer Polytechnic Institute. A historian and anthropologist of the life sciences, he has covered the policy, scientific, and social history of the Human Genome Project in the United States, the history of biotechnology, and the growth of commercial genomics and bioinformatics in the speculative economies of the 1990s. His current research focuses on the genomics of asthma, and he contributes to *The Asthma Files*, a collaborative and experimental digital ethnographic project. He is also researching the truth- and subject-producing practices in contemporary genomics, which he calls the "care of the data." His most recent book is *Promising Genomics: Iceland and deCODE Genetics in a World of Speculation* (2008).

DONNA J. HARAWAY is a professor in the history of consciousness department at the University of California, Santa Cruz, where she teaches feminist theory, science studies, and animal studies. Her books include *Crystals, Fabrics, and Fields: Metaphors that Shape Embryos* (1976), *Primate Visions: Gender, Race, and Nature in the World of Modern Science* (1989), *Simians, Cyborgs, and Women: The Reinvention of Nature* (1991), *Modest_Witness@Second_Millennium.FemaleMan©_Meets_OncoMouse™: Feminism and Technoscience* (1997); and *The Companion Species Manifesto: Dogs, People, and Significant Otherness* (2003). Under the title "Staying with the Trouble," her current work inhabits the relational labor of human and nonhuman animals in urban and periurban agriculture.

SHEILA JASANOFF is Pforzheimer Professor of Science and Technology Studies at Harvard University's John F. Kennedy School of Government. Her work explores the role of science and technology in the law, politics, and policy of modern democracies. Among her books are *The Fifth Branch: Science Advisers as Policymakers* (1990), *Science at the Bar: Law, Science, and Technology in America* (1995), and *Designs on Nature: Science and Democracy in Europe and the United States* (2005). She was founding chair of the department of science and technology studies at Cornell University and has held distinguished visiting appointments at numerous institutions, including MIT, the University of Cambridge, Kyoto University, the University of Vienna, and the Wissenschaftskolleg Berlin.

WEN-HUA KUO is an associate professor at National Yang-Ming University, Taiwan, where he holds joint appointments at the Institute of Science, Technology and Society and the Institute of Public Health. His research concerns health governance in the transformation of East Asian states, with thematic focuses on Cold War epidemic control and debates on the harmonization of clinical-trial regulations. His publications include "The Voice on the Bridge: Taiwan's Regulatory Engagement with Global Pharmaceuticals" (2009), which won the David Edge Prize of the Society for Social Studies of Science.

ANDREW LAKOFF is an associate professor of anthropology, sociology, and communication at the University of Southern California. He is the author of *Pharmaceutical Reason: Knowledge and Value in Global Psychiatry* (2005) and coeditor of *Global Pharmaceuticals: Ethics, Markets, Practices* (with Arthur Kleinman and Adriana Petryna, 2006) and *Biosecurity Interventions: Global Health and Security in Question* (with Stephen Collier, 2008). His current research concerns the articulation of public-health and national-security expertise around the problem of emerging infectious disease.

KRISTIN PETERSON is an assistant professor of anthropology at the University of California, Irvine. Her work broadly concerns science, political economy, and Africa. She is currently pursuing several projects: a study on the relationship between global drug marketing and pharmaceutical markets in Nigeria; health and security logics as they relate to AIDS policies and politics; and a collaborative study (with Morenike Ukpong), analyzing the ethics, structural inequities, and biosocialities of HIV-related clinical trials in Nigeria and Malawi.

CHLOE SILVERMAN is an assistant professor of science, technology, and society at Pennsylvania State University. She is the author of *Understanding Autism: Parents, Doctors, and the History of a Disorder* (2011). Her work describes the role of affect in scientific knowledge production, with a focus on biomedical social movements and the relationships between scientists and their research subjects.

ELTA SMITH is a senior consultant at GHK, London. She received her doctorate in public policy from the John F. Kennedy School of Government, at Harvard University, specializing in the political economy of agricultural biotechnology. She works on agriculture, food-chain, and sustainability-related projects primarily for the British Government and the European Commission.

KAUSHIK SUNDER RAJAN is an associate professor of anthropology at the University of Chicago. He is the author of *Biocapital: The Constitution of Postgenomic Life* (2006). His work explores the relationships between the life sciences and global capital, with a specific empirical focus on the United States and India. He is currently working on a number of projects relating to various aspects of pharmaceutical development in the Indian context, such as global clinical trials, intellectual-property regimes, and translational research.

TRAVIS J. TANNER is a postdoctoral teaching fellow in the department of English at Tulane University. He received his doctorate in comparative literature from the University of Califonia, Irvine, specializing in Native American literature and critical theory. He has published an essay on the Acoma poet Simon Ortiz in *The Kenyon Review*, and is revising his manuscript "X-Communicated Subjects in Native American Literature" for publication. His work explores how literature reconfigures modes of subjectivity and politics in situations of postcolonial subjection.

antibiotics, 60, 380, 400

antidepressants, 46, 260, 285

antihypertensive, 400

antipsychotics, 267, 368, 383

antiretrovirals, 239, 244

antithrombin 3, 389, 393, 395

antitrust law, 170

anti-tumor agent, 389

Applied Behavioral Analysis (ABA), 361

Argentina, 31, 150, 252–56, 259–64, 267, 270–78, 438, 443

Arizona Genomics Institute, 209

arthritis, 417

Asia, 129, 143, 201, 209, 281–82, 285–89, 292–97, 301–5, 419, 423–26, 435, 443

Asian Pacific region, 132, 423, 426

Asia-Pacific Economic Cooperation (APEC), 288–89, 291, 294, 302–5

Asperger's Syndrome, 383

assemblage, 38, 104, 133, 182, 252, 274, 430–31

Association of Southeast Asian Nations (ASEAN), 289, 294, 304

asthma, 72, 262, 275, 316–19, 325

AstraZeneca, 302

Atlanta, 433

Atlas Ventures, 336

attention-deficit disorder, 277

Austin, J. L., 331–33, 341, 350

Australia, 73, 132, 144, 301; Therapeutic Goods Administration in, 304

authorship, 392

autism, 35–36, 354–71, 373–84, 438, 446, 448

Autism Biomedical Discussion Group, 366, 383

Autism Network for Dietary Intervention, 382

Autism One, 361, 381

Autism Research Institute, 361–62, 366, 368, 382–83; conferences, 381

Autism Society of America, 361

autistic enterocolitis, 372

automated sequencing technologies, 193, 209

Babandiga, Ibrahim, 245

Bank of Iceland, 347

BASF, 309, 323

Basic Local Alignment Sequence Tool (BLAST), 416

basic research, 204, 208, 415

Baton Rouge, 315

Bayh-Dole Act, 2–3, 37, 196, 415

behavioral genetics, 105, 118–19; therapies of, 378, 383

Beijing Genomics Institute (BGI), 201–2, 206, 210, 421

Belgium, 396

Belmont Report, 445

bench to clinic, 420

benign prostatic hypertrophy, 75

Bermuda Principles, 198–99, 210

Berne Convention, 208

Beth Israel Deaconess Medical Center, 414

Bettelheim, Bruno, 355, 361

Bhopal, 306, 313, 315–16, 323

Bidil, 298

Bight of Benin, 241

bioavailability, 82, 91

biobricks, 436

biocapital, 23–24, 29, 39, 90, 94–95, 102, 114, 387, 439, 445–46, 449–50

biochemistry, 365, 369, 378, 384

biocolonialism, 214

biodiversity, 322, 324

bioethics, 29, 102, 182, 261, 276, 326, 444–45, 448

Biogen, 335

bioinformatics, 253, 258, 442

biomarkers, 56–57, 74, 79, 85, 88, 146, 371

BioMarx, 50–53, 55, 57, 61–63, 69, 76, 79, 81–88, 92, 152

biomedical mode of reproduction, 112

biomedical platform, 277

Biopolis, 115, 305, 419, 423–24, 442

biopolitics, biopolitical, 18, 26, 33, 41, 91, 108–9, 231, 280, 306, 311, 323, 326, 441, 447, 450

biopower, 94, 107

biorepository, 169

biosociality, biosocialities, 278, 439, 450

biostatistics (tician), 303, 435

biotechnology industry, 2, 5, 172

Biotek Invest, 348

bipolar disorder, 150, 252–60, 263–78, 439

black market, 234

Blair, Tony, 184

blockbuster drugs, 64, 81, 434

Boehringer-Ingelheim, 247

Bombay, 450

Boston, 390, 393, 403–4, 433

Boyer, Herbert, 2–3, 5, 37

Brahmachari, Samir, 424, 434

brain drain, 419

Brazil, 198, 209

Brazilian Rice Genome Initiative, 209

breathers, 128, 142

breedwealth, 102, 113

bridging studies, 287–93, 296, 302–3

Brigham and Women's Hospital, 417

Bristol-Myers Squibb, 247, 302, 396

British Medical Journal, 72, 85

British Petroleum (BP), 323

Broad Institute, 105

Buenos Aires, 252, 256–62, 270, 274–76

Burroughs Wellcome, 68

Bush, George W., 159–60, 175, 427, 436; administration of, 239, 241

Business Communications Company, 96, 98

California, 135–36, 145–46, 342, 362, 432; Supreme Court of, 168, 171

Cambodia, 425

Campomar Institute, 257, 260

Canada, 73, 140, 173, 176, 182, 262, 438; parliament of, 180; Patent Act of, 173, 183; Supreme Court of, 27, 158, 173, 176, 178–81, 183

cancer, 99, 104, 108–9, 118–19, 172, 177, 307–10, 316, 321, 330, 383, 389, 401–5; breast, 56, 329, 400, 403–4, 406; colon and colorectal, 56, 90, 329, 404–5; genomics and, 119; ovarian, 405; prostate, 47, 54, 169, 404; war on, 2

Canine Genome Project. *See* dog genome projects

capital: fixed, 62; flexible, 126, 228; flight of, 233, 246; markets of, 342

Capital (Marx), 90; Vol. 1, 12–13, 24, 48–53, 58, 65, 76, 213, 217, 219, 221, 450; Vol. 3, 439

carcinogens, 310

Catholic Church, 277

cDNA library, 256

Celebrex, 61

Celera Genomics, 203, 337, 339

Cell, 400, 418

cell dogs, 110, 112

cell line, 160, 169, 181, 386–87

cellular signaling, 371

Center for Autism and Related Disorders, 381

Center for Drug Development, Tufts University, 283, 299

Central Bank of Iceland, 345

central dogma of molecular biology, 39

Cereon, 420

Chad Basin, 245

chelating agents, 364, 382

chelation, 364–65

chemical, chemicals: industry, 314; security of, 314–15, 325; toxicity of, 321

Chemical Week, 311

chemotherapy, 396

Chernobyl, 152

Children's Hospital, 394

Chile, 262

China, 32, 61, 114–15, 124, 129, 132, 142–44, 147, 195, 198, 201–2, 209–10, 253, 275, 295, 324, 385, 418–25, 434

Chinese Academy of Sciences, 209, 421

Chinese University of Hong Kong (CUHK), 129, 150

cholesterol, 46–47, 52–58, 69–73, 88, 90–91

chronic diarrhea, 372–73

chronic illness, 80–82, 359, 357

Cipla, 239, 434

397, 402, 426, 433–36; exclusivity, 243; management, 22, 299; mining, 22, 425; sharing, agreements for, 196–97, 205

databanks, 291, 386, 426, 429, 433

databases, 105, 191–206, 210, 242, 253–54, 257, 276–77, 309, 311, 316, 320, 324, 338–39, 345, 427

Da Vinci Surgical System, 432

debt-worthiness, 246

Declaration of Independence, 164

DeCode Genetics, 34, 253, 263, 276–77, 330–31, 335–39, 346–50, 352

deconstruction, 220

deep play, 296, 386, 429

deep venous thrombosis (DVT), 77–78

Defeat Autism Now! (DAN!), 361–70, 372–76, 378, 381–84

Delaware, 337

Deleuze, Gilles, 430–31

Democratic Party, 173

depression, 54, 56

Derrida, Jacques, 151, 312, 326, 331–33, 426–30, 436, 444–50

determination, determinations, 2, 9, 11, 15; in last instance, 10, 39; multiple, 18; over-, 10, 19, 449

diabetes, 106, 264, 329

Diamond v. Chakrabarty, 3, 37, 167, 172–74, 176–81, 196, 415

differential reproduction, 351

digitalis, 432

diphtheria, pertussis, and tetanus vaccine, 434

direct-to-consumer advertising, 54

Disney. See Walt Disney Corporation

dispossession, 16, 29–30, 36, 213–14, 224, 228–33, 238–39, 244–45, 438–41, 444

DNA Databank of Japan (DDBJ), 199, 202, 210

DNA sequence, sequencing, 197, 199, 210; technology of, 253

dog genome projects, 105, 118

Dolly, 102–3, 118, 160, 176, 277, 424

dot-coms, 338

double binds, 133, 331, 335

Dred Scott v. Sandford, 163–64, 182

drugs: delivery of, 386; distribution systems for, 234; labeling for, 232; manufacturing of, 229–30, 232–33, 245; markets for, 232–38, 242

DSM, 266, 274, 277

DSM-III, 265, 271

DSM-IV, 268

due diligence, 344–45

Duke University, 188

East Asia, 151, 281, 293, 302

economic crisis, 234, 238

Edinburgh School, 20

Egypt, 245

Eighth Circuit Court of Appeals, 169

Eisner, Michael, 122–23, 129

electrical engineers, electrical engineering, 415

Eli Lilly, 54, 302, 394

embodiment, 55

eminent domain, 155–56

Enbrel, 417, 419

enclosure movement, 13

encounter value, 25, 95, 109–10, 112, 379, 440–41

endocrine system, 370, 384

endostatin, 394, 400, 402–3

Engels, Frederich, 10–11, 38

Enlightenment, 17, 427, 448

Enron, 349

EntreMed, 394, 400, 410

entrepreneurial university, 3–4, 6

Environmental Defense, 311, 319, 323

environmentalists, environmentalism, 17, 33–34, 306, 311, 442; activists and, 123; burdens of, 309; economics of, 128; ethics of, 321, 325–26; informatics and, 131, 146; information systems and, 311, 313, 321–22, 324–25; justice and, 306, 317–18; medicine and, 364; politics and, 313, 321–22, 447; regulations of, 311; risk and, 313–15; science and, 437; scientists and, 320; toxins and, 376

epidemiology, 57–58, 75–81, 133, 254, 264–66, 271, 435

epigenetics, 94

epistemology, epistemologies, 7–9, 13, 15, 17, 31–36, 47, 133, 194, 225, 252, 265, 267, 271, 274, 358, 361, 365, 379–80, 438–40, 443, 445, 447, 450

erythropoietin (EPO), 421

ethical plateaus, 422, 429

ethnography, 124, 150, 211, 298, 333, 430, 438

eugenics, 101, 261

Europe, Europeans, 61, 66, 73, 77, 131, 173, 145, 209, 242–43, 253–54, 262–63, 274–75, 292, 389, 394, 400–404, 423, 428, 433, 435, 438

European Agency for the Evaluation of Medicinal Products, 304

European Community, 300

European Economic Area, 345

European Federation of Pharmaceutical Industries and Associations, 300

European Free Trade Association, 300

European Institute of Bioinformatics, 201

European Molecular Biology Organization, 199, 202

European Union, 280, 286–87

evangelical Christianity, 363

evidence-based medicine, 274

exchange value, 25, 54, 93, 95, 97, 112

Executive Committee for Research (ECOR), 414–15

experiential knowledge, 358

experimentality, 240

experimental psychology, 363

experimental subjects, 61

experimental systems, 23, 324, 326, 351, 386, 396

exploitation, 79, 213–14, 219, 221, 223

Express Scripts, 80, 82

expropriation, 24, 213, 216, 440, 444, 450

factory farming, 97

Fanon, Frantz, 232

felt-illness, 53, 56–59

feminists, feminism, 20–21, 25, 39

fetishism, 221

finance capital, 228

financial markets, 245, 331

fluorescent tagging, 413

Folkman, Judah, 36, 385–92, 396, 401–4, 408–9, 411, 417, 419, 426–31, 433

food: riots for, 234; security of, 28–29, 183, 185

Forbes, 82

foreign direct investment, 232, 246

foreign exchange, 245

forward-looking statements, 34, 335, 339–44

Foucault, Michel, 17–18, 41, 91, 107–8, 280, 295, 433, 442

Framingham Heart Study, 57

France, 145, 174, 198, 209–10, 276, 409, 434, 443

Frankfurt School, 325

Fred Hutchinson Cancer Research Center, 119

Freeport, 309

free-trade agreements, 229, 243

Freud, Sigmund, 11

functional medicine, 363–64, 377–78, 380, 384

Fundación Bipolares de Argentina (FUBIPA), 260, 276

futures markets, 255, 345

Gabon, 262

Galveston County, Texas, 306–9, 321

gastroenterology, 374

gastrointenstinal pathologies, 365

GenBank, 197, 199–204, 210

gene-expression studies, 39

Genentech, 174–79, 330, 352, 395, 401, 406

gene patents, gene patenting, 38, 184, 263

General Electric, 77, 167

general public license (GPL), 208

generic drugs, 233, 244, 289; industry, 229–30, 239, 242–43, 434; market for, 442

model animal. *See* model organism

model cereal, 198, 205

model organism, 28, 192

modernity, 125, 213, 222, 227, 299, 441

modes of production, 125, 213, 312, 360, 440

molecular biology, 22, 269, 415, 433, 436

molecular genetics, 32, 269

molecular medicine, 443

molecular profiling, 291

Mongolia, 425

monopoly rights, 166, 254

Monsanto, 170–71, 181, 194–98, 201, 206, 406, 420, 422

moral economy, 158, 350

Morgan Stanley, 349–50, 352–53

Mount Sinai Hospital, 317

mouse genome, 210

mouse models, 420

multi-locale ethnographic strategy. *See* multisited ethnography

multiculturalism, 423

multidisciplinary, 387, 414–15, 429

multilateral organizations, 243

multinational corporations, 237; pharmaceutical companies as, 38, 241, 262, 434; software companies as, 441

multiple chemical sensitivity, 357

multisited ethnography, 21, 430

multispecies, 94, 97, 99, 106, 114

Myriad Genetics, 195

Nader, Ralph, 243

NASDAQ, 336, 345, 348–49

National Autism Association, 361

National Bank of Iceland, 347

National Cancer Institute (NCI), 37, 108, 119, 393, 398, 423–24

National Center for Biotechnology Information, 210

National Center for Gene Research, 209

National Center for Genetic Engineering and Biotechnology, 209

National Geographic Society, 227

National Health Service (NHS), 74

National Heart, Lung and Blood Institute, 46, 416–17

National Human Genome Research Institute (NHGRI), 105, 119

National Institute of Agrobiological Sciences, 209

National Institutes of Health (NIH), 2, 192, 199, 210, 386, 388, 399, 414–16, 433, 436

nationalism, nationalisms, 116, 282, 296, 298–99

National Science Foundation, 434

national security, 232, 240, 316

National Society for Autistic Children, 361–62

nation-states, 78, 237–38, 295, 299, 301, 336, 442–43

Nature, 160, 210, 434

Nature Medicine, 408

NeoGenesis, 416–17

neoliberalism, 23, 30, 32, 126, 212, 218, 230, 238, 244–46, 271, 434, 443, 446; economics of, 7, 9

neuroinflammation, 365

neurology, 374

New Deal, 340

New Jersey, 315, 337

New Orleans, 315

New Testament, 384

New York City, 318

New York State: Department of Health of, 317; Metropolitan Transportation Authority of, 317

New York Times, 172, 314, 325, 359, 405

New Zealand, 29, 73–74, 211, 213, 223

Nigeria, 30–31, 228–35, 239–47; civil war in, 234; Intellectual Property Law Association in, 242; National Agency for Drug Administration and Control in, 234–35, 238; National Drug Policy of, 234; Patent Office of, 242; pharmaceutical industry in, 238

nonexclusive licensing, 4–5

non-governmental organizations (NGOs), 240, 243, 313, 324

non-tariff barrier, 281, 302

North America, 158, 178, 209, 253–54, 259, 263, 271–72, 274, 438
Northwestern University, 169
Novartis, 195, 302
Novo Nordisk, 302
NSAIDS, 61
nuclear magnetic resonance, 413
number needed to treat (NNT), 59, 70–71, 73, 76, 80, 85
nutritional biochemistry, 363

Obama administration, 241
obesity, 47, 97, 262
objectivity, 360, 376–77
obligatory point of passage, 11
off-label uses, 366
Oil, Chemical and Atomic Workers union, 323
oil extraction, 231, 241, 246
Old Testament, 384
oncogene, 177–78, 406
oncology, 386
OncoMouse, 27–28, 172–73, 177, 179, 181
ontological choreography, 114
ontological surgery, 162
open-access publication, 189, 208
open-source forms, 436
open-source software, 189, 311
oppression, 232
optical technology, 414
organs: donation of, 276; trade of, 275; transplantation of, 450
orphan drug laws, 89
orthomolecular medicine, 362–63
orthomolecular psychiatry, 363
Osmania University, 434
osteoporosis, 79
overdetermination, 10, 15, 19, 449. See also determination, determinations
ownership, 132, 155–56, 169, 175, 182, 186–90, 193, 197, 200, 205, 207, 427
oxidative stress, 364–65

Panama, 348
Pan-Asian SNP initiative, 426, 429
panopticon, 24

parallel importation, 241
parent advocacy and advocates, 361, 439, 448
parent-scientists, 365
Paris Club, 231, 246
Parsons, Talcott, 430
partial perspective, 360
particularity, particularities, particularism, 125, 127, 131, 148–50, 438, 449
Partners Health Care, 417
Pasteur, Louis, 168
Pasteur Institute, 257
patent, patents, patenting, 3–5, 27–28, 38, 41, 66, 71, 85, 89, 108, 158, 166–82, 185–88, 190, 192, 194–200, 202–6, 210, 214, 242, 252–54, 257, 261–63, 277, 283, 330, 336, 390, 392, 395, 409, 425–27, 434
patent attorneys and law, 158, 166, 170–71, 178, 180, 182, 390
patient advocacy, 38, 437, 439
patients-in-waiting, 47, 56, 78
Pauling, Linus, 362
penicillin, 410
People's Business Commission (PBC), 174–76, 179
peopling of technoscience, 36
performativity, 55, 331–33, 335, 339
Perry, Rick, 307
personalized medicine, 119, 419
pet-food market, 97
Pfizer, 56, 234, 302, 424
pharmaceutical capital, 229, 238
pharmaceutical industry, 45, 49, 51–52, 61–67, 75, 88–89, 229, 241–42, 247, 280, 283, 294, 299, 441, 447; manufacturing and, 233
pharmaceutical marketing, 34, 45, 49, 52, 74, 78, 146, 438, 446
Pharmaceutical Research and Manufacturers of America (PhRMA), 299–302
Pharmaceuticals and Medical Devices Agency (PMDA), 304
Pharmacists Council of Nigeria, 236
pharmacoeconomics, 68
pharmacogenomics, 251, 291, 293

single nucleotide polymorphism (SNP), 258, 276–77, 291, 303, 435; chip, 426, 435; consortium, 276, 436
situated knowledge, 357
Snuppy, 102–3, 105, 118
social capital, 122
social contract, 230
social medicine, 432
social responsibility, 195, 207
social theory, 20, 438
Sociology of Scientific Knowledge (SSK), 20, 39
software company, 325
somatic cell nuclear transfer cloning, 102, 118
South Africa, 450; Medicines Act in, 241
South America, 209, 438
South Asia, 438
South China Morning Post, 140
Southeast Asia, 151, 294, 297, 302, 305, 438
Southern California, 77
South Korea, 118, 130, 132, 198, 209–10, 295, 302, 425
sovereignty, 124
Specific Carbohydrate Diet (SCD), 365, 382
speculative capital, 34, 439–40
speculative economy, 330
speech acts, 331–32
Stanford University, 3–8, 37–38, 269, 415, 420, 434
state privatization, 238
statin, 25, 47, 74, 86, 97
Stefansson, Kari, 336, 349, 352
stem-cell (research/therapy), 102–5, 118, 304, 424, 428, 436
Sterling Chemicals, 309
stock market, 64
streptococcus, 410
Structural Adjustment Programs (SAPS), 30, 230–34, 237, 239, 245–46
structural biology, structural biologists, 411–12, 415
structure, 12–13

Subaltern Studies Collective, 217, 310
subjectivity, subjectivities, 32, 40, 220, 230, 240, 311, 449
subsumption, 127, 217–18
superstructure, 11
supplementarity, 312
surplus health, 24, 45, 49, 52, 69, 71, 76, 84–86
surplus labor, 24–25, 51, 62, 82–83
surplus population, 65
surplus value, 13–14, 24, 62, 66, 71, 103, 126, 218–19, 221; absolute, 12; relative, 12
surrogacy, 171
Swiss Federal Institute of Technology, 167
Switzerland, 194, 337
symbolic capital, 387
symptom fetishism, 52, 57
symptoms, 53–57, 74, 81, 83, 86, 152, 237, 356, 359, 364, 367–68, 372–73, 376, 380
Syngenta, 185, 194–97, 201–4, 206, 210
synthetic biology, 387, 430–31, 436
systems biology, 414, 416, 430–31

tacit knowledge, 411, 429, 433
Tagamet, 410
Taiwan, 32, 198, 209–10, 281, 286–89, 292–97, 301–5, 425, 435, 438, 443; Bureau of Pharmaceutical Affairs of, 301; Center for Drug Evaluation of, 287–88, 292, 297, 302, 304; Department of Health of, 302
Tanzania, 247
Taxol, 396, 403, 405
T-cells, 411–12; T-cell receptors and, 418
technology: licensing office for, 3–4, 6; transfer of, 6, 390
terra nullius, 229
testing device, 324
Texas, 306–7, 309
Texas City, 306, 309–10, 320–21, 323
Thailand, 198, 209–10, 294, 425; Food and Drug Administration of, 302
thalidomide, 283

The Institute for Genomic Research
(TIGR), 198, 209
thematic centers, 414–15, 426
therapeutic economy, 233
therapeutic marketing. *See* pharmaceutical marketing
Thoughtful House, 381
tissue engineering, 386
Torrey Mesa Research Institute (TMRI), 194–95
toxicogenomics, 324
Toxic Release Inventory (TRI), 309–10, 321
trade barriers, 294
Trade Related Aspects of Intellectual Property Rights, 208, 242, 244, 434
trade secrets, 185, 188, 194, 196–97, 202–6
trading zones, 298
Traditional Chinese Medicine (TCM), 135–36
traditional medicine, 233
transgenic goats, 395
translational research, 36–37, 385–86, 414, 419, 424
transnationalism, 245; transnational capital and, 244
trans-species, 94–95, 441, 447
treatment maximization, 51
Tristan da Cunha, 262
tropical medicines, 67
tuberculosis, 68

ulcerative colitis, 382
UNESCO, 276
Union Carbide, 306, 309–10, 320–21
United Kingdom, 65, 73, 160, 185, 192, 198, 209–10, 410, 434; Parliament, 161
United Nations, 208, 284, 435; Accelerated Access Initiative (AAI), 241–42; Children's Fund (UNICEF), 434; Program on HIV/AIDS (UNAIDS), 241–42
United States, 46, 60–61, 66, 73–74, 77, 89, 91, 94–97, 102–5, 109, 114–17,

123, 144, 146, 156, 158, 162–65, 168–73, 176–82, 185, 192–95, 198–202, 209, 231–32, 239–44, 247, 254, 262–64, 272, 276, 278, 280–81, 285–87, 292, 298, 303, 306–7, 310–12, 321, 334, 337, 345–46, 355, 389, 394, 401, 405, 410, 419–23, 428, 433–35, 438, 441, 443, 445, 450; Africa Command (AFRICOM), 240; Agency for International Development, 240, 242–43; Census, 317; Center for Disease Control and Prevention, 90, 317; Chemical Safety and Hazard Investigation Board, 323; Chemical Safety Information, Site Security and Fuels Regulatory Act, 313–14; Clean Air Act, 307, 313–14; Community Right-to-Know Act, 313, 325; Constitution, 155, 163–65, 180; Department of Commerce, 242; Department of Defense, 240, 247, 435; Department of Energy, 196, 436; Department of Homeland Security, 314; Department of State, 240; Department of Transportation, 317; Environmental Protection Agency, 307–9, 311, 314–16, 323, 325; Global AIDS Coordinator, 240; National Library of Medicine, 323; Occupational Safety and Health Administration, 323; Patent Act, 165–66, 178, 180; Patent and Trademark Office, 168, 172–73, 183, 196, 243, 263; Plant Patent Act, 166, 196; Plant Variety Protection Act, 166, 196; President's Emergency Program for HIV/AIDS Relief, 231, 239–41, 247; Secretary of State, 240; Securities and Exchange Commission, 17, 34–35, 335–41, 343, 345, 347–53, 410; Superfund Amendments and Reauthorization Act, 313; trade representative, 301–2
U.S. Congress, 160, 163, 175, 178–79, 309, 335, 342–45, 349; House of Representatives, 339
U.S. Food and Drug Administration (FDA), 59, 65–66, 265, 277, 287,

298–300, 303–4, 390, 395, 401–8,
435; Center for Drug Evaluation and
Research, 61
U.S. Supreme Court, 3, 156, 158, 167, 170,
172, 174, 179–81, 196, 263, 415
universality, universalism, 125–27, 148–
50
University of Arizona, 209
University of California, Berkeley, 415
University of California, Davis, 99
University of California, M.I.N.D. Insti-
tute, 381; Los Angeles, 168–69
University of California, San Diego, 424
University of California, San Francisco,
269, 392, 415
University of California, Santa Cruz, 119,
436
University of California, School of Veteri-
nary Medicine, 105
University of California system, 37, 51
University of Detroit, Mercy, 64
University of Hong Kong (HKU), 129–30,
132–33, 150
University of North Carolina, Chapel
Hill, 106
University of Pittsburgh, 102
University of Washington, 194
University of Wisconsin, 209, 420
use value, 25, 54, 93, 95, 97, 112
utopian socialists, 10

vaccine, vaccines, 364, 376, 382, 384
value, 5–8, 12–15, 19, 24–29, 34, 48, 52,
59, 62–64, 67–72, 75, 77, 80–83, 87,
91, 93, 95, 99–100, 103–4, 108–9,
114, 117, 124, 126, 143, 156–58, 166,
169, 177–81, 190, 198, 212–14, 217–
20, 226, 252–58, 262–63, 275–76,
279, 297, 304, 325, 332, 337, 373, 375,
379, 415, 439, 442, 445, 450
vancomycin, 380
variable capital, 62
Velasquez, 432–33
venture capital, venture capitalists, 2, 17,
89, 264, 335–37, 352, 387, 406, 415,
420, 434

Viagra, 66
Vietnam, 424–25
Villanova University, 188
Vioxx, 61
Virginia, 166
vitamin B6, 371

wage labor, wage laborers, 86, 113, 232
Waitts Foundation, 227
Wall Street Journal, 405
Walt Disney Corporation, 122–24, 141,
146
Washington University, 169, 209
wealth extraction, 229–30, 233
Weber, Max, 9–11, 38, 430
Weissman, August, 412
welfare state, 271
Wellcome Trust, 210, 276–77
Wellington Public Hospital, 213
West Africa, 237, 241
West Harlem Environmental Action
(We Act), 316–17, 320
Wheat Commission, 340
White House, 342
whole genome shotgun [sequencing],
195, 201
William J. Clinton Foundation, 434
Wisconsin, 404
Wistar Institute, 276
World Bank, 30, 231, 246
World Health Organization (WHO), 146,
152, 241–42, 264–65, 284, 299–300
World Intellectual Property Organiza-
tion, 208
World Trade Organization (WTO), 208,
229, 238, 241–42, 434
worst-case scenarios, 313–14, 325

xenotransplantation, 177

Yahoo, 383
Yahoo Finance, 410
Yale University, 434
Young Hegelians, 10

Zimbabwe, 245

Library of Congress Cataloging-in-Publication Data
Lively capital : biotechnologies, ethics, and
governance in global markets /
Kaushik Sunder Rajan, ed.
p. cm. — (Experimental futures)
Includes bibliographical references and index.
ISBN 978-0-8223-4820-7 (cloth : alk. paper)
ISBN 978-0-8223-4831-3 (pbk. : alk. paper)
1 Biotechnology—Social aspects. 2. Biotechnology—
Political aspects. 3. Biotechnology industries—
Finance. I. Sunder Rajan, Kaushik. II. Series:
Experimental futures.
TP248.23.I58 2012
660.6—dc23 2011027558